Essentials of Geology

SIXTH EDITION

Essentials of Geology

Frederick K. Lutgens

Edward J. Tarbuck

Illinois Central College

Illustrated by
Dennis Tasa

Prentice Hall
Upper Saddle River, New Jersey 07458

Library of Congress Cataloging-in-Publication Data

Lutgens, Frederick K.
   Essentials of geology / Frederick K. Lutgens, Edward J. Tarbuck
 -6th ed. / illustrated by Dennis Tasa
     p.   cm.
   Includes index.
   ISBN 0-13-752510-9
   1. Geology.   I. Tarbuck, Edward J.   II. Title.
  QE26.2.L87 1998   97-24128
  550—dc21   CIP

Acquisition Editor: *Robert A. McConnin*
Editor in Chief: *Paul F. Corey*
Editorial Director: *Tim Bozik*
Assistant Vice President of Production and Manufacturing: *David W. Riccardi*
Executive Managing Editor: *Kathleen Schiaparelli*
Assistant Managing Editor: *Shari Toron*
Production Editor: *Edward Thomas*
Development Editor: *Fred Schroyer*
Editor in Chief of Development: *Ray Mullaney*
Associate Editor in Chief of Development: *Carol Trueheart*
Marketing Manager: *Leslie Cavaliere*
Creative Director: *Paula Maylahn*
Art Director: *Joseph Sengotta*
Art Manager: *Gus Vibal*
Photo Editors: *Lorinda Morris-Nantz and Melinda Reo*
Photo Researcher: *Barbara Scott*
Copy Editor: *James Tully*
Editorial Assistants: *Grace Anspake and Nancy Gross*
Production Assistant: *Adam Velthaus*
Interior Designer: *Thomas Nery*
Cover Designer: *Amy Rosen*
Manufacturing Manager: *Trudy Pisciotti*
Text Composition: *Molly Pike, Lido Graphics*
Cover & Title Page Photo: *White Island Volcano, Bay of Plenty, on New Zealand's North Island. Photo by Paul Kenward/Tony Stone Images.*

Printed in the United States of America
10 9 8 7 6 5 4 3 2 1

ISBN 0-13-752510-9

Prentice-Hall International (UK) Limited, *London*
Prentice-Hall of Australia Pty. Limited, *Sydney*
Prentice-Hall Canada Inc., *Toronto*
Prentice-Hall Hispanoamericana, S.A., *Mexico*
Prentice-Hall of India Private Limited, *New Delhi*
Prentice-Hall of Japan, Inc., *Tokyo*
Simon & Schuster Asia Pte. Ltd., *Singapore*
Editora Prentice-Hall do Brasil, Ltda., *Rio de Janeiro*

Brief Contents

Contents

5

Weathering and Soils   93

6

Sedimentary Rocks   114

7

Metamorphic Rocks   137

8

Mass Wasting   155

9

Running Water   169

10

Groundwater   191

11

Glaciers and Glaciation   210

12

Deserts and Wind   232

Preface

Earth is a *very* small part of a vast universe, but it is our home. It provides the resources that support our modern society and the ingredients necessary to maintain life. Therefore, a knowledge and understanding of our planet is critical to our social well being and indeed, vital to our survival.

In recent years, media reports have made us increasingly aware of the geological forces at work in our physical environment. News stories graphically portray the violent force of a volcanic eruption, the devastation generated by a strong earthquake, and the large numbers left homeless by mudflows and flooding. Such events, and many others as well are destructive to life and property, and we must be better able to understand and deal with them. To comprehend and prepare for such events requires an awareness of how science is done and the scientific principles that influence our planet, its rocks, mountains, atmosphere, and oceans.

Essentials of Geology is a college-level text designed for students taking their first and perhaps only course in geology. The book is intended to be a meaningful nontechnical survey for people with little background in science. Usually students are taking this class to meet a portion of their college's or university's general requirements.

In addition to being informative and up-to-date, a major goal of *Essentials of Geology* is to meet the need of beginning students for a readable and user-friendly text, a book that is a highly usable "tool" for learning the basic principles and concepts of geology. To accomplish this we have incorporated the following features.

Readability

The language of this book is straightforward and *written to be understood*. Clear, readable discussions with a minimum of technical language are the rule. When new terms are introduced, they are placed in **boldface** and defined. The frequent headings and subheadings help students follow discussions and identify the important ideas presented in each chapter. In the sixth edition, improved readability was achieved by reducing sentence and paragraph length, omitting unnecessary details, examining chapter organization and flow, and writing in a more personal style. Large portions of the text were substantially rewritten in an effort to make the material more understandable.

Illustrations and Photographs

Geology is highly visual. Therefore, photographs and artwork are a very important part of an introductory book. *Essentials of Geology, Sixth Edition*, contains dozens of new high-quality photographs that were carefully selected to aid understanding, add realism and heighten the interest of the reader.

The illustrations in each new edition of *Essentials of Geology* keep getting better and better. In the sixth edition *every* piece of existing art was redesigned. The new art illustrates ideas and concepts more clearly and realistically than ever before. The new art program was carried out by Dennis Tasa, a gifted artist and respected Earth science illustrator.

Focus on Learning

New to the sixth edition: To assist student learning, every chapter now opens with a series of questions. Each question alerts the reader to an important idea or concept in the chapter. When a chapter has been completed, four useful devices help students review. First, a helpful summary—*The Chapter In Review*—recaps all of the major points. Next is a checklist of *Key Terms* with page references. Learning the language of geology helps students learn the material. This is followed by *Questions For Review* that help students examine their knowledge of significant facts and ideas. Finally, a twenty-item chapter test—*Testing What You Have Learned*—wraps up the chapter-end review.

Earth as a System

An important occurrence in modern science has been the realization that Earth is a giant multidimensional system. Our planet consists of many separate but interacting parts. A change in any one part can produce changes in any or all of the other parts—

often in ways that are neither obvious nor immediately apparent. Although it is not possible to study the entire system at once, it is possible to develop an awareness and appreciation for the concept and for many of the system's important interrelationships. With this as a goal, a number of new headings and boxes have been added to the sixth edition of *Essentials of Geology*.

Environmental Issues

Because knowledge about our planet and how it works is necessary to our survival and well-being, the treatment of environmental and resource topics has always been an important part of *Essentials of Geology*. These issues serve to illustrate the relevance and application of geologic knowledge. The text integrates a great deal of information about the relationship between people and the physical environment and explores applications of geology to understanding and solving problems that arise from these interactions.

Maintaining a Focus on Basic Principles and Instructor Flexibility

Although new topical issues are treated in *Essentials of Geology, Sixth Edition*, it should be emphasized that the main focus of this new edition remains the same as its predecessors—to foster student understanding of basic geological principles. Whereas student use of the text is a primary concern, the books adaptability to the needs and desires of the instructor is equally important.

The organization of the text remains intentionally traditional. Following the overview of geology in Chapter 1, we turn to a discussion of Earth materials and the related processes of volcanism and weathering. Next, we explore the geological work of gravity, water, wind, and ice in modifying and sculpting landscapes. After this look at external processes, we examine Earth's internal structure and the processes that deform rocks and give rise to mountains. Finally, the text concludes with chapters on geologic time and Earth history. This organization accommodates the study of minerals and rocks in the laboratory, which usually comes early in the course.

Realizing that some instructors may prefer to structure their courses somewhat differently, we made each of the chapters self-contained so that they may be taught in a different sequence. Thus, the instructor who wishes to discuss earthquakes, plate tectonics, and mountain building prior to dealing with erosional processes may do so without difficulty. We also chose to provide a brief overview of plate tectonics in Chapter 1 so that this important theory could be incorporated in appropriate places throughout the text.

Supplements

In addition to the text itself, we have worked with a number of talented people to produce an excellent supplements package. This package includes the traditional supplements that students and professors have come to expect from authors and publishers, as well as some new kinds of supplements that involve electronic media. We always strive to produce the highest quality materials, and hope the supplements described here will make the study of geology more fulfilling for both students and instructors.

- *GEODe* (Geologic Explorations on Disk): This fully interactive CD-ROM was developed by the authors in collaboration with Dennis Tasa. It includes excellent coverage of Earth materials, external processes, internal processes, geologic time, landforms of the United States, and more. Animation and interactive exercises reinforce key geologic concepts. GEODe is available for student purchase for both Macintosh and Windows CD-ROM compatible computers—and at a special price for adopters of *Essentials of Geology 6/e*. (ISBN 0-13-516545-8)

- *Instructor's Resource Manual with Tests*: In this manual Ken Pinzke gives new instructors the benefit of his years of teaching experience and experienced instructors a ready source of new ideas to complement their teaching style. The manual contains a variety of lecture outlines, teaching tips, advice on how to integrate visual supplements, and many suggested test questions. (ISBN 0-13-775677-1)

- *Prentice Hall Custom Test:* Based on the powerful technology developed by Engineering Software Associates, Inc., this supplement, available for both Mac and Windows (IBM), allows instructors to tailor exams to their own needs. With the *On-line Testing* option, exams can be administered on-line so data can be automatically transferred for evaluation.
  A comprehensive desk reference guide is included, along with on-line assistance. (ISBN for IBM 0-13-680679-1; ISBN for Mac 0-13-680711-9)

- *Transparencies:* More than one hundred full color acetates of illustrations from the text. (ISBN 0-13-775693-3)

- *Slides:* One hundred fifty full-color slides of figures from the text and from other sources. (ISBN 0-13-775685-2)
- *The New York Times Themes of the Times—Changing Earth:* This unique newspaper-format supplement features recent articles on dynamic geological applications from the pages of the *New York Times*. This free supplement, available in quantity from your local Prentice Hall representative, encourages students to make connections between the classroom and the world around them. (ISBN 0-13-757980-2)
- *Geoscience on the Internet: A Student Guide:* Written by Andrew Stull of California State University-Fullerton and Duane Griffin of the University of Wisconsin-Madison, this guide to the internet helps geology students use the internet to enrich their studies. It includes general information on internet and WWW basics, as well as specific material about how to use the internet for the Geosciences. *Geoscience on the Internet* is free to qualified adopters of *Essentials of Geology.* Please contact your local Prentice Hall representative for details. (ISBN 0-13-890138-4)
- *Prentice Hall Geodisc:* This laser disk allows instructors to bring a wealth of images into the classroom. It contains more than 50 minutes of motion video and more than 900 still images and animations. The disk is accompanied by a bar code manual to access images easily. (ISBN 0-13-304163-8)

Acknowledgments

Writing a college textbook requires the talents and cooperation of many individuals. Once again we benefited from the skills of our Developmental Editor Fred Schroyer. He helped us make the Sixth Edition a more readable, user-friendly text. We are also grateful to Professor Kenneth Pinzke at Belleville Area College. Ken prepared the chapter-opening questions and the chapter-end summaries and tests that help make the book an even more effective tool for beginning students. Professor Pinzke is also responsible for preparing the *Instructor's Resource Manual* for the text. Working with Dennis Tasa, who is responsible for all of the outstanding illustrations, is always special for us. We not only value his outstanding artistic talents and imagination, but his friendship as well.

Special thanks goes to those colleagues who prepared in-depth reviews. Their critical comments and thoughtful input helped guide our work and clearly strengthened the text. We wish to thank:

John L. Berkley, SUNY College at Fredonia
Michael Bikerman, University of Pittsburgh
Richard B. Bonnett, Marshall University
Sherman Clebnik, Eastern Connecticut State University
Edward Heideman, Mercy College, NY
Betsy D. Torrez, Sam Houston State University

In addition we would like to acknowledge the aid of our students. Their comments and criticism sometimes challenge us to search for a deeper understanding and always help us maintain our focus on readability.

We also want to acknowledge the team of professionals at Prentice Hall. Thanks to Editor-in-Chief Paul Corey. We sincerely appreciate his continuing strong support for excellence and innovation. The production team led by Ed Thomas has done an outstanding job. They are true professionals with whom we are very fortunate to be associated. Finally, we would like to acknowledge Robert McConnin, the executive editor for geology at Prentice Hall. This is the last project that we will work on together. Over the past 23 years, we have had *many* editors, but none finer than Bob McConnin. He is a first rate professional who we are happy to call our friend. We will miss his talents on future projects and wish him the very best.

Fredrick K. Lutgens
Edward J. Tarbuck

This text is dedicated to Bob McConnin, a good friend and one of the finest editors in college publishing.

An Introduction to Geology

Chapter 1

Fisher Towers with LaSal Mountains, Utah, in background. (Photo by Tom Till)

Focus on Learning

To assist you in learning the important concepts in this chapter, you will find it helpful to focus on the following questions:

- What is the fundamental difference between the doctrine of uniformitarianism and the doctrine of catastrophism?

- What is relative dating? What are some principles of relative dating?

- How does a scientific hypothesis differ from a scientific theory?

- What are the principal divisions of Earth's interior?

- What is the theory of plate tectonics? How do the three types of plate boundaries differ?

- What is the rock cycle? Which geologic interrelationships are illustrated by the cycle?

The spectacular eruption of a volcano, the terror brought by an earthquake, the magnificent scenery of a mountain valley, the destruction created by a landslide—all are subjects for the geologist (Figure 1.1). The study of geology deals with many fascinating and practical questions about our physical environment. What forces produce mountains? What was the Ice Age like? Will there be another? What created this cave and the stone icicles hanging from its ceiling? Can we find water here? Is strip mining practical in this area? Will oil be found if a well is drilled at this location? What will result if the landfill is located in the old quarry?

The Science of Geology

This book introduces you to the science of **geology,** a word that literally means "the study of Earth." Unraveling Earth's secrets is not an easy task because our planet is not a static, unchanging mass of rock. It is a dynamic body possessing a long and complex history.

The science of geology is divided into two broad areas—physical and historical. **Physical geology** examines Earth's rocks and minerals and seeks to understand the hundreds of processes that operate beneath or upon its surface.

The aim of **historical geology,** on the other hand, is to understand Earth's origin and how it changed through time. Historical geology strives to establish the chronology of physical and biological changes of the past 4.6 billion years.

The study of physical geology logically precedes the study of Earth history because we must first understand how Earth works before attempting to unravel its past. Thus, the first seventeen chapters of this book introduce physical geology while the last two chapters present Earth's remarkable history.

Historical Notes About Geology

The nature of our Earth—its materials and processes—has been a focus of study since early times. Writings about fossils, gems, earthquakes, and volcanoes date back to the Greeks, more than 2300 years ago. The most influential Greek philosopher was Aristotle, but unfortunately his explanations of the natural world were not based on keen observations and experiments, as modern science is. Instead, they were his opinions, based on the limited knowledge of his day. Aristotle believed that rocks were created under the "influence" of the stars and that

FIGURE 1.1 Erosional processes create many of Earth's varied landscapes. (Photo by Carr Clifton)

earthquakes occurred when air in the ground was heated by central fires and escaped explosively! When confronted with a fossil fish, he explained that "a great many fishes live in the earth motionless and are found when excavations are made." Although Aristotle's explanations may have been adequate for his day, they unfortunately continued to be believed for many centuries, thus thwarting the acceptance of better ideas that were based on observations.

Catastrophism

During the 1600s and 1700s, the doctrine of **catastrophism** strongly influenced explanations of Earth dynamics. Catastrophists believed that Earth's landscapes had been shaped primarily by great catastrophes. They felt that mountains and canyons resulted from sudden, often worldwide disasters produced by unknowable causes that no longer operate. Catastrophism tried to fit the *rate* of Earth processes to then-current ideas of Earth's age.

In the mid-seventeenth century, James Ussher, an Anglican Archbishop in Ireland and a respected Bible scholar, constructed a chronology of human and Earth history. He calculated that Earth was only a few thousand years old, having been created in 4004 B.C. Ussher's treatise earned widespread acceptance among scientific and religious leaders alike, and his chronology was soon printed in the margins of the Bible itself.

The relationship between catastrophism and the age of Earth has been summarized nicely:

That the earth had been through tremendous adventures and had seen mighty changes during its obscure past was plainly evident to every inquiring eye; but to concentrate these changes into a few brief millenniums required a tailor-made philosophy, a philosophy whose basis was sudden and violent change.*

The Birth of Modern Geology

Against this backdrop of Aristotle's views and an Earth created in 4004 B.C., a Scottish physician and gentleman farmer named James Hutton published *Theory of the Earth* in 1795 (Figure 1.2). In it he put forth the doctrine of **uniformitarianism.** Today, uniformitarianism is a fundamental concept in geology. It states that the *physical, chemical, and biological laws that operate today also operated in the geologic past.* In other words, the forces and

*H. E. Brown, V. E. Monnett, and J. W. Stovall, *Introduction to Geology* (New York: Blaisdell, 1958).

FIGURE 1.2 James Hutton, Scottish geologist of the 1700s, who is often called the "father of modern geology." (Photo courtesy of the Natural History Museum, London)

processes that we observe shaping our planet today have been at work for a very long time. Thus, to understand ancient rocks, we must first understand present-day processes and their results. This idea is commonly stated as *the present is the key to the past.*

Prior to Hutton's *Theory of the Earth,* no one had effectively demonstrated that geological processes continue over extremely long periods of time. Hutton persuasively argued that even weak, slow processes could, over long spans of time, produce effects just as great as those resulting from sudden catastrophic events. Unlike his predecessors, Hutton cited verifiable observations to support his ideas.

Hutton's literary style was cumbersome and difficult, so his work was not widely read nor easily understood. Fortunately, a more readable English geologist, Charles Lyell, advanced the basic principles of modern geology. Between 1830 and 1872, Lyell produced eleven editions of his great work, *Principles of Geology.* As was customary in those days, Lyell's book had a long subtitle that outlined the main theme of the book: *Being an Attempt to Explain the Former Changes of the Earth's Surface, by Reference to Causes Now in Operation.* He painstakingly illustrated the concept of the uniformity of nature through time and was able to show more convincingly than Hutton that those geologic processes observed today most likely operated in the past. Although uniformitarianism did not originate with Lyell, he was most successful in interpreting and publicizing it for society at large.

Today uniformitarianism is just as viable as in Lyell's day. We realize more strongly than ever that the present gives us insight into the past and that the physical, chemical, and biological laws that govern geological processes remain unchanging through time. However, we also understand that the doctrine should not be taken too literally. To say that geological processes in the past were the same as those occurring today is not to suggest that they always had the same *relative importance* or operated at the *same rate*. Although these processes have prevailed through time, their rates have undoubtedly varied.

Although processes vary in their intensity, they still take a very long time to create or destroy major landscape features. For example, rocks containing fossils of organisms that lived in the sea more than 15 million years ago are now part of mountains that stand 3000 meters (9800 feet) above sea level. This means that the mountains were uplifted 3000 meters during the past 15 million years—an average rate of only 0.2 millimeter per year! Rates of erosion are equally slow, explaining why Earth's landscapes appear to be permanent (Figure 1.3). Estimates indicate that the North American continent is being eroded just 3 centimeters every 1000 years. Thus, as you can see, it takes tens of millions of years for nature to build mountains and wear them down again. But even these time spans are relatively short on the time scale of Earth's 4.6-billion-year history, for the rock record contains evidence that shows Earth has experienced many cycles of mountain-building and erosion.

It is important to remember that although many features of our physical landscape may seem to be unchanging in terms of the decades over which we observe them, they are nevertheless changing, but on time scales of hundreds, thousands, or even many millions of years. As James Hutton said, "We find no vestige of a beginning, no prospect of an end."

Geologic Time

Although Hutton, Lyell, and others recognized that geologic time is exceedingly long, they had no methods to accurately determine the age of Earth (see Box 1.1). However, in 1896, radioactivity was discovered. Using radioactivity for dating was first attempted in 1905 and has been refined ever since. Geologists are

FIGURE 1.3 Geologic processes usually act so slowly that changes may not be visible during an entire human lifetime. Today, Monument Valley looks much the same as it did when first encountered by explorers. (Photo by Carr Clifton)

Box 1.1 — The Magnitude of Geologic Time

omeone who is 100 years old seems *very old* to us. A 1000-year-old artifact we call *ancient.* But in geology, we routinely deal with truly vast time periods—a million years (14,000 human lifetimes) or a billion years (which is a thousand million). When viewed in the context of Earth's incredible 4.6-billion-year history, geologists often characterize an event that occurred 1 million years ago as "recent" and a rock sample that has been dated at 10 million years as "young."

Your appreciation for the magnitude of geologic time is essential, because many geologic processes are so gradual that vast spans of time are needed for a cumulative change to become noticeable or important. Hendrick Van Loon tries to convey the idea of immense geologic time in this passage:

High up in the North in the land called Svithjod, there stands a rock. It is a hundred miles high and a hundred miles wide. Once every thousand years a little bird comes to

this rock to sharpen its beak. When the rock has thus been worn away, then a single day of eternity will have gone by.*

Although the preceding quotation is an exaggeration, it makes the point that geologic time is vast and that Earth's "everlasting landscapes" are indeed slowly changing. Thus, over millions of years, mountains rise and are eroded to hills, and rivers excavate deep canyons.

How long is 4.6 billion years? If you were to begin counting at the rate of one number per second and continued 24 hours a day, 7 days a week, and never stopped, it would take about two lifetimes (150 years) to reach 4.6 billion! Another interesting way to illustrate Earth's age:

Compress…the entire 4.6 billion years of geologic time into a single year. On that scale, the oldest rocks we know date from about mid-March. Living things first appeared in the sea in May. Land plants and animals emerged in late

November…. Dinosaurs became dominant in mid-December, but disappeared on the 26th, at about the time the Rocky Mountains were first uplifted. Manlike creatures appeared sometime during the evening of December 31st, and the most recent continental ice sheets began to recede from the Great Lakes area and from northern Europe about 1 minute and 15 seconds before midnight on the 31st. Rome ruled the Western world for 5 seconds from 11:59:45 to 11:59:50. Columbus discovered America 3 seconds before midnight, and the science of geology was born with the writings of James Hutton just slightly more than one second before the end of our eventful year of years.**

*Hendrick Van Loon (1951), *The Story of Mankind,* ed. Black and Gold (New York: Liveright).

**Don L. Eicher, *Geologic Time,* 2d ed. (Englewood Cliffs, New Jersey: Prentice Hall, 1978), pp. 18–19. Reprinted by permission.

now able to assign fairly accurate dates to events in Earth history. For example, we know that the dinosaurs died out about 65 million years ago.

Before such *absolute* dates were established, the best that geologists could do was to put the events of Earth history in the proper order, called **relative dating.** They determined, for example, that the ancestors of modern shellfish appeared in the oceans before those of fish. This was done by applying several principles. One is the **law of superposition**—in layers of sedimentary rocks or lava flows, the youngest layer is on top, and the oldest is on the bottom (assuming that nothing has turned the layers upside down, which sometimes happens). Arizona's Grand Canyon provides a fine example where the oldest rocks are located in the inner gorge while the youngest rocks are found on the rim (Figure 1.4). So the law of superposition establishes the *sequence* of rock layers—but not, of course, their absolute ages.

Fossils, the remains or traces of prehistoric life, were also essential to the development of a geologic time scale. Fossils are the basis for the **principle of fossil succession**—fossil organisms succeed one another in a definite and determinable order, and, therefore, any time period can be recognized by its fossil content. This principle was laboriously worked out over decades by collecting fossils from countless rock layers around the world. Once established, it allowed geologists to identify rocks of the same age in widely separated places and to build the *geologic time scale* shown in Figure 1.5. We will consider the geologic time scale in more detail in Chapter 18.

The Nature of Scientific Inquiry

All science is based on the assumption that the *natural world behaves in a consistent and predictable manner.* This implies that the physical laws governing the smallest atomic particles also operate in the

FIGURE 1.4 By applying the law of superposition, can you determine the relative ages (which is older, which is younger) of these undeformed layers in the Grand Canyon? (Photo by Tom Till)

largest, most distant galaxies. Evidence for the existence of these underlying patterns is abundant in the physical and biological worlds. For example, the same biochemical processes and the same genetic codes that are found in bacterial cells are also found in human cells. The overall goal of science is to discover the underlying patterns in the natural world and then to use this knowledge to predict what will or will not happen, given certain facts or circumstances.

Collecting Facts

The development of new scientific knowledge involves some basic, logical processes that are universally accepted. To determine what is occurring in the natural world, scientists collect *facts* through observation and measurement (Figure 1.6). These data are the raw material from which scientific theories and laws are formulated.

Hypothesis

Once facts have been gathered and principles have been formulated to describe a natural phenomenon, investigators try to explain how or why things happen in the manner observed. They can do this by constructing a preliminary, untested explanation,

which we call a scientific **hypothesis**. Often, several different hypotheses are advanced to explain the same facts and observations.

For example, there are currently five major hypotheses that have been proposed to explain the origin of Earth's moon. Until recently, the most widely held hypothesis argued that the Moon and Earth formed simultaneously from the same cloud of nebular dust and gases. A newer hypothesis suggests that a Mars-sized body impacted Earth. The collision ejected a huge quantity of material into Earth's orbit that eventually accumulated into the Moon. Both hypotheses will be submitted to rigorous testing that will undoubtedly result in the modification or rejection of one, or perhaps both.

The history of science is littered with discarded hypotheses. One of the best known is the idea that Earth was at the center of the universe, a proposal that was supported by the apparent daily motion of the Sun, Moon, and stars around Earth.

Theory

After a hypothesis has survived exhaustive scrutiny, and when competing hypotheses have been eliminated, a hypothesis may be elevated to the status of a scientific **theory**. In everyday language, we may

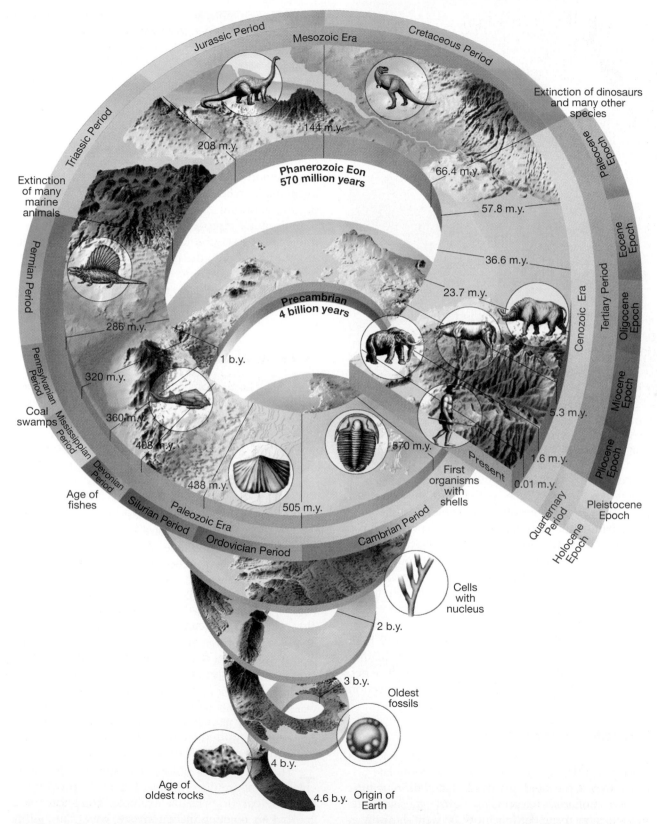

FIGURE 1.5 The geological time scale. Numbers on the time scale represent time in billions and millions of years before the present. These dates were added long after the time scale had been established using relative dating techniques. The Precambrian accounts for more than 85 percent of geologic time. (Data from Geological Society of America)

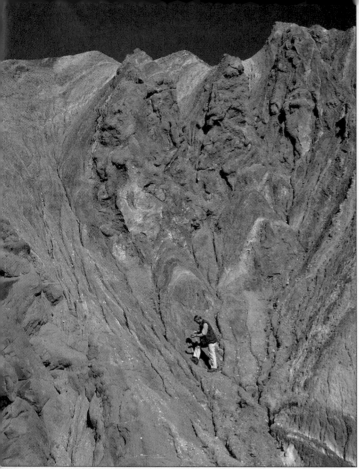

FIGURE 1.6 Geologist examining deformed rocks of Cockscomb Grand Staircase, Escalante National Monument, Utah. (Photo by Tom Bean)

say "that's only a theory." But a scientific theory is a well-tested and widely accepted view that scientists agree best explains certain observable facts.

Scientific theories, like scientific hypotheses, are accepted only provisionally. It is always possible that a theory which has withstood previous testing may eventually be disproven. As theories survive more testing, they are regarded with higher levels of confidence. Theories that have withstood extensive testing, such as the theory of plate tectonics or the theory of evolution, are held with a very high degree of confidence. Nevertheless, as the mathematician Jacob Bronowski so aptly stated, "Science is the acceptance of what works and the rejection of what does not."

Scientific Methods

The processes just described, in which scientists gather facts through observations and formulate scientific hypotheses and theories is called the *scientific method*. Contrary to popular belief, the scientific method is not a standard recipe that scientists apply in a routine manner to unravel the secrets of our natural world. Neither is this process a haphazard one.

Some scientific knowledge is gained through the following steps: (1) collection of scientific facts (data) through observation and measurement (Figure 1.7); (2) development of a working hypothesis to explain these facts; (3) construction of experiments to validate the hypothesis; and (4) acceptance, modification, or rejection of the hypothesis on the basis of extensive testing.

Other scientific discoveries represent purely theoretical ideas, which then stood up to extensive examination. Still other scientific advancements have been made when a totally unexpected happening occurred during an experiment. These so-called serendipitous discoveries are more than pure luck, for as Louis Pasteur said, "In the field of observation, chance favors only the prepared mind."

Because scientific knowledge is acquired through several avenues, it might be best to describe the nature of scientific inquiry as the *methods* of science, rather than *the* scientific method.

A View of Earth

A view of Earth from space gives us a unique perspective of our planet (Figure 1.8). At first, it may strike us that Earth is a fragile-appearing sphere surrounded by the blackness of space. In fact, it is just a speck of matter in a vast universe. As we look more closely at our planet from space, it becomes apparent that Earth is much more than rock and soil. Indeed, the most conspicuous features are not the continents but the swirling clouds suspended above the surface and the vast global ocean. These features emphasize the importance of water to our planet.

From such a vantage point we can appreciate why Earth's physical environment is traditionally divided into three major parts: the solid Earth; the water portion of our planet, the hydrosphere; and Earth's gaseous envelope, the atmosphere. It should be emphasized that our environment is highly integrated and is not dominated by rock, water, or air alone. Rather, it is characterized by continuous interactions as air comes in contact with rock, rock with water, and water with air. Moreover, the biosphere, the totality of life-forms on our planet, extends into each of the three physical realms and is an equally integral part of Earth.

The interactions among the spheres of Earth's environment are uncountable. Figure 1.9 provides us with one easy-to-visualize example. The shoreline is an obvious meeting place for rock, water, and air. In this scene, ocean waves that were created by the drag of air moving across the water are breaking

FIGURE 1.7 Temperature probe of active lava flow, Kilauea Caldera. (Photo by Norman Banks, U.S.G.S. Hawaiian Volcano Observatory)

against the rocky shore. The force of the water can be powerful and the erosional work that is accomplished can be great.

Hydrosphere

Earth is sometimes called the *blue* planet. Water more than anything else makes Earth unique. The **hydrosphere** is a dynamic mass of liquid that is continually on the move, from the oceans to the atmosphere, to the land, and back again. The global ocean is certainly the most prominent feature of the hydrosphere, blanketing nearly 71 percent of Earth's surface and accounting for about 97 percent of Earth's water. However, the hydrosphere also includes the freshwater found in streams, lakes, and glaciers, as well as that found underground.

Although these latter sources constitute just a tiny fraction of the total, they are much more important than their meager percentage indicates. In addition to providing the freshwater that is so vital to life on the continents, streams, glaciers, and groundwater are responsible for sculpturing and creating many of our planet's varied landforms.

Atmosphere

Earth is surrounded by a life-giving gaseous envelope called the **atmosphere**. This thin blanket of air is an integral part of the planet. It not only provides the air that we breathe but also acts to protect us from the Sun's intense heat and dangerous radiation. The energy exchanges that continually occur between the atmosphere and the surface and between the atmosphere and space produce the effects we call weather. If, like the Moon, Earth had no atmosphere, our planet would not only be lifeless but also many of the processes and interactions that make the surface such a dynamic place could not operate. Without weathering and erosion, the face of our planet might more closely resemble the lunar surface, which has not changed appreciably in nearly three billion years.

Lithosphere

Lying beneath the atmosphere and the ocean is the solid Earth. It is divided into three principal units: the dense **core**; the less dense **mantle**; and the **crust**, which is the light and very thin outer skin of Earth (Figure 1.10). The term **lithosphere** refers to a rigid outer layer of Earth, which includes the crust and part of the upper mantle. The crust is not a layer of uniform thickness; rather, it is characterized by many irregularities. It is thinnest beneath the oceans and thickest where continents exist.

A very important zone exists within the mantle and deserves special mention. This region, called the **asthenosphere**, is located between the depths of 100 and 350 kilometers and may extend down to 700 kilometers. The asthenosphere is a hot, weak zone

A.

FIGURE 1.8 A. View of Earth that greeted the *Apollo 8* astronauts as their spacecraft came from behind the Moon. **B.** A closer view of Earth from *Apollo 17*. Africa and Arabia are prominent in this image. The dark blue of the oceans and the cloud patterns remind us of the importance of the oceans and atmosphere. Antarctica, a continent covered by glacial ice, is visible at the South Pole. (Both photos courtesy of NASA)

B.

that is capable of gradual flow. As we shall see in the next section, the weak asthenosphere permits the gradual movement of solid lithosphere located above.

Biosphere

The **biosphere** includes all life on Earth and penetrates parts of the solid Earth, hydrosphere, and atmosphere. Plants and animals depend on the physical environment for the basics of life. However, it is important to note that organisms do more than just respond to their physical environment. Indeed, through countless interactions, life-forms help maintain and alter their physical environment. Without life, the makeup and nature of the lithosphere, hydrosphere, and atmosphere would be very different. Indeed, Earth would be a very different planet.

FIGURE 1.9 The shoreline is one obvious meeting place for the hydrosphere, atmosphere, and solid Earth. In this scene, along California's Big Sur coastline, ocean waves that were created by the force of moving air break against the rocky shore. The force of the water can be powerful, and the erosional work that is accomplished can be great. (Photo by Darrell Gulin/DRK Photo)

Dynamic Earth

Earth is a dynamic planet! If we could go back in time a billion years or more, we would find a planet with a surface dramatically different from what it is today. There would be no Grand Canyon, Rocky Mountains, or Appalachian Mountains. Moreover, we would find continents with different shapes and located in different positions than today.

In contrast, a billion years ago the Moon's surface was almost the same as we now find it. In fact, if viewed telescopically from Earth, perhaps only a few craters would be missing. Thus, when compared to Earth, the Moon is a static, lifeless body wandering through space and time.

The processes that alter Earth's surface can be divided into two categories—destructive and constructive. *Destructive processes* are those that wear away the land, including weathering and erosion. Unlike the Moon, where weathering and erosion

progress at infinitesimally slow rates, these processes daily alter Earth's landscape. In fact, these destructive forces would have leveled the continents long ago had it not been for opposing constructional processes. Included among *constructional processes* are volcanism and mountain-building, which increase the average elevation of the land. As you shall see, these constructive processes depend upon Earth's internal heat as their source of energy.

Plate Tectonics

Within the past few decades, a great deal has been learned about the workings of our dynamic planet. In fact, many have called this period an unequalled revolution in our knowledge about Earth. This revolution began about 1915 with the radical proposal of *continental drift,* the idea that the continents moved about the face of the planet. This proposal contradicted the established view that the continents

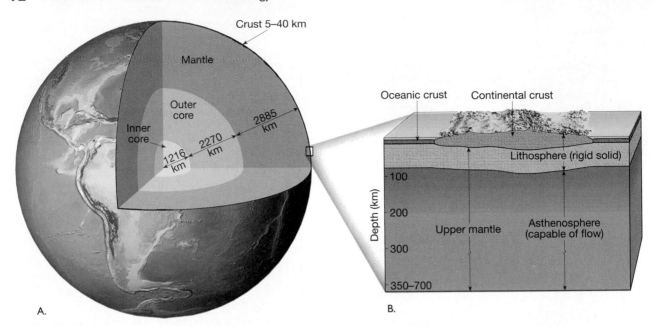

FIGURE 1.10 View of Earth's layered structure. **A.** The inner core, outer core, and mantle are drawn to scale, but the thickness of the crust is exaggerated by about five times. **B.** A blowup of Earth's outer shell. It shows the two types of crust (oceanic, continental), the solid lithosphere, and plastic asthenosphere.

and ocean basins are permanent and stationary features. For that reason, the notion was received with great skepticism and even ridicule. More than 50 years passed before enough data were gathered to transform this controversial hypothesis into a sound theory that weaved together the basic processes known to operate on Earth. The theory that finally emerged, called **plate tectonics,** provided geologists with the first comprehensive model of Earth's internal workings. (*Tectonics* is the study of large-scale deformations of Earth's lithosphere that result in major features such as mountains and ocean basins.)

According to the plate tectonics model, Earth's rigid outer shell, the lithosphere, is broken into vast slabs called **plates.** In Figure 1.11, each plate is tinted a different color to make them easier to see. Note that several plates include an entire continent plus a large area of seafloor (for example, the South American plate).

We know that these rigid plates are slowly and continually in motion. This movement is driven by a thermal engine, the result of an unequal distribution of heat within Earth. As hot material gradually moves up from deep within Earth and spreads laterally in the plastic asthenosphere, the plates are set in motion. Ultimately, this titanic, grinding movement of Earth's lithospheric plates generates earthquakes, causes volcanic activity, and deforms large masses of rock into mountains.

Plate Boundaries

Because each plate moves as a distinct unit, all interaction among individual plates occurs along their *boundaries*. In fact, the first attempts to outline plate boundaries were made using points of earthquake and volcanic activity. Later work revealed three distinct types of plate boundaries, which are differentiated by the type of movement they exhibit. They are depicted at the bottom of Figure 1.11 and are briefly described here:

1. **Divergent boundaries**—where plates move apart, leaving a gap between them (Figure 1.11A).
2. **Convergent boundaries**—where plates move together, forcing one of the slabs of lithosphere to descend beneath the other (Figure 1.11B).
3. **Transform boundaries**—where plates grind past each other, scraping and deforming as they pass (Figure 1.11C).

Study each plate and you can see that it is bounded by a combination of these boundaries. Movement along one boundary requires that adjustments be made at the others. We will examine this further in Chapter 16, but here is a little more about these important boundaries.

Divergent Boundaries

As plates separate, the gap created is immediately filled with molten rock that wells up from the hot asthenosphere (Figure 1.12). This fresh material

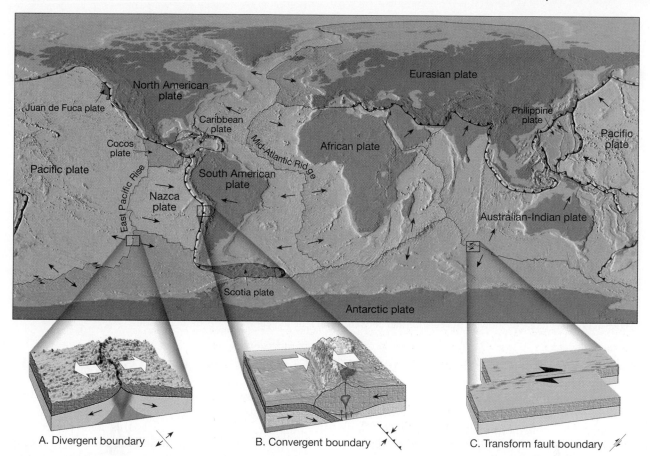

A. Divergent boundary   B. Convergent boundary   C. Transform fault boundary

FIGURE 1.11 Mosaic of rigid plates that constitute Earth's outer shell. (After W. B. Hamilton, U.S. Geological Survey)

slowly cools to hard rock, producing a new sliver of seafloor. As this happens again and again over millions of years, thousands of square kilometers of seafloor are added. This mechanism has created the floor of the Atlantic Ocean during the past 160 million years and is appropriately called **sea-floor spreading.** The typical rate of sea-floor spreading is estimated to be 5 centimeters (2 inches) per year, although it varies considerably from one location to another. This extremely slow rate of movement is nevertheless rapid enough so that all of Earth's ocean basins could have been generated within the last 5 percent of geologic time (about 200 million years).

Further, along divergent boundaries where molten rock emerges, the ocean floor is elevated. Worldwide, this ridge extends for 64,000 kilometers through all major ocean basins. You can see parts of these mid-ocean ridges in Figure 1.13.

Convergent Boundaries

Although Earth's crust is constantly being added at the oceanic ridges, the planet is not growing in size—its total surface area remains constant. Therefore, old crust must be disappearing somewhere. Indeed it is,

along *convergent boundaries.* As two plates slowly collide, the leading edge of one slab is bent downward, allowing it to slide beneath the other. This is shown in Figure 1.12. The subsiding plate pushes down into the asthenosphere. The surface expression of this is an ocean *trench,* like the Peru–Chile trench in Figure 1.13.

The regions where oceanic crust is being consumed are called **subduction zones.** Here, as the solid plates move downward, they enter a high-pressure, high-temperature environment. Some subducted material melts and migrates upward into the overriding plate (Figure 1.12). Occasionally this molten rock may reach the surface, where it gives rise to explosive volcanic eruptions like Mount St. Helens in 1980 and Mount Pinatubo in 1991.

Transform Fault Boundaries

Transform fault boundaries are located where plates grind past each other without either generating new crust or consuming old crust. These faults form in the direction of plate movement and were first discovered in association with offsets in the oceanic ridges (Figure 1.12). Although most transform faults are located along

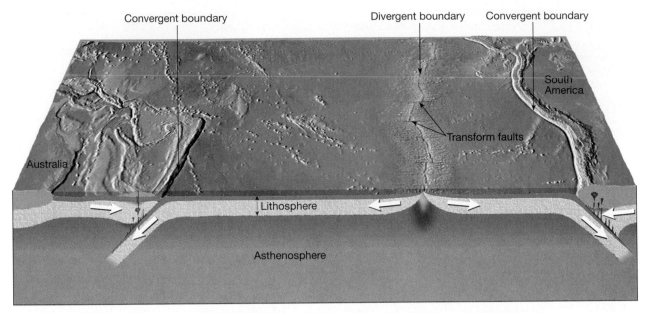

FIGURE 1.12 View of Earth showing the relationship between divergent and convergent plate boundaries.

mid-ocean ridges, a few slice through the continents. The earthquake-prone San Andreas fault of California is a famous example. Along this fault the Pacific plate is moving toward the northwest, past the North American plate. The movement along this boundary does not go unnoticed. As these plates pass, strain builds in the rocks on opposite sides of the fault. Occasionally the rocks adjust, releasing energy in the form of a great earthquake of the type that devastated San Francisco in 1906 and caused extensive damage in the 1990s.

Interaction of plates along their boundaries initiates most of our planet's volcanism, earthquakes, and mountain building. It also can form a new ocean basin. For example, a divergent boundary that runs through eastern Africa is developing Africa's Rift Valley. If spreading continues there, Africa will split into two continents separated by a new ocean basin. At other locations, continents are presently moving toward each other and may eventually join into a "supercontinent." When continents collide, their edges are gradually thrust into majestic mountain ranges.

As long as temperatures deep within our planet remain significantly higher than those near the surface, the material within Earth will continue to move. This internal flow, in turn, will keep the rigid outer shell of Earth in motion. Thus, while Earth's internal heat engine is operating, the positions and shapes of the continents and ocean basins will change, and Earth will remain a dynamic planet.

In the rest of the book we will examine in more detail the workings of our dynamic planet in light of the plate tectonics theory.

Earth as a System: The Rock Cycle

Earth is a system (see Box 1.2). This means that our planet consists of many interacting parts that form a complex whole. Nowhere is this idea better illustrated than when we examine the rock cycle (Figure 1.14). The **rock cycle** allows us to view many of the interrelationships among different parts of the Earth system. It helps us understand the origin of igneous, sedimentary, and metamorphic rocks and to see that each type is linked to the others by the processes that act upon and within the planet. Learn the rock cycle well; you will be examining its interrelationships in greater detail throughout the book.

The Basic Cycle

Let us begin at the top of Figure 1.14. **Magma** is molten material that forms inside Earth. Eventually magma cools and solidifies. This process, called *crystallization*, may occur either beneath the surface or, following a volcanic eruption, at the surface. In either situation, the resulting rocks are called **igneous rocks**.

If igneous rocks are exposed at the surface, they will undergo *weathering*, in which the day-in and day-out influences of the atmosphere slowly disintegrate and decompose rocks. The materials that result are often moved downslope by gravity before being picked up and transported by any of a number of erosional agents—running water, glaciers, wind, or waves. Eventually these particles and dissolved substances, called **sediment**, are deposited. Although

FIGURE 1.13 Major physical features of the continents and ocean basins. Note the underwater mountain chains that mark diverging plate boundaries in the Atlantic (Mid-Atlantic Ridge) and Pacific (East Pacific Rise). Also note the trenches created where two plates converge, like the Peru–Chile trench.

most sediment ultimately comes to rest in the ocean, other sites of deposition include river floodplains, desert basins, swamps, and dunes.

Next the sediments undergo *lithification*, a term meaning "conversion into rock." Sediment is usually lithified into **sedimentary rock** when compacted by the weight of overlying layers or when cemented as percolating water fills the pores with mineral matter.

If the resulting sedimentary rock is buried deep within Earth and involved in the dynamics of mountain building, or intruded by a mass of magma, it will be subjected to great pressures and/or intense heat. The sedimentary rock will react to the changing environment and turn into the third rock type, **metamorphic rock**. When metamorphic rock is subjected to additional pressure changes or to still higher temperatures, it will melt, creating magma, which will eventually crystallize into igneous rock.

Processes driven by heat from Earth's interior are responsible for creating igneous and metamorphic rocks. Weathering and erosion, external processes powered by energy from the Sun, produce the sediment from which sedimentary rocks form.

| Box 1.2 | Earth as a System |

A *system* is a group of interrelated, interacting, or interdependent parts that form a complex whole. Most of us hear and use the term frequently. We may service our car's cooling *system*, make use of the city's transportation *system*, and be a participant in the political *system*. A news report may inform us of an approaching weather *system*.

We know that Earth is just a small part of a large system known as the *solar system*. As we study Earth, it also becomes clear that our planet can be viewed as a system with many separate but interacting parts or subsystems. The hydrosphere, atmosphere, biosphere, and solid Earth and all of their components can be studied separately. However, the parts are not isolated. Each is related in some way to the others to produce a complex and continuously interacting whole that we call the *Earth system*.

The parts of the Earth system are linked so that a change in one part can produce changes in any or all of the other parts. For example, when a volcano erupts, lava from Earth's interior may flow out at the surface and block a nearby valley. This new obstruction influences the region's drainage system by creating a lake or causing streams to change course. The large quantities of volcanic ash and gases that can be emitted during an eruption may be blown high into the atmosphere and influence the amount of solar energy that can reach the surface. The result could be a drop in air temperatures over the entire hemisphere. Where the surface is covered by lava flows or a thick layer of volcanic ash, existing soils are buried. This causes the soil-forming processes to begin anew to transform the new surface material

FIGURE I.A When Mount St. Helens erupted in May 1980, the area shown here was buried by a volcanic mudflow. Now, plants are reestablished and new soil is forming. (Photo by Jack Dykinga and Associates)

into soil (Figure 1.A). The soil that eventually forms will reflect the interaction among many parts of the Earth system. Of course, there would also be significant changes in the biosphere. Some organisms and their habitats would be eliminated by the lava and ash, while new settings for life, such as the lake, would be created. The potential climate change could also impact sensitive life-forms.

The Earth system is powered by energy from two sources. The Sun drives external processes that occur in the atmosphere, hydrosphere, and at Earth's surface. Weather and climate, ocean circulation, and erosional processes are driven by energy from the Sun. Earth's interior is the second source of energy. Heat remaining from when our planet formed and heat that is continuously generated

by radioactive decay powers the internal processes that produce volcanoes, earthquakes, and mountains.

Humans are *part of* the Earth system, a system in which the living and nonliving components are entwined and interconnected. Therefore, our actions produce changes in all of the other parts. When we burn gasoline and coal, build breakwaters along the shoreline, dispose of our wastes, and clear the land, we cause other parts of the system to respond, often in unforeseen ways. Throughout this book you will learn about many of Earth's subsystems: the hydrologic system, the tectonic (mountain-building) system, and the rock cycle, to name a few. Remember that these components *and we humans* are all part of the complex interacting whole we call the Earth system.

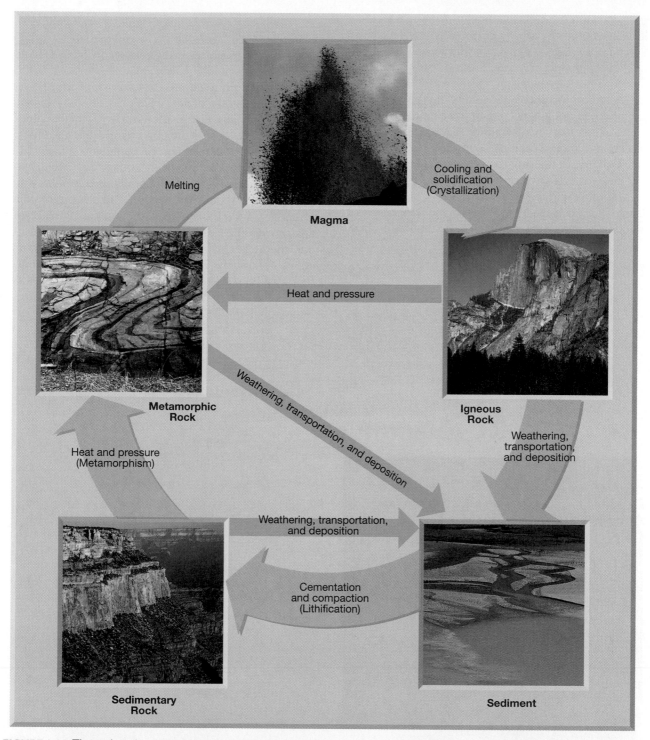

Magma

Melting

Cooling and solidification (Crystallization)

Heat and pressure

Metamorphic Rock

Igneous Rock

Weathering, transportation, and deposition

Weathering, transportation, and deposition

Heat and pressure (Metamorphism)

Weathering, transportation, and deposition

Cementation and compaction (Lithification)

Sedimentary Rock

Sediment

FIGURE 1.14 The rock cycle is one way of viewing many of the interrelationships among different parts of the Earth system. It shows us that Earth's materials (boxes) are all linked to each other by processes (arrows). Every rock contains clues to the processes that formed it. [Photos by J. D. Griggs, U.S.G.S. (top), E. J. Tarbuck (top right, bottom right, bottom left), and Phil Dombrowski (top left).]

Alternative Paths

The paths shown in the basic cycle are not the only ones that are possible. To the contrary, other paths are just as likely to be followed as those described in the preceding section. These alternatives are indicated by the blue arrows in Figure 1.14.

Igneous rocks, rather than being exposed to weathering and erosion at Earth's surface, may remain deeply buried. Eventually these masses may be subjected to the strong compressional forces and high temperatures associated with mountain building. When this occurs, they are transformed directly into metamorphic rocks.

Metamorphic and sedimentary rocks, as well as sediment, do not always remain buried. Rather, overlying layers may be stripped away, exposing the once-buried rock. When this happens, the material is attacked by weathering processes and turned into new raw materials for sedimentary rocks.

Although rocks may seem to be unchanging masses, the rock cycle shows that they are not. The changes, however, take time—great amounts of time.

The Rock Cycle and Plate Tectonics

When the rock cycle was first proposed by James Hutton, very little was actually known about the processes by which one rock was transformed into another; only evidence for the transformation existed. In fact, it was not until the development of

the theory of plate tectonics that a relatively complete picture of the rock cycle emerged.

Figure 1.15 illustrates the rock cycle in terms of the plate tectonics model. According to this model, weathered material from elevated landmasses is transported to the continental margins, where it is deposited in layers that collectively are thousands of meters thick. Once lithified, these sediments create a thick wedge of sedimentary rocks flanking the continents.

Eventually the relatively quiescent activity of sedimentation along a continental margin may be interrupted if the region becomes a convergent plate boundary. When this occurs, the oceanic lithosphere adjacent to the continent begins to inch downward into the asthenosphere beneath the continent. Along active continental margins such as this, convergence deforms them into linear belts of metamorphic rocks.

As the oceanic plate descends, some of the overlying sediments that were not crumpled into mountains are carried downward into the hot asthenosphere, where they too undergo metamorphism. Eventually some of these metamorphic rocks

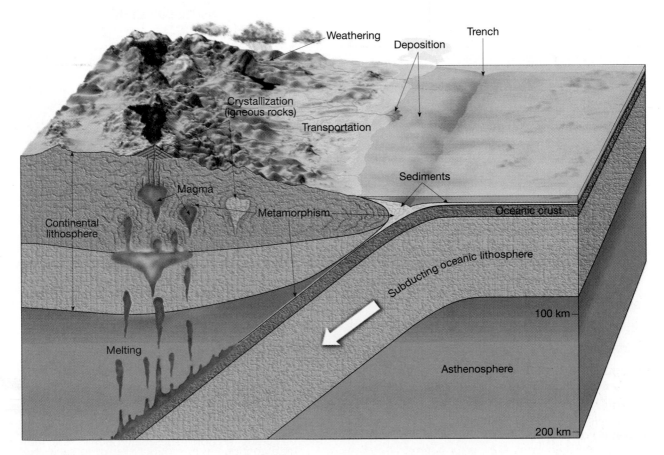

FIGURE 1.15 The rock cycle as it relates to the plate tectonics model.

are transported to depths where the conditions are conducive to melting. The newly formed magma will then migrate upward through the underlying lithosphere to produce igneous rocks. Some will crystallize prior to reaching the surface and the remainder will erupt and solidify at the surface. When igneous rocks are exposed at the surface, they are immediately attacked by the processes of weathering. Thus, the rock cycle is ready to begin anew.

The Chapter in Review

The following statements are intended to help you review the primary objectives presented in this chapter.

- *Geology* literally means "the study of Earth." The two broad areas of the science of geology are 1) *physical geology*, which examines the materials composing Earth and the processes that operate beneath and upon its surface; and 2) *historical geology*, which seeks to understand the origin of Earth and its development through time.

- During the seventeenth and eighteenth centuries the *doctrine of catastrophism* influenced the formulation of explanations about the dynamics of Earth. Catastrophism states that Earth's landscapes have been developed primarily by great catastrophes. On the other hand, the doctrine of *uniformitarianism*, one of the fundamental principles of modern geology put forth by *James Hutton* in his published work *Theory of the Earth* in the late 1700s, states that the physical, chemical, and biological laws that operate today have also operated in the geologic past. The idea is often summarized as "the present is the key to the past." Hutton argued that processes that appear to be slow-acting could, over long spans of time, produce effects that were just as great as those resulting from sudden catastrophic events. *Sir Charles Lyell* (mid–1800s) is given the most credit for advancing the basic principles of modern geology with the publication of the eleven editions of his great work, *Principles of Geology*.

- Using the principles of *relative dating*, the placing of events in their proper sequence or order without knowing their absolute age in years, scientists developed a geologic time scale during the nineteenth century. Relative dates can be established using the *law of superposition, fossils*, and the *principle of fossil succession*.

- All science is based on the assumption that the natural world behaves in a consistent and predictable manner. The process by which scientists gather facts from observations and formulate scientific *hypotheses* and *theories* is called the *scientific method*. To determine what is occurring in the natural world, scientists often 1) collect facts, 2) develop a scientific hypothesis, 3) construct experiments to validate the hypothesis, and 4) accept, modify, or reject the hypothesis on the basis of extensive testing. Other discoveries represent purely theoretical ideas which have stood up to extensive examination. Still other scientific advancements have been made when a totally unexpected happening occurred during an experiment.

- Earth's physical environment is traditionally divided into several parts. The *hydrosphere* is the dynamic mass of water that is continually on the move from the oceans to the air, to the land, and back again. The *atmosphere* is the life-giving gaseous envelope surrounding Earth. The solid Earth is divided into the dense *core*, less dense *mantle*, and *crust*. Existing within the mantle between the depths of approximately 100 and 700 kilometers is a very important region called the *asthenosphere*, a hot, weak zone that is capable of gradual flow. Above the asthenosphere is the *lithosphere*, the cool and rigid outer layer of Earth which includes the crust and uppermost mantle. The *biosphere* includes all life on Earth and penetrates parts of the lithosphere, hydrosphere, and atmosphere.

- The theory of *plate tectonics*, which provides geologists with a comprehensive model of Earth's internal workings, holds that Earth's rigid outer shell, the lithosphere, consists of several segments called *plates* that are slowly and continually in motion relative to each other. Most of Earth's earthquakes, volcanic activity, and mountain building are associated with the movements of these plates.

- The three distinct types of plate boundaries are 1) *divergent boundaries*—where plates move apart, leaving a gap between them; 2) *convergent boundaries*—where plates move together, causing one to go beneath the other, or where plates collide, which occurs when the leading edges are made of continental crust; and 3) *transform boundaries*—where plates slide past each other.
- Earth is a system which consists of many interacting parts that form a complex whole. The

rock cycle—perhaps the best illustration of this idea—is a means of viewing many of the interrelationships of geology. It illustrates the origin of the three basic rock types and the role of various geologic processes in transforming one rock type into another. *Igneous rock* forms from *magma* that cools and solidifies in a process called *crystallization*. *Sedimentary rock* forms when the products of *weathering*, called *sediment*, undergo *lithification*. *Metamorphic rock* forms from rock that has been subjected to great pressure and heat in a process called *metamorphism*.

Key Terms

asthenosphere (p. 9)
atmosphere (p. 9)
biosphere (p. 10)
catastrophism (p. 3)
convergent boundary (p. 12)
core (p. 9)
crust (p. 9)
divergent boundary (p. 12)

fossil (p. 5)
fossil succession, principle of (p. 5)
geology (p. 2)
historical geology (p. 2)
hydrosphere (p. 9)
hypothesis (p. 6)
igneous rock (p. 14)
lithosphere (p. 9)
magma (p. 14)

mantle (p. 9)
metamorphic rock (p. 15)
physical geology (p. 2)
plate (p. 12)
plate tectonics (p. 12)
relative dating (p. 5)
rock cycle (p. 14)
sea-floor spreading (p. 13)

sediment (p. 14)
sedimentary rock (p. 15)
subduction zone (p. 13)
superposition, law of (p. 5)
theory (p. 6)
transform boundary (p. 12)
uniformitarianism (p. 3)

Questions for Review

1. Geology is traditionally divided into two broad areas. Name and describe these two subdivisions.
2. Briefly describe Aristotle's influence on the science of geology.
3. How did the proponents of catastrophism perceive the age of Earth?
4. Describe the doctrine of uniformitarianism.
5. How did the advocates of uniformitarianism view the age of Earth?
6. Briefly describe the contributions of Hutton and Lyell.
7. How old is Earth currently thought to be?

8. The geologic time scale was established without the aid of radiometric dating. What principles were used to develop the time scale?
9. How is a scientific hypothesis different from a scientific theory?
10. List and briefly describe the four "spheres" that constitute our environment.
11. Describe the asthenosphere.
12. With which type of plate boundary is each of the following associated: subduction zone, San Andreas fault, sea-floor spreading, and Mount St. Helens?
13. Using the rock cycle, explain the statement "one rock is the raw material for another."

Testing What You Have Learned

To test your knowledge of the material presented in this chapter, answer the following questions:

Multiple-Choice Questions

1. The mechanism which produced the floor of the Atlantic Ocean during the past 200 million years is called _____.
   - **a.** subduction
   - **b.** convergence
   - **c.** succession
   - **d.** lithification
   - **e.** seafloor spreading

2. Which zone of Earth's interior includes the crust and uppermost mantle?
   - **a.** asthenosphere
   - **b.** outer core
   - **c.** inner core
   - **d.** lithosphere
   - **e.** tectosphere

3. The word that literally means "the study of Earth" is _____.
   - **a.** astronomy
   - **b.** tectonics
   - **c.** science
   - **d.** geology
   - **e.** chronology

4. Which force(s) wear(s) away the land?
   - **a.** weathering
   - **b.** volcanism
   - **c.** erosion
   - **d.** both a. and b.
   - **e.** both a. and c.

5. Placing events in their proper sequence or order without knowing their absolute age in years is referred to as _____ dating.
   - **a.** chronological
   - **b.** relative
   - **c.** absolute
   - **d.** estimated
   - **e.** comparative

6. Which number best represents the age of Earth?
   - **a.** 6 thousand
   - **b.** 4.6 million
   - **c.** 12.8 million
   - **d.** 1.2 billion
   - **e.** 4.6 billion

7. James _____, an eighteenth century Scottish physician and gentleman farmer, is often called the "father of modern geology."
   - **a.** Smith
   - **b.** Lyell
   - **c.** Steno
   - **d.** Hutton
   - **e.** Adams

8. The doctrine of _____ states that the physical, chemical, and biological laws that operate today have also operated in the geologic past.
   - **a.** uniformitarianism
   - **b.** succession
   - **c.** relative dating
   - **d.** catastrophism
   - **e.** superposition

9. Which rock type(s) originate(s) when molten material called magma cools and solidifies?
   - **a.** igneous
   - **b.** metamorphic
   - **c.** sedimentary
   - **d.** both b. and c.
   - **e.** both a. and c.

10. A scientific _____ is a well-tested and widely accepted view that scientists agree best explains certain observable facts.
    - **a.** determination
    - **b.** estimate
    - **c.** hypothesis
    - **d.** fact
    - **e.** theory

Fill-in Questions

11. The process in which scientists gather facts through observations and formulate scientific hypotheses and theories is called the _____ _____.

12. _____ is the study of large-scale deformation of Earth's lithosphere that results in the formation of major structural features such as those associated with mountains.

13. The water portion of our planet is called the _____.

14. The term _____ is used to describe the conversion of sediment into sedimentary rock.

15. Earth's rigid outer shell, the lithosphere, is broken into several large individual pieces called _____.

True/False Questions

16. The regions where oceanic lithosphere is being consumed are called subduction zones. ___

17. Earth's inner core is a molten metallic zone some 2270 kilometers thick. ___

18. The full rock cycle does not always take place. ___

19. Transform boundaries are zones where plates move apart, leaving a gap between them. ___

20. The three major divisions of Earth's physical environment are the atmosphere, the hydrosphere, and the solid Earth. ___

Answers

1.e; 2.d; 3.d; 4.e; 5.b; 6.e; 7.d; 8.a; 9.a; 10.e; 11. scientific method; 12. Tectonics; 13. hydrosphere; 14. lithification; 15. plates; 16.T; 17.F; 18.T; 19.F; 20.T

Minerals:
Building Blocks
of Rocks

Focus on Learning

To assist you in learning the important concepts in this chapter, you will find it helpful to focus on the following questions:

- What are minerals and how are they different from rocks?

- What is the basic structure of an atom? How do atoms bond together?

- How do isotopes of the same element vary from each other and why are some isotopes radioactive?

- What are some of the physical and chemical properties of minerals? How can these properties be used to distinguish one mineral from another?

- What is the most abundant mineral group? What do all minerals within this group have in common?

- What are some important nonsilicate minerals?

- When is the term *ore* used with reference to a mineral?

Fluorite (light purple) and gypsum crystals from Coahuila, Mexico. (Photo by Breck P. Kent)

Earth's crust is the source of a wide variety of useful and essential minerals. In fact, practically every manufactured product contains materials obtained from minerals. Most people are familiar with the common uses of many basic metals, including aluminum in beverage cans, copper in electrical wiring, gold in jewelry, and silicon in computer chips. But fewer are aware that pencil "lead" does not contain the metal lead but is really made of the greasy-feeling mineral graphite. Baby powder is ground-up rock made of the mineral talc, with perfume added. Drill bits impregnated with pieces of diamond (a mineral) are used to cut through rock layers in search of oil. The common mineral quartz is the main ingredient in glass, and so on and so on. As the material demands of modern society grow, the need to locate additional supplies of useful minerals also grows, and becomes more challenging as the easily mined sources become depleted.

In addition to the economic uses of rocks and minerals, every process studied by geologists is in some way dependent upon the properties of these basic Earth materials. Events such as volcanic eruptions, mountain building, weathering and erosion, and even earthquakes involve rocks and minerals. Consequently, a basic knowledge of materials in the Earth is essential to the understanding of all geologic phenomena.

Minerals Versus Rocks

Rocks to many people are nondescript, hard, often dirty objects. Minerals are considered by many to be part of the human diet ("vitamins and minerals"), or possibly rare ores or precious gems. However, these common perceptions are far from reality.

Figure 2.1 shows the relation between rocks and minerals. Minerals are the building blocks that make up rocks. Thus, **rock** can be defined simply as an aggregate of minerals. Here, the term *aggregate* implies that the minerals occur together as a *mixture*; in other words, each mineral retains its distinctive properties.

Although most rocks are composed of more than one mineral, certain minerals are commonly found by themselves in large, impure quantities. In these

Granite
(Rock)

Quartz
(Mineral)

Hornblende
(Mineral)

Feldspar
(Mineral)

FIGURE 2.1 Rocks are aggregates of one or more minerals. (All photos by E. J. Tarbuck)

instances they are considered to be a rock. A common example is the mineral *calcite*, which frequently is the dominant constituent in large rock units, where it is given the name *limestone.*

Minerals, the building blocks of rocks, have a very precise, specific definition. A **mineral** is defined as a naturally occurring inorganic solid that possesses a definite chemical structure, which gives it a unique set of physical properties. Thus, for any Earth material to be considered a mineral, it must exhibit the following characteristics:

1. It must be naturally occurring.
2. It must be inorganic.
3. It must be a solid.
4. It must possess a definite chemical structure.

When the term *mineral* is used by geologists, only those substances that fulfill these precise conditions are meant. Consequently, synthetic diamonds, although chemically the same as natural diamonds, are not considered minerals. Further, oil and natural gas, which are not inorganic solids, and coal, which is produced from decayed organic matter, are also not classified as minerals by geologists. In the fields of business and economics, however, a "mineral" is anything of commercial value that is extracted from Earth. Thus, people in other professions refer to coal and petroleum as "mineral resources."

Composition and Structure of Minerals

Each of Earth's nearly 4000 minerals is a unique chemical compound, defined by its chemical composition and internal structure. What elements make up a particular mineral? In what pattern are the elements bonded together?

At this point, let us briefly review the basic building blocks of minerals, the elements. At present, over 100 elements are known, although a dozen and a half of them have been produced only in the laboratory (see Periodic Table in Appendix B). Some minerals, such as gold and sulfur, are made entirely from one element. But most minerals are a combination of two or more elements, joined to form a chemically stable compound. To better understand how elements combine to form compounds, we must first consider the **atom**, the smallest part of matter that still retains the characteristics of an element, because it is this extremely small particle that does the combining.

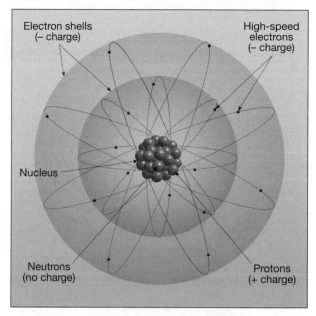

FIGURE 2.2 Simplified model of an atom. An atom consists of a central nucleus composed of protons and neutrons that is encircled by electrons.

Atomic Structure

A simplified model of an atom is shown in Figure 2.2. Each atom has a central region, called the **nucleus**, which contains very dense **protons** (particles with positive electrical charges) and equally dense **neutrons** (particles with neutral electrical charges). Orbiting the nucleus are **electrons**, which have negative electrical charges. For convenience, we often diagram atoms to show the electrons orbiting the nucleus, like the orderly orbiting of the planets around the Sun. However, electrons move so rapidly that they create a sphere-shaped negative zone around the nucleus. Hence, a more realistic picture of the positions of electrons can be obtained by envisioning a cloud of negatively charged electrons surrounding the nucleus. It is also known that individual electrons are located at given distances from the nucleus in regions called **energy-levels**, or **shells**. As we shall see, an important fact about these shells is that each can have only a specific number of electrons.

The number of protons found in an atom's nucleus determines the **atomic number** and name of the element. For example, all atoms with six protons are carbon atoms, all those with eight protons are oxygen atoms, and so forth. Since "free" atoms (those not combined with other atoms) have the same number of electrons as protons, the atomic number also equals the number of electrons surrounding the

TABLE 2.1 Atomic number and distribution of electrons

Element	Symbol	Atomic Number	Number of Electrons in Each Shell			
			1st	2nd	3rd	4th
Hydrogen	H	1	1			
Helium	He	2	2			
Lithium	Li	3	2	1		
Beryllium	Be	4	2	2		
Boron	B	5	2	3		
Carbon	C	6	2	4		
Nitrogen	N	7	2	5		
Oxygen	O	8	2	6		
Fluorine	F	9	2	7		
Neon	Ne	10	2	8		
Sodium	Na	11	2	8	1	
Magnesium	Mg	12	2	8	2	
Aluminum	Al	13	2	8	3	
Silicon	Si	14	2	8	4	
Phosphorus	P	15	2	8	5	
Sulfur	S	16	2	8	6	
Chlorine	Cl	17	2	8	7	
Argon	Ar	18	2	8	8	
Potassium	K	19	2	8	8	1
Calcium	Ca	20	2	8	8	2

nucleus (Table 2.1). Moreover, since neutrons have no charge, the positive charge of the protons is exactly balanced by the negative charge of the electrons. Consequently, uncombined atoms are neutral electrically and have no overall electrical charge. Thus, an **element** is a large collection of electrically neutral atoms, all having the same atomic number.

The simplest element, hydrogen, is composed of atoms that have only one proton in the nucleus and one electron orbiting the nucleus. Each successively heavier atom has one more proton and one more electron, plus neutrons (Table 2.1). Studies of electron configurations have shown that, as elements increase in complexity, each electron is added in a systematic fashion to a particular energy level or shell. In general, electrons enter higher energy levels only after lower energy levels have been filled to capacity. The innermost shell can have a maximum of two electrons, while additional shells hold eight or more electrons. However, the outermost shell may contain a maximum of only eight electrons. As we shall see, these outermost electrons are the ones generally involved in bonding the elements together to form compounds.

How Atoms Bond Together

Elements combine with each other to form a wide variety of more complex substances called compounds. A **compound** is a substance composed of two or more elements bonded together in definite proportions. When the elements separate, the bonds are broken and the compound no longer exists; only its component elements do. Through experimentation it has been learned that the forces bonding the atoms together are electrical in nature. Further, it is known that chemical bonding results in a change in the arrangement of electrons of the bonded atoms.

Importantly, most elements have fewer than the maximum number of electrons in their outermost shells. However, every atom seeks a full outer shell to become chemically stable. This means that the atoms of most elements are "chemically active," seeking partner atoms to bond with, which explains the great variety and occurrence of minerals.

An atom can either gain, lose, or share electrons with one or more other atoms. The result of this process is "electrical gluing" that bonds atoms together. The electrons involved in bonding are called **valance electrons**. The number of valence electrons that an element has determines the number of bonds it will form. For example, the element silicon has four valence electrons and forms four bonds in the process of completing its outer shell. On the other hand, oxygen forms two bonds, and hydrogen forms only one.

Ionic Bonds Perhaps the easiest type of bond to visualize is an **ionic bond**. In ionic bonding, one or more valence electrons are transferred from one atom to another. One atom becomes stable by giving up its valence electrons and the other uses them to complete its outer shell, making itself stable. A common example of ionic bonding is sodium (Na) and chlorine (Cl) bonding to produce sodium chloride (common table salt). This is shown in Figure 2.3. Notice that sodium loses its single outer electron to chlorine. As a result, sodium acquires a stable configuration by having eight electrons in its outermost shell. By adding an electron, a chlorine atom fills its outermost shell and becomes stable, too.

You might think that this makes the atoms electrically neutral, but this is not so. Although the sodium and chlorine atoms form a stable unit *together*, their union unbalances the atoms *internally*. The sodium atom changes from a balanced 11 protons/11 electrons to an unbalanced 11 protons/10 electrons, giving it a *net positive charge*. The chlorine atom changes from a balanced 17 protons/17 electrons to an unbalanced 17 protons/18 electrons, giving it a *net negative charge*. Atoms such as these, which have an

electrical charge because of the unequal numbers of electrons and protons, are called **ions**.

We know that particles (ions) with like charges repel and those with unlike charges attract. Thus, an ionic bond is one in which oppositely charged ions attract one another to produce a chemical compound with a neutral charge. Figure 2.4 illustrates the arrangement of sodium and chloride ions in ordinary table salt. Notice that salt consists of alternating sodium and chloride ions, positioned in such a manner that each positive ion is attracted to and surrounded on all sides by negative ions, and vice versa. This arrangement maximizes the attraction between ions with unlike electrical charges while minimizing the repulsion between ions with like charges. Thus, *ionic compounds consist of an orderly arrangement of oppositely charged ions assembled in a definite ratio that provides overall electrical neutrality.*

The properties of a chemical compound are dramatically different from the properties of the elements composing it. For example, chlorine is a

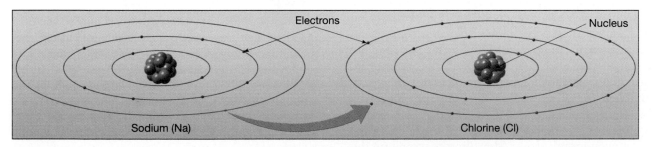

FIGURE 2.3 Chemical bonding of sodium and chlorine atoms to produce sodium chloride. Through the transfer of one electron in the outer shell of a sodium atom to the outer shell of a chlorine atom, the sodium becomes a positive ion and chlorine a negative ion.

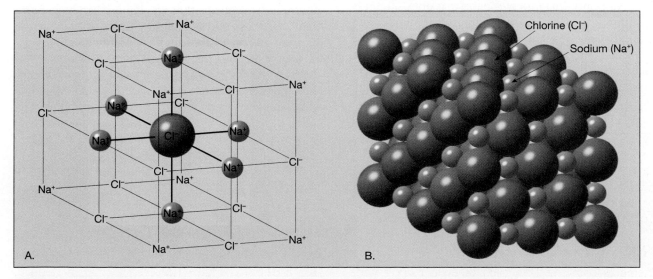

FIGURE 2.4 Schematic diagrams illustrating the arrangement of sodium and chloride ions in table salt. **A.** Structure has been opened up to show arrangement of ions. **B.** Actual ions are closely packed.

green, poisonous gas that is so toxic it was used as a weapon during World War I. Sodium is a soft, silvery metal that reacts vigorously with water and, if held in your hand, could burn it severely. Together, however, these atoms produce the compound sodium chloride (table salt), a clear crystalline solid that is essential for human life. This example also illustrates an important difference between a rock and a mineral. Most *minerals* are chemical compounds with unique properties that are very different from the elements which make them up. A *rock*, on the other hand, is a *mixture* of minerals, with each mineral retaining its own identity.

Covalent Bonds Not all atoms combine by transferring electrons to form ions. Other atoms *share* electrons. For example, the gaseous elements oxygen (O_2), hydrogen (H_2), and chlorine (Cl_2) exist as stable molecules consisting of two atoms bonded together, without the complete transfer of electrons.

Figure 2.5 illustrates the sharing of a pair of electrons between two chlorine atoms to form a molecule of chlorine gas (Cl_2). By overlapping their outer shells, each chlorine atom shares an electron in its outer shell, which has seven electrons. Thus, each chlorine atom has acquired, through cooperative action, the needed electron to complete its outer shell. The bond produced by the sharing of electrons to acquire this stable arrangement is called a **covalent bond**. The most common mineral group, the silicates, contains the element silicon, which readily forms covalent bonds with oxygen.

A common analogy may help you visualize a covalent bond. Imagine two people at opposite ends of a dimly lit room, each reading under a separate lamp. By moving the lamps to the center of the room, they are able to combine their light sources so each can see better, sharing light. Just as the overlapping light beams meld, shared electrons are indistinguishable from each other.

As you might suspect, many chemical bonds are actually a blend, consisting to some degree of electron sharing, as in covalent bonding, and to some degree of electron transfer, as in ionic bonding. Furthermore, both ionic and covalent bonds may occur within the same compound. This occurs in many silicate minerals, where silicon and oxygen atoms are covalently bonded to form the basic building block common to all silicates. These structures in turn are ionically bonded to metallic ions, producing various electrically neutral chemical compounds.

There is another chemical bond in which mobile valence electrons serve as the "electrical glue." This type of electron sharing, found in metals such as copper, gold, aluminum, and silver, is called **metallic bonding**. Metallic bonding accounts for the high electrical conductivity of metals, the ease with which metals are reshaped, and numerous other special properties of metals.

Isotopes and Radioactivity

Subatomic particles, such as protons, are so incredibly small that a special unit was devised to express their mass. A proton or a neutron has a mass just slightly more than one *atomic mass unit*, whereas an electron is only about one two-thousandth of an atomic mass unit. Thus, although electrons play an active role in chemical reactions, they do not contribute significantly to the mass of an atom.

The **mass number** of an atom is simply the total of its neutrons and protons in the nucleus. Atoms of the same element always have the same number of protons, but commonly have varying numbers of neutrons. This means that an element can have more than one mass number. These variants of the same element are called **isotopes** of that element.

For example, carbon has two well-known isotopes, one having a mass number of 12 (carbon-12), the other a mass number of 14 (carbon-14). All atoms of the same element must have the same number of

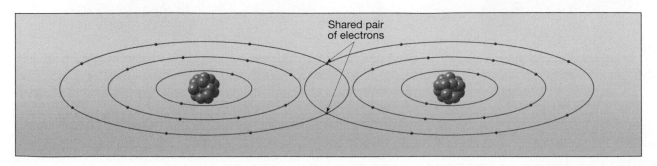

FIGURE 2.5 Illustration of the sharing of a pair of electrons between two chlorine atoms to form a chlorine molecule.

protons (atomic number), and carbon always has six. Hence, carbon-12 must have six protons plus six neutrons to give it a mass number of 12, whereas carbon-14 must have six protons plus *eight* neutrons to give it a mass number of 14. The *average* atomic mass of any random sample of carbon is much closer to 12 than 14, because carbon-12 is the more common isotope. This average is called *atomic weight.**

Note that in a chemical sense all isotopes of the same element are nearly identical. To distinguish among them would be like trying to differentiate individual members from a group of similar objects, all having the same shape, size, and color, with only some being slightly heavier. Different isotopes of an element can become parts of the same mineral.

Although the vast majority of atoms are stable, many elements do have isotopes that are unstable. Unstable isotopes, such as carbon-14, disintegrate naturally through a process called **radioactivity**. It is so called because a disintegrating atom actively radiates energy and particles. Radioactivity occurs when the forces that bind the nucleus are not strong enough to keep it together. The rate at which the unstable nuclei break apart (decay) is steady and measurable, thus making such isotopes useful

------

*The term *weight* as used here is a misnomer that has been sanctioned by long use. The correct term is atomic *mass*.

"clocks" for dating the events of Earth history. A discussion of radioactivity and its applications in dating past geologic events can be found in Chapter 18.

Properties of Minerals

Minerals occur in a fascinating palette of colors, shapes, and lusters. They vary in hardness and may even have a distinctive taste or scent. We will now look at these and other properties that are used to help identify minerals in the field.

Crystal Form

Crystal form is the external expression of the orderly internal arrangement of atoms. Generally, when a mineral forms without space restrictions, it will develop individual crystals with well-formed crystal faces. Figure 2.6A illustrates this for pyrite, an iron-and-sulfur mineral with cubic crystals. Figure 2.6B shows the distinctive hexagonal crystals of quartz that grow when space and time permit. However, most of the time, crystal growth is severely constrained. It is stunted because of competition for space, resulting in an intergrown mass of small, jammed crystals, none of which exhibits its crystal form. Thus, most inorganic solid objects are composed of crystals, but they are not clearly visible to the unaided eye.

A.

B.

FIGURE 2.6 Crystal form is the external expression of a mineral's orderly internal structure. **A.** Pyrite, commonly known as "fool's gold," often forms cubic crystals that may contain parallel lines called striations. (Photo by E. J. Tarbuck) **B.** Quartz sample that exhibits well-developed hexagonal (six-sided) crystals with pyramidal-shaped ends. (Photo by Breck P. Kent)

Luster

Luster is the appearance or quality of light reflected from the surface of a mineral. Minerals that have the appearance of metals, regardless of color, are said to have a *metallic luster* like the pyrite crystals in Figure 2.6A. Minerals with a *nonmetallic luster* are described by various adjectives, including vitreous (glassy), pearly, silky, resinous, and earthy (dull). Some minerals appear somewhat metallic in luster and are said to be *submetallic*.

Color

Although **color** is an obvious feature of a mineral, it is often an unreliable diagnostic property. Slight impurities in the common mineral quartz, for example, give it a variety of colors, including pink, purple (amethyst), white, and even black (see Figure 2.18).

Streak

Streak is the color of a mineral in its powdered form, which is a much more consistent indication of color. Streak is obtained by rubbing the mineral across a piece of unglazed porcelain termed a *streak plate*. Whereas the color of a mineral may vary from sample to sample, the streak usually does not, and is therefore the more reliable property. Streak can also be an aid in distinguishing minerals with metallic lusters from those having nonmetallic lusters. Metallic minerals generally have a dense, dark streak, whereas minerals with nonmetallic lusters do not.

Hardness

One of the most useful diagnostic properties is **hardness**, a measure of the resistance of a mineral to abrasion or scratching. This property is determined by rubbing the mineral to be identified against another mineral of known hardness, or vice versa. One will scratch the other (unless they have the same hardness).

Geologists use a standard hardness scale, called the **Mohs scale**. It consists of ten minerals arranged in order from 1 (softest) to 10 (hardest) as shown in Table 2.2. Any mineral of unknown hardness can be rubbed against these to determine its hardness. In the field, other handy objects work, too. For example, your fingernail has a hardness of 2.5, a copper penny 3, and a piece of glass 5.5. The mineral gypsum, which has a hardness of 2, can be easily scratched with your fingernail. On the other hand, the similar-looking mineral calcite, which has a hardness of 3, will scratch your fingernail but will not scratch glass.

TABLE 2.2 Mohs scale of mineral hardness

Relative Scale		Mineral	Hardness of Some Common Objects
Hardest	10	Diamond	
	9	Corundum	
	8	Topaz	
	7	Quartz	
	6	Potassium Feldspar	5.5 Glass, Pocketknife
	5	Apatite	
	4	Fluorite	3.5 Copper Penny
	3	Calcite	2.5 Fingernail
	2	Gypsum	
Softest	1	Talc	

Quartz, the hardest of the common minerals, will scratch a glass plate. You can see that by rubbing an unknown mineral against objects of known hardness, you can quickly narrow its hardness range.

Cleavage

In the crystal structure of a mineral, some bonds may be weaker than others. These bonds are where a mineral will break when it is stressed. **Cleavage** is the tendency of a mineral to break along planes of weak bonding. Not all minerals have definite planes of weak bonding, but those that possess cleavage can be identified by the distinctive smooth surfaces which are produced when the mineral is broken.

The simplest type of cleavage is exhibited by the micas (Figure 2.7). Because the micas have weak bonds in one direction, they cleave to form thin, flat sheets. Some minerals have several cleavage planes which produce smooth surfaces when broken, while

FIGURE 2.7 The thin sheets shown here were produced by splitting a mica (muscovite) crystal parallel to its perfect cleavage. (Photo by Breck P. Kent)

others exhibit poor cleavage, and still others have no cleavage at all. When minerals break evenly in more than one direction, cleavage is described by the *number of planes* exhibited and the *angles at which they meet* (Figure 2.8).

Do not confuse cleavage with crystal form. When a mineral exhibits cleavage, it will break into pieces *that have the same shape as the original sample.* By contrast, the quartz crystals shown in Figure 2.6B do not have cleavage. If broken, they fracture into shapes that do not resemble each other or the original crystals.

Fracture

Minerals that do not exhibit cleavage when broken, such as quartz, are said to **fracture**. Even fracturing has variety: minerals that break into smooth curved surfaces like those seen in broken glass have a *conchoidal fracture* (Figure 2.9). Others break into splinters or fibers, but most minerals fracture irregularly.

Specific Gravity

Specific gravity compares the weight of a mineral to the weight of an equal volume of water. For example, if a mineral weighs three times as much as an equal volume of water, its specific gravity is 3. With a little practice, you can estimate the specific gravity of minerals by hefting them in your hand. For example, if a mineral feels as heavy as the common rocks you have handled, its specific gravity will probably be somewhere between 2.5 and 3. Some metallic minerals have a specific gravity two or three times

FIGURE 2.9 Conchoidal fracture. The smooth curved surfaces result when minerals break in a glasslike manner. (Photo by E. J. Tarbuck)

that of common rock-forming minerals. Galena, which is an ore of lead, has a specific gravity of roughly 7.5, whereas the specific gravity of 24-karat gold is approximately 20 (Figure 2.10).

Other Properties of Minerals

In addition to the properties already discussed, some minerals can be recognized by other distinctive properties. For example, halite is ordinary salt, so it is quickly identified with your tongue. Thin sheets of mica will bend and elastically snap back. Gold is malleable and can be easily shaped. Talc and graphite both have distinctive feels; talc feels soapy and graphite feels greasy. A few minerals, such as

FIGURE 2.8 Smooth surfaces produced when a mineral with cleavage is broken. The sample on the left (fluorite) exhibits four planes of cleavage (eight sides), whereas the other two samples exhibit three planes of cleavage (six sides). Also notice that the mineral in the center (halite) has cleavage planes that meet at 90-degree angles, whereas the mineral on the right (calcite) has cleavage planes that meet at 75-degree angles. (Photo by E. J. Tarbuck)

FIGURE 2.10 Galena is lead sulfide. Like other metallic ores, it has a relatively high specific gravity and therefore feels heavy when hefted. (Photo by E. J. Tarbuck)

magnetite, have a high iron content and can be picked up with a magnet, while some varieties (lodestone) are natural magnets and will pick up small iron-based objects such as pins and paper clips. Some minerals exhibit special optical properties. For example, when a transparent piece of calcite is placed over printed material, the letters appear twice. This optical property is known as *double refraction*. In addition, the streak of many sulfur-bearing minerals smells like rotten eggs.

One very simple chemical test involves placing a drop of dilute hydrochloric acid from a dropper bottle on a freshly broken mineral surface. Certain minerals, called carbonates, will effervesce (fizz) with hydrochloric acid. This test is useful in identifying the mineral calcite, which is a common carbonate mineral.

In summary, a number of special physical and chemical properties are useful in identifying certain minerals. These include taste, smell, elasticity, malleability, feel, magnetism, double refraction, and chemical reaction with hydrochloric acid. Remember that every one of these properties depends on the composition (elements) of a mineral and its structure (how the atoms are arranged).

Mineral Groups

Nearly 4000 minerals have been named and about 40 to 50 new ones are being identified each year.* Fortunately, for students who are beginning to study minerals, no more than a few dozen are abundant!

---

*Appendix C describes some of the common minerals of Earth's crust.

Collectively, these few make up most of the rocks of Earth's crust and, as such, are classified as the *rock-forming minerals*. It is also interesting to note that *only eight elements* compose the bulk of these minerals and represent over 98 percent (by weight) of the continental crust (Table 2.3).

The two most abundant elements are silicon and oxygen, which combine to form the framework of the most common mineral group, the **silicates**. Perhaps the next most common mineral group is the carbonates, of which calcite is the most prominent member. Other common rock-forming minerals include gypsum and halite.

We will first discuss the most common mineral group, the silicates, and then consider some of the other prominent mineral groups.

Silicate Structures

Every silicate mineral contains the elements oxygen and silicon. Moreover, except for a few minerals such as quartz, every silicate mineral includes one or more additional elements that are needed to produce electrical neutrality. These additional elements give rise to the great variety of silicate minerals and their varied properties.

All silicates have the same fundamental building block, the **silicon-oxygen tetrahedron**. This structure consists of four oxygen atoms surrounding a much smaller silicon atom (Figure 2.11). The silicon-oxygen tetrahedron is not a compound, because it is "incomplete" and unstable. Rather, it is an ion (SiO_4^{4-}) with a charge of -4. This negative charge results because each of the four oxygen ions contributes a charge of -2, for a total of -8, whereas the one silicon atom has a charge of $+4$ (Figure 2.12). In nature, one of the simplest ways in which these tetrahedra become neutral compounds is

TABLE 2.3 Relative abundance of the most common elements in Earth's crust (Source: Data from Brian Mason.)

Element	Approximate Percentage by Weight
Oxygen (O)	46.6
Silicon (Si)	27.7
Aluminum (Al)	8.1
Iron (Fe)	5.0
Calcium (Ca)	3.6
Sodium (Na)	2.8
Potassium (K)	2.6
Magnesium (Mg)	2.1
All others	1.5
Total	100

FIGURE 2.11 Two representations of the silicon-oxygen tetrahedron. **A.** The four large spheres represent oxygen atoms and the blue sphere represents a silicon atom. The spheres are drawn in proportion to the radii of the atoms. **B.** A model of the tetrahedron using rods to depict the bonds that connect the atoms.

the tetrahedra may be joined to form single chains, double chains, or sheet structures, as shown in Figure 2.13. The joining of tetrahedra in each of these configurations results from the sharing of oxygen atoms between silicon atoms in adjacent tetrahedra.

To better understand how this sharing takes place, select one of the silicon atoms (small blue spheres) near the middle of the single chain structure shown in Figure 2.13. Notice that this silicon atom is completely surrounded by four larger oxygen atoms (you are looking *through* one of the four, to see the blue silicon atom). Also notice that, of the four oxygen atoms, two are joined to other silicon atoms, whereas the other two are not shared in this manner. *It is the linkage across the shared oxygen atoms that joins the tetrahedra into a chain structure.* Now, examine a silicon atom near the middle of the sheet structure and count the number of shared and unshared oxygen atoms surrounding it. The increase in the degree of sharing accounts for the sheet structure. Other silicate structures exist, and the most common has all of the oxygen atoms shared to produce a complex three-dimensional framework (not shown because of its complexity).

By now you can see that the ratio of oxygen atoms to silicon atoms differs in each of the silicate structures. In the isolated tetrahedron there are 4 oxygen atoms for every silicon atom. In the single chain, the oxygen-to-silicon ratio is 3 to 1, and in the three-dimensional framework this ratio is 2 to 1. Consequently, as more of the oxygen atoms are shared, the percentage of silicon in the structure increases. Silicate minerals are therefore described as having a "high" or "low" silicon content, based on their ratio of oxygen to silicon. This difference in silicon content is important, as we shall see in a later chapter when we consider the formation of igneous rocks.

through the addition of positively charged ions (Figure 2.12). In this way a chemically stable structure is produced, consisting of individual tetrahedra linked together by positively charged ions.

In addition to positive ions acting as the "glue" to bind the tetrahedra, the tetrahedra themselves may be linked in a variety of configurations. For example,

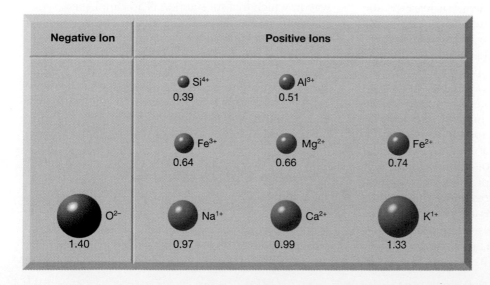

Negative Ion	Positive Ions		
	Si^{4+} 0.39	Al^{3+} 0.51	
	Fe^{3+} 0.64	Mg^{2+} 0.66	Fe^{2+} 0.74
O^{2-} 1.40	Na^{1+} 0.97	Ca^{2+} 0.99	K^{1+} 1.33

FIGURE 2.12 Relative sizes and electrical charges of the eight most common elements in Earth's crust. These are the most common ions in rock-forming minerals. Ionic radii are expressed in angstroms (1 angstrom equals 10^{-8} cm).

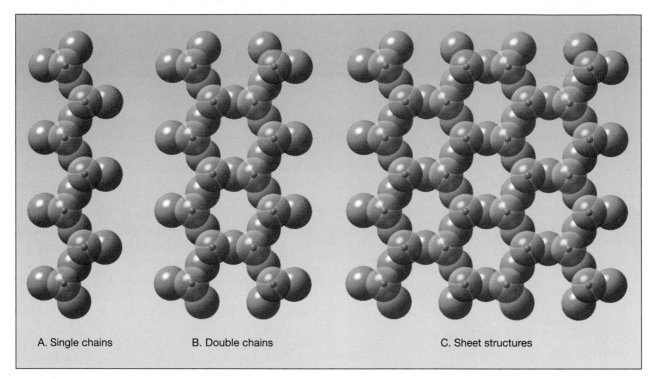

A. Single chains   B. Double chains   C. Sheet structures

FIGURE 2.13 Three types of silicate structures. **A.** Single chains. **B.** Double chains. **C.** Sheet structures.

Most silicate structures, like those shown in Figure 2.13, are not neutral chemical compounds. Thus, like the individual tetrahedra, they all are neutralized by the inclusion of charged metallic ions that bond them together into a variety of crystalline configurations. The ions that most often link silicate structures are those of the elements iron (Fe), magnesium (Mg), potassium (K), sodium (Na), aluminum (Al), and calcium (Ca).

Notice in Figure 2.12 that each of these positive ions has a particular atomic size and a particular charge. Generally, ions of approximately the same size are able to freely substitute for one another. For instance, the ions of iron (Fe^{2+}) and magnesium (Mg^{2+}) are nearly the same size and substitute for each other without altering the mineral structure. This also holds true for calcium and sodium, which can occupy the same site in a crystalline structure, and aluminum (Al), which substitutes for silicon in the silicon-oxygen tetrahedron.

Because of the ability of mineral structures to readily accommodate different ions at a given bonding site, individual specimens of a particular mineral may contain varying amounts of certain elements. A mineral of this type is often expressed by a chemical formula that uses parentheses to show the variable component. A good example is the mineral olivine, $(Mg, Fe)_2SiO_4$, which is magnesium/iron silicate. As you can see from the formula, it is the iron (Fe^{2+}) and magnesium (Mg^{2+}) ions in olivine that freely

substitute for each other. At one extreme, olivine may contain iron without any magnesium (Fe_2SiO_4, or iron silicate) and at the other, iron is totally lacking (Mg_2SiO_4, or magnesium silicate). Between these end members, any ratio of iron to magnesium is possible. Thus olivine, as well as many other silicate minerals, is actually a *family* of minerals with a range of composition between the two end members.

In certain substitutions, the ions that interchange do not have the same electrical charge. For instance, when calcium (Ca^{2+}) substitutes for sodium (Na^{1+}), the structure gains a positive charge. In nature, one way in which this substitution is accomplished, while still maintaining overall electrical neutrality, is the simultaneous substitution of aluminum (Al^{3+}) for silicon (Si^{4+}). This particular double substitution occurs in a mineral called plagioclase feldspar. It is a member of the most abundant family of minerals found in Earth's crust. The end members of this particular feldspar series are a calcium-aluminum silicate (anorthite, $CaAl_2Si_2O_8$), and sodium-aluminum silicate (albite, $NaAlSi_3O_8$).

Common Silicate Minerals

To reiterate, silicate minerals are those that have the silicate ion (SiO_4^{4-}) as their basic building block. Major groups of silicate minerals and common examples are given in Figure 2.14. The feldspars are by far the most plentiful group, comprising over 50 percent

Mineral		Idealized Formula	Cleavage	Silicate Structure	
Olivine		(Mg, Fe)$_2$SiO$_4$	None	Single tetrahedron 	
Pyroxene group (Augite)		(Mg,Fe)SiO$_3$	Two planes at right angles	Single chains 	
Amphibole group (Hornblende)		Ca$_2$(Fe,Mg)$_5$Si$_8$O$_{22}$(OH)$_2$	Two planes at 60° and 120°	Double chains 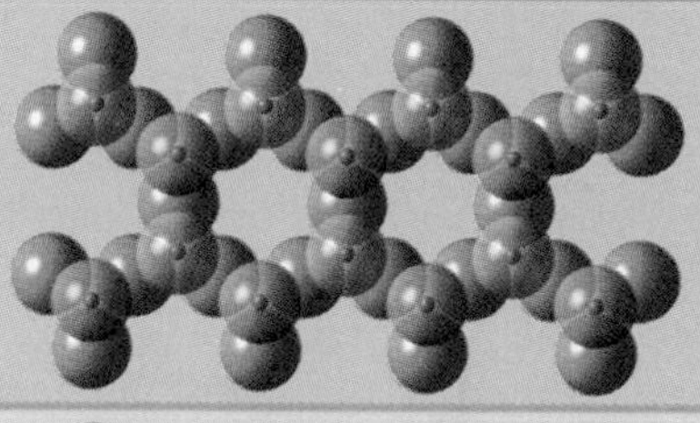	
Micas	Biotite	K(Mg,Fe)$_3$AlSi$_3$O$_{10}$(OH)$_2$	One plane	Sheets 	
	Muscovite	KAl$_2$(AlSi$_3$O$_{10}$)(OH)$_2$			
Feld-spars	Orthoclase	KAlSi$_3$O$_8$	Two planes at 90°	Three-dimensional networks	
	Plagioclase	(Ca,Na)AlSi$_3$O$_8$			
Quartz		SiO$_2$	None	(Expanded view)	

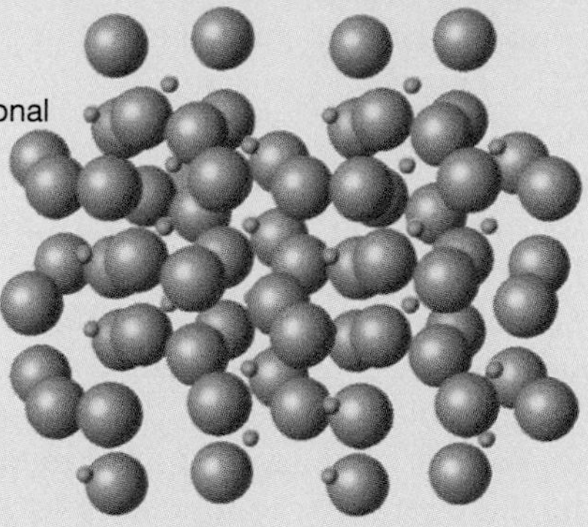

FIGURE 2.14 Common silicate minerals. Note that the complexity of the silicate structure increases down the chart.

of Earth's crust. Quartz, the second most abundant mineral in the continental crust, is the only common mineral made completely of silicon and oxygen.

Notice in Figure 2.14 that each mineral *group* has a particular silicate *structure*, and that a relationship exists between the internal structure of a mineral and the *cleavage* it exhibits. Because the silicon-oxygen bonds are strong, silicate minerals tend to cleave between the silicon-oxygen structures rather than across them. For example, the micas have a sheet structure and thus tend to cleave into flat plates (see Figure 2.7). Quartz, which has equally strong silicon-oxygen bonds in all directions, has no cleavage, but fractures instead.

Most silicate minerals form (crystallize) as molten rock is cooling. This cooling can occur at or near Earth's surface (low temperature and pressure) or at great depths (high temperature and pressure). The environment during crystallization and the chemical composition of the molten rock determine to a large degree which minerals are produced. For example, the silicate mineral olivine crystallizes at high temperatures, whereas quartz crystallizes at much lower temperatures.

In addition, some silicate minerals form at Earth's surface from the weathered products of older silicate minerals. Still other silicate minerals are formed under the extreme pressures associated with mountain-building. Each silicate mineral, therefore, has a structure and a chemical composition that *indicate the conditions under which it formed.* Thus, by carefully examining the mineral constituents of rocks, geologists can often determine the circumstances under which the rocks formed.

We will now examine some of the most common silicate minerals, which we divide into two major groups on the basis of their chemical makeup.

The Dark Silicate Minerals The *dark* (or *ferromagnesian*) *silicates* are those minerals containing ions of iron (iron = *ferro*) and/or magnesium in their structure. Because of their iron content, ferromagnesian silicates are dark in color and have greater specific gravity, between 3.2 and 3.6, than do nonferromagnesian silicates. The most common dark silicate minerals are olivine, the pyroxenes, the amphiboles, and the dark mica, biotite.

Olivine is black to olive green in color and has a glassy luster and a conchoidal fracture. Although not abundant in continental rocks, olivine is thought to be a major component of Earth's upper mantle.

The *pyroxenes* are actually a group of chemically complex minerals. The most common member, *augite*, is a black, opaque mineral that is found in *basalt*, a common igneous rock of the oceanic crust and volcanic areas on the continents.

Hornblende is the most common member of a group of silicate minerals called amphiboles (Figure 2.15). Hornblende is usually dark green to black in color and very similar in appearance to augite. Predominately found in continental rocks of igneous origin, hornblende makes up the dark portion of an otherwise light-colored rock.

Biotite is the dark iron-rich member of the mica family. Like other micas, biotite possesses a sheet structure that gives it excellent cleavage in one direction. Biotite also has a shiny black appearance that helps distinguish it from the other dark ferromagnesian minerals. Like hornblende, biotite is a common constituent of continental rocks, including the igneous rock granite.

The Light Silicate Minerals As the name implies, the *light* (or *nonferromagnesian*) *silicates* are generally light in color and have a specific gravity of about 2.7, which is considerably less than the

FIGURE 2.15 Hornblende amphibole. Hornblende is a common dark silicate material having two cleavage directions that intersect at roughly 60° and 120°. (Photo by E. J. Tarbuck)

ferromagnesian silicates. As indicated earlier, these differences are mainly attributable to the presence or absence of iron and magnesium. The light silicates contain varying amounts of aluminum, potassium, calcium, and sodium, rather than iron and magnesium.

The most common mineral group found in Earth's crust are the nonferromagnesians known as the *feldspars*. Because the feldspars form (crystallize) under a wide range of pressures and temperatures, they are a major constituent of most igneous rocks. All of the feldspars have similar physical properties. They have two planes of cleavage meeting at or near 90 degree angles, are relatively hard (6 on the Mohs scale), and have a luster that ranges from glassy to pearly.

Two different feldspar structures exist. *Orthoclase feldspar* is a common member of a group of feldspar minerals that contains potassium ions in its

FIGURE 2.16 Sample of orthoclase feldspar. (Photo by E. J. Tarbuck)

structure. The other group, called *plagioclase feldspar*, contains both sodium and calcium ions that freely substitute for one another depending on the environment during crystallization.

Orthoclase feldspar is usually light cream to salmon pink in color (Figure 2.16). The plagioclase feldspars, on the other hand, range in color from white to medium gray. However, color should not be used to distinguish these groups. The only sure way to physically distinguish the feldspars is to look for a multitude of fine parallel lines, called *striations*. Striations are found on some cleavage planes of plagioclase feldspar, but are not present on orthoclase feldspar (Figure 2.17).

Quartz is the only common silicate mineral consisting entirely of silicon and oxygen. As such, the term *silica* is applied to quartz, which has the chemical formula SiO_2. Quartz is hard, resistant to weathering, and does not have cleavage. When broken, quartz generally exhibits conchoidal fracture. In a pure form, quartz is clear and if allowed to solidify without interference, will form hexagonal crystals that develop pyramid-shaped ends (see Figure 2.6B). However, like most other clear minerals, quartz is often colored (Figure 2.18) by the inclusion of various ions (impurities) and forms without developing good crystal faces. The most common varieties of quartz are milky (white), smoky (gray), rose (pink), amethyst (purple), and rock crystal (clear).

Muscovite is a common member of the mica family. It is light in color and has a pearly luster. Like other micas, muscovite has excellent cleavage in one direction (see Figure 2.7). In thin sheets muscovite is clear, a property that accounts for its use as window "glass" during the Middle Ages. Because muscovite is very shiny, it can often be identified by the sparkle it gives a rock.

FIGURE 2.18 Quartz. Some minerals, such as quartz, occur in a variety of colors. These samples include crystal quartz (colorless), amethyst (purple), citrine (yellow), and smoky quartz (gray to black). (Photo by E. J. Tarbuck)

Important Nonsilicate Minerals

Other mineral groups can be considered scarce when compared to the silicates, although many are important economically. Table 2.4 lists examples of oxides, sulfides, sulfates, halides, native elements, and carbonates of economic value. A discussion of a few of the more common nonsilicate, rock-forming minerals follows.

The carbonate minerals are much simpler structurally than the silicates. This mineral group is composed of the carbonate ion (CO_3^{2-}), and one or more kinds of positive ions. The two most common carbonate minerals are *calcite*, $CaCO_3$ (calcium carbonate), and *dolomite*, $CaMg(CO_3)_2$ (calcium/magnesium carbonate). Because these minerals are similar both physically and chemically, they are difficult to distinguish from one another. Both have a vitreous luster, a hardness between 3 and 4, and nearly perfect rhombic cleavage (see Figure 2.8, right side). They can, however, be distinguished by using dilute hydrochloric acid. Calcite reacts vigorously with this acid, whereas dolomite reacts much more slowly. Calcite and dolomite are usually found together as the primary constituents in the sedimentary rocks limestone and dolostone. When calcite is the dominant mineral, the rock is called *limestone*, whereas *dolostone* results from a predominance of dolomite. Limestone has numerous economic uses, including uses as road aggregate, as building stone, and as the main ingredient in portland cement.

Two other nonsilicate minerals frequently found in sedimentary rocks are *halite* and *gypsum*. Both minerals are commonly found in thick layers, which

FIGURE 2.17 These parallel lines, called striations, are a distinguishing characteristic of the plagioclase feldspars. (Photo by E. J. Tarbuck)

TABLE 2.4 Common nonsilicate mineral groups

Mineral Group	Name	Chemical Formula	Economic Use
Oxides	Hematite	Fe_2O_3	Ore of iron, pigment
	Magnetite	Fe_3O_4	Ore of iron
	Corundum	Al_2O_3	Gemstone, abrasive
	Ice	H_2O	Solid form of water
	Uraninite	UO_2	Ore of uranium
Sulfides	Galena	PbS	Ore of lead
	Sphalerite	ZnS	Ore of zinc
	Pyrite	FeS_2	Sulfuric acid production
	Chalcopyrite	$CuFeS_2$	Ore of copper
	Cinnabar	HgS	Ore of mercury
Sulfates	Gypsum	$CaSO_4 \cdot 2H_2O$	Plaster
	Anhydrite	$CaSO_4$	Plaster
	Barite	$BaSO_4$	Drilling mud
Native elements	Gold	Au	Trade, jewelry
	Copper	Cu	Electrical conductor
	Diamond	C	Gemstone, abrasive
	Sulfur	S	Sulfa drugs, chemicals
	Graphite	C	Pencil lead, dry lubricant
	Silver	Ag	Jewelry, photography
	Platinum	Pt	Catalyst
Halides	Halite	NaCl	Common salt
	Fluorite	CaF_2	Used in steel making
	Sylvite	KCl	Fertilizer
Carbonates	Calcite	$CaCO_3$	Portland cement, lime
	Dolomite	$CaMg(CO_3)_2$	Portland cement, lime

are the last vestiges of ancient seas that have long since evaporated (Figure 2.19). Like limestone, both are important nonmetallic resources. Halite is the mineral name for common table salt (NaCl). Gypsum ($CaSO_4 \cdot 2H_2O$, which is calcium sulfate with water bound into the structure) is the mineral of which plaster and other similar building materials are composed.

In addition, a number of other minerals are prized for their economic value (see Table 2.4). Included in this group are the ores of metals, such as hematite (iron), sphalerite (zinc), and galena (lead); the native (free-occurring, not in compounds) elements, including gold, silver, carbon (diamonds); and a host of others, such as fluorite, corundum, and uraninite (a uranium source).

Mineral Resources

Earth's crust and oceans are the source of a wide variety of useful and essential substances (see Box 2.1). In fact, practically every manufactured product contains materials obtained from minerals. Table 2.4 lists important mineral groups in this category.

Mineral resources are the endowment of useful minerals ultimately available commercially. Resources include already identified deposits from which minerals can be extracted profitably, called **reserves**, as well as known deposits that are not yet economically or technologically recoverable. Deposits inferred to exist, but not yet discovered, are also considered as mineral resources. In addition, the term **ore** is used to denote those useful metallic minerals that can be mined at a profit. In common usage, the term *ore* is also applied to some nonmetallic minerals such as fluorite and sulfur. However, materials used for such purposes as building stone, road aggregate, abrasives, ceramics, and fertilizers are not usually called ores; rather, they are classified as industrial rocks and minerals.

Recall that more than 98 percent of Earth's crust is composed of only eight elements and, except for oxygen and silicon, all other elements make up a relatively small fraction of common crustal rocks (see Table 2.3). Indeed, the natural concentrations of many elements are exceedingly small. A deposit containing the average percentage of a valuable element

FIGURE 2.19 Thick beds of halite (salt) are being drilled at an underground mine near Grand Saline, Texas. (Courtesy of Morton International, Inc.)

It is important to realize that a deposit may become profitable to extract or lose its profitability because of economic changes. If demand for a metal increases and prices rise, the status of a previously unprofitable deposit changes, and it becomes an ore. The status of unprofitable deposits may also change if a technological advance allows the useful element to be extracted at a lower cost than before. Conversely, changing economic factors can turn a once-profitable ore deposit into an unprofitable deposit that can no longer be called an ore. This situation was illustrated at the copper mining operation located at Bingham Canyon, Utah, the largest open-pit mine on Earth.* Mining was halted there in 1985 because outmoded equipment had driven the cost of extracting the copper beyond the current selling price. The owners responded by replacing an antiquated 1000-car railroad with conveyor belts and pipelines for transporting the ore and waste. These devices achieved a cost reduction of nearly 30 percent and returned this mining operation to profitability.

Over the years, geologists have been keenly interested in learning how natural processes produce localized concentrations of essential minerals. One well-established fact is that occurrences of valuable mineral resources are closely related to the rock cycle. That is, the mechanisms that generate igneous, sedimentary, and metamorphic rocks, including the processes of weathering and erosion, play a major role in producing concentrated accumulations of useful elements. Moreover, with the development of the theory of plate tectonics, geologists have added another tool for understanding the processes by which one rock is transformed into another. As these rock-forming processes are examined in the following chapters, we will consider their role in producing some of our important mineral resources.

———

*The huge Bingham Canyon mine is described in Box 3.1.

is worthless if the cost of extracting it greatly exceeds the value of the material recovered. To be considered of value, an element must be concentrated above the level of its average crustal abundance. Generally, the lower the crustal abundance, the greater the concentration must be.

For example, copper makes up about 0.0135 percent of the crust. However, for a material to be considered as copper ore, it must contain a concentration that is about 100 times this amount. Aluminum, on the other hand, represents 8.13 percent of the crust and must be concentrated to only about four times its average crustal percentage before it can be extracted profitably.

The Chapter in Review

The following statements are intended to help you review the primary objectives presented in this chapter.

- A *mineral* is a naturally occurring inorganic solid that possesses a definite chemical structure which gives it a unique set of physical properties. Most *rocks* are aggregates composed of two or more minerals.

- The building blocks of minerals are *elements*. An *atom* is the smallest particle of matter that still retains the characteristics of an element. Each atom has a *nucleus* which contains *protons* (particles with positive electrical charges) and *neutrons* (particles with neutral electrical charges). Orbiting the nucleus of an atom in regions called *energy levels*, or *shells*, are *electrons*,

| Box 2.1 | ## Asbestos: What are the Risks? |

Are the following statements true or false? "Asbestos" is a single mineral; exposure to a single asbestos fiber can cause cancer; all asbestos should be removed from buildings. (Answer: all are false.)

All three statements are common misconceptions. Why do the public and some government and industry workers hold such misconceptions? As often happens, some nonscientists have misunderstood what they have heard. Let us look at the story of asbestos.

Asbestos mineral fibers are valuable because they are flexible, heat-resistant, chemically inert, and insulating (Figure 2.A). An estimated 30 million tons of asbestos minerals have been mined during the past century to strengthen concrete and floor tiles, to make fireproof fabrics, to insulate boilers and hot pipes, and to make automotive brake linings. During the 1950s and 1960s, wall coatings strengthened with asbestos fibers were used extensively, including in schools.

FIGURE 2.A Asbestos. This sample is a fibrous form of the mineral serpentine. (Photo by E. J. Tarbuck)

The first misconception is that "asbestos" is a single mineral. Rather, it is a group of silicate minerals that readily separate into thin, strong fibers. Most asbestos is the mineral chrysotile ("white asbestos"). It is a fibrous form of the mineral serpenti-nite, and is the only type mined in North America. The other asbestos minerals, mined in other countries, are "blue asbestos" and "brown asbestos."

"Blue" asbestos fibers definitely are a health hazard. Prolonged exposure in an unregulated workplace allows these very

which have negative electrical charges. The number of protons in an atom's nucleus determines its *atomic number* and the name of the element. An element is a large collection of electrically neutral atoms, all having the same atomic number.

- Atoms combine with each other to form more complex substances called *compounds*. Atoms bond together by either gaining, losing, or sharing electrons with other atoms. In an *ionic bond*, one or more electrons are transferred from one atom to another, giving the atom a net positive or negative charge. The resulting electrically charged atom is called an *ion*. Ionic compounds consist of an orderly arrangement of oppositely charged ions assembled in a definite ratio that provides overall electrical neutrality. Another type of bond, the *covalent bond*, is produced when atoms share electrons.

- *Isotopes* are variants of the same element, but with a different *mass number* (the total number of neutrons plus protons found in an atom's nucleus). Some isotopes are unstable and disintegrate naturally through a process called *radioactivity*.

- The properties of minerals include *crystal form, luster, color, streak, hardness, cleavage, fracture,* and *specific gravity*. In addition, a number of special physical and chemical properties (*taste, smell, elasticity, malleability, feel, magnetism, double refraction,* and *chemical reaction to hydrochloric acid*) are useful in identifying certain minerals. Each mineral has a unique set of properties which can be used for identification.

- Of the nearly 4000 minerals, no more than a few dozen make up most of the rocks of Earth's crust and as such, are classified as *rock-forming minerals*. Eight elements (oxygen, silicon, aluminum, iron, calcium, sodium,

thin, rodlike particles to be inhaled. They are not expelled easily nor break down chemically, and they can remain in the lungs for life. Three diseases may result: *asbestosis*, a lung tissue scarring that reduces oxygen absorption; *mesothelioma*, a rare cancer of the lung lining; or *lung cancer*. Studies of "blue" asbestos mine and mill workers in South Africa and Australia revealed an extremely high incidence of mesothelioma, sometimes following less than a year's exposure.

Such studies led in 1986 to the EPA's Asbestos Hazard Emergency Response Act. It required inspection of all U.S. public and private schools. This aroused public concern of an "asbestos health risk" to school children. It led people to believe our second misconception, the "one-fiber theory" that a single asbestos fiber could cause cancer. Billions of dollars have been spent on asbestos testing and removal, and many classrooms have been unusable for months during removal.

Which leads to our third misconception about asbestos. Just as not all snakes are poisonous, not all asbestos fibers are a health risk. The asbestos in U. S. construction is chrysotile, or "white" asbestos. Its curly fibers are more readily expelled from the lungs. Although extreme exposure in the workplace could cause disease, the United States Geological Survey and others conclude that most asbestos exposure in the United States is harmless. Studies of "white" asbestos workers in Canada and Italy show mesothelioma and lung cancer rates differ very little from the general public.

So, does asbestos present a risk to U. S. school children? With few exceptions, the asbestos in schools is "white" asbestos. Further, asbestos-particle levels in schools are approximately 0.01 of the level that is permissible in the workplace. Indoor concentrations are comparable to outdoor concentrations. Table 2.A compares the asbestos risk in schools to other hazards. These data clearly demonstrate that the concern over asbestos, and the large expenditure by school districts to purge their classrooms of asbestos particles, has been unwarranted. Unfortunately, the "asbestos panic" of the late 1980s established a prosperous asbestos-removal industry. Geologic misconceptions can be expensive!

TABLE 2A Estimates of risk from asbestos exposure in schools in comparison with other risks

Cause	Annual Death Rate (per million)
Asbestos exposure in schools	0.005–0.093
Whooping cough vaccination (1970–80)	1–6
Aircraft accidents (1979)	6
High school football (1970–80)	10
Drowning (ages 5–14)	27
Motor vehicle accident, Pedestrian (ages 5–14)	32
Home accidents (ages 1–14)	60
Long-term smoking	1200

potassium, and magnesium) compose the bulk of these minerals and represent over 98 percent (by weight) of Earth's continental crust.

- The most common mineral group is the *silicates.* All silicate minerals have the negatively charged *silicon-oxygen tetrahedron* as their fundamental building block. In some silicate minerals the tetrahedra are joined in chains (the pyroxene and amphibole groups); in others, the tetrahedra are arranged into sheets (the micas, biotite and muscovite), or three-dimensional networks (the feldspars and quartz). Binding the tetrahedra and various silicate structures are often the positive ions of iron, magnesium, potassium, sodium, aluminum, and calcium. Each silicate mineral has a structure and a chemical composition that indicates the conditions under which it was formed.

- The *nonsilicate* mineral groups, which contain several economically important minerals, include the *oxides* (e.g. the mineral hematite, mined for iron), *sulfides* (e.g. the mineral sphalerite, mined for zinc, and the mineral galena, mined for lead), *sulfates, halides*, and *native elements* (e.g. gold and silver). The more common nonsilicate rock-forming minerals include the *carbonate minerals*, calcite and dolomite. Two other nonsilicate minerals frequently found in sedimentary rocks are halite and gypsum.

- *Mineral resources* are the endowment of useful minerals ultimately available commercially. Resources include already identified deposits from which minerals can be extracted profitably, called *reserves*, as well as known deposits that are not yet economically or technologically recoverable. Deposits inferred to exist, but not yet discovered, are also considered as mineral resources. The term *ore* is used to denote those useful metallic minerals that can be mined for a profit, as well as some nonmetallic minerals, such as fluorite and sulfur, that contain useful substances.

Key Terms

atom (p. 25)	energy-level (p. 25)	mineral (p. 25)	reserve (p. 38)
atomic number (p. 25)	fracture (p. 31)	mineral resource (p. 38)	rock (p. 24)
cleavage (p. 30)	hardness (p. 30)	Mohs scale (p. 30)	shells (p. 25)
color (p. 30)	ion (p. 27)	neutron (p. 25)	silicates (p. 32)
compound (p. 26)	ionic bond (p. 26)	nucleus (p. 25)	silicon-oxygen tetrahe-
covalent bond (p. 28)	isotope (p. 28)	ore (p. 38)	dron (p. 32)
crystal form (p. 29)	luster (p. 30)	proton (p. 25)	specific gravity (p. 31)
electron (p. 25)	mass number (p. 28)	radioactivity (p. 29)	streak (p. 30)
element (p. 26)	metallic bond (p. 28)		valence electron (p. 26)

Questions for Review

1. Define the term *rock*.

2. List the three main particles of an atom and explain how they differ from one another.

3. If the number of electrons in an atom is 35 and its mass number is 80, calculate the following:
   a. The number of protons.
   b. The atomic number.
   c. The number of neutrons.

4. What is the significance of valence electrons?

5. Briefly distinguish between ionic and covalent bonding.

6. What occurs in an atom to produce an ion?

7. What is an isotope?

8. Although all minerals have an orderly internal arrangement of atoms (crystalline structure), most mineral samples do not demonstrate their crystal form. Why?

9. Why might it be difficult to identify a mineral by its color?

10. If you found a glassy-appearing mineral while rock hunting and had hopes that it was a diamond, what simple test might help you make a determination?

11. Explain the use of corundum as given in Table 2.4 in terms of Mohs hardness scale.

12. Gold has a specific gravity of almost 20. If a 25-liter pail of water weighs about 25 kilograms, how much would a 25-liter pail of gold weigh?

13. Explain the difference between the terms *silicon* and *silicate*.

14. What do ferromagnesian minerals have in common? List examples of ferromagnesian minerals.

15. What do muscovite and biotite have in common? How do they differ?

16. Should color be used to distinguish between orthoclase and plagioclase feldspar? What is the best means of distinguishing between the two types of feldspar?

17. Each of the following statements describes a silicate mineral or mineral group. In each case, provide the appropriate name.
    a. The most common member of the amphibole group.
    b. The most common nonferromagnesian member of the mica family.
    c. The only silicate mineral made entirely of silicon and oxygen.
    d. A high-temperature silicate with a name that is based on its color.
    e. Characterized by striations.
    f. Originates as a product of chemical weathering.

18. What simple test can be used to distinguish calcite from dolomite?

19. Contrast resource and reserve.

20. What might cause a mineral deposit that had not been considered an ore to be reclassified as an ore?

Testing What You Have Learned

To test your knowledge of the material presented in this chapter, answer the following questions:

Multiple-Choice Questions

1. Most rock-forming minerals are _____.
   a. oxides
   b. silicates
   c. carbonates
   d. sulfides
   e. sulfates

2. Which of the following is NOT one of the eight most abundant elements found in Earth's continental crust?
 a. carbon   **c.** oxygen   **e.** silicon

 b. potassium   **d.** iron

3. A naturally occurring inorganic solid that possesses a definite chemical structure, which gives it a unique set of physical properties, is called a _____.
 a. rock   **c.** glass   **e.** gas

 b. liquid   **d.** mineral

4. When an atom combines chemically with another atom, it either gains, loses, or shares _____.
 a. protons   **c.** electrons   **e.** nuclei

 b. neutrons   **d.** compounds

5. Variations of the same element, but with different mass numbers, are referred to as _____.
 a. atoms   **c.** neutrons   **e.** isotopes

 b. minerals   **d.** silicates

6. Which one of the following is NOT a physical property of minerals?
 a. hardness   **c.** luster   **e.** chemical composition

 b. cleavage   **d.** streak

7. The fundamental building block of all silicate minerals is a tetrahedron composed of _____.
 a. carbon-oxygen   **c.** iron-zinc   **e.** calcium-carbon

 b. silicon-oxygen   **d.** iron-oxygen

8. Each silicate mineral group has its own particular silicate _____.
 a. structure   **c.** color   **e.** taste

 b. luster   **d.** atoms

9. The most common silicate mineral group found in Earth's crust is the _____.
 a. micas   **c.** feldspars   **e.** amphiboles

 b. sulfides   **d.** pyroxenes

10. A metallic mineral that can be mined at a profit is called a(n) _____.
 a. ion   **c.** rock   **e.** ore

 b. proton   **d.** isotope

Fill-in Questions

11. Most _____ are aggregates of minerals.
12. Atomic particles called _____ have positive charges, whereas _____ have negative charges.
13. A(n) _____ is a substance composed of two or more elements bonded together in definite proportions.
14. The external expression of the orderly internal arrangement of atoms in a mineral is called the _____ _____.
15. The tendency of a mineral to break along planes of weak bonding is called _____.

True/False Questions

16. An atom is the smallest part of matter that still retains the characteristics of an element. ___
17. Protons orbit the nucleus of an atom. ___
18. The nuclei of some isotopes disintegrate naturally in a process called radioactivity. ___
19. One of the most common and useful diagnostic physical properties of minerals is double refraction. ___
20. The carbonates are an important nonsilicate rock-forming mineral group. ___

Answers

1.b; 2.a; 3.d; 4.c; 5.e; 6.e; 7.b; 8.a; 9.c; 10.e; 11. rocks; 12. protons, electrons; 13. compound; 14. crystal form; 15. cleavage; 16.T; 17.F; 18.T; 19.F; 20.T

Chapter 3

Igneous Rocks

Focus on Learning

To assist you in learning the important concepts in this chapter, you will find it helpful to focus on the following questions:

- How are igneous rocks formed?
- How does magma differ from lava?
- What two criteria are used to classify igneous rocks?
- How does the rate of cooling of magma influence the crystal size of minerals in igneous rocks?
- How is the mineral makeup of an igneous rock related to Bowen's reaction series?
- In what ways are granitic rocks different from basaltic rocks?
- How are economic deposits of gold, silver, and many other metals formed?

Sentinel Rock, Yosemite National Park, California. (Photo by Larry Ulrich/DRK Photo)

In our discussion of the rock cycle, it was pointed out that igneous rocks form when **magma** cools and solidifies. This molten rock, which originates at depths as great as 200 kilometers beneath the surface, consists primarily of the elements found in silicate minerals, along with some gases, particularly water vapor, which are confined within the magma by the surrounding rocks. Because the magma body is less dense than the surrounding rocks, it works its way toward the surface, and on occasion breaks through, producing a volcanic eruption (Figure 3.1). The spectacular explosions that sometimes accompany an eruption are produced by the gases (volatiles) escaping as the confining pressure lessens near the surface. Sometimes blockage of the vent, coupled with surface water seepage into the magma chamber to produce steam, can produce catastrophic explosions. Along with ejected rock fragments, a volcanic eruption can generate extensive lava flows. **Lava** is magma that has reached the surface. The rocks that result when lava solidifies are classified as **extrusive**, or **volcanic**. The magma not able to reach the surface eventually crystallizes at depth. Igneous rocks produced in this manner are termed **intrusive**, or **plutonic**, and would never be observed if not for the processes of erosion stripping away the overlying rocks.

Crystallization of Magma

Magma is a hot fluid that may contain suspended crystals and a gaseous component. The liquid portion of a magma body is composed of ions that move about freely. However, as magma cools, the random movements of the ions slow and the ions begin to arrange themselves into orderly patterns. This process is called **crystallization**. Before we examine crystallization in more detail, let us first examine how a simple crystalline solid melts. In a crystalline solid, the ions are arranged in a closely packed regular pattern. However, they are not without some motion. They exhibit a sort of restricted vibration about fixed points. As the temperature rises, the ions vibrate more and more rapidly and consequently collide with ever-increasing vigor with their neighbors. Continued heating causes the ions to occupy additional space. This results in expansion of the solid and greater distance between ions. When the ions become far enough apart and are vibrating rapidly enough to overcome the force of

FIGURE 3.1 Lava from Hawaii's Kilauea Volcano explosively enters the Pacific Ocean. (Photo by Philip Rosenberg/Pacific Stock)

the chemical bonds which had joined them, the solid begins to melt. At this stage, the ions are able to slide past one another, and their orderly crystalline structure disintegrates. Thus, what was once a solid has become a liquid composed of unordered ions moving randomly about.

In the process of crystallization, cooling reverses the events of melting. As the temperature of the liquid drops, the ions pack closer together and begin to lose their freedom of movement. When cooling is sufficient, the force of the chemical bonds will again confine the atoms to an orderly crystalline arrangement. Usually, all of the molten material does not solidify at the same time. Rather, as it cools, numerous embryo crystals develop. In a systematic fashion, ions are added to these centers of crystal growth. When the crystals grow large enough that their edges meet, their growth ceases and crystallization continues elsewhere. Eventually, all of the liquid is transformed into a solid mass of interlocking crystals (Figure 3.2).

The rate of cooling strongly influences the crystallization process, in particular the size of the crystals. When a magma cools slowly, relatively few centers of crystal growth develop. Slow cooling also allows ions time to migrate over relatively great distances. Consequently, slow cooling results in the formation of rather large crystals, from millimeters to meters in length. On the other hand, when cooling occurs rapidly, the ions quickly lose their motion and readily combine. This results in the development of large numbers of nuclei which all compete for the available ions. The result is a solid mass formed of small intergrown crystals, sometimes microscopic.

When the molten material is quenched instantly, there is not sufficient time for the ions to arrange themselves into a crystalline network. Therefore, the solids produced in this manner consist of randomly distributed ions. Rocks that consist of unordered atoms are referred to as **glass** and are similar to ordinary manufactured glass.

The crystallization of a magma, although more complex, occurs in a manner similar to that just described. Rather than being composed of only one or two elements, most magma consists of the eight elements that are the primary constituents of the silicate minerals (see Table 2.3). These are silicon, oxygen, aluminum, sodium, potassium, calcium, iron, and magnesium. In addition, trace amounts of many other elements, as well as volatiles, particularly water and carbon dioxide, are also found in magma. A *volatile* is a material that is commonly a gas at temperatures and pressures existing at Earth's surface.

When magma cools, it is generally the silicon and oxygen atoms that link together first, to form silicon-oxygen tetrahedra. As cooling continues, the tetrahedra join with each other and with other ions to form crystal nuclei of the various silicate minerals. Each crystal nucleus grows as ion after ion is added to the crystalline network in an unchanging pattern. However, minerals do not all form at the same time or under the same conditions. As we shall see, certain minerals crystallize at much higher temperatures than others. Consequently, magmas often consist of solid crystals surrounded by molten material.

In addition to the rate of cooling, the composition of a magma and the amount of volatile material influence the crystallization process. Because magmas differ in each of these aspects, the physical appearance and mineral composition of igneous rocks vary widely. Nevertheless, it is possible to classify igneous rocks based on their mode of origin and mineral constituents. The environment during

A.

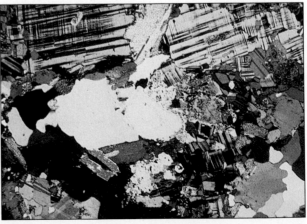

B.

FIGURE 3.2 A. Close-up of interlocking crystals in a coarse-grained igneous rock. The largest crystals are about 1 centimeter in length. **B.** Photomicrograph of interlocking crystals in a coarse-grained igneous rock. (Photos by E. J. Tarbuck)

crystallization can be inferred from the size and arrangement of the mineral grains, a property called texture. Consequently, *igneous rocks are most often classified by their texture and mineral composition.* We will consider both of these rock characteristics in the following sections.

Igneous Rock Textures

The term *texture*, when applied to an igneous rock, is used to describe the overall appearance of the rock based on the size and arrangement of its interlocking crystals (Figure 3.3). Texture is an important characteristic because it reveals a great deal about the environment in which the rock formed. This fact allows geologists to make inferences about a rock's origin while working in the field where sophisticated equipment is not available.

The most important factor affecting the texture of a rock is the *rate at which the magma cooled.* From our discussion of crystallization, we learned that rapid cooling produces small crystals, whereas very slow cooling results in the formation of much larger crystals. As we might expect, the rate of cooling is quite slow in magma chambers lying deep within the crust, whereas a thin layer of lava extruded upon the surface may chill in a matter of hours, and small molten blobs ejected into the air during a violent eruption can solidify almost instantly.

Igneous rocks that form at the surface or as small masses within the upper crust possess a very fine-grained texture termed **aphanitic**. By definition, the grains in aphanitic rocks are too small for individual minerals to be distinguished with the unaided eye (Figure 3.3A). Mineral identification is not possible, so we commonly characterize fine-grained rocks as

A.

B.

C.

D.

FIGURE 3.3 Igneous rock textures. **A.** Aphanitic (fine-grained). **B.** Phaneritic (coarse-grained). **C.** Porphyritic (large crystals embedded in a matrix). **D.** Glassy (cooled too rapidly to form crystals). (Photos by E. J. Tarbuck)

being light, intermediate, or dark in color. Using this system of grouping, light-colored aphanitic rocks are those composed primarily of light-colored nonferromagnesian silicate materials, and so forth.

Commonly seen in many aphanitic rocks are the voids left by gas bubbles that escape as the magma solidifies (Figure 3.4). These spherical or elongated openings are called **vesicles** and are limited to the upper portion of lava flows (Figure 3.5). It is in the upper zone of a lava flow that cooling occurs rapidly enough to "freeze" the lava, thereby preserving the openings produced by the expanding gas bubbles.

FIGURE 3.4 Scoria is a volcanic rock that exhibits a vesicular texture. The vesicles were gas bubbles preserved near the top of a solidified lava flow. (Photo by E. J. Tarbuck)

FIGURE 3.5 Upper layer of a lava flow with characteristic voids called vesicles. (Photo by E. J. Tarbuck)

When large masses of magma slowly solidify far below the surface, they form igneous rocks that exhibit a coarse-grained texture described as **phaneritic**. These coarse-grained rocks consist of a mass of intergrown crystals, which are roughly equal in size and large enough so that the individual minerals can be identified with the unaided eye (Figure 3.3B). Because phaneritic rocks form deep within the crust, their exposure at the surface results only after erosion removes the overlying rocks that once surrounded the magma chamber.

A large mass of magma located at depth may require tens of thousands, even millions, of years to solidify. Because all minerals within a magma do not crystallize at the same rate or at the same time during cooling, it is possible for some crystals to become quite large before others even start to form. If magma containing some large crystals should change environments, by erupting at the surface, for example, the molten portion of the lava would cool quickly. The resulting rock, which has large crystals embedded in a matrix of smaller crystals, is said to have a **porphyritic texture** (Figure 3.3C). The large crystals in such a rock are referred to as **phenocrysts**, while the matrix of smaller crystals is called **groundmass**. A rock that has such a texture is called a **porphyry**.

During some volcanic eruptions, molten rock is ejected into the atmosphere where it is quenched quickly. Rapid cooling of this type may generate rock with a **glassy texture**. As was indicated earlier, glass results when the ions have not been permitted the time to unite into an orderly crystalline structure. *Obsidian*, a common type of natural glass, is similar in appearance to a dark chunk of manufactured glass (Figure 3.3D).

Although the rate of cooling is the major factor that determines the texture of an igneous rock, other factors are also important. In particular, the composition of the magma influences the resulting texture. For example, basaltic magma, which is very fluid, will usually generate crystalline rocks when cooled quickly in a thin lava flow. Under the same conditions, granitic magma, which is quite viscous (resists flow), is much more likely to produce a rock with a glassy texture. Consequently, most of the lava flows which are composed of volcanic glass are granitic in composition. However, the surface of basaltic lava may be quenched rapidly enough to form a thin, glassy skin. Moreover, Hawaiian volcanoes sometimes generate lava fountains which spray basaltic lava tens of meters into the air. Such activity may produce strands of volcanic glass called *Pele's hair*, after the Hawaiian goddess of volcanoes.

Some igneous rocks are formed from the consolidation of individual rock fragments that are ejected during a violent eruption. The ejected particles may be very fine ash, molten blobs, or large angular blocks which are torn from the walls of the vent during the eruption. Igneous rocks composed of these rock fragments are said to have a **pyroclastic texture** (see Figure 3.14).

A common type of pyroclastic rock is composed of glass shards (thin strands) that remained hot enough during their flight to fuse together upon impact. Other pyroclastic rocks are composed of fragments that solidified before impact and became cemented together at some later time. Because pyroclastic rocks are made of individual rock fragments rather than interlocking crystals, their overall textures are often more similar to sedimentary rocks than to other igneous rocks.

Igneous Rock Composition

The mineral makeup of an igneous rock is ultimately determined by the chemical composition of the magma from which it crystallizes. Such a large variety of igneous rocks exists that it is logical to assume that an equally large variety of magmas must also exist. However, geologists have found that various eruptive stages of the same volcano often extrude lavas or pyroclastic material exhibiting somewhat different compositions, particularly if an extensive period of time separated the eruptions. Data of this type led them to look into the possibility that a single magma might produce rocks of varying mineral content.

Bowen's Reaction Series

A pioneering investigation into the crystallization of magma was carried out by N. L. Bowen in the first quarter of the twentieth century. Bowen discovered that, as magma cools in the laboratory, those minerals with higher melting points crystallize before minerals with lower melting points. As shown in Figure 3.6, the first mineral to crystallize from a typical magma is the ferromagnesian mineral olivine. As the magma cools further, the minerals pyroxene and calcium-rich plagioclase begin to form. During the crystallization process, the composition of the melt (liquid portion of a magma, excluding any solid material) continually changes. For example, at the stage when about 50 percent of the magma has solidified, the melt will be depleted of iron, magnesium, and calcium because these elements all went into the earliest-formed minerals. Consequently, the melt is enriched in aluminum, sodium, and potassium. Further, the silica component of the melt becomes enriched toward the latter stages of crystallization.

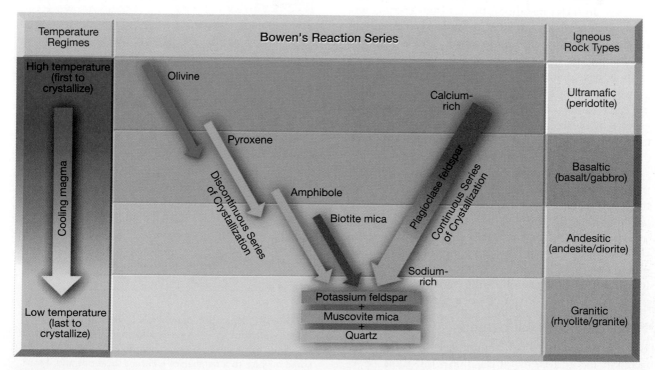

FIGURE 3.6 Bowen's reaction series shows the sequence in which minerals crystallize from a magma. Compare this figure to the mineral composition of the rock groups in Table 3.1. Note that each rock group consists of minerals that crystallize at about the same temperature.

Bowen also demonstrated that, as a magma cooled, its solid components (minerals) would react with the remaining melt and produce the next mineral in the sequence shown in Figure 3.6. For this reason, this arrangement of minerals became known as **Bowen's reaction series**. On the upper left branch of this reaction series, olivine, the first mineral to form, will react with the remaining melt to form pyroxene. As the magma body cools further, the pyroxene crystals will in turn react with the melt to generate amphibole. This reaction will continue until the last mineral in the series, biotite, is formed. The left branch of Bowen's reaction series is called a *discontinuous reaction series* because each mineral has a different crystalline structure. Olivine is composed of single tetrahedra, whereas the other minerals in this sequence are composed of single chains, double chains, and sheet structures, respectively. Ordinarily, these reactions do not run to completion, so that various amounts of each of these minerals may exist at any given time.

The right branch of the reaction series, called the *continuous reaction series*, demonstrates that calcium-rich feldspar crystals react with the sodium ions contained in the melt to become progressively more sodium-rich. Oftentimes, the rate of cooling occurs rapidly enough to prohibit the reaction between early formed, calcium-rich plagioclase crystals and the melt. In these instances, the feldspar crystals will have calcium-rich interiors surrounded by zones that are progressively richer in sodium.

During the last stage of crystallization, after most magma has solidified, the minerals muscovite and potassium feldspar are generated. Finally, the remaining melt (if any) will have a high silica content that eventually crystallizes as quartz. Although these latter-formed minerals crystallize in the order shown, they do not react with the melt to produce other minerals.

Bowen's reaction series illustrates the sequence in which minerals crystallize from a magma. One of the consequences of this crystallization scheme is that minerals that form in the same general temperature regime are found together in the same igneous rock. For example, in Figure 3.6, notice that the minerals potassium feldspar, muscovite, and quartz are found in the same region of the diagram and are the major constituents of the igneous rock granite.

Magmatic Differentiation

Bowen demonstrated that minerals crystallize from magma in a systematic fashion. But how does Bowen's reaction series account for the great diversity of igneous rocks? It has been shown that, at one

or more stages in the crystallization process, a separation of the solid and liquid components of a magma can occur. This happens, for example, if the earlier-formed minerals are denser (heavier) than the liquid portion and settle to the bottom of the magma chamber as shown in Figure 3.7A. This process, called **crystal settling**, is thought to occur frequently with the dark silicates, such as olivine and pyroxene. When the remaining melt solidifies, either in place or in a new location if it migrates out of the magma chamber, it will form a rock with a chemical

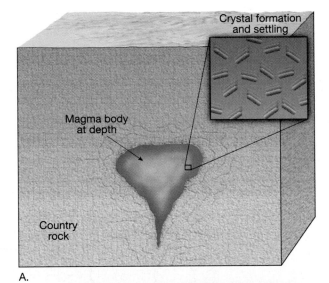

A.

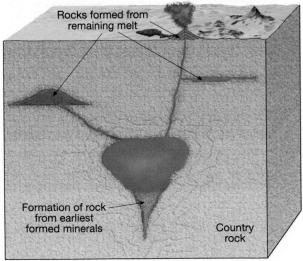

B.

FIGURE 3.7 Separation of minerals by crystal settling. **A.** Illustration of how the earliest-formed minerals can be separated from a magma by settling. **B.** The remaining melt could migrate to a number of different locations and, upon further crystallization, generate rocks having a composition much different from that of the parent magma.

composition much different from the overall parent magma (Figure 3.7B). The process of developing more than one rock type from a common magma is called **magmatic differentiation**.

A classic example of magmatic differentiation is found in the Palisades Sill, which is a 300-meter-thick tabular mass of dark igneous rock exposed along the west shore of the Hudson River. Because of its great thickness and subsequent slow rate of crystallization, olivine, the first mineral to crystallize, sank toward the bottom and makes up about 25 percent of the lower portion of the sill. By contrast, near the top of this igneous body, olivine represents only 1 percent of the rock mass.*

At any stage in the evolution of a magma, it can be separated into two or more chemically distinct components. Further, each of these components may undergo further segregation by a variety of processes. Consequently, magmatic differentiation can produce a variety of igneous rocks having a wide range of compositions.

Assimilation and Magma Mixing

Bowen successfully demonstrated that, through magmatic differentiation, a single magma can generate several different igneous rocks. However, more recent work indicates that this process alone cannot account for the great quantities of the rock types known to exist. Apparently, other mechanisms also generate magmas of varied chemical compositions.

Once a magma body forms, its composition can change through the incorporation of foreign material. For example, as magma migrates upward it may incorporate some of the surrounding country rock, a process called **assimilation** (Figure 3.8). One type of assimilation operates in a near-surface environment where rocks are brittle. As the magma pushes upward, stress causes numerous cracks in the overlying rock. The force of the injected magma is strong enough to dislodge blocks of the surrounding rock and incorporate them into the magma body. In other environments, the magma may simply melt and assimilate some of the country rock.

Another means by which the composition of a magma body is altered is **magma mixing**. Simply, this process is believed to occur during ascent, when one magma body overtakes another (Figure 3.8).

---

*Recent studies indicate that this igneous body was produced by multiple injections of magma and represents more than just a simple case of crystal settling.

Once contact occurs, convective flow and other mechanisms mix the two magmas, generating a fluid with a different composition. Both magma mixing and the assimilation of country rock are similar in that the magma body is contaminated through the incorporation of foreign material.

Naming Igneous Rocks

As was stated previously, igneous rocks are most often classified, or grouped, on the basis of their texture and mineral composition. The various igneous textures result from different cooling histories, whereas the mineral composition of an igneous rock is the consequence of the chemical makeup of the parent magma and the environment of crystallization. As we might expect from the results of Bowen's work, minerals that crystallize under similar conditions are most often found together comprising the same igneous rock. Hence, the classification of igneous rocks closely corresponds to Bowen's reaction series (see Figure 3.6).

The first minerals to crystallize—calcium feldspar, pyroxene, and olivine—are high in iron, magnesium, or calcium, and low in silicon. Because basalt is a common rock with this mineral makeup, the term *basaltic* is often used to denote any rock having a similar mineral composition. Moreover, because basaltic rocks contain a high percentage of ferromagnesian minerals, geologists may also refer to them as *mafic* rocks (from *mag*nesium and *Fer*ric, the Latin name for iron). Because of their iron content, mafic rocks are typically darker and slightly denser than other igneous rocks commonly found at the surface.

Among the last minerals to crystallize are potassium feldspar and quartz, the primary components of the abundant rock granite. Igneous rocks in which these two minerals predominate are said to have a *granitic* composition. Geologists also refer to granitic rocks as being *felsic*, a term derived from *fel*dspar and *si*lica (quartz).

Intermediate igneous rocks contain minerals found near the middle of Bowen's reaction series. Amphibole and the intermediate plagioclase feldspars are the main constituents of this compositional group. We will refer to rocks that have a mineral makeup between that of granite and basalt as being *andesitic*, after the rock andesite.

Although the rocks in each of these basic categories consist mainly of minerals located in a specific region of Bowen's reaction series, other constituents are usually present in lesser amounts. For example,

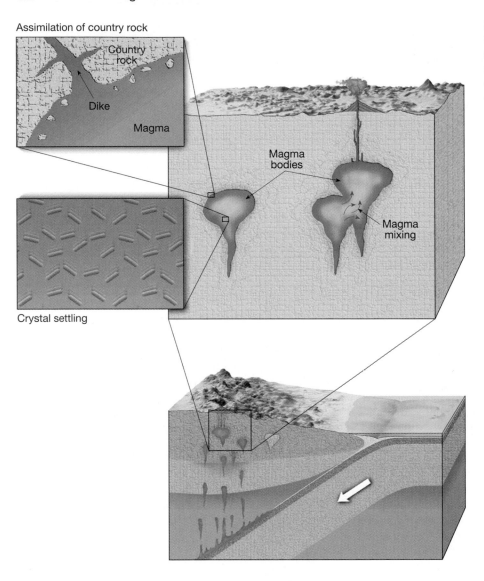

Assimilation of country rock

Crystal settling

FIGURE 3.8 This illustration shows three ways that the composition of a magma body may be altered: magma mixing; assimilation of country rock; and crystal settling (magmatic differentiation).

granitic rocks are composed primarily of quartz and potassium feldspar (K feldspar), but may also contain muscovite, biotite, amphibole, and sodium feldspar (Na feldspar). See Table 3.1.

Thus far this discussion has focused on only three mineral compositions, yet it is important to note that gradations among these types also exist (Figure 3.9). For example, an abundant intrusive igneous rock called *granodiorite* has a mineral composition between that of granitic rocks and those with an andesitic composition. Another important igneous rock, *peridotite*, contains mostly olivine and pyroxene and thus falls near the very beginning of Bowen's reaction series.

Because peridotite is composed almost entirely of ferromagnesian minerals, its chemical composition is often referred to as *ultramafic*. Although ultramafic rocks are rarely observed on Earth's surface, peridotite is believed to be a major constituent of the upper mantle.

An important aspect of the chemical composition of igneous rocks is silica (SiO_2) content. Recall that most of the minerals in igneous rocks contain some silica. Typically, the silica content of crustal rocks ranges from a low of 50 percent in basaltic rocks to a high of over 70 percent in granitic rocks. The percentage of silica in igneous rocks actually varies in a systematic manner that parallels the abundance of the other elements. For example, rocks low in silica contain large amounts of calcium, iron, and magnesium. Consequently, the chemical makeup of an igneous rock can be inferred directly from its silica content. Further, the amount of silica present in magma strongly influences its behavior. Granitic magma, which has a high silica content, is viscous and exists as a fluid at temperatures as low as 800°C. On the other hand, basaltic magmas are low in silica and generally more fluid. Further, basaltic magmas are largely crystalline below 950°C.

TABLE 3.1 Classification of igneous rocks.

	Granitic (Felsic)	Andesitic (Intermediate)	Basaltic (Mafic)	Ultramafic
Phaneritic (coarse-grained)	Granite	Diorite	Gabbro	Peridotite
Aphanitic (fine-grained)	Rhyolite	Andesite	Basalt	Komatiite (rare)
Mineral Composition	Quartz Potassium feldspar Sodium feldspar	Amphibole Intermediate plagioclase	Calcium feldspar Pyroxene	Olivine Pyroxene
Minor Mineral Constituents	Muscovite Biotite Amphibole	Pyroxene Amphibole Biotite	Olivine Amphibole	Calcium feldspar
Rock color Based on % dark (mafic) minerals	Light-colored Less than 15% dark minerals	Medium-colored 15–40% dark minerals	Dark gray to black More than 40% dark minerals	Dark-green to black Nearly 100% dark minerals

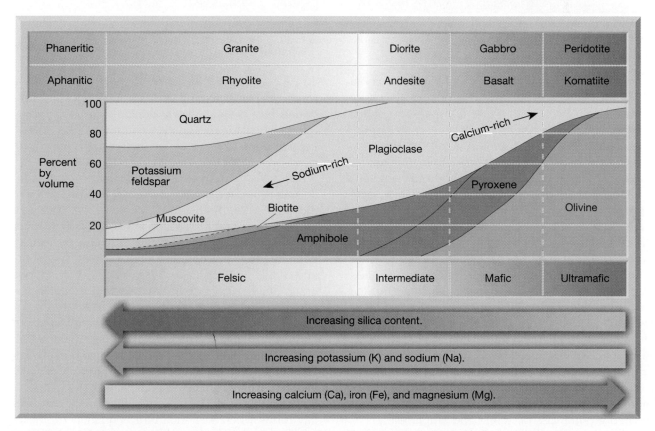

FIGURE 3.9 Mineralogy of the common igneous rocks. Phaneritic (coarse-grained) rocks are plutonic, solidifying deep underground. Aphanitic (fine-grained) rocks are volcanic, or solidify near Earth's surface. (After Dietrich)

Because igneous rocks are classified on the basis of their mineral composition and texture, two rocks may have the same mineral constituents but have different textures and hence different names. For example, the coarse-grained intrusive rock granite has a fine-grained volcanic equivalent called rhyolite (Figure 3.10). Although these rocks are mineralogically the same, they have different textures and do not look at all alike.

Granitic Rocks

Granite is perhaps the best known of all igneous rocks (Figure 3.10A). This is partly because of its natural beauty, which is enhanced when it is polished, and partly because of its abundance. Slabs of polished granite are commonly used for tombstones and monuments and as building stones.

A.

B.

FIGURE 3.10 A. Granite, one of the most common phaneritic igneous rocks. **B.** Rhyolite, the aphanitic equivalent of granite, is less abundant. (Photos by E. J. Tarbuck)

Granite is a phaneritic rock composed of about 25 to 35 percent quartz and over 50 percent potassium feldspar and sodium-rich feldspar. The quartz crystals, which are roughly spherical in shape, are most often clear to light gray in color. In contrast to quartz, the feldspar crystals in granite are not as glassy, but rectangular in shape and generally salmon pink to white in color. Other common constituents of granite are muscovite and the dark silicates, particularly biotite and amphibole. Although the dark components of granite make up less than 20 percent of most samples, dark minerals appear to be more prominent than their percentage would indicate. In some granites, K feldspar is dominant and dark pink in color, so that the rock appears almost reddish. This variety is popular as a building stone. However, most often the feldspar grains are white, so that when viewed at a distance granite appears light gray in color. Granite may also have a porphyritic texture, in which feldspar crystals of a centimeter or more in length are scattered among the coarse-grained groundmass of quartz and amphibole.

Granite is often produced by the processes that generate mountains. Because granite is a by-product of mountain building and is very resistant to weathering and erosion, it frequently forms the core of eroded mountains. For example, Pikes Peak in the Rockies, Mount Rushmore in the Black Hills, the White Mountains of New Hampshire, Stone Mountain in Georgia, and Yosemite National Park in the Sierra Nevada are all areas where large quantities of granite are exposed at the surface (Figure 3.11). As we can see from these examples, granite is a very abundant rock. However, it has become common

practice among geologists to apply the term *granite* to any coarse-grained intrusive rock composed predominantly of light silicate minerals. We will follow this practice for the sake of simplicity. You should keep in mind that this use of the term *granite* covers rocks having a range of mineral compositions.

Rhyolite is the volcanic equivalent of true granite. Like granite, rhyolite is composed primarily of the light-colored silicates (Figure 3.10B). This fact accounts for its color, which is usually buff to pink or occasionally very light gray. Rhyolite is usually aphanitic and frequently contains glassy fragments and voids indicating rapid cooling in a surface environment. In those instances when rhyolite contains phenocrysts, they are usually small and composed of either quartz or potassium feldspar. In contrast to granite, rhyolite is rather uncommon. Yellowstone Park is one well-known exception. Here rhyolitic lava flows and ash deposits of similar composition are widespread.

Obsidian is a dark-colored, glassy rock that usually forms when silica-rich lava is quenched quickly (Figure 3.12A). In contrast to the orderly arrangement of ions that is characteristic of minerals, the ions in glass are unordered. Consequently, glassy rocks like obsidian are not composed of minerals in the same sense as most other rocks.

Although usually black or reddish-brown in color, obsidian has a high silica content (Figure 3.12A). Thus, its composition is more akin to the light igneous rocks such as granite than to the dark rocks of basaltic composition. By itself, silica is clear like window glass; the dark color results from the presence of metallic ions. If you examine a thin edge

FIGURE 3.11 El Capitan, a large igneous monolith located in Yosemite National Park, California. Granite bedrock is exposed over much of the park. (Photo by Lewis Kemper/DRK Photo)

A.

B.

FIGURE 3.12 Igneous rocks that exhibit a glassy texture. **A.** Obsidian, a glassy volcanic rock. **B.** Pumice, a glassy rock containing numerous vesicles. (Photos by E. J. Tarbuck)

of a piece of obsidian, it will be nearly transparent. Because of its excellent conchoidal fracture and ability to hold a sharp, hard edge, obsidian was a prized material from which Native Americans chipped arrowheads and cutting tools.

Pumice is a volcanic rock that, like obsidian, has a glassy texture. Usually found with obsidian, pumice forms when large amounts of gas escape through lava to generate a gray, frothy mass (Figure 3.12B). In some samples, the voids are quite noticeable, whereas in others, the pumice resembles fine shards of intertwined glass. Because of the large percentage of voids, many samples of pumice will float when placed in water. Oftentimes, flow lines are visible in pumice, indicating that some movement occurred before solidification was complete. Moreover, pumice and obsidian often form in the same rock mass, where they exist in alternating layers.

Andesitic Rocks

Andesite is a medium gray, aphanitic rock of volcanic origin. Its name comes from the Andes Mountains where numerous volcanoes are composed of this rock type. In addition to the volcanoes of the Andes, many of the volcanic structures encircling the Pacific Ocean are of andesitic composition. Andesite quite commonly exhibits a porphyritic texture (Figure 3.13). In these cases, the phenocrysts are often light, rectangular crystals of plagioclase feldspar or black, elongated hornblende crystals.

Diorite is a phaneritic intrusive rock that looks somewhat similar to gray granite. However, it can be distinguished from granite by the absence of visible quartz crystals. The mineral makeup of diorite is primarily sodium-rich plagioclase and amphibole, with lesser amounts of biotite. Because the white feldspar grains and dark amphibole crystals are roughly equal in abundance, diorite has a "salt and pepper" appearance.

Basaltic Rocks

Basalt is a very dark green to black, fine-grained volcanic rock composed primarily of pyroxene and calcium-rich feldspar, with lesser amounts of olivine and amphibole present. When porphyritic, basalt commonly contains small, light-colored calcium feldspar phenocrysts or glassy-appearing olivine phenocrysts embedded in a dark groundmass.

Basalt is the most common extrusive igneous rock. Many volcanic islands, such as the Hawaiian Islands and Iceland, are composed mainly of basalt. Further, the upper layers of the oceanic crust consist of basalt. In the United States, large portions of central Oregon and Washington were the sites of extensive basaltic outpourings (see Figure 4.20). At some locations these once-fluid basaltic flows have accumulated to thicknesses approaching 3 kilometers.

Gabbro is the intrusive equivalent of basalt. Like basalt, it is very dark green to black in color and composed primarily of pyroxene and calcium-rich plagioclase. Although gabbro is not a common constituent of the continental crust, it undoubtedly makes up a significant percentage of the oceanic crust. Here large portions of the magma found in underground reservoirs that once fed basalt flows eventually solidified at depth to form gabbro.

Pyroclastic Rocks

Pyroclastic rocks form from fragments ejected during a volcanic eruption. One of the most common pyroclastic rocks, called *tuff*, is composed of tiny ash-sized fragments which were later cemented together

A.

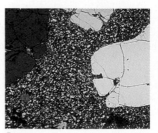

B.

FIGURE 3.13 Andesite porphyry. **A.** Hand sample of andesite porphyry, a common volcanic rock. **B.** Photomicrograph of a thin section of andesite porphyry to illustrate texture. Notice that the few large crystals (phenocrysts) are surrounded by much smaller crystals (groundmass). (Photos by E. J. Tarbuck)

FIGURE 3.14 Outcrop of welded tuff interbedded with obsidian (black) near Shoshone, California. As the close-up view shows, tuff is composed mainly of ash-sized particles and may contain larger fragments of pumice or other volcanic rocks. (Outcrop photo by Breck P. Kent, close-up by E. J. Tarbuck)

(Figure 3.14). In situations where the ash particles remained hot enough to fuse, the rock is generally called *welded tuff*. Although welded tuffs consist mostly of glass shards, they frequently contain pieces of obsidian or other rock fragments. Further, a microscope is often required to distinguish tuffs from other fine-grained igneous rocks. Deposits of partially welded tuffs are easily quarried and used as a durable building material. Several villages in Cappadocia in central Turkey, which date to the fourth century A.D., have been carved into vertical cliffs composed of this material.

Pyroclastic rocks composed mainly of particles larger than ash are called *volcanic breccia*. The particles in volcanic breccia can consist of streamlined fragments that solidified in air, blocks broken from the walls of the vent, crystals, and glass fragments. Unlike the other igneous rock names, the terms *tuff* and *volcanic breccia* do not denote mineral composition.

Mineral Resources and Igneous Processes

Some of the most important accumulations of metals, such as gold, silver, copper, mercury, lead, platinum, and nickel, are produced by igneous processes (Table 3.2). These mineral resources, like most others, result from processes that concentrate desirable materials to the extent that extraction is economically feasible.

The igneous processes that generate some of these metal deposits are quite straightforward. For example, as a large magma body cools, the heavy minerals that crystallize early tend to settle to the lower portion of the magma chamber. This type of magmatic segregation is particularly active in large basaltic magmas where chromite (ore of chromium), magnetite, and platinum are occasionally generated. Layers of chromite, interbedded with other heavy minerals, are mined from such deposits in the Stillwater Complex of Montana. Another example is the Bushveld Complex in South Africa, which contains over 70 percent of the world's known reserves of platinum.

Magmatic segregation is also important in the late stages of the magmatic process. This is particularly true of granitic magmas in which the residual melt can become enriched in rare elements and heavy metals. Further, because water and other volatile substances do not crystallize along with the bulk of the magma body, these fluids make up a high percentage of the melt during the final phase of solidification. Crystallization in a fluid-rich environment, where ion migration is enhanced, is believed to result in the formation of crystals several centimeters, or even a few meters, in length. The resulting rocks, called **pegmatites**, are composed of these unusually large crystals (Figure 3.15).

Feldspar masses the size of houses have been quarried from a pegmatite located in North Carolina. Gigantic hexagonal crystals of muscovite measuring a few meters across have been found in Ontario, Canada. In the Black Hills, spodumene crystals as large as telephone poles have been mined. The largest of these was more than 12 meters (40 feet) long. Not all pegmatites contain such large crystals, but these examples emphasize the special conditions that must exist during their formation.

TABLE 3.2 Occurrences of metallic minerals.

Metal	Principal Ores	Geological Occurrences
Aluminum	Bauxite	Residual product of weathering
Chromium	Chromite	Magmatic segregation
Copper	Chalcopyrite Bornite Chalcocite	Hydrothermal deposits; contact metamorphism; enrichment by weathering processes
Gold	Native gold	Hydrothermal deposits; placers
Iron	Hematite Magnetite Limonite	Banded sedimentary formations; magmatic segregation
Lead	Galena	Hydrothermal deposits
Magnesium	Magnesite Dolomite	Hydrothermal deposits
Manganese	Pyrolusite	Residual product of weathering
Mercury	Cinnabar	Hydrothermal deposits
Molybdenum	Molybdenite	Hydrothermal deposits
Nickel	Pentlandite	Magmatic segregation
Platinum	Native platinum	Magmatic segregation; placers
Silver	Native silver Argentite	Hydrothermal deposits; enrichment by weathering processes
Tin	Cassiterite	Hydrothermal deposits; placers
Titanium	Ilmenite Rutile	Magmatic segregation; placers
Tungsten	Wolframite Scheelite	Pegmatites; contact metamorphic deposits; placers
Uranium	Uraninite (pitchblende)	Pegmatites; sedimentary deposits
Zinc	Sphalerite	Hydrothermal deposits

Most pegmatites are granitic in composition and consist of unusually large crystals of quartz, feldspar, and muscovite. Feldspar is used in the production of ceramics and muscovite is used for isinglass, electrical insulation, and glitter. Further, pegmatites often contain some of the least abundant elements. Thus, in addition to the common silicates, some pegmatites include semiprecious gems such as beryl, topaz, and tourmaline. Moreover, minerals containing the elements lithium, cesium, uranium, and the rare earths[*] are occasionally found. Most pegmatites are located within large igneous masses or as dikes or veins that cut into the country rock that surrounds the magma chamber (Figure 3.16).

Not all late-stage magmas produce pegmatites, nor do all have a granitic composition. Rather, some magmas become enriched in iron or occasionally copper. For example, at Kirava, Sweden, magma composed of over 60 percent magnetite solidified to produce one of the largest iron deposits in the world.

FIGURE 3.15 Granite pegmatite. The pinkish mineral is feldspar and the glassy gray-to-black mineral is quartz. Quarter for scale. (Photo by Sherman Clebnik)

[*]The rare earths are a group of 15 elements (atomic numbers 57 through 71) that possess similar properties. They are useful catalysts in petroleum refining and are used to improve color retention in television picture tubes.

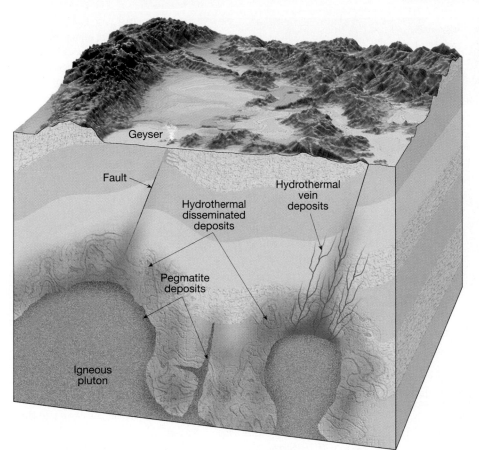

FIGURE 3.16 Illustration of the relationship between a parent igneous body and the associated pegmatite and hydrothermal deposits.

Another economically important mineral with an igneous origin is diamond. Although best known as gems, diamonds are used extensively as abrasives. Diamonds are thought to originate at depths of nearly 200 kilometers, where the confining pressure is great enough to generate this high-pressure form of carbon. Once crystallized, the diamonds are carried upward through pipe-shaped conduits that increase in diameter toward the surface. In diamond-bearing pipes, nearly the entire pipe contains diamond crystals that are disseminated throughout an ultramafic rock called *kimberlite*. The most productive kimberlite pipes are those found in South Africa. The only equivalent source of diamonds in the United States is located near Murfreesboro, Arkansas, but this deposit is exhausted and serves today merely as a tourist attraction.

Among the best-known and most important ore deposits are those generated from **hydrothermal** (hot-water) **solutions**. Included in this group are the gold deposits of the Homestake mine in South Dakota, the lead, zinc, and silver ores near Coeur d'Alene, Idaho, the silver deposits of the Comstock Lode in Nevada, and the copper ores of the Keweenaw Peninsula in Michigan.

The majority of hydrothermal deposits are thought to originate from hot, metal-rich fluids that are remnants of late-stage magmatic processes. During solidification, liquids plus various metallic ions accumulate near the top of the magma chamber. Because of their mobility, these ion-rich solutions can migrate great distances through the surrounding rock before they are eventually deposited, usually as sulfides of various metals (Figure 3.16). Some of this fluid moves along openings such as fractures or bedding planes, where it cools and precipitates the metallic ions to produce *vein deposits* (Figure 3.17). Many of the most productive deposits of gold, silver, and mercury occur as hydrothermal vein deposits.

Another important type of accumulation generated by hydrothermal activity is called a *disseminated deposit*. Rather than being concentrated in narrow veins and dikes, these ores are distributed as minute masses throughout the entire rock mass. Much of the world's copper is extracted from disseminated deposits, including those at Chuquicamata, Chile, and

FIGURE 3.17 Light-colored vein deposits emplaced along a series of fractures in dark-colored igneous rock. (Photo by James E. Patterson)

FIGURE 3.18 Geysers along Firehole River, Yellowstone National Park. (Photo by Carr Clifton)

the huge Bingham Canyon copper mine in Utah (see Box 3.1). Because these accumulations contain only 0.4 to 0.8 percent copper, between 125 and 250 kilograms of ore must be mined for every kilogram of metal recovered. The environmental impact of these large excavations, including the problems of waste disposal, is significant.

Some hydrothermal deposits have been generated by the circulation of ordinary groundwater in regions where magma was emplaced near the surface. The Yellowstone National Park area is a modern example of such a situation (Figure 3.18). When groundwater invades a zone of recent igneous activity its temperature rises, greatly enhancing its ability to dissolve minerals. These migrating hot waters remove metallic ions from the intrusive igneous rocks and carry them upward, where they may be deposited as an ore body. Depending on the conditions, the resulting accumulations may occur as vein deposits, as disseminated deposits, or, where hydrothermal solutions reach the surface in the form of hot springs or geysers, as surface deposits.

Box 3.1   # The Largest Open-pit Mine

In the photo, a mountain once stood where there is now a huge pit (Figure 3.A). This is the world's largest open-pit mine, Bingham Canyon copper mine, about 40 kilometers (25 miles) southwest of Salt Lake City, Utah. The rim is nearly 4 kilometers (2.5 miles) across and covers almost 8 square kilometers (3 square miles). Its depth is 900 meters (3000 feet). If a steel tower were erected at the bottom, it would have to be five times taller than the Eiffel Tower to reach the top of the pit!

It began in the late 1800s as an underground mine for veins of silver and lead. Later, copper was discovered. Similar deposits occur at several sites in the American Southwest and in a belt from southern Alaska to northern Chile.

As in other places in this belt, the ore at Bingham Canyon is disseminated throughout *porphyritic* igneous rocks, hence the name *porphyry copper deposits*. The deposit formed after magma was intruded to shallow depths. Following this, shattering created extensive fractures that were penetrated by hydrothermal solutions from which the ore minerals precipitated.

Although the percentage of copper in the rock is small, the total volume of

FIGURE 3.A Aerial view of Bingham Canyon copper mine near Salt Lake City, Utah. Although the amount of copper in the rock is less than 1 percent, the huge volumes of material removed and processed each day (about 200,000 tons) yield enough metal to be profitable. (Photo by Michael Collier)

copper is huge. Ever since open-pit operations started in 1906, some 5 billion tons of material have been removed, yielding more than 12 million tons of copper. Significant amounts of gold, silver, and molybdenum have also been recovered.

The ore body is far from exhausted today. Over the next 25 years, plans call for an additional 3 billion tons of material to be removed and processed. This largest of artificial excavations has generated most of Utah's mineral production for more

than 80 years and has been called the "richest hole on Earth."

Like many older mines, the Bingham pit was unregulated during most of its history. Development occurred prior to the present-day awareness of the environmental impacts of mining and prior to effective environmental legislation. Today, problems of groundwater and surface water contamination, air pollution, solid and hazardous wastes, and land reclamation are receiving long overdue attention at Bingham Canyon.

The Chapter in Review

The following statements are intended to help you review the primary objectives presented in this chapter.

- *Igneous rocks* form when *magma* cools and solidifies. *Extrusive*, or *volcanic*, igneous rocks result when *lava*, which is similar to magma except that most of the gaseous component has escaped, cools at the surface. Magma that solidifies at depth produces *intrusive*, or *plutonic*, igneous rocks.

- As magma cools, the ions that compose it arrange themselves into orderly patterns during a process called *crystallization*. Slow cooling results in the formation of rather large crystals. On the other hand, when cooling occurs rapidly, the outcome is a solid mass consisting of tiny intergrown crystals. When molten material is quenched instantly, a mass of unordered atoms, referred to as *glass*, forms.

- Igneous rocks are most often classified by their *texture* and *mineral composition*.

- The texture of an igneous rock refers to the overall appearance of the rock based on the size and arrangement of its interlocking crystals. The most important factor affecting texture is the rate at which magma cools. Common igneous rock textures include *aphanitic*, with grains too small to be distinguished with the unaided eye; *phaneritic*, with intergrown crystals that are roughly equal in size and large enough to be identified with the unaided eye; *porphyritic*, which has large crystals interbedded in a matrix of smaller crystals; and *glassy*.

- The mineral makeup of an igneous rock is ultimately determined by the chemical composition of the magma from which it crystallizes. N.L. Bowen discovered that as magma cools in the laboratory, those minerals with higher melting points crystallize before minerals with lower melting points. *Bowen's reaction series* illustrates the sequence of mineral formation within magma. In the *discontinuous reaction series*, the left side of Bowen's reaction series, each mineral has a different crystalline structure that forms as the solid components (minerals) react with the remaining melt (liquid portion of a magma, excluding any solid material) and produce the next mineral in the sequence. The right branch of Bowen's reaction series, called the *continuous reaction series*, demonstrates that calcium-rich feldspar crystals react with the sodium ions contained in the melt to become progressively more sodium rich. During the last stage of crystallization, after most of the magma has solidified, the minerals muscovite, potassium feldspar, and quartz are generated.

- During the crystallization of magma, if the earlier-formed minerals are denser than the liquid portion, they will settle to the bottom of the magma chamber during a process called *crystal settling*. Owing to the fact that crystal settling removes ions from the magma, as the remaining melt solidifies, it will form a rock much different from the parent magma. The process of developing more than one rock type from a common magma is called *magmatic differentiation*.

- Once a magma body forms, its composition can change through the incorporation of foreign material during a process termed *assimilation*, or by *magma mixing*.

- The mineral composition of an igneous rock is the consequence of the chemical makeup of the parent magma and the environment of crystallization. Hence, the classification of igneous rocks closely corresponds to Bowen's reaction series. *Granitic rocks* (e.g., the rocks granite and rhyolite), also referred to as being *felsic*, form from the last minerals to crystallize, potassium feldspar and quartz, and are light-colored. *Andesitic rocks* (e.g., the rocks andesite and diorite) are intermediate and form from plagioclase feldspar and amphibole minerals. *Basaltic rocks* (e.g., the rocks basalt and gabbro), also referred to as *mafic* rocks, form from the first minerals to crystallize—calcium feldspar, pyroxene, and olivine—are high in iron, magnesium, and calcium, low in silicon, and dark-gray to black.

- Some of the most important accumulations of metals, such as gold, silver, lead, and copper, are produced by igneous processes. The best-known and most important ore deposits are generated from *hydrothermal* (hot-water) *solutions*. Hydrothermal deposits are thought to originate from hot, metal-rich fluids that are remnants of late-stage magmatic processes. These ion-rich solutions move along fractures or bedding planes, cool, and precipitate the metallic ions to produce *vein deposits*. In a *disseminated deposit* (e.g., much of the world's copper deposits) the ores from hydrothermal solutions are distributed as minute masses throughout the entire rock mass.

Key Terms

aphanitic texture (p. 47)
assimilation (p. 51)
Bowen's reaction series (p. 50)
crystal settling (p. 50)

crystallization (p. 45)
extrusive (p. 45)
glass (p. 46)
glassy texture (p. 48)
groundmass (p. 48)

hydrothermal solutions (p. 59)
intrusive (p. 45)
lava (p. 45)
magma (p. 45)

magma mixing (p. 51)
magmatic differentiation (p. 51)
pegmatites (p. 57)
phaneritic texture (p. 48)

phenocryst (p. 48)   porphyry (p. 48)   vesicle (p. 48)

plutonic (p. 45)   pyroclastic texture (p. 49)   volcanic (p. 48)

porphyritic texture (p. 48)

Questions for Review

1. How does lava differ from magma?

2. How does the rate of cooling influence crystallization?

3. In addition to the rate of cooling, what other factors influence crystallization?

4. The classification of igneous rocks is based largely upon two criteria. Name these criteria.

5. The statements that follow relate to terms describing igneous rock textures. For each statement, identify the appropriate term.

  a. Openings produced by escaping gases.
  b. Obsidian exhibits this texture.
  c. A matrix of fine crystals surrounding phenocrysts.
  d. Crystals are too small to be seen with the unaided eye.
  e. A texture characterized by two distinctively different crystal sizes.
  f. Coarse grained, with crystals of roughly equal size.

6. What does a porphyritic texture indicate about an igneous rock?

7. Relate the classification of igneous rocks to Bowen's reaction series.

8. How are granite and rhyolite different? In what way are they similar?

9. Why are the crystals in pegmatites so large?

10. Compare and contrast each of the following pairs of rocks:

  a. Granite and diorite.
  b. Basalt and gabbro.
  c. Andesite and rhyolite.

11. How do tuff and volcanic breccia differ from other igneous rocks such as granite and basalt?

12. List two general types of hydrothermal deposits.

Testing What You Have Learned

To test your knowledge of the material presented in this chapter, answer the following questions:

Multiple-Choice Questions

1. Igneous rocks form when _____ within Earth, or lava on the surface, cools and solidifies.
  a. water   **c.** carbonates   **e.** organic compounds
  b. magma   **d.** sediments

2. Which one of the following metals is NOT associated with an ore that occurs as a hydrothermal deposit?
  a. gold   **c.** lead   **e.** zinc
  b. tin   **d.** aluminum

3. The rate at which a magma cools strongly influences the _____ of the mineral crystals in an igneous rock.
  a. color   **c.** shape   **e.** size
  b. hardness   **d.** luster

4. The last mineral to crystallize from a silica-rich magma is _____.
  a. olivine   **c.** pyroxene   **e.** sodium-rich plagioclase
  b. quartz   **d.** biotite mica

5. In the _____ reaction series of Bowen's reaction series, each mineral has a different crystalline structure.
   a. discontinuous   **c.** continuous   **e.** tertiary
   b. middle   **d.** secondary

6. Which one of the following igneous rocks has an aphanitic texture?
   a. granite   **c.** gabbro   **e.** basalt
   b. peridotite **d.** diorite

7. An igneous rock that has large crystals embedded in a matrix of smaller crystals is said to have a(n) _____ texture.
   a. phaneritic **c.** pyroclastic **e.** aphanitic
   b. glassy   **d.** porphyritic

8. The process of magmatic _____ is responsible for developing more than one type of igneous rock from a common magma.
   a. isolation   **c.** lithification **e.** differentiation
   b. termination **d.** deposition

9. Basaltic rocks that contain a high percentage of ferromagnesian minerals are referred as _____ rocks.
   a. pegmatic **c.** felsic **e.** secondary
   b. mafic   **d.** glassy

10. Potassium feldspar and quartz are the primary components of the abundant igneous rock called _____.
    a. basalt   **c.** granite   **e.** peridotite
    b. obsidian **d.** andesite

Fill-In Questions

11. Igneous rocks that result when lava solidifies are classified as _____, or volcanic, rocks.
12. The two criteria used to classify an igneous rock are _____ and mineral _____.
13. _____, the most common extrusive igneous rock, is composed primarily of pyroxene and calcium-rich feldspar.
14. Among the best-known and most important ore deposits are veins generated from _____ solutions.
15. As magma migrates, it may incorporate some of the surrounding rock during a process called _____.

True/False Questions

16. Bowen's reaction series illustrates the sequence in which minerals crystallize from a magma. ___
17. Slow cooling of a magma results in tiny mineral crystals. ___
18. Basaltic magma has a higher silica content than granitic magma. ___
19. Igneous rocks formed from the consolidation of individual rock fragments that are ejected during a volcanic eruption are said to have a pyroclastic texture. ___
20. During the process of crystallization, the composition of the liquid portion of a magma remains the same. ___

Answers

1.b; 2.d; 3.e; 4.b; 5.a; 6.e; 7.d; 8.e; 9.b; 10.c; 11. extrusive; 12. texture, composition; 13. Basalt; 14. hydrothermal; 15. assimilation; 16.T; 17.F; 18.F; 19.T; 20.F

Volcanoes and Other Igneous Activity

Chapter 4

Focus on Learning

To assist you in learning the important concepts in this chapter, you will find it helpful to focus on the following questions:

- What primary factors determine the nature of volcanic eruptions? How do these factors affect a magma's viscosity?

- What materials are associated with a volcanic eruption?

- What are the eruptive patterns and basic characteristics of the three types of volcanoes generally recognized by volcanologists?

- What criteria are used to classify intrusive igneous bodies? What are some of these features?

- What are the roles of temperature, pressure, and partial melting in the formation of magma?

- What is the relation between volcanic activity and global tectonics?

Eruption of Mount Augustine Volcano, 1986, Cook Inlet, Alaska. (Photo by Steve Kaufman/DRK Photo)

On Sunday, May 18, 1980, the largest volcanic eruption to occur in North America in historic times transformed a picturesque volcano into a decapitated remnant (compare Figures 4.1 and 4.2). On this date in southwestern Washington State, Mount St. Helens erupted with tremendous force. The blast blew out the entire north flank of the volcano, leaving a gaping hole. In one brief moment, a prominent volcano whose summit had been more than 2900 meters (9500 feet) above sea level was lowered by more than 400 meters (1350 feet).

The event devastated a wide swath of timber-rich land on the north side of the mountain. Trees within a 400-square-kilometer area lay intertwined and flattened, stripped of their branches and appearing from the air like toothpicks strewn about. The accompanying mudflows carried ash, trees, and water-saturated rock debris 29 kilometers down the Toutle River. The eruption claimed 59 lives, some dying from the intense heat and the suffocating cloud of ash and gases, others from being hurled by the blast, and still others from entrapment in the mudflows.

The eruption ejected nearly one cubic kilometer of ash and rock debris (Figure 4.3). Following the devastating explosion, Mount St. Helens continued to emit great quantities of hot gases and ash. The force of the blast was so strong that some ash was propelled more than 18,000 meters (over 11 miles) into the stratosphere. During the next few days, this very fine-grained material was carried around Earth by strong upper-air winds. Measurable deposits were reported in Oklahoma and Minnesota, with crop damage into central Montana. Meanwhile, ash fallout in the immediate vicinity exceeded 2 meters in depth. The air over Yakima, Washington, (130 kilometers to the east) was so filled with ash that residents experienced midnight-like darkness at noon.

Anatomy of the Eruption

The events leading to the Mount St. Helens eruption began about two months earlier as a series of minor Earth tremors centered beneath the awakening

FIGURE 4.1 Mount St. Helens before the May 18, 1980, eruption. Compare this scene to Figure 4.2, which shows the volcano after the eruption. (Photo by Stephen Trimble)

FIGURE 4.2 Mount St. Helens as it appeared shortly after the May 18, 1980 eruption. (Photo by C. C. Lockwood/DRK Photo)

mountain (Figure 4.4A). The tremors were caused by the upward movement of magma within the mountain. The first volcanic activity took place a week later, when a small amount of ash and steam rose from the summit. Over the next several weeks, sporadic eruptions of varied intensity occurred. Prior to the main eruption, the primary concern had been the potential hazard of mudflows. These moving lobes of saturated soil and rock are created as ice and snow melt from the heat emitted from magma within the volcano.

The only warning of a potential eruption was a bulge on the volcano's north flank (Figure 4.4B). Careful monitoring of this dome-shaped structure indicated a very slow but steady growth rate of a few meters per day. If the growth rate of the bulge changed appreciably, an eruption might quickly follow. Unfortunately, no such variation was detected prior to the explosion. In fact, the seismic activity decreased during the two days preceding the huge blast.

Dozens of scientists were monitoring the mountain when it exploded. "Vancouver, Vancouver, this is it!" was the only warning—and last words from one

scientist—that preceded the unleashing of tremendous quantities of pent-up gases. The trigger was a medium-sized earthquake. Its vibrations sent the north slope of the cone plummeting into the Toutle River, removing the overburden that had trapped the magma below (Figure 4.4C). With the pressure reduced, the water in the magma vaporized and expanded, causing the mountainside to rupture like an overheated steam boiler. Because the eruption originated around the bulge, several hundred meters below the summit, the initial blast was directed laterally rather than vertically. Had the full force of the eruption been upward, far less destruction would have occurred.

Mount St. Helens is one of 15 large volcanoes and innumerable smaller ones that comprise the Cascade Range, which extends from British Columbia to northern California. Eight of the largest volcanoes have been active in the past few hundred years. Of the remaining seven "active" volcanoes, the most likely to erupt again are Mount Baker and Mount Rainier in Washington, Mount Shasta and Lassen Peak in California, and Mount Hood in Oregon.

FIGURE 4.3 In May 1980, Washington state's Mount St. Helens erupted explosively, sending huge quantities of volcanic ash and debris into the atmosphere. (Photo by R. Hoblitt, U.S. Geological Survey)

Not All Volcanoes Are Alike

Not all volcanic eruptions are as violent as the 1980 Mount St. Helens event. Some volcanoes, such as Hawaii's Kilauea volcano, generate relatively quiet outpourings of fluid lavas (Figure 4.5). These "gentle" eruptions are not without some fiery displays; occasionally fountains of incandescent lava spray hundreds of meters into the air. Such events, however, are typically short-lived and harmless, and the lava generally falls back into a lava pool.

Testimony to the quiet nature of Kilauea's eruptions is the fact that the Hawaiian Volcanoes Observatory has operated on its summit since 1912. This, despite the fact that Kilauea has had more than 50 eruptive phases since record keeping began in 1823. Further, the longest and largest of Kilauea's eruptions began in 1983 and remains active, although it has received only modest media attention.

Why do volcanoes like Mount St. Helens erupt explosively, whereas others like Kilauea are relatively quiet? Why do volcanoes occur in chains like the Aleutian Islands or the Cascade Range? Why do some volcanoes form on the ocean floor, while others occur on the continents? This chapter will deal with these and other questions as we explore the formation and movement of magma.

The Nature of Volcanic Eruptions

Volcanic activity is commonly perceived as a process that produces a picturesque, cone-shaped structure that periodically erupts in a violent manner, like Mount St. Helens. But, as we pointed out, some eruptions are very explosive, whereas many others are not. What determines whether a volcano extrudes magma violently or "gently"? The primary factors are the magma's *composition*, its *temperature*, and the amount of *dissolved gases* it contains. To varying degrees these factors affect the magma's **viscosity**. The more viscous ("thicker") the material, the greater its resistance to flow. For example, syrup is more viscous than water.

Factors Affecting Viscosity

The effect of temperature on viscosity is easily visualized. Just as heating syrup makes it more fluid (less viscous), the mobility of lava is strongly influenced by temperature changes. As a lava flow cools and begins to congeal, its mobility decreases and eventually the flow halts.

But a more significant factor influencing volcanic behavior is the chemical composition of magmas. This was discussed in Chapter 2 along with the classification of igneous rocks. Recall that a major difference among various igneous rocks is their silica (SiO_2) content. The same is true of the magma from which rocks form. Magmas that produce basaltic rocks contain about 50 percent silica, whereas magmas that produce granitic rocks contain over 70 percent silica (Table 4.1).

Note that *a magma's viscosity is directly related to its silica content*. In general, the more silica in magma, the greater is its viscosity. The flow of magma is impeded because silicate structures link together into long chains, even before crystallization begins. Consequently, because of high silica content, granitic lavas are very viscous and tend to form comparatively short, thick flows. By contrast, basaltic lavas, which contain less silica, tend to be more fluid and have been known to travel distances exceeding 150 kilometers (90 miles) before congealing (Figure 4.5).

In Hawaiian eruptions, the magmas are hot and basaltic, so they are extruded with ease. By contrast, highly viscous granitic magmas are more difficult to force through a vent. The vent may become plugged with viscous magma, causing a buildup of gases and

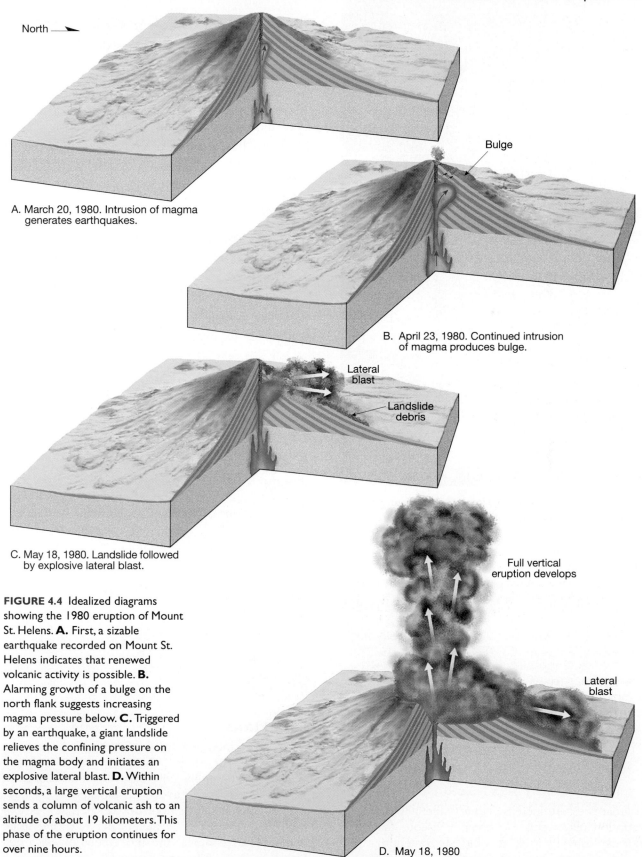

North ━━━▶

A. March 20, 1980. Intrusion of magma
   generates earthquakes.

Bulge

B. April 23, 1980. Continued intrusion
   of magma produces bulge.

Lateral
blast

Landslide
debris

C. May 18, 1980. Landslide followed
   by explosive lateral blast.

Full vertical
eruption develops

Lateral
blast

FIGURE 4.4 Idealized diagrams
showing the 1980 eruption of Mount
St. Helens. **A.** First, a sizable
earthquake recorded on Mount St.
Helens indicates that renewed
volcanic activity is possible. **B.**
Alarming growth of a bulge on the
north flank suggests increasing
magma pressure below. **C.** Triggered
by an earthquake, a giant landslide
relieves the confining pressure on
the magma body and initiates an
explosive lateral blast. **D.** Within
seconds, a large vertical eruption
sends a column of volcanic ash to an
altitude of about 19 kilometers. This
phase of the eruption continues for
over nine hours.

D. May 18, 1980

FIGURE 4.5 Fluid lava emitted from a flank eruption on Kilauea, Hawaii. (Photo by Paul Chesley/Tony Stone Images)

TABLE 4.1 Variations in properties among magmas of different composition.

Property	Basaltic magma	Andesitic magma	Granitic magma
Silica content	Least (about 50%)	Intermediate (about 60%)	Most (about 70%)
Viscosity	Least ("thinnest")	Intermediate	Greatest ("thickest")
Tendency to form lavas	Highest	Intermediate	Least
Tendency to form pyroclastics	Least	Intermediate	Greatest
Melting temperature	Highest	Intermediate	Lowest

a great pressure increase, so a potentially explosive eruption may result. However, a viscous magma is not explosive by itself. It is the gas content that puts the "bang" into a violent eruption.

Importance of Dissolved Gases in Magma

Dissolved gases in magma provide the force that extrudes molten rock from the vent. These gases are mostly water vapor and carbon dioxide. As magma moves into a near-surface environment, such as within a volcano, the confining pressure in the uppermost portion of the magma body is greatly reduced. This allows dissolved gases to be released suddenly, just as opening a soda bottle allows dissolved carbon dioxide gas to bubble out of the soda.

In the high-temperature magma and low near-surface pressure, these gases expand to hundreds of times their original volume. Very fluid basaltic magmas allow the expanding gases to bubble upward and escape from the vent with relative ease. As they escape, the gases will often propel incandescent lava hundreds of meters into the air, producing lava fountains, spectacular but harmless. Thus, eruptions of these fluid basaltic lavas, such as those that occur in Hawaii, are relatively quiet.

At the other extreme, highly viscous magmas impede the upward migration of expanding gases. The gases collect in bubbles and pockets that increase in size and pressure until they explosively eject the semimolten rock from the volcano. The result is a Mount St. Helens or Mt. Pinatubo (Philippines).

To summarize, the viscosity of magma, plus the quantity of dissolved gases and the ease with which they can escape, determines the nature of a volcanic eruption. You can now understand the "gentle" volcanic eruptions of hot, fluid lavas in Hawaii and the explosive, violent, dangerous eruptions of viscous lavas in volcanoes such as Mount St. Helens.

What is Extruded During Eruptions?

Volcanoes extrude lava, large volumes of gas, and pyroclastics (broken rock, lava "bombs," fine ash, and dust). In this section we will examine each of these materials.

Lava Flows

Because of their low silica content, basaltic lavas are usually very fluid. They flow in thin, broad sheets or streamlike ribbons. On the island of Hawaii, such lavas have been clocked at 30 kilometers (20 miles) per hour down steep slopes, but flow rates of 10 to 300 meters per hour are more common. In contrast, the movement of silica-rich lava may be too slow to perceive.

When fluid basaltic lavas of the Hawaiian type congeal, they commonly form a relatively smooth skin that wrinkles as the still-molten subsurface lava continues to advance (Figure 4.6A). These are known as **pahoehoe flows** (pronounced *pah-hoy-hoy*) and resemble the twisted braids in ropes.

Another common basaltic lava has a surface of rough, jagged blocks with dangerously sharp edges and spiny projections, called **aa** (pronounced *ah-ah*) (Figure 4.6B). Active aa flows are relatively cool and thick and advance at 5 to 50 meters per hour, depending on the slope. Further, escaping gases fragment the cool surface and produce numerous voids and sharp spines in the congealing lava. As the molten interior advances, the outer crust is broken further, giving the flow the appearance of an advancing mass of lava rubble.

Gases

Magmas contain varied amounts of dissolved gases held in the molten rock by confining pressure, just as carbon dioxide is held in soft drinks. As with soft drinks, as soon as the pressure is reduced, the gases begin to escape. Obtaining gas samples from an erupting volcano is difficult and dangerous, so geologists usually only estimate the amount of gas originally contained within the magma.

The gaseous portion of most magmas is believed to make up 1 to 5 percent of the total weight, with most of this being water vapor. Although the percentage may be small, the actual quantity of emitted gas can exceed thousands of tons per day.

The composition of volcanic gases is important because these gases are thought to be the original source of the water in the oceans. Further, volcanic eruptions have contributed much of the gases that compose the atmosphere. Analysis of samples taken during Hawaiian eruptions indicates that the gases are about 70 percent water vapor, 15 percent carbon dioxide, 5 percent nitrogen, 5 percent sulfur, and lesser amounts of chlorine, hydrogen, and argon. Sulfur compounds are easily recognized by their pungent odor and because they readily form sulfuric acid. Volcanoes are a natural source of air pollution, and can cause serious problems.

Pyroclastics

When basaltic lava is extruded, the dissolved gases escape quite freely and continually. These gases propel incandescent blobs of lava to great heights. Some ejected material may land near the vent and build a cone structure, whereas smaller particles will be carried great distances by the wind. In addition, the gases in highly viscous magmas become superheated, and upon release they expand a thousandfold as they blow pulverized rock, lava, and glass fragments from the vent. The particles produced by these processes are called **pyroclastics** (meaning "fire fragments"). These ejected lava fragments range in size from very fine dust and sand-sized volcanic ash, to large pieces (Figure 4.3).

Ash particles are produced when the extruded lava contains so many gas bubbles that it resembles the froth flowing from a newly opened bottle of

A.

B.

FIGURE 4.6 A. Typical pahoehoe (ropy) lava flow, Kilauea, Hawaii. **B.** Typical slow-moving aa flow. (Photos by J. D. Griggs/U.S. Geological Survey)

champagne. As the hot gases expand explosively, the lava is disseminated into very fine glassy fragments. When the hot ash falls, the glassy shards often fuse to form *welded tuff*, which covers vast portions of the western United States. Sometimes the frothlike lava is ejected as *pumice*, a material having so many voids (air spaces) that it often floats in water.

Also common are walnut-sized pyroclastics called *lapilli* ("little stones") and pea-sized particles called *cinders*. Particles larger than lapilli are called *blocks* when made of hardened lava and *bombs* when they are ejected as incandescent lava. Because bombs are semimolten upon ejection, they often take on a streamlined shape as they hurtle through the air (Figure 4.7).

Volcano Types

Successive eruptions from a central vent build a mountainous accumulation we call a **volcano**. Located at the summit of many volcanoes is a steep-walled **crater**. The crater is connected to the underground magma chamber via a pipelike **vent**. Some volcanoes have unusually large craters—exceeding 1 kilometer in diameter—and are known as **calderas**.

When fluid Hawaiian-type lava leaves a conduit, it is often stored in the crater or caldera until it overflows. On the other hand, viscous lava forms a plug in the pipe. It rises slowly or is blown out, often enlarging the crater.

Lava does not always issue from a central crater. Sometimes it is easier for the magma or escaping gases to push through fissures on the volcano's

FIGURE 4.7 Volcanic bombs. Ejected lava fragments take on a streamlined shape as they sail through the air. This bomb is approximately 10 centimeters in length. (Photo by E. J. Tarbuck)

flanks. Mount Etna in Italy, for example, has more than 200 secondary vents. Some of these emit only gases and are appropriately called *fumaroles*.

The eruptive history of each volcano is unique, so volcanoes vary in form and size. Nevertheless, volcanologists recognize three general eruptive patterns and characteristic forms: shield volcanoes, cinder cones, and composite cones. These three types are shown together for comparison in Figure 4.8. Note the massiveness of the shield volcano (Mauna Loa in Hawaii, Earth's largest active volcano). A large composite cone, Washington State's Mt. Rainier is dwarfed by Mauna Loa, and cinder cones are tiny in proportion to the other types.

We now will look at examples of the three types.

Shield Volcanoes

When fluid Hawaiian-type lava is extruded, the volcano takes the shape of a broad, slightly domed structure called a **shield volcano** (Figure 4.9). They are so-called because they roughly resemble the shape of a warrior's shield. Shield volcanoes are built primarily of basaltic lava flows and contain only a small percentage of pyroclastic material.

Mauna Loa is one of five shield volcanoes that together make up the island of Hawaii. Its base rests on the ocean floor 5000 meters below sea level, and its summit is 4170 meters above the water, giving it a total height approaching 6 miles, greater than the height of Mount Everest. Nearly one million years and numerous eruptive cycles built this truly gigantic pile of volcanic rock. Many other volcanic structures, including Midway Island and the Galapagos Islands, have been built in a similar manner from the ocean's depths.

Perhaps the most active and intensively studied shield volcano is Kilauea, also on the island of Hawaii, on the flank of the larger Mauna Loa. Kilauea has erupted more than 50 times in recorded history and is still active today. Several months before an eruptive phase, Kilauea's summit inflates as magma rises from its source 60 kilometers or more below the surface. This molten rock gradually works its way upward and accumulates in smaller reservoirs 3 to 5 kilometers below the summit. For up to 24 hours in advance of each eruption, swarms of small earthquakes warn of the impending activity.

The longest and largest rift eruption ever recorded at Kilauea began in 1983 and continues as this is written. The eruption began along a 6.5-kilometer (4-mile) fissure in an inaccessible forested area east of the summit caldera (Figure 4.10). By the summer of 1986, the eruptions shifted 3 kilometers

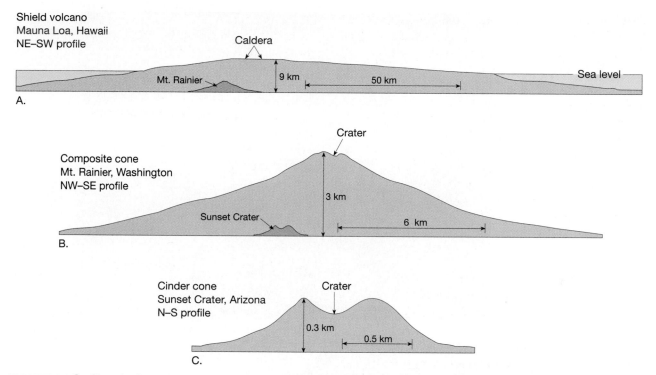

FIGURE 4.8 Profiles of volcanic landforms. **A.** Profile of Mauna Loa, Hawaii, the largest shield volcano in the Hawaiian chain. Note size comparison with Mt. Rainier, Washington, a large composite cone. **B.** Profile of Mt. Rainier, Washington. Note how it dwarfs a typical cinder cone. **C.** Profile of Sunset Crater, Arizona, a typical steep-sided cinder cone.

downslope. Here, smooth-surfaced pahoehoe lava formed a lava lake. Eventually, it overflowed, and the fast-moving pahoehoe destroyed nearly a hundred rural homes, covered a major roadway, and eventually flowed into the sea. Lava has been intermittently pouring into the ocean since that time, adding new land to the island.

Interestingly, shield volcanoes exist elsewhere in the solar system. Olympus Mons, a huge shield volcano on Mars, is the largest yet located. Its summit caldera is nearly large enough to contain the above-water portion of Mauna Loa.

Cinder Cones

As the name suggests, **cinder cones** are built from ejected lava fragments. Loose pyroclastic material has a high angle of repose (between 30 and 40 degrees), the steepest angle at which the material remains stable. Thus, volcanoes of this type have very steep slopes (Figure 4.11). Cinder cones are rather small, usually less than 300 meters (1000 feet) high, often forming near larger volcanoes, and often in groups.

One of the very few volcanoes observed by geologists from beginning to end is the cinder cone called Parícutin about 200 miles west of Mexico City. In 1943, it erupted in a cornfield owned by Dionisio

Pulido, who witnessed the event as he prepared the field for planting. For 2 weeks prior to the first eruption, numerous Earth tremors caused apprehension in the village of Parícutin about 3.5 kilometers away. Then sulfurous smoke began billowing from a small hole that had been in the cornfield for as long as Señor Pulido could remember. During the night, hot, glowing rock fragments thrown into the air from the hole produced a spectacular fireworks display. In one day the cone grew to 40 meters (130 feet) and by the fifth day it was over 100 meters (330 feet) high. Explosive eruptions threw hot fragments 1000 meters (3300 feet) above the crater rim. Larger fragments fell near the crater, some remaining incandescent as they rolled down the slope. These built an aesthetically pleasing cone, while finer ash fell over a much larger area, burning and eventually covering the village of Parícutin. Within 2 years the cone attained its final height of about 400 meters (1300 feet).

The first lava flow came from a fissure that opened just north of the cone, but after a few months flows began to emerge from the base of the cone itself. In June 1944, a clinkery aa flow 10 meters thick moved over much of the village of San Juan Parangaricutiro, leaving only the church steeple exposed (Figure 4.12). After 9 years, the activity ceased almost as

FIGURE 4.9 Mauna Loa is one of five shield volcanoes that together make up the island of Hawaii. Shield volcanoes are built primarily of fluid basaltic lava flows and contain only a small percentage of pyroclastic materials. (Photo by Greg Vaughn)

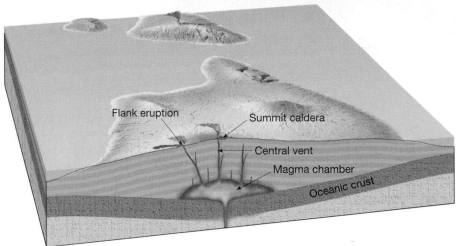

quickly as it had begun. Today, Parícutin is just another one of the numerous quiet cinder cones dotting the landscape in this region of Mexico. Like the others, it will probably not erupt again.

Composite Cones

Earth's most picturesque volcanoes are composite cones. Most active composite cones are in a narrow zone that encircles the Pacific Ocean, appropriately named the *Ring of Fire*. In this region are Fujiyama (Mt. Fuji) in Japan, Mount Mayon in the Philippines, and the picturesque volcanoes of the Cascade Range in the northwestern United States, including Mount St. Helens, Mount Rainier, and Mount Shasta (Figure 4.13).

A **composite cone** or **stratovolcano** is a large, nearly symmetrical structure composed of alternating lava flows and pyroclastic deposits, emitted mainly from a central vent. Just as shield volcanoes owe their shape to the highly fluid nature of the extruded lavas, so too do composite cones reflect the nature of the erupted material.

Composite cones are produced when relatively viscous lavas of andesitic composition are extruded. A composite cone may extrude viscous lava for long periods. Then, suddenly, the eruptive style changes and the volcano violently ejects pyroclastic material. Most of it falls near the summit, building a steep-sided mound of cinders. In time, this debris becomes covered by new lava. Occasionally, both activities occur simultaneously. The resulting cone consists of alternating layers (strata) of lava and pyroclastics, giving rise to the name *stratovolcano*. Two of the most perfect cones, Mount Mayon in the Philippines and Fujiyama in Japan, exhibit the classic form of the stratovolcano with its steep summit and more gently sloping flanks.

FIGURE 4.10 Lava extruded along the East Rift Zone, Kilauea, Hawaii. (Photo by Greg Vaughn)

Composite cones represent the most violent volcanic activity. Their eruption can be unexpected and devastating, as was the A.D. 79 eruption of the Italian volcano we now call Vesuvius. Prior to this eruption, Vesuvius had been dormant for centuries. Although minor earthquakes probably warned of the events to follow, Vesuvius was covered with dense vegetation and hardly looked threatening. On August 24, however, the tranquility ended, and in the next 3 days the city of Pompeii (near Naples) and more than 2000 of its 20,000 residents were buried. They remained so for nearly 17 centuries, until the city was rediscovered and excavated.

Nuée Ardente

Although the destruction of Pompeii was catastrophic, even more devastating eruptions occur when a volcano ejects hot gases infused with incandescent ash. This produces a fiery cloud called a **nuée ardente**. Also referred to as *glowing avalanches*, these turbulent steam clouds and companion ash flows race down steep volcanic slopes at speeds that can approach 200 kilometers (125 miles) per hour (Figure 4.14). The ground-hugging portions of glowing avalanches are rich in particulate matter that is suspended by hot, buoyant gases. Thus, these flows, which can include larger rock fragments in addition to ash, travel downslope in a nearly frictionless environment cushioned by expanding volcanic gases. This explains why some nuée ardente deposits extend more than 100 kilometers (60 miles) from their source.

In 1902, a nuée ardente from Mount Pelée, a small volcano on the Caribbean island of Martinique, destroyed the port town of St. Pierre. The destruction happened in moments and was so devastating that almost all of St. Pierre's 28,000 inhabitants were killed. Only a prisoner protected in a dungeon, a shoemaker, and a few people on ships in the harbor were spared (Figure 4.15).

It is interesting to contrast the destruction of St. Pierre with that of Pompeii. In 3 days, Pompeii was completely buried, whereas St. Pierre was destroyed in moments and its remains were mantled by only a thin layer of volcanic debris. The structures of Pompeii remained intact, except for roofs that collapsed under the weight of the ash. In St. Pierre, masonry walls nearly a meter thick were knocked over like dominoes; large trees were uprooted and cannons were torn from their mounts. Clearly, volcanic hazards vary from one volcano to another.

Lahar

In addition to their violent eruptions, volcanoes offer other hazards. Large composite cones often generate a mudflow called by its Indonesian name **lahar**. These

Pyroclastic material

Crater

Central vent filled
with rock fragments

FIGURE 4.11 SP Crater, a cinder cone north of Flagstaff, Arizona. Note the road across the lava flow for scale. Cinder cones are built from ejected lava fragments and are usually less than 300 meters (1000 feet) high. (Photo by Michael Collier)

destructive flows occur when volcanic ash and debris become saturated with water and flow down steep volcanic slopes, generally following stream valleys. Some lahars are produced when rainfall saturates volcanic deposits, whereas others are triggered as large volumes of ice and snow melt during an eruption.

When Mount St. Helens erupted in 1980, several lahars formed. The flows and accompanying floods

FIGURE 4.12 The village of San Juan Parangaricutiro engulfed by lava from Parícutin, shown in the background. Only the church towers remain. (Photo by Tad Nichols)

raced down the valleys of the north and south forks of the Toutle River at speeds exceeding 30 kilometers per hour. Water levels reached 4 meters above flood stage, and nearly all the homes and bridges along the river were destroyed or severely damaged (Figure 4.16). Fortunately, the affected area was not densely populated. This was not the case in 1985, when Nevado del Ruiz, a volcano in the Andes, erupted and generated a lahar that killed nearly 20,000 people (see the section entitled "Mudflow" in Chapter 8).

FIGURE 4.14 Nuée ardente races down the slope of Mount St. Helens on August 7, 1980, at speeds in excess of 100 kilometers (60 miles) per hour. (Photo by Peter W. Lipman, U.S. Geological Survey)

FIGURE 4.13 Mount Shasta, California, one of the largest composite cones in the Cascade Range. (Photo by John S. Shelton)

FIGURE 4.15 St. Pierre as it appeared shortly after the eruption of Mount Pelée, 1902. (Reproduced from the collection of the Library of Congress)

FIGURE 4.16 A house damaged by debris from lahars that flowed along the Toutle River, west-northwest of Mount St. Helens. The end section of the house was torn free and lodged against trees. (Photo by D. R. Crandell, U.S. Geological Survey)

Volcanic Landforms

The most obvious volcanic landform is a cone. But other distinctive landforms are associated with volcanic activity.

Volcanic Pipes and Necks

Most volcanoes are fed by conduits called **pipes** or **vents**, which are connected to a magma source near the surface. Some pipes are thought to extend tube-like directly into the asthenosphere, perhaps more than 100 kilometers. Consequently, the materials in these vents are considered to be samples of the asthenosphere that have undergone very little alteration during their ascent. Geologists thus consider pipes to be "windows" into Earth.

Occasionally a volcanic pipe will reach the surface, but the eruptive phase will cease before any lava is extruded. The upper portions of these pipes often contain a jumble of lava fragments and fragments that were torn from the walls of the vent. The best-known of these structures are the diamond-bearing pipes of South Africa. Here the rocks filling the pipes are thought to have originated at depths of about 200 kilometers, where pressures are great enough to generate diamonds. The diamonds crystallized at depth and were carried upward by the still-molten portion of the magma.

Volcanoes on land are continually being lowered by weathering and erosion. Cinder cones are easily eroded, because they are composed of unconsolidated materials. As erosion progresses, the rock occupying the vent is often more resistant and may remain standing above the surrounding terrain long after most of the cone has vanished. Ship Rock, New Mexico, is such a feature, called a **volcanic neck** (Figure 4.17). This structure, higher than many skyscrapers, is but one of many such landforms that protrude conspicuously from the red desert landscapes of the Southwest. Ultimately, even these resistant necks will succumb to relentless erosion over geologic time.

Craters and Calderas

As noted, most volcanoes have a steep-walled *crater*. A crater is called a *caldera* when it exceeds 1 kilometer in diameter. Calderas are usually circular, with rather flat floors and steep walls. Most calderas probably form when the summit of a volcanic structure collapses into the partially emptied magma chamber below (Figure 4.18).

Crater Lake in Oregon, located in such a caldera, is 10 kilometers (6 miles) at its widest and 1175 meters (over 3800 feet) deep. The formation of Crater Lake began about 7000 years ago when the volcano, later to be named Mount Mazama, violently extruded 50 to 70 cubic kilometers of volcanic material. With

FIGURE 4.17 Ship Rock, New Mexico, is a volcanic neck. This structure, over 420 meters (a quarter mile) high, consists of igneous rock that crystallized in the vent of a volcano. The volcano has long since been eroded away, leaving the harder rock of the neck. The tabular structure in the foreground is a dike that fed lava flows along the flanks of the once-active volcano. (Photo by Tom Till)

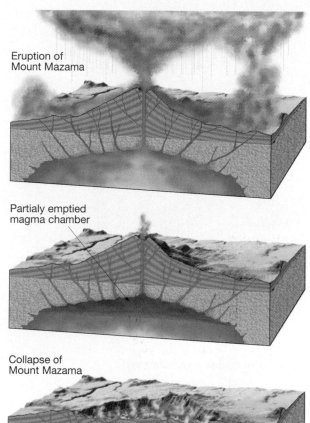

Eruption of Mount Mazama

Partialy emptied magma chamber

Collapse of Mount Mazama

Formation of Crater Lake and Wizard Island

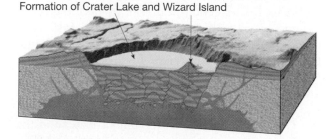

FIGURE 4.18 Sequence of events that formed Crater Lake, Oregon. About 7000 years ago, a violent eruption partly emptied the magma chamber, causing the summit of former Mount Mazama to collapse. Rainfall and groundwater contributed to form Crater Lake, the deepest lake in the United States. Subsequent eruptions produced the cinder cone called Wizard Island. (After H. Williams, *The Ancient Volcanoes of Oregon*, p. 47. Courtesy of the University of Oregon)

the loss of support, 1500 meters (nearly a mile) of this once-prominent 3600-meter cone collapsed. After the collapse, rainwater filled the caldera. Later volcanic activity built a small cinder cone in the lake called Wizard Island, which today provides a mute reminder of past activity (Figure 4.19).

Not all calderas are produced by explosive eruptions. For example, Hawaii's active shield volcanoes, Mauna Loa and Kilauea, have large calderas (3 to 5 kilometers or 2 to 3 miles across and nearly 200 meters, or 650 feet deep). They formed by collapse as magma slowly drained from the summit magma chambers during flank eruptions.

FIGURE 4.19 Crater Lake in Oregon occupies a caldera about 6 miles in diameter. (Photo by Greg Vaughn/Tom Stack and Associates)

Fissure Eruptions and Lava Plateaus

We think of volcanic eruptions as building a cone or mountain from a central vent. But by far the greatest volume of volcanic material is extruded from fractures in the crust called **fissures**. Rather than building a cone, these long, narrow cracks pour forth a low-viscosity Hawaiian-type lava, blanketing a wide area. This process occurs on both the seafloor (at spreading centers, Chapters 14 and 16) and on land.

On land, the extensive Columbia Plateau in the northwestern United States was formed this way (Figure 4.20). Here, numerous fissure eruptions extruded very fluid basaltic lava. Successive flows, some 50 meters thick, buried the existing landscape as they built a lava plateau that is nearly a mile thick in places (Figure 4.21). The fluidity is evident, because some lava remained molten long enough to flow 150 kilometers (90 miles) from its source. The term **flood basalts** appropriately describes these flows.

Pyroclastic Flows

When silica-rich magma is extruded, **pyroclastic flows** of ash and pumice fragments usually result. They are propelled away from the vent at high speeds and may blanket extensive areas before coming to rest. Once deposited, the pyroclastic materials may closely resemble lava flows.

Extensive pyroclastic flow deposits exist in many parts of the world, most associated with large calderas. Perhaps best known is the Yellowstone Plateau in northwestern Wyoming. Here a large magma body, rich in silica, still exists a few kilometers below the surface. Several times over the past two million years, fracturing of the rocks overlying the magma chamber has resulted in huge outpourings, accompanied by the formation of calderas. In Yellowstone National Park, numerous buried fossil forests have been discovered. Following volcanic activity, a forest would develop on a newly formed volcanic surface, only to be covered by ash from the next eruptive phase. Fortunately, no eruption of this type has occurred in historic times.

Volcanoes and Climate

The idea that explosive volcanic eruptions change Earth's climate was first proposed many years ago and still appears to explain some aspects of climatic variability. Explosive eruptions emit huge quantities

FIGURE 4.20 Volcanic areas in the northwestern United States. The Columbia River basalts cover an area of nearly 200,000 square kilometers (80,000 square miles). Activity here began about 17 million years ago as lava began to pour out of large fissures, eventually producing a basalt plateau with an average thickness of more than 1 kilometer. (After U.S. Geological Survey)

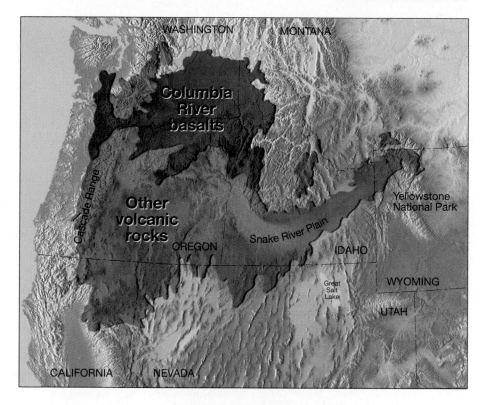

FIGURE 4.21 Basalt flows near Idaho Falls. (Photo by John S. Shelton)

of gases and fine-grained debris into the atmosphere. The greatest eruptions inject material high into the stratosphere, where it spreads around the globe and remains for months or even years. The basic premise is that this suspended volcanic material (most importantly, droplets of sulfuric acid) will filter out a portion of the incoming solar radiation, and this, in turn, will lower air temperatures worldwide.

Perhaps the most notable cool period linked to a volcanic event is the "year without a summer" that followed the 1815 eruption of Mount Tambora in Indonesia. The abnormally cold spring and summer of 1816 in many parts of the Northern Hemisphere, including New England, are believed to have been caused by the cloud of volcanic debris and gases ejected from Tambora (see Box 4.1).

When Mount St. Helens erupted in 1980, there was almost immediate speculation: Can an eruption such as this change our climate? Although spectacular, a single explosive volcanic eruption of the magnitude of Mount St. Helens occurs somewhere in the world every 2 to 3 years. Studies of these events indicate that a very slight cooling of the lower atmosphere does occur. However, it is believed that the cooling is so slight, less than one-tenth of one degree Celsius, as to be inconsequential.

Box 4.1 The Year Without A Summer

In Figure 4.A, compare the volume of volcanic debris extruded during some well-known eruptions, beginning with Mount Vesuvius in A.D. 79. The eruption of Tambora is clearly the largest in modern times. In 1815, this nearly 4000-meter Indonesian volcano violently ejected an estimated 30 cubic kilometers of volcanic debris. That is 30 times more than was emitted from Mount St. Helens in 1980.

Although the Tambora eruption occurred in an isolated part of the world, circulation high in the atmosphere spread its influence far and wide. The impact of its dust and gases on climate is believed to have been widespread in the Northern Hemisphere. One researcher noted, "The extreme cold that prevailed during the spring and summer of 1816 in some regions of the world represents one of the most unusual climatic episodes that has occurred since the advent of instrumental weather observations."* The effects were especially severe in New England, where 1816 has come to be known as the "*year without a summer.*"

From May through September 1816, unprecedented cold affected the Northeast and adjacent Canada. The result was a late spring, a cold summer,

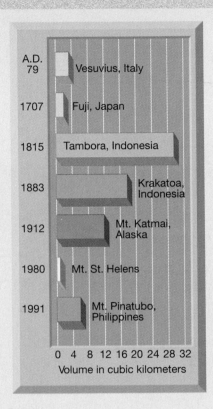

FIGURE 4.A
Approximate volume of volcanic debris emitted during some well-known eruptions. The 1815 eruption of Tambora, the largest-known eruption in historic time, ejected over 30 times more ash than did Mount St. Helens in 1980.

and an early fall. Heavy snow fell in June and frosts occurred in July and August, killing crops. Temperatures in New England averaged 3.5°C below normal in June and 1–2°C below normal in August. Although temperatures this cold had occurred before, there has been no other such protracted span of cold since record keeping

began. The reduced averages may seem modest to you, but they took place in a region where even a small drop can mean severe frost.

The chill was widespread. "Although the New England farmer considered it a local tragedy, the abnormal weather was widespread throughout the Northern Hemisphere. In England it was almost as cold as in the United States, and 1816 was a famine year there, as it was in France and Germany."*

The unusual meteorological events of 1816 that followed the massive 1815 eruption of Tambora are regarded by many as a spectacular example of the influence of explosive volcanism on climate.

* *American Weather Stories* (Washington, DC: National Oceanic and Atmospheric Administration.

The huge eruption of the Philippines' Mount Pinatubo in 1991 gave scientists another opportunity to test the connection between volcanism and climatic change. In 1991, this volcano emitted huge quantities of ash and 25 million to 30 million tons of sulfur dioxide gas. The gaseous component turned into a haze of tiny sulfuric acid droplets. Because this material was superhot, it rose to heights of 20 to 30 kilometers and quickly encircled the entire globe. This Sun-blocking haze was probably the most massive since the eruption of the Indonesian volcano Krakatau in 1883. Studies indicated that the average global surface temperature had declined by about 0.6°C (1°F) when

compared to the average of the year before. Scientists hope to learn more about the volcano/climate connection as subsequent data are analyzed.

Intrusive Igneous Activity

Volcanic eruptions can be among the most violent and spectacular events in nature and thus receive much scientific attention. Yet most magma is deep underground, intruding into existing rocks. An understanding of this intrusive igneous activity is therefore as important to geologists as is the study of volcanic events on the surface. In mythology, Pluto was lord of the underworld. Thus, underground igneous rock bodies are called *plutons*.

Intrusive Igneous Bodies

Figure 4.22 shows intrusive igneous bodies (plutons) that form when magma crystallizes within Earth's crust. Notice that some of these structures are flat, whereas others are bulky masses. Some cut through existing structures, such as layers of sedimentary rocks, while others form when magma is injected between sedimentary layers. Owing to these differences, plutons are generally classified according to their shape as either *tabular* (flat) or *massive*, and by their orientation with respect to the host rock. Intrusive igneous bodies are said to be *discordant* if they cut across existing sedimentary beds, and *concordant* if they form parallel to the existing sedimentary beds.

Dikes are tabular plutons produced when magma is injected into fractures that cut across rock layers. They range in thickness from less than 1 centimeter to more than 1 kilometer. The largest are hundreds of kilometers long. Dikes are often oriented vertically, forming in pathways followed by molten rock that fed ancient lava flows. Frequently, dikes are more resistant to weathering than the surrounding rock. When exposed, such dikes have the appearance of a wall, as shown in Figures 4.17 and 4.22B.

Sills are tabular plutons formed when very fluid basaltic magma is injected along sedimentary bedding surfaces (Figure 4.23). Horizontal sills are the most common; hence the name, like a window sill. Injection of a sill raises the overlying sedimentary rock to a height equal to the thickness of the sill. Consequently, sills can form only at relatively shallow depths where the pressure exerted by the weight of overlying strata is relatively low.

One of the largest and best-known sills in the United States is the Palisades Sill, which is exposed along the Hudson River near New York City. This resistant, 300-meter-thick sill has formed an imposing cliff easily seen from across the river.

Laccoliths form the same way as sills, but from more viscous magma, which collects as a lens-shaped mass that arches the overlying strata upward. Consequently, a laccolith can occasionally be detected because of the dome it creates at the surface (Figure 4.22B).

Batholiths are by far the largest intrusive igneous bodies. The Idaho batholith, for example, encompasses more than 40,000 square kilometers. Indirect evidence from gravitational studies indicates that batholiths are also very thick, possibly extend-

ing dozens of kilometers into the crust. Geologists define a batholith as an intrusive body with a surface exposure of more than 100 square kilometers (40 square miles). Similar but smaller intrusive bodies are termed *stocks*. Many stocks appear to be portions of batholiths that are not yet fully exposed. Both a batholith and a stock are illustrated in Figure 4.22C.

Batholiths frequently form the cores of mountain systems. Here, uplift and erosion have removed the surrounding rock to expose the resistant igneous batholith. Some of the highest mountain peaks, such as Mount Whitney in the Sierra Nevada, are being eroded from such a granitic mass. Large batholiths form over millions of years. The intrusive activity that created the Sierra Nevada batholith, for example, occurred nearly continuously over a 130-million-year span (Figure 4.24).

Large expanses of granitic rock are also exposed in the stable interiors of the continents, such as the Canadian Shield of North America. These relatively flat outcrops are believed to be the remnants of ancient mountains that erosion has long since leveled.

Igneous Activity and Plate Tectonics

The origin of magma has been controversial in geology almost from the beginning of the science. How do magmas of different compositions form? Why do volcanoes in the deep-ocean basins primarily extrude basaltic lava (Hawaii), whereas those adjacent to oceanic trenches extrude mainly andesitic lava (Mount St. Helens)? Why does an area of igneous activity, commonly called the *Ring of Fire*, surround the Pacific Ocean? Insights from the theory of plate tectonics are providing answers. We will first examine the origin of magma and then look at the global distribution of volcanic activity as viewed by the plate tectonics model.

Origin of Magma

Based on available scientific evidence, the crust and mantle are composed primarily of solid rock, not a molten liquid. The outer core is in a fluid state, but this material is very dense and remains deep within Earth. So what could be the source of magma that produces volcanic activity?

Temperature and Magma Generation Geologists conclude that magma must originate from essentially solid rock in the crust and mantle. The most obvious way to generate magma from solid rock is to raise the temperature above the rock's

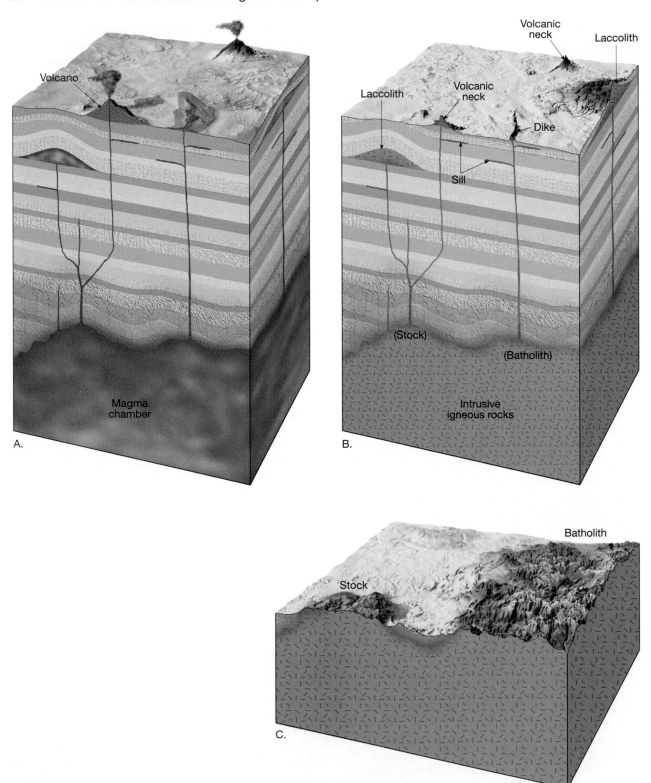

FIGURE 4.22 Illustrations showing basic igneous structures. **A.** This cross-sectional view shows the relationship between volcanism and intrusive igneous activity. **B.** This view illustrates the basic intrusive igneous structures, some of which have been exposed by erosion long after their formation. **C.** After millions of years of uplifting and erosion a stock and batholith are exposed at the surface.

FIGURE 4.23 Salt River Canyon, Arizona. The dark, essentially horizontal band is a sill of basaltic composition that intruded into horizontal layers of sedimentary rock. (Photo by E. J. Tarbuck)

recognized that temperatures increase as they descend to greater depths. A rule of thumb is that temperature increases 20–30°C (35–55°F) per kilometer depth in the upper crust.

This gradual increase in temperature with depth is thought to contribute to magma production (rock-melting) in two ways. First, at deep-ocean trenches, slabs of cool oceanic lithosphere descend into the hot mantle. Heat supplied by the surrounding rocks is thought to melt some of the subducting oceanic crust and overlying sediments. Second, heat drives water and other fluids from the subducted rocks into the overlying mantle. These substances then act as a flux does at a foundry, causing melting in the mantle at reduced temperatures. The result of these processes is the generation of basaltic magma.

Once a hot magma body forms as just described, the molten rock could migrate upward to the base of the crust and intrude granitic rocks. Because granitic rocks have melting temperatures well below those of basalt, heat derived from the hotter basaltic magma could melt the already warm crustal rocks. The volcanic activity that produced the vast ash flows in Yellowstone National Park is believed to have resulted from such activity. Here basaltic magma from the mantle transported heat to the crust, where the melting of silica-rich rocks generated explosive outflows of pyroclastics.

melting point. In a near-surface environment, silica-rich rocks of granitic composition begin to melt about 750°C (1400°F), whereas basaltic rocks must reach temperatures above 1000°C (1850°F) before melting commences.

What is the source of heat sufficient to melt this rock? One source is the heat liberated during the decay of radioactive elements found in the mantle and crust. Workers in underground mines have long

FIGURE 4.24 Half Dome in Yosemite National Park, California. This feature is just a tiny portion of the Sierra Nevada batholith, a huge structure that extends for approximately 400 kilometers. (Photo by Lewis Kemper/DRK Photo)

Role of Pressure If temperature were the only factor that determined whether or not rock melts, Earth would be a molten ball covered with a thin, solid outer shell. This, of course, is not the case. The reason is that pressure also increases with depth.

In general, an increase in the confining pressure increases a rock's melting temperature. (Molten rock requires more space than solid rock, so under great pressure rocks stay solid, even well above their melting temperature at surface pressures.) By contrast, reducing confining pressure lowers a rock's melting temperature. Consequently, a drop in confining pressure can lower the melting temperature of rock sufficiently to trigger melting. This occurs when rock ascends, thereby moving into zones of lower pressures.

Partial Melting An important difference exists between the melting of a substance that consists of a single compound, such as ice, and the melting of igneous rocks, which are mixtures of several different minerals. Whereas ice melts at a definite temperature, most igneous rocks melt over a temperature range of a few hundred degrees (recall Bowen's Reaction Series in Chapter 3). As a rock is heated, the first melt to form will contain a higher percentage of the low-melting-temperature minerals than the original rock. Should melting continue, the composition of the melt will steadily approach the overall composition of the rock from which it is derived as more of its minerals melt. Most often, however, melting is not complete. This process, known as **partial melting**, produces most, if not all, magma.

Thus, an important consequence of partial melting is the production of a melt with a higher silica content than the parent rock. Recall that basaltic rocks have a relatively low silica content and that granitic rocks have a much higher silica content. Consequently, magmas generated by partial melting are more granitic than the parent material from which they formed. As you shall see, this will help us understand the global distribution of the different types of volcanic activity.

Distribution of Igneous Activity

Earth has more than 600 active volcanoes and most are near convergent plate margins. Further, extensive volcanic activity (fissure eruptions) occurs along spreading centers of the oceanic ridge system, hidden from easy view by the world's ocean. In this section we will examine three zones of volcanic activity and relate all of them to global tectonic activity. These active areas are along oceanic ridges

(spreading centers), adjacent to ocean trenches (subduction zones), and within the plates themselves (intraplate). These three settings are shown in Figure 4.25, which highlights two examples of each. Please refer to this illustration as you read the following.

Spreading Center Volcanism The greatest volume of volcanic rock is produced along the oceanic ridge system, where sea-floor spreading is active (Figure 4.25A). As the rigid lithosphere pulls apart, the pressure on the underlying rocks is lessened. This reduced pressure, in turn, lowers the melting temperature of the mantle rocks. Partial melting of these rocks (primarily peridotite) generates large quantities of basaltic magma that moves upward to fill the newly formed cracks.

Some of the molten basalt reaches the ocean floor, where it produces extensive lava flows or occasionally grows into a volcanic pile. Sometimes this activity produces a volcanic cone that rises above sea level, as the island of Surtsey did near the coast of Iceland in 1963 (Figure 4.26). Numerous submerged volcanic cones also dot the flanks of the ridge system and the adjacent deep-ocean floor (see Figure 14.4). Many of these formed along the ridge crests and were moved away as new oceanic crust was created by the process of sea-floor spreading.

Spreading center volcanism also occurs less often on the continents. The classic example is the African Rift Zone shown in Figure 4.25A (lower right). If this rift continues to spread, Africa eventually will be divided into two continents.

Subduction Zone Volcanism Recall that ocean trenches are sites where slabs of oceanic crust are bent and move downward into the upper mantle (Figure 4.25B). When the descending oceanic plate reaches a depth of about 100 kilometers (60 miles), partial melting of the water-rich ocean crust and the overlying mantle rocks takes place. The partial melting of these materials is thought to produce basaltic and andesitic magmas. When enough magma has accumulated, the molten material slowly migrates upward toward the surface. It moves upward because it is less dense than the surrounding solid rock.

When subduction zone volcanism occurs in the ocean, a chain of volcanoes called an *island arc* is produced. Examples are numerous in the Pacific and include the Tonga Islands, Mariana Islands, and Aleutian Islands (Figure 4.25B, upper left).

When subduction occurs beneath continental crust, the magma generated is often altered before it solidifies. Assimilation of crustal fragments into the

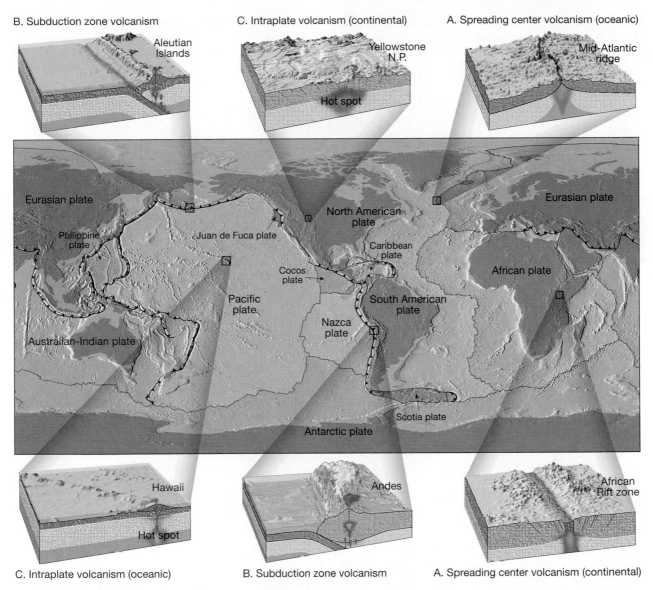

B. Subduction zone volcanism — Aleutian Islands

C. Intraplate volcanism (continental) — Yellowstone N.P. — Hot spot

A. Spreading center volcanism (oceanic) — Mid-Atlantic ridge

C. Intraplate volcanism (oceanic) — Hawaii — Hot spot

B. Subduction zone volcanism — Andes

A. Spreading center volcanism (continental) — African Rift zone

FIGURE 4.25 Three settings where volcanism occurs (two examples of each). **A.** Spreading centers are along divergent plate boundaries. **B.** Subduction zones are along convergent plate boundaries. **C.** Intraplate volcanism occurs in isolated places within plates, away from boundaries.

ascending magma body can create a melt with a composition that is andesitic to granitic. The volcanoes of the Andes Mountains, from which andesite gets its name, are examples of this mechanism at work (Figure 4.25B, lower center).

Many volcanoes border the Pacific Basin, giving this region the name *Ring of Fire*. Volcanism here is associated with subduction and partial melting of the Pacific seafloor as it is shoved beneath the continents that surround the Pacific. As these oceanic plates sink, they carry with them sediments and oceanic crust containing abundant water. Because water reduces the melting point of rock, it aids the melting process that generates fresh magma for volcanoes.

The presence of water contributes to the high gas content (water vapor) and explosive nature of volcanoes that make up the *Ring of Fire*. The volcanoes of the Cascade Range in the northwestern United States (including Mount St. Helens, Mount Rainier, and Mount Shasta), as well as Mount Pinatubo, Tambora, and Krakatau, are all of this type. The Cascade volcanoes are shown in their tectonic setting in Figure 4.27.

Intraplate Volcanism Volcanic activity occurs along plate boundaries for the reasons just described. But eruptions also occur in the *middle* of plates, both continental and oceanic. Why?

FIGURE 4.26 The volcanic island of Surtsey emerged from the ocean just south of Iceland in 1963. (Courtesy of P. Vauthey/Sygma)

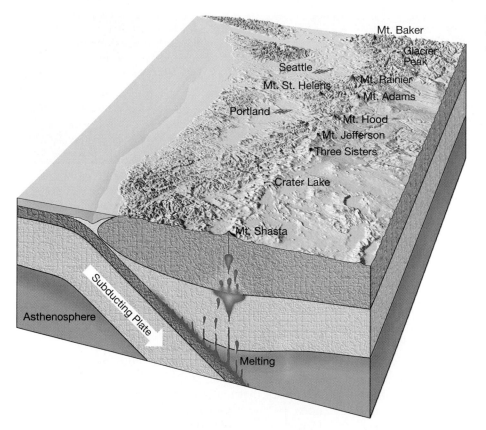

FIGURE 4.27 Locations of several of the larger composite cones that comprise the Cascade Range.

The processes that trigger volcanic activity within a rigid plate are difficult to establish. Consider this: Activity in the Yellowstone region produced silica-rich lava, pumice, and ash flows, whereas extensive basaltic flows have formed the nearby Columbia Plateau. Yet these rocks of greatly varying compositions actually overlie one another in several locations.

Because basaltic volcanism occurs on the continents as well as within the ocean basins, it may have a common source. Most likely the source is the partial melting of mantle rocks. One proposal suggests that some intraplate basaltic magma comes from rising plumes of hot mantle material. These hot plumes, which may extend to the core-mantle boundary, produce **hot spots**, local volcanic regions only a few hundred kilometers across. Most hot spots appear to have persisted for a few tens of millions of years. One hot spot is located beneath the island of Hawaii

(Figure 4.25C, lower left). Another may have produced the large outpourings of lava that make up the Columbia Plateau.

Generally, lavas and ash of granitic composition are extruded from vents located inland from the continental margins. This suggests that remelting of the continental crust may form these silica-rich magmas. But what mechanism causes large quantities of continental material to melt? One idea is that a thick segment of continental crust occasionally becomes situated over a rising plume of hot mantle material. Rather than producing vast outpourings of basaltic lava, as occurs at oceanic sites such as Hawaii, magma from the rising plume is emplaced at depth. Here the incorporation and melting of surrounding rock forms a secondary, silica-rich, viscous magma, which slowly migrates upward. Continued hot spot activity supplies heat to the rising mass, thereby aiding its ascent. This may explain the activity in the Yellowstone region (Figure 4.25C, upper center).

Although the theory of plate tectonics has answered many questions that plagued volcanologists for decades, new questions have arisen: Why does sea-floor spreading occur in some areas and not in others? How do hot spots originate? These and other questions are the subject of continuing geologic research.

The Chapter in Review

The following statements are intended to help you review the primary objectives presented in this chapter.

- The primary factors that determine the nature of volcanic eruptions include the magma's *composition,* its *temperature,* and the *amount of dissolved gases* it contains. As lava cools, it begins to congeal, and, as *viscosity* increases, its mobility decreases. *The viscosity of magma is directly related to its silica content.* Granitic lava, with its high silica content (over 70 percent), is very viscous and forms short, thick flows. Basaltic lava, with a lower silica content (about 50 percent), is more fluid and may travel a long distance before congealing. Dissolved gases tend to increase the fluidity of magma and, as they expand, provide the force that propels molten rock from the vent of a volcano.

- The materials associated with a volcanic eruption include (1) *lava flows* (*pahoehoe* flows, which resemble twisted braids, and *aa* flows, consisting of rough jagged blocks, both form from basaltic lavas); (2) *gases* (primarily in the form of *water vapor*); and (3) *pyroclastic material* (pulverized rock and lava fragments blown from the volcano's vent, which include *ash, pumice, lapilli, cinders, blocks,* and *bombs*).

- Successive eruptions of lava from a central vent result in a mountainous accumulation of material known as a *volcano.* Located at the summit of many volcanoes is a steep-walled depression called a *crater. Shield cones* are broad, slightly domed volcanoes built primarily of fluid, basaltic lava. *Cinder cones* have very steep slopes composed of pyroclastic material. *Composite cones,* or *stratovolcanoes,* are large, nearly symmetrical structures built of interbedded lavas and pyroclastic deposits. Composite cones represent the most violent type of volcanic activity. Often associated with a violent eruption is a *nuée ardente,* a fiery cloud of hot gases infused with incandescent ash that races down steep volcanic slopes. Large composite cones may generate a type of mudflow known as a *lahar.*

- Most volcanoes are fed by conduits called *pipes* or *vents.* As erosion progresses, the rock occupying the vent is often more resistant and may remain standing above the surrounding terrain as a *volcanic neck.* The summits of some volcanoes have large, nearly circular depressions that exceed 1 kilometer in diameter called *calderas.* Although volcanic eruptions from a central vent are the most familiar, by far the largest amounts of volcanic material are extruded from cracks in the crust called *fissures.* The term *flood basalts* describes the fluid, water-like, basaltic lava flows that cover an extensive region in the northwestern United States known as the Columbia Plateau. When silica-rich magma is extruded, *pyroclastic flows* consisting largely of ash and pumice fragments usually result.

- Explosive volcanic eruptions are regarded as an explanation for some aspects of Earth's climatic variability. The basic premise is that suspended volcanic material will filter out a portion of the incoming solar radiation, which, in turn, will lower air temperature in the lower atmosphere.

- Intrusive igneous bodies are classified according to their *shape* and by their *orientation with respect to the host rock*, generally sedimentary rock. The two general shapes are *tabular* (sheetlike) and *massive*. Intrusive igneous bodies that cut across existing sedimentary beds are said to be *discordant*; those that form parallel to existing sedimentary beds are *concordant*.

- *Dikes* are tabular, discordant igneous bodies produced when magma is injected into fractures that cut across rock layers. Tabular, concordant bodies, called *sills*, form when magma is injected along the bedding surfaces of sedimentary rocks. In many respects sills closely resemble buried lava flows. *Laccoliths* are similar to sills but form from less-fluid magma that collects as a lens-shaped mass that arches the overlying strata upward. *Batholiths*, the largest intrusive igneous bodies with surface exposures of more than 100 square kilometers (40 square miles), frequently make up the cores of mountains.

- *Temperature, pressure,* and *partial melting all influence the formation of magma.* By raising its temperature, solid rock will melt and generate magma. One source of heat is that which is released by the decay of radioactive elements found in Earth's mantle and crust. Second, a drop in confining pressure can lower the melting temperature of rock sufficiently to trigger melting. Third, igneous rock, which contains several different minerals, melts over a range of temperatures, with the lower temperature minerals melting first. This process, known as *partial melting*, produces most, if not all, magma. Partial melting often results in the production of a melt with a higher silica content than the parent rock.

- *Most active volcanoes are associated with plate boundaries.* Active areas of volcanism are found along the oceanic ridges (*spreading center volcanism*), adjacent to ocean trenches (*subduction zone volcanism*), as well as the interiors of plates themselves (*intraplate volcanism*).

Key Terms

aa flow (p. 71)
batholith (p. 83)
caldera (p. 72)
cinder cone (p. 73)
composite cone (strato-
    volcano) (p. 74)
crater (p. 72)

dike (p. 83)
fissure (p. 80)
flood basalt (p. 80)
hot spot (p. 88)
laccolith (p. 83)
lahar (p. 75)
nuée ardente (p. 75)

pahoehoe flow (p. 71)
partial melting (p. 86)
pipe (p. 78)
pyroclastic flow (p. 80)
pyroclastics (p. 71)
shield volcano (p. 72)
sill (p. 83)

vent (p. 72)
viscosity (p. 68)
volcanic neck (p. 78)
volcano (p. 72)

Questions for Review

1. What triggered the 1980 eruption of Mount St. Helens?

2. Name five volcanoes in the western United States that geologists believe will erupt again.

3. What is the difference between magma and lava?

4. What three factors determine the nature of a volcanic eruption? What role does each play?

5. Why is a volcano fed by highly viscous magma likely to be a greater threat than a volcano supplied with very fluid magma?

6. Describe pahoehoe and aa lava.

7. List the main gases released during a volcanic eruption. Why are gases important in eruptions?

8. Analysis of samples taken during Hawaiian eruptions on the island of Hawaii indicate that _____ was the most abundant gas released.

9. Describe each type of pyroclastic material.

10. Compare and contrast the main types of volcanoes (size, shape, eruptive style, and so forth).

11. Name one example of each of the three types of volcanoes.

12. Compare the formation and size of Mauna Loa and Parícutin.

13. How is a caldera different from a crater?

14. Describe the formation of Crater Lake. Compare it to the caldera formed during the eruption of Kilauea.

15. Briefly describe the mechanism by which explosive volcanic eruptions are thought to influence Earth's climate.

16. What is Ship Rock, New Mexico, and how did it form?

17. Describe each of the four intrusive features discussed in the text.

18. Why might a laccolith be detected at Earth's surface before being exposed by erosion?

19. What is the largest of all intrusive igneous bodies? Is it tabular or massive? Concordant or discordant?

20. Explain how most magma is thought to originate.

21. Spreading center volcanism is associated with which rock type? What causes rocks to melt in regions of spreading center volcanism?

22. What is the *Ring of Fire?*

23. Are volcanic eruptions in the *Ring of Fire* generally quiet or violent? Name a volcano that would support your answer.

24. The Hawaiian Islands and Yellowstone are thought to be associated with which of the three zones of volcanism?

25. Volcanic islands in the deep ocean are composed primarily of what igneous rock type?

26. Explain why andesitic and granitic rocks are confined to the continents and oceanic margins but are absent from deep-ocean basins.

Testing What You Have Learned

To test your knowledge of the material presented in this chapter, answer the following questions:

Multiple-Choice Questions

1. Which one of the following provides the force that propels molten rock from the vent of a volcano?
  a. large lapilli   **c.** subduction   **e.** crystalline silica
  b. fluid basalt   **d.** escaping gases

2. Igneous intrusive bodies with a surface exposure of more than 100 square kilometers (40 square miles) that frequently compose the cores of mountain systems are referred to as _____.
  a. dikes   **c.** sills   **e.** laccoliths
  b. batholiths   **d.** stocks

3. Subduction zone volcanism in the ocean produces a chain of volcanoes called an island _____.
  a. arc   **c.** group   **e.** system
  b. reef   **d.** string

4. Intrusive igneous bodies that cut across sedimentary beds are said to be _____.
  a. concordant   **c.** homogenous   **e.** discordant
  b. composite   **d.** disinclined

5. What is the most abundant gas found in magma?
  a. oxygen   **c.** methane   **e.** water vapor
  b. hydrogen   **d.** nitrogen

6. The Hawaiian Islands consist primarily of what type of volcanic cone?
  a. shield   **c.** pyroclastic   **e.** laccolithic
  b. composite   **d.** cinder

7. The process by which igneous rocks melt over a temperature range of a few hundred degrees is known as _____ melting.
  a. subduction   **c.** zonal   **e.** fractional
  b. partial   **d.** separation

8. The area commonly called the *Ring of Fire* encircles the _____.
  a. United States   **c.** Pacific Ocean   **e.** Mediterranean Sea
  b. Atlantic Ocean   **d.** Indian Ocean

9. Nuée ardente and lahar are phenomena associated with:
   a. batholiths   **c.** fissure eruptions   **e.** all of these
   b. composite cones **d.** shield volcanoes

10. Which of the following is NOT a primary factor that affects the viscosity of magma?
    a. composition   **c.** temperature   **e.** both a. and c.
    b. age   **d.** dissolved gases

Fill-Ins Questions

11. Basaltic lava with a surface of rough, jagged blocks is called _____.
12. Although volcanic eruptions from a central vent are most familiar, by far the largest amounts of volcanic material are extruded from fractures in the crust called _____.
13. _____ are sheetlike intrusive igneous bodies that form when magma is injected into fractures that cut across rock layers.
14. Most of the more than 600 active volcanoes that have been identified are located in the vicinity of _____ plate boundaries.
15. Rising plumes of hot mantle material produce intraplate volcanic regions a few hundred kilometers across called _____ _____.

True/False Questions

16. The greatest volume of volcanic rock is produced along the oceanic ridge system. ___
17. Shield cones represent the most violent type of volcanic activity. ___
18. Pyroclastic flows consist largely of ash and pumice fragments produced when silica-rich magma is extruded. ___
19. Within Earth there is a gradual temperature increase with depth. ___
20. Partial melting often produces a melt with a lower silica content than the parent rock. ___

Answers

1.d; 2.b; 3.a; 4.e; 5.e; 6.a; 7.b; 8.c; 9.b; 10.b; 11. aa; 12. fissures; 13. Dikes; 14. convergent; 15. hot spots; 16.T; 17.F; 18.T; 19.T; 20.F

Weathering and Soils

Delicate Arch, Arches National Park, Utah. (Photo by Stephen Trimble)

Focus on Learning

To assist you in learning the important concepts in this chapter, you will find it helpful to focus on the following questions:

- What are Earth's external processes and what roles do they play in the rock cycle?

- What are the two types of weathering? In what ways are they different?

- What factors determine the rate at which rock weathers?

- What is soil? What are the factors that control soil formation?

- How is weathering related to the formation of ore deposits?

Earth's surface is constantly changing. Rock is disintegrated and decomposed, moved to lower elevations by gravity, and carried away by water, wind, or ice. In this manner Earth's physical landscape is sculptured. This chapter focuses on the first step of this never-ending process—weathering. What causes solid rock to crumble and why does the type and rate of weathering vary from place to place? Soil, an important product of the weathering process and a vital resource, is also examined.

Earth's External Processes

Weathering, mass wasting, and erosion are called *external processes* because they occur at or near Earth's surface and are powered by energy from the Sun. External processes are a basic part of the rock cycle because they are responsible for transforming solid rock into sediment.

To the casual observer, the face of Earth may appear to be without change, unaffected by time. In fact, less than 200 years ago, most people believed that mountains, lakes, and deserts were permanent features of an Earth that was thought to be no more than a few thousand years old. Today, however, we know that mountains eventually succumb to weathering and erosion, lakes fill with sediment and vegetation or are drained by streams, and deserts come and go as relatively minor climatic changes occur.

Earth is a dynamic body. Some parts of Earth's surface are gradually elevated by mountain building and volcanic activity. Meanwhile opposing processes are continually removing materials from higher elevations and transporting them to lower elevations (Figure 5.1). The latter processes include:

1. **Weathering**—disintegration and decomposition of rock at or near Earth's surface.
2. **Mass wasting**—transfer of rock material downslope under the influence of gravity.
3. **Erosion**—incorporation and transportation of material by a mobile agent, usually water, wind, or ice.

We will first turn our attention to the process of weathering and the products generated by this activity. However, weathering cannot be easily separated from the other two processes because, as weathering breaks rocks apart, it facilitates the movement of rock debris by mass wasting and erosion. On the other hand, the transport of material by mass wasting and erosion furthers the disintegration and decomposition of rock.

Weathering

All materials are susceptible to weathering. Consider, for example, the synthetic rock we call concrete, which closely resembles the sedimentary rock called conglomerate. A newly poured concrete sidewalk has a smooth, fresh look. However, not many years later, the same sidewalk will appear chipped, cracked, and rough, with pebbles exposed at the surface. If a tree is nearby, its roots may grow under the concrete, heaving and buckling it. The same natural processes that eventually break apart a concrete sidewalk also act to disintegrate natural rocks, regardless of their type or strength.

Why does rock weather? Simply, weathering is the response of Earth materials to a changing environment. For instance, after millions of years of uplift and erosion, the rocks overlying a large body of intrusive igneous rock may be removed. This exposes the rock to a whole new environment at the surface. The mass of crystalline rock, which formed deep below ground where temperatures and pressures are much greater than at the surface, is now subjected to very different and comparatively hostile surface conditions. In response, this rock mass will gradually change until it is once again in equilibrium, or balance, with its new environment. Such transformation of rock is what we call *weathering*.

In the following sections we will discuss two kinds of weathering—mechanical and chemical. Mechanical weathering is the physical breaking up of rocks into smaller pieces. Chemical weathering actually alters a rock's chemistry, changing it into different substances. Although we will consider these two processes separately, keep in mind that they usually work simultaneously in nature.

Mechanical Weathering

When a rock undergoes **mechanical weathering** it is broken into smaller and smaller pieces, each retaining the characteristics of the original material. The end result is many small pieces from a single large one. Figure 5.2 shows that breaking a rock into smaller pieces increases the surface area available for chemical attack. An analogous situation occurs when a sugar cube is added to water. A cube of sugar will dissolve much more slowly than will an equal volume of loose granules because of the vast difference in surface area. Hence, by breaking rocks into smaller pieces, mechanical weathering increases the amount of surface area available for chemical weathering.

FIGURE 5.1 Slopes here, as elsewhere, are places where materials are continually moving from higher to lower elevations. Weathering begins the process by attacking the solid rock exposed at the surface. Next, gravity moves the weathered debris downslope. This step, termed mass wasting, may range from a slow and gradual creep to a thundering landslide. Eventually, the material that was once high up the slope reaches the stream at the bottom. The moving water then transports the debris away. Colorado River in Marble Canyon, Grand Canyon National Park, Arizona. (Photo by Tom Bean/DRK Photo)

FIGURE 5.2 Chemical weathering can occur only to those portions of a rock that are exposed to the elements. Mechanical weathering breaks rock into smaller and smaller pieces, thereby increasing the surface area available for chemical attack.

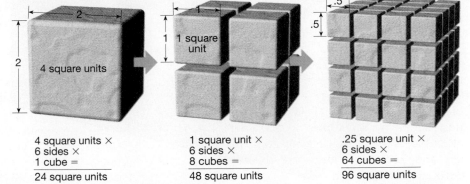

4 square units ×
6 sides ×
1 cube =
_____
24 square units

1 square unit ×
6 sides ×
8 cubes =
_____
48 square units

.25 square unit ×
6 sides ×
64 cubes =
_____
96 square units

In nature, four important physical processes break rocks into smaller fragments: frost wedging, expansion resulting from unloading, thermal expansion, and biological activity.

Frost Wedging

Alternate freezing and thawing of water is one of the most important processes of mechanical weathering. Water has the unique property of expanding about 9 percent when it freezes. This increase in volume occurs because, as ice forms, the water molecules arrange themselves into a very open crystalline structure. As a result, when water freezes, it expands and exerts a tremendous outward force. This can be verified by filling a container with water and freezing it. The formation of ice will rupture the container.

In nature, water works its way into every crack or void in rock and, upon freezing, expands and enlarges the opening. After many freeze-thaw cycles, the rock is broken into pieces. This process is appropriately called **frost wedging** (Figure 5.3). Frost wedging is most pronounced in mountainous regions in the middle latitudes where a daily freeze-thaw cycle often exists. Here, sections of rock are

wedged loose and may tumble into large piles called **talus slopes** that often form at the base of steep rock outcrops (Figure 5.3).

Unloading

When large masses of igneous rock, particularly those composed of granite, are exposed by erosion, slabs begin to break loose like the layers of an onion. The process is called **sheeting** and is thought to occur, at least in part, because of the great reduction in pressure when the overlying rock is eroded away. Accompanying this unloading, the outer layers expand more than the rock below, and thus separate from the rock body. Continued weathering eventually causes the slabs to separate and spall off, creating **exfoliation domes**. Excellent examples of exfoliation domes include Stone Mountain, Georgia, and Half Dome (Figure 5.4) and Liberty Cap in Yosemite National Park.

Deep underground mining provides us with another example of how rocks behave once the confining pressure is removed. Large rock slabs sometimes explode off the walls of newly cut mine tunnels because of the abruptly reduced pressure. Evidence of this type, plus the fact that fracturing

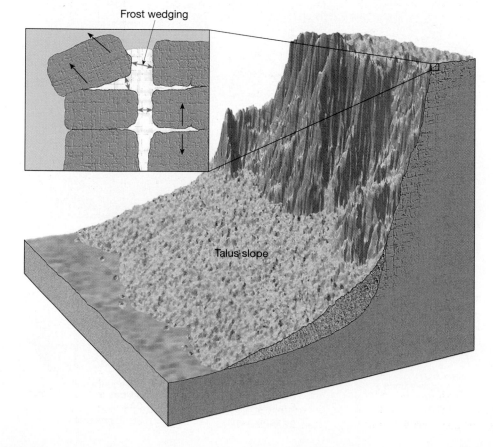

Frost wedging

Talus slope

FIGURE 5.3 Frost wedging. As water freezes it expands to 109 percent of its original volume, exerting a force great enough to break rock. When frost wedging occurs in a setting such as this, the broken rock fragments fall to the base of the cliff and create a cone-shaped accumulation known as talus.

FIGURE 5.4 Summit of Half Dome, an exfoliation dome in Yosemite National Park. (Photo by Breck P. Kent)

occurs parallel to the floor of a quarry when large blocks of rock are removed, strongly supports the process of unloading as the cause of sheeting.

Although many fractures are created by expansion, others are produced by contraction during the crystallization of magma, and still others by tectonic forces during mountain building. Fractures produced by these activities generally form a definite pattern and are called *joints*. Joints are important rock structures that allow water to penetrate to depth and start the process of weathering long before the rock is exposed at the surface.

Thermal Expansion

The daily cycle of temperature change is thought to weaken rocks, particularly in deserts where daily variations may exceed 30°C (54°F). Heating a rock causes it to expand, and cooling causes it to contract. Repeated swelling and shrinking of minerals that have different expansion and contraction rates should exert some stress on the rock's outer shell. However, laboratory experiments have not substantiated this. In one test, unweathered rocks were

heated much hotter than is normally experienced on Earth's surface, and then cooled. This procedure was repeated many times to simulate hundreds of years of weathering, but the rocks showed little apparent change. Nevertheless, in desert areas pebbles do exhibit unmistakable evidence of shattering from what appear to be temperature changes.

Biological Activity

Weathering is also accomplished by the activities of organisms, including plants, burrowing animals, and humans. Plant roots in search of minerals and water grow into fractures, and as the roots grow they wedge the rock apart (Figure 5.5). Burrowing animals further break down the rock by moving fresh material to the surface, where physical and chemical processes can more effectively attack it. Decaying organisms also produce acids, which contribute to chemical weathering. Where rock has been blasted in search of minerals or for road construction, the impact is quite noticeable, but on a worldwide scale human activity probably ranks behind burrowing animals in Earth-moving accomplishments.

FIGURE 5.5 Root wedging widens fractures in rock and aids the process of mechanical weathering. (Photo by Tom Bean/DRK Photo)

Although usually considered separately from mechanical weathering, the activities of the erosional agents—wind, water, and glaciers—are nonetheless important. For as these mobile agents move rock debris, they relentlessly disintegrate the Earth materials they carry.

Chemical Weathering

Chemical weathering involves the complex processes that alter the internal structures of minerals by removing and/or adding elements. During this transformation, the original rock decomposes into substances that are stable in the surface environment. Consequently, the products of chemical weathering will remain essentially unchanged as long as they remain in an environment similar to the one in which they formed.

Water and Carbonic Acid

Water is by far the most important agent of chemical weathering. Although pure water is nonreactive, a small amount of dissolved material is generally all that is needed to activate it. Oxygen dissolved in water will *oxidize* some materials. For example, when an iron nail is found in moist soil, it will have a coating of rust (iron oxide), and if the time of exposure has been long, the nail will be so weak that it can be broken as easily as a toothpick. When rocks containing iron-rich minerals oxidize, a yellow to reddish-brown rust will appear on the surface.

Carbon dioxide (CO_2) dissolved in water (H_2O) forms carbonic acid (H_2CO_3), the same weak acid produced when soft drinks are carbonated. Rain dissolves some carbon dioxide as it falls through the atmosphere, and additional amounts released by decaying organic matter are acquired as the water percolates through the soil. Carbonic acid ionizes to form the very reactive hydrogen ion (H^+) and the bicarbonate ion (HCO_3^-).

How Granite Weathers

To illustrate how rock chemically weathers when attacked by carbonic acid, we will consider the weathering of granite, the most abundant continental rock. Recall that granite consists mainly of quartz and potassium feldspar. The weathering of the potassium feldspar component of granite takes place as follows:

$$2KAlSi_3O_8 + 2(H^+ + HCO_3^-) + H_2O \rightarrow$$
potassium   carbonic acid   water
feldspar

$$Al_2Si_2O_5(OH)_4 + 2KHCO_3 + 4SiO_2$$
clay mineral   potassium   silica
bicarbonate

In this reaction, the hydrogen ions (H^+) attack and replace potassium ions (K^+) in the feldspar structure, thereby disrupting the crystalline network. Once removed, the potassium is available as a nutrient for plants. Or it becomes potassium bicarbonate ($KHCO_3$), a soluble salt that may be incorporated into other minerals or carried to the ocean in dissolved form by streams.

The most abundant products of the chemical breakdown of feldspar are residual clay minerals. Because clay minerals are the end product of weathering, they are very stable under surface conditions. Consequently, clay minerals make up a high percentage of the inorganic material in soils. Moreover, the most abundant sedimentary rock, shale, also contains a high proportion of clay minerals.

In addition to the formation of clay minerals during this reaction, some silica is removed from the feldspar structure and is carried away by groundwater (water beneath Earth's surface). This dissolved silica will eventually precipitate to produce nodules of chert or flint, fill in the port spaces between sediment grains, or be carried to the ocean, where microscopic animals will remove it to build hard silica shells.

To summarize, the weathering of potassium feldspar generates a residual clay mineral, a soluble salt (potassium bicarbonate), and some silica which enters into solution.

Quartz, the other main component of granite, is very resistant to chemical weathering; hence, it remains substantially unaltered when attacked by weakly acidic solutions. As a result, when granite weathers, the feldspar crystals dull and slowly turn to clay, releasing the once-interlocked quartz grains, which still retain their fresh, glassy appearance. Although some quartz remains in the soil, much is transported to the sea or to other sites of deposition, where it becomes the main constituent of such features as sandy beaches and sand dunes. In time it may become lithified to form the sedimentary rock *sandstone*.

Weathering of Silicate Minerals

Table 5.1 lists the weathered products of some of the most common silicate minerals. Remember that silicate minerals make up most of Earth's crust and that these minerals are composed essentially of only eight elements. When chemically weathered, silicate minerals yield sodium, calcium, potassium, and magnesium ions. These form soluble products that may be removed by groundwater. The element iron combines with oxygen, producing relatively insoluble iron oxides, which give soil a reddish-brown or yellowish color. Under most conditions the three

TABLE 5.1 Products of weathering.

Mineral	Residual Products	Material in Solution
Quartz	Quartz grains	Silica
Feldspars	Clay minerals	Silica, K^+, Na^+, Ca^{2+}
Amphibole (hornblende)	Clay minerals Limonite Hematite	Silica, Ca^{2+}, Mg^{2+}
Olivine	Limonite Hematite	Silica, Mg^{2+}

remaining elements, aluminum, silicon, and oxygen, join with water to produce residual clay minerals. However, even the highly insoluble clay minerals are very slowly removed by subsurface water.

Spheroidal Weathering

In addition to altering the internal structure of minerals, chemical weathering causes physical changes as well. For instance, when angular rock masses are attacked by water that enters along joints, the rocks tend to take on a spherical shape. Gradually the corners and edges of the angular blocks become more rounded. The corners are attacked most readily because of their greater surface area, as compared to the edges and faces. This process, called **spheroidal weathering**, gives the weathered rock a more rounded or spherical shape (Figure 5.6).

Sometimes during the formation of spheroidal boulders, successive shells separate from the rock's main body. Eventually the outer shells spall off, allowing the chemical weathering activity to penetrate deeper into the boulder. This spherical scaling

results because, as the minerals in the rock weather to clay, they increase in size through the addition of water to their structure. This increased bulk exerts an outward force that causes concentric layers of rock to break loose and fall off. Hence, chemical weathering does produce forces great enough to cause mechanical weathering.

This type of spheroidal weathering, in which shells spall off, should not be confused with the phenomenon of sheeting discussed earlier. In sheeting, the fracturing occurs as a result of unloading and the rock layers which separate from the main body are largely unaltered at the time of separation.

Rates of Weathering

Several factors influence the type and rate of rock weathering. We have already seen how mechanical weathering affects the rate of weathering. By breaking rock into smaller pieces, the amount of surface area exposed to chemical weathering is increased. The presence or absence of joints (cracks) can be significant because they influence the ability of water to penetrate the rock (Figure 5.6). Other important factors include the mineral makeup of rocks and climate.

Mineral Makeup

The variations in weathering rates, attributable to the mineral constituents, can be demonstrated by comparing old headstones made from different rock types. Headstones of granite, which is composed of silicate minerals, are relatively resistant to chemical

FIGURE 5.6 Spheroidal weathering is evident in this exposure of granite in California's Joshua Tree National Monument. Because the rocks are attacked more vigorously on the corners and edges, they take on a spherical shape. The lines visible in the rock are called *joints*. Joints are important rock structures that allow water to penetrate and start the weathering process long before the rock is exposed. (Photo by E. J. Tarbuck)

weathering. We can see this by examining the inscriptions on the headstones shown in Figure 5.7. This is not true of the marble headstone, which shows signs of extensive chemical alteration over a relatively short period. Marble is composed of calcite (calcium carbonate), which readily dissolves even in a weakly acidic solution.

The silicates, the most abundant mineral group, weather in essentially the same order as their order of crystallization. The minerals that crystallize first form under much higher temperatures than those that crystallize last. Consequently, the early formed minerals are not as stable at Earth's surface, where the temperature and pressure are drastically different from the environment in which they formed. By examining Bowen's reaction series (see Figure 3.7, p. 50), we see that olivine crystallizes first and is therefore the least resistant to chemical weathering, whereas quartz, which crystallizes last, is the most resistant.

Climate

Climatic factors, particularly temperature and moisture, are crucial to the rate of rock weathering. These climatic elements largely determine the weathering rate and strongly influence the kind and amount of vegetation present. Regions with lush vegetation generally have a thick mantle of soil rich in decayed organic matter from which chemically active fluids such as carbonic and humid acids are derived.

The optimum environment for chemical weathering is a combination of warm temperatures and abundant moisture. In polar regions chemical weathering is ineffective because frigid temperatures keep the available moisture locked up as ice, whereas in arid regions there is insufficient moisture to foster rapid chemical weathering.

The sum of these factors determines the type and rate of rock weathering for a given place. However, there is generally enough variation, even within a relatively small area, for the rocks to exhibit some differential weathering. **Differential weathering** simply relates to the fact that rocks exposed at Earth's surface usually do not weather at the same rate. Because of variations in such factors as mineral makeup, degree of jointing, and exposure to the elements, significant differences occur. Consequently, differential weathering and subsequent erosion create many unusual and often spectacular rock formations and landforms. Included are features such as the natural bridges found in Arches National Park (see chapter-opening photo) and the sculptured rock pinnacles found in Bryce Canyon National Park (Figure 5.8).

Soil

Soil covers most land surfaces. Along with air and water, it is one of our most indispensable resources (Figure 5.9). Also like air and water, soil is taken for granted by many of us. The following quote helps put this vital layer in perspective.

Science, in recent years, has focused more and more on the Earth as a planet, one that for all we know is unique—where a thin blanket of air, a thinner film of water, and the thinnest veneer of soil combine to support a web of life of wondrous diversity in continuous change. [*]

———

[*]Jack Eddy. "A Fragile Seam of Dark Blue Light," in *Proceedings of the Global Change Research Forum*. U.S. Geological Survey Circular 1086, 1993, p. 15.

FIGURE 5.7 An examination of headstones reveals the rate of chemical weathering on diverse rock types. The granite headstone (left) was erected three years after the marble headstone (right). The inscription date of 1885 on the marble monument is nearly illegible. (Photos by E. J. Tarbuck)

FIGURE 5.8 Differential weathering is illustrated by these sculpted rock pinnacles in Bryce National Park, Utah. (Photo by Tom Till)

Soil has accurately been called "the bridge between life and the inanimate world." All life—the entire biosphere—owes its existence to a dozen or so elements that must ultimately come from Earth's crust. Once weathering and other processes create soil, plants carry out the intermediary role of assimilating the necessary elements and making them available to animals including humans.

An Interface in the Earth System

When Earth is viewed as a system, soil is referred to as an *interface*—a common boundary where different parts of a system interact. This is an appropriate designation because soil forms where the solid Earth, the atmosphere, the hydrosphere, and the biosphere meet. Soil is a material that develops in response to complex environmental interactions among different parts of the Earth system. Over time, soil gradually evolves to a state of equilibrium or balance with the environment. Soil is dynamic and sensitive to almost every aspect of its surroundings. Thus, when environmental changes occur, such as climate, vegetative

cover, and animal (including human) activity, the soil responds. Any such change produces a gradual alteration of soil characteristics until a new balance is reached. Although thinly distributed over the land surface, soil functions as a fundamental interface; providing an excellent example of the integration among many parts of the Earth system.

What Is Soil?

With few exceptions, Earth's land surface is covered by **regolith**, the layer of rock and mineral fragments produced by weathering. Some would call this material soil, but soil is more than an accumulation of weathered debris. **Soil** is a combination of mineral and organic matter, water, and air—that portion of the regolith that supports the growth of plants. Although the proportions of the major components in soil vary, the same four components always are present to some extent (Figure 5.10). About one-half of the total volume of a good quality surface soil is a mixture of disintegrated and decomposed rock (mineral matter) and **humus**, the decayed remains of animal and plant

FIGURE 5.9 Soil is an essential resource that we often take for granted. Soil is not a living entity, but it contains a great deal of life. Moreover, this complex medium supports nearly all plant life, which, in turn, supports animal life. (Photo by James E. Patterson)

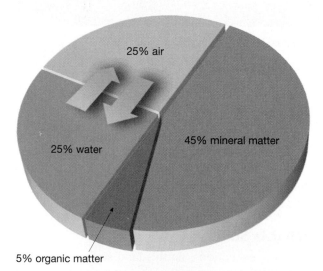

FIGURE 5.10 Composition (by volume) of a soil in good condition for plant growth. Although the percentages vary, each soil is composed of mineral and organic matter, water, and air.

life (organic matter). The remaining half consists of pore spaces among the solid particles where air and water circulate.

Although the mineral portion of the soil is usually much greater than the organic portion, humus is an essential component. In addition to being an important source of plant nutrients, humus enhances the soil's ability to retain water. Because plants

require air and water to live and grow, the portion of the soil consisting of pore spaces that allow for the circulation of these fluids is as vital as the solid soil constituents.

Soil water is far from "pure" water; instead, it is a complex solution containing many soluble nutrients. Soil water not only provides the necessary moisture for the chemical reactions that sustain life; it also supplies plants with nutrients in a form they can use. The pore spaces not filled with water contain air. This air is the source of necessary oxygen and carbon dioxide for most microorganisms and plants that live in the soil.

Controls of Soil Formation

Soil is the product of the complex interplay of several factors. The most important of these are parent material, time, climate, plants, animals, and slope. Although all of these factors are interdependent, their roles will be examined separately.

Parent Material

The source of the weathered mineral matter from which soils develop is called the **parent material** and is a major factor influencing a newly forming soil. Gradually it undergoes physical and chemical changes as the processes of soil formation progress. Parent material may be the underlying bedrock or it can be a layer of unconsolidated deposits, as in a stream valley. When the parent material is bedrock, the soils are termed *residual soils*. By contrast, those developed on unconsolidated sediment are called *transported soils* (Figure 5.11). Note that transported soils form *in place* on parent materials that have been carried from elsewhere and deposited by gravity, water, wind, or ice.

The nature of the parent material influences soils in two ways. First, the type of parent material affects the rate of weathering, and thus the rate of soil formation. (Consider the weathering rates of granite versus marble.) Also, because unconsolidated deposits are already partly weathered and provide more surface area for chemical weathering, soil development on such material usually progresses more rapidly. Second, the chemical makeup of the parent material affects the soil's fertility. This influences the character of the natural vegetation the soil can support.

At one time the parent material was believed to be the primary factor causing differences among soils. Today soil scientists realize that other factors, especially climate, are more important. In fact, it has been found that similar soils are often produced from different parent materials and that dissimilar soils

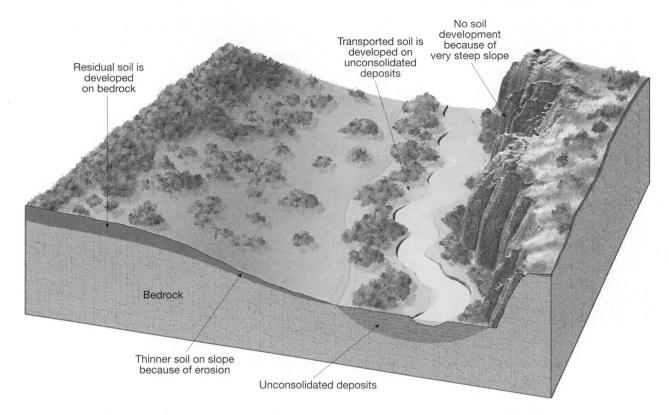

FIGURE 5.11 The parent material for residual soils is the underlying bedrock, whereas transported soils form on unconsolidated deposits. Also note that as slopes become steeper, soil becomes thinner.

have developed from the same parent material. Such discoveries reinforce the importance of the other soil-forming factors.

Time

Time is an important component of every geological process, and soil formation is no exception. The nature of soil is strongly influenced by the length of time processes have been operating. If weathering has been going on for a comparatively short time, the parent material determines to a large extent the characteristics of the soil. As weathering processes continue, the influence of parent material on soil is overshadowed by the other soil-forming factors, especially climate. The amount of time required for various soils to evolve cannot be specified because the soil-forming processes act at varying rates under different circumstances. However, as a rule the longer a soil has been forming, the thicker it becomes and the less it resembles the parent material.

Climate

Climate is the most influential control of soil formation. Just as temperature and precipitation are the climatic elements that influence people the most, so too are they the elements that exert the strongest

impact on soil formation. Variations in temperature and precipitation determine whether chemical or mechanical weathering predominates. They also greatly influence the rate and depth of weathering. For instance, a hot, wet climate may produce a thick layer of chemically weathered soil in the same amount of time that a cold, dry climate produces a thin mantle of mechanically weathered debris. Also, the amount of precipitation influences the degree to which various materials are removed (leached) from the soil, thereby affecting soil fertility. Finally, climatic conditions are an important control of the type of plant and animal life present.

Plants and Animals

Plants and animals play a vital role in soil formation. The types and abundance of organisms present have a strong influence on the physical and chemical properties of a soil. In fact, for well-developed soils in many regions, the significance of natural vegetation is frequently implied in the description used by soil scientists. Such phrases as *prairie soil, forest soil,* and *tundra soil* are common.

The chief function of plants and animals is to furnish organic matter to the soil. Certain bog soils are composed almost entirely of organic matter, whereas desert soils may contain as little as a small fraction of 1 percent. Although the quantity of organic matter varies substantially among soils, it is the rare soil that completely lacks it.

The primary source of organic matter is plants, although animals and the uncountable microorganisms also contribute. When organic matter decomposes, important nutrients are supplied to plants, as well as to animals and microorganisms living in the soil. Consequently, soil fertility depends in part on the amount of organic matter present. Furthermore, the decay of plant and animal remains causes the formation of various organic acids. These complex acids hasten the weathering process. Organic matter also has a high water-holding ability and thus aids water retention in a soil.

Microorganisms, including fungi, bacteria, and single-celled protozoa, play an active role in the decay of plant and animal remains. The end product is *humus*, a material that no longer resembles the plants and animals from which it is formed. In addition, certain microorganisms aid soil fertility because they have the ability to *fix* (alter) atmospheric nitrogen gas into soil nitrogen compounds.

Earthworms and other burrowing animals act to mix the mineral and organic portions of a soil. Earthworms, for example, feed on organic matter and thoroughly mix soils in which they live, often moving and enriching many tons per acre each year. Burrows and holes also aid the passage of water and air through the soil.

Slope

The slope of the land can vary greatly over short distances. Such variations, in turn, can lead to widely varied local soil types. Many of the differences exist because slope has a significant impact on the amount of erosion and the water content of soil.

On steep slopes soils are often poorly developed. In such situations little water can soak in, and as a result, soil moisture may be insufficient for vigorous plant growth. Further, because of accelerated erosion on steep slopes, the soils are thin, or nonexistent (Figure 5.11).

On the other hand, poorly drained and waterlogged soils in bottomlands have a much different character. Such soils are usually thick and dark. The dark color results from generous organic matter that accumulates because saturated conditions retard the decay of vegetation. The optimum terrain for soil development is a flat-to-undulating upland surface. Here we find good drainage, minimum erosion, and sufficient infiltration of water into the soil.

Slope orientation, the direction the slope is facing, also is significant. In the mid-latitudes of the Northern Hemisphere, a south-facing slope receives a great deal more sunlight than a north-facing slope. In fact, a steep north-facing slope may receive no direct sunlight at all. The difference in the amount of solar radiation received causes substantial differences in soil temperature and moisture, which in turn may influence the nature of the vegetation and the character of the soil.

Although we have dealt separately with each of the soil-forming factors, remember that *all work together* to form soil. No single factor is responsible for a soil being as it is. Rather, it is the combined influence of parent material, time, climate, plants, animals, and slope that determines a soil's character.

The Soil Profile

It is important to realize that soil-forming processes *operate from the surface downward*. Thus, variations in composition, texture, structure, and color gradually evolve at varying depths. These vertical differences, which usually become more pronounced as time passes, divide the soil into zones or layers that soil scientists call **horizons**. If you were to dig a trench in soil, you would see that its walls are layered. Such a vertical section through all of the soil horizons constitutes the **soil profile** (Figure 5.12).

Figure 5.13 presents an idealized view of a well-developed soil profile in which five horizons are identified. From the surface downward, they are designated as *O, A, E, B,* and *C,* respectively. These five horizons are common to soils in temperate regions. The characteristics and extent of development of horizons vary in different environments. Thus, different localities exhibit soil profiles that can contrast greatly with one another.

The *O* horizon consists largely of organic material, unlike the layers beneath it that consist mainly of mineral matter. The upper portion of the *O* horizon is primarily plant litter such as loose leaves and other organic debris that are still recognizable. By contrast, the lower portion of the *O* horizon is made up of partly decomposed organic matter (humus) in which plant structures can no longer be identified. In addition to plants, the *O* horizon is teeming with microscopic life including bacteria, fungi, algae, and insects. All of these organisms contribute oxygen, carbon dioxide, and organic acids to the developing soil.

Underlying the organic-rich *O* horizon is the *A* horizon. This zone is largely mineral matter, yet biological activity is high and humus is generally present—up to 30 percent in some instances. Together

FIGURE 5.12 A soil profile is a vertical cross section from the surface down to the parent material. Well-developed soils show distinct layers called horizons. (Photo by E. J. Tarbuck)

the *O* and *A* horizons make up what is commonly called *topsoil*. Below the *A* horizon, the *E* horizon is a light-colored layer that contains little organic material. As water percolates downward through this zone, finer particles are carried away. This washing out of the fine soil components is termed **eluviation**. Water percolating downward also dissolves soluble inorganic soil components and carries them to deeper zones. This depletion of soluble materials from the upper soil is termed **leaching**.

Immediately below the *E* horizon is the *B* horizon, or *subsoil*. Much of the material removed from the *E* horizon by eluviation is deposited in the *B* horizon, which is often referred to as the *zone of accumulation*. The accumulation of the fine clay particles enhances water retention in the subsoil. However, in extreme cases, clay accumulation can form a very compact and impermeable layer called *hardpan*. The *O, A, E,* and *B* horizons together constitute the **solum**, or "true soil." It is in the solum that the soil-forming processes are active and that living roots and other plant and animal life are largely confined.

Below the solum and above the unaltered parent material is the *C* horizon, a layer characterized by partially altered parent material. Whereas the *O, A, E,*

and *B* horizons bear little resemblance to the parent material, it is easily identifiable in the *C* horizon. Although this material is undergoing changes that will eventually transform it into soil, it has not yet crossed the threshold that separates regolith from soil.

Soil characteristics and development can vary greatly in different environments. The boundaries between soil horizons may be very distinct, as in Figures 5.12 and 5.13, or the horizons may blend gradually from one to another. Consequently, a well-developed soil profile indicates that environmental conditions have been relatively stable over an extended time span and that the soil is *mature*. By contrast, some soils lack horizons altogether.

Such soils are called *immature*, because soil building has been going on for only a short time. Immature soils are also characteristic of steep slopes where erosion continually strips away the soil, preventing full development.

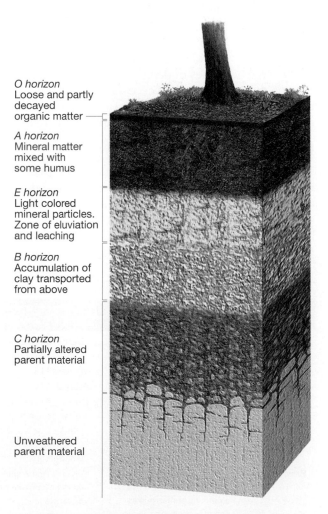

O horizon
Loose and partly decayed organic matter

A horizon
Mineral matter mixed with some humus

E horizon
Light colored mineral particles. Zone of eluviation and leaching

B horizon
Accumulation of clay transported from above

C horizon
Partially altered parent material

Unweathered parent material

FIGURE 5.13 Idealized soil profile from a humid climate in the middle latitudes.

Soil Types

There are hundreds of soil types and subtypes worldwide, far too many to consider here. Instead, we shall look at just three very generic types—pedalfers, pedocals, and laterites. As you read, notice that the characteristics of each soil primarily reflect the prevailing climatic conditions. A summary of the characteristics is provided in Table 5.2.

Pedalfer

The term **pedalfer** is derived from the Greek **ped***on*, meaning "soil," and the chemical symbols **Al** (aluminum) and **Fe** (Iron). Pedalfers are characterized by an accumulation of iron oxides and aluminum-rich clays in the *B* horizon. In mid-latitude areas where the annual rainfall exceeds 63 centimeters (25 inches), most of the soluble materials, such as calcium carbonate, are leached from the soil and carried away by underground water. The less soluble iron oxides and clays are carried from the *E* horizon and deposited in the *B* horizon, giving it a brown to red-brown color. These soils are best developed under forest vegetation where large quantities of decomposing organic matter provide the acid conditions necessary for leaching. In the United States pedalfers are found east of a line from northwestern Minnesota to south-central Texas.

Pedocal

The term **pedocal** is derived from the Greek **ped***on*, meaning "soil," and the first three letters of **cal***cite* (calcium carbonate). As the name implies, pedocals are characterized by an accumulation of calcium carbonate. This soil type is found in the drier western United States in association with grassland and brush vegetation. Because chemical weathering is less intense in drier areas, pedocals generally contain a smaller percentage of clay materials than pedalfers.

In the arid and semiarid western states, a calcite-enriched layer called *caliche* may be present in the soils. In these areas little of the rain that falls penetrates to great depths. Rather, it is held by the soil particles near the surface until it evaporates. As a result, the soluble materials, chiefly calcium carbonate, are removed from the uppermost layer and redeposited below, forming the caliche layer.

Laterite

In hot, wet tropical climates, soils called **laterites** may develop. Chemical weathering is intense under such climate conditions, so these soils are usually deeper than soils developing over a similar period in the mid-latitudes. Not only does leaching remove the soluble materials such as calcite, but the great quantities of percolating water also remove much of the silica, with the result that oxides of iron and aluminum become concentrated in the soil. The iron gives the soil a distinctive red color.

Because bacterial activity is great in the tropics, laterites contain practically no humus. This fact, coupled with the highly leached and bricklike nature of these soils, makes laterites poor for growing crops. Their infertility has been demonstrated repeatedly in tropical countries where cultivation has been attempted (see Box 5.1).

In sharp contrast, cold or dry climates generally have very thin and poorly developed soils. The reasons for this are fairly obvious. Chemical weathering progresses very slowly in such climates, and the scant plant life yields very little organic matter.

TABLE 5.2 Summary of soil types.

Climate	Temperate humid (>63 cm rainfall)	Temperate dry (<63 cm rainfall)	Tropical (heavy rainfall)	Extreme arctic or desert
Vegetation	Forest	Grass and brush	Grass and trees	Almost none, so no humus develops
Typical Area	Eastern U.S.	Western U.S.		
Soil Type	Pedalfer	Pedocal	Laterite	
Topsoil	Sandy, light colored; acid	Commonly enriched in calcite; whitish color	Zones not developed. Enriched in iron (and aluminum); brick red color	No real soil forms because there is no organic material. Chemical weathering is very slow.
Subsoil	Enriched in aluminum, iron and clay; brown color	Enriched in calcite; whitish color	All other elements removed by leaching	
Remarks	Extreme development in conifer forests, because abundant humus makes groundwater very acidic. Produces light gray soil because of removal of iron	*Caliche* is name applied to the accumulation of calcite	Apparently bacteria destroy humus, so no acid is available to remove iron	

Box 5.1 **Laterites and the Clearing of the Rain Forest**

aterites are thick red soils that form in the wet tropics and subtropics and are the end product of extreme chemical weathering. Because lush tropical rain forests have lateritic soils, it might seem as though they would have great potential for agriculture. However, this is not the case. In fact, just the opposite is true; laterites are among the poorest soils for farming. Why is this the case?

Because laterites develop under conditions of high temperatures and heavy rainfall, they are severely leached. Leaching leads to low fertility because plant nutrients are removed by the large volume of downward-per-colating water. Therefore, even though the vegetation may be dense and luxu-riant, the soil itself contains few available nutrients. Most of the nutrients that support the rain forest are locked up in the trees themselves. The forest is maintained by the decay of plant material. That is, as vegetation dies and decomposes, the roots of the rainfor-est trees quickly absorb the nutrients before they are leached from the soil. The nutrients are continuously recy-cled as trees die and decompose.

Therefore, when forests are cleared to provide land for farming or to harvest the timber, most of the nutrients are removed as well (Figure 5.A). What remains is a soil that con-tains little to nourish crop growth.

The clearing of rain forests not only removes the supply of plant nutrients, but also leads to accelerated erosion. When vegetation is present, its roots anchor the soil while its leaves and branches provide a canopy that protects the ground from the full force of the fre-quent heavy rains. The removal of the vegetation also exposes the ground to strong direct sunlight. When baked by the sun, laterites can harden to a brick-like consistency and become practically

FIGURE 5.A Clearing the Amazon rain forest in Surinam. The thick lateritic soil is highly leached. (Photo by Wesley Bocxe/Photo Researchers)

FIGURE 5.B This ancient temple at Angor Wat, Cambodia, was built of bricks made of laterite. (Photo by R. Ian Lloyd/The Stock Market)

impenetrable to water and crop roots. In only a few years, lateritic soils in a freshly cleared area may no longer be cultivable.

The term *laterite* is derived from the Latin word *latere*, meaning "brick," and was first applied to the use of this material for brick making in India and Cambodia. Laborers simply excavated the soil, shaped it, and allowed it to harden in the sun. Ancient but still well-preserved structures built of lat-erite remain standing today in the wet tropics (Figure 5.B). Such structures have withstood centuries of weather-ing because all of the original soluble

materials were already removed from the soil by chemical weathering. Lat-erites are therefore virtually insoluble and thus very stable.

In summary, we have seen that lat-erites are highly leached soils that are the products of extreme chemical weathering in the warm, wet tropics. Although they may be associated with lush tropical rain forests, these soils are unproductive when vegetation is removed. Moreover, when cleared of plants, laterites are subject to acceler-ated erosion and can be baked to bricklike hardness by the sun.

Soil Erosion

Soils are just a tiny fraction of all Earth materials, yet they are a vital resource. Because soils are necessary for the growth of rooted plants, they are the very foundation of the human life-support system. Just as human ingenuity can increase the agricultural productivity of soils through fertilization and irrigation, soils can be damaged or destroyed by careless activities. Despite their basic role in providing food, fiber, and other basic materials, soils are among our most abused resources.

Perhaps this neglect and indifference has occurred because a substantial amount of soil seems to remain even where soil erosion is serious. Nevertheless, although the loss of fertile topsoil may not be obvious to the untrained eye, it is a growing problem as human activities expand and disturb more and more of Earth's surface.

How Soil Is Eroded

Soil erosion is a natural process; it is part of the constant recycling of Earth materials that we call the *rock cycle*. Once soil forms, erosional forces, especially water and wind, move soil components from one place to another. Every time it rains, raindrops strike the land with surprising force (Figure 5.14). Each drop acts like a tiny bomb, blasting movable soil particles out of their positions in the soil mass. Then, water flowing across the surface carries away the dislodged soil particles. Because the soil is moved by thin sheets of water, this process is termed *sheet erosion*.

After flowing as a thin, unconfined sheet for a relatively short distance, threads of current typically develop and tiny channels called *rills* begin to form. Still deeper cuts in the soil, known as *gullies*, are created as rills enlarge (Figure 5.15). When normal farm cultivation cannot eliminate the channels, we know the rills have grown large enough to be called gullies. Although most dislodged soil particles move only a short distance during each rainfall, substantial quantities eventually leave the fields and make their way downslope to a stream. Once in the stream channel, these soil particles, which can now be called *sediment*, are transported downstream and eventually deposited.

Rates of Erosion

We know that soil erosion is the ultimate fate of practically all soils. In the past, erosion occurred at slower rates than it does today because more of the land surface was covered and protected by trees,

FIGURE 5.14 When it is raining, millions of water drops are falling at velocities approaching 10 meters per second (35 kilometers per hour). When water drops strike an exposed surface, soil particles may splash as high as 1 meter into the air and land more than a meter away from the point of raindrop impact. Soil dislodged by splash erosion is more easily moved by sheet erosion. (Photo courtesy of U.S. Department of Agriculture)

FIGURE 5.15 Gully erosion in poorly protected soil. (Photo by John C. Coulter/Visuals Unlimited)

shrubs, grasses, and other plants. However, human activities such as farming, logging, and construction, which remove or disrupt the natural vegetation, have greatly accelerated the rate of soil erosion. Without the stabilizing effect of plants, the soil is more easily swept away by the wind or carried downslope by sheet wash.

Natural rates of soil erosion vary greatly from one place to another and depend on soil characteristics as well as such factors as climate, slope, and type of vegetation. Over a broad area, erosion caused by surface runoff may be estimated by determining the sediment loads of the streams that drain the region. When studies of this kind were made on a global scale they indicated that, prior to the appearance of humans, sediment transport by rivers to the ocean amounted to just over 9 billion metric tons per year. By contrast, the amount of material currently transported to the sea by rivers is about 24 billion metric tons per year, or more than two and one-half times the earlier rate.

It is more difficult to measure the loss of soil due to wind erosion. However, the removal of soil by wind is generally much less significant than erosion by flowing water except during periods of prolonged drought. When dry conditions prevail, strong winds can remove large quantities of soil from unprotected fields. Such was the case in the 1930s in the portions of the Great Plains that came to be called the Dust Bowl (see Box 12.1, p. 235).

In many regions the rate of soil erosion is significantly greater than the rate of soil formation. This means that a renewable resource has become nonrenewable in these places. At present, it is estimated that topsoil is eroding faster than it forms on more than one-third of the world's croplands. The result is lower productivity, poorer crop quality, reduced agricultural income, and an ominous future.

Sedimentation and Chemical Pollution

Another problem related to excessive soil erosion involves the deposition of sediment. Each year in the United States hundreds of millions of tons of eroded soil are deposited in lakes, reservoirs, and streams. The detrimental impact of this process can be significant. For example, as more and more sediment is deposited in a reservoir, the capacity of the reservoir is diminished, limiting its usefulness for flood control, water supply, and/or hydroelectric power generation. In addition, sedimentation in streams and other waterways can restrict navigation and lead to costly dredging operations.

In some cases soil particles are contaminated with pesticides used in farming. When these chemicals are introduced into a lake or reservoir, the quality of the water supply is threatened and aquatic organisms may be endangered. In addition to pesticides, nutrients found naturally in soils as well as those added by agricultural fertilizers make their way into streams and lakes, where they stimulate the growth of plants. Over a period of time, excessive nutrients accelerate the

process by which plant growth leads to the depletion of oxygen and an early death of the lake.

The availability of good soils is critical if the world's rapidly growing population is to be fed. On every continent unnecessary soil loss is occurring because appropriate conservation measures are not being used. Although it is a recognized fact that soil erosion can never be completely eliminated, soil conservation programs can substantially reduce the loss of this basic resource. Windbreaks (rows of trees), terracing, and plowing along the contours of hills are some of the effective measures, as are special tillage practices and crop rotation.

Weathering Creates Ore Deposits

Weathering creates many important mineral deposits. It does so by concentrating minor amounts of metals that are scattered through unweathered rock into economically valuable concentrations. Such a transformation is often termed **secondary enrichment** and takes place in one of two ways. In one situation, chemical weathering coupled with downward-percolating water removes undesired materials from decomposing rock, leaving the desired elements enriched in the upper zones of the soil. The second way is basically the reverse of the first. That is, the desirable elements that are found in low concentrations near the surface are removed and carried to lower zones, where they are redeposited and become more concentrated.

FIGURE 5.16 Bauxite is the principal ore of aluminum and forms as a result of weathering processes under tropical conditions. Its color varies from red or brown to nearly white. (Photo by E. J. Tarbuck)

Bauxite

The formation of *bauxite*, the principal ore of aluminum, is one important example of an ore created as a result of enrichment by weathering processes (Figure 5.16). Although aluminum is the third most abundant element in Earth's crust, economically valuable concentrations of this important metal are not common, because most aluminum is tied up in silicate minerals, from which it is extremely difficult to extract.

Bauxite forms in rainy tropical climates in association with laterites. (In fact, bauxite is sometimes called aluminum laterite.) When aluminum-rich source rocks are subjected to the intense and prolonged chemical weathering of the tropics, most of the common elements, including calcium, sodium, and silicon, are removed by leaching. Because aluminum is extremely insoluble, it becomes concentrated in the soil (as bauxite, a hydrated aluminum oxide). Thus, the formation of bauxite depends on climatic conditions in which chemical weathering and leaching are pronounced, plus of course the presence of aluminum-rich source rock. Important deposits of nickel and cobalt are also found in laterite soils that develop from igneous rocks rich in other silicate minerals.

Other Deposits

Many copper and silver deposits result when weathering processes concentrate metals that are dispersed through a low-grade primary ore. Usually such enrichment occurs in deposits containing pyrite (FeS_2), the most common and widespread sulfide mineral. Pyrite is important because when it chemically weathers, sulfuric acid forms, which enables percolating waters to dissolve the ore metals. Once dissolved, the metals gradually migrate downward through the primary ore body until they are precipitated. Deposition takes place because of changes that occur in the chemistry of the solution when it reaches the groundwater zone (the zone beneath the surface where all pore spaces are filled with water). In this manner, the small percentage of dispersed metal can be removed from a large volume of rock and redeposited as a higher-grade ore in a smaller volume of rock.

The Chapter in Review

The following statements are intended to help you review the primary objectives presented in this chapter.

- Earth's external processes that are continually removing materials from higher elevations and moving them to lower elevations include (1) *weathering*—the disintegration and decomposition of rock at or near the surface; (2) *mass wasting*—the transfer of rock material downslope under the influence of gravity; and (3) *erosion*—the incorporation and transportation of material by a mobile agent, usually water, wind, or ice.

- *Mechanical weathering* is the physical breaking up of rock into smaller pieces. Rocks can be broken into smaller fragments by *frost wedging* (where water works its way into cracks or voids in rock, and, upon freezing, expands and enlarges the openings), *unloading* (expansion and breaking due to a great reduction in pressure when the overlying rock is eroded away), *thermal expansion* (weakening of rock as the result of expansion and contraction as it heats and cools), and *biological activity* (by humans, burrowing animals, plant roots, etc.).

- *Chemical weathering* alters a rock's chemistry, changing it into different substances. Water is by far the most important agent of chemical weathering. Oxygen dissolved in water will *oxidize* iron-rich minerals, while carbon dioxide (CO_2) dissolved in water forms *carbonic acid*, which attacks and alters rock. The chemical weathering of silicate minerals frequently produces (1) soluble products containing sodium, calcium, potassium, and magnesium ions, and silica in solution; (2) insoluble iron oxides, including limonite and hematite; and (3) clay minerals.

- The rate at which rock weathers depends on such factors as (1) *particle size*—small pieces generally weather faster than large pieces; (2) *mineral makeup*—calcite readily dissolves in mildly acidic solutions, and silicate minerals that form first from magma are least resistant to chemical weathering; and (3) *climatic factors*, particularly temperature and moisture. Frequently, rocks exposed at Earth's surface do not weather at the same rate. This *differential weathering* of rocks is influenced by such factors as mineral makeup, degree of jointing, and exposure to the elements.

- *Soil* is a combination of mineral and organic matter, water, and air—that portion of the *regolith* (the layer of rock and mineral fragments produced by weathering) that supports the growth of plants. About one-half of the total volume of a good quality soil is a mixture of disintegrated and decomposed rock (mineral matter) and *humus* (the decayed remains of animal and plant life); the remaining half consists of pore spaces, where air and water circulate. The most important factors that control soil formation are *parent material, time, climate, plants* and *animals,* and *slope.*

- Soil-forming processes operate from the surface downward and produce zones or layers in the soil that are called *horizons.* From the surface downward, the soil horizons are respectively designated as *O* (largely organic matter), *A* (largely mineral matter), *E* (where the fine soil components and soluble materials have been removed by *eluviation* and *leaching*), *B* (or *subsoil,* often referred to as the *zone of accumulation*), and *C* (partially altered parent material). Together the *O* and *A* horizons make up what is commonly called the *topsoil.* Although there are hundreds of soil types and subtypes worldwide, the three very generic types are (1) *pedalfer*—characterized by an accumulation of iron oxides and aluminum-rich clays in the *B* horizon; (2) *pedocal*—characterized by an accumulation of

calcium carbonate; and (3) *laterite*—deep soils that develop in the hot, wet tropics that are poor for growing because they are highly leached.

- Soil erosion is a natural process; it is part of the constant recycling of Earth materials that we call the rock cycle. Once in a stream channel, soil particles, which can now be called *sediment,* are transported downstream and eventually deposited. *Rates of soil erosion* vary from one place to another and depend on the soil's characteristics as well as such factors as climate, slope, and type of vegetation.

- Weathering creates ore deposits by concentrating minor amounts of metals into economically valuable deposits. The process, often called *secondary enrichment,* is accomplished by either (1) removing undesirable materials and leaving the desired elements enriched in the upper zones of the soil, or (2) removing and carrying the desirable elements to lower soil zones where they are redeposited and become more concentrated. *Bauxite,* the principal ore of aluminum, is one important ore created as a result of enrichment by weathering processes. In addition, many copper and silver deposits result when weathering processes concentrate metals that were formerly dispersed through low-grade primary ore.

Key Terms

chemical weathering (p. 98)
differential weathering (p. 100)
eluviation (p. 105)
erosion (p. 94)
exfoliation dome (p. 96)
frost wedging (p. 96)

horizon (p. 104)
humus (p. 101)
laterite (p. 106)
leaching (p. 105)
mass wasting (p. 94)
mechanical weathering (p. 94)
parent material (p. 102)

pedalfer (p. 106)
pedocal (p. 106)
regolith (p. 101)
secondary enrichment (p. 109)
sheeting (p. 96)
soil (p. 101)
soil profile (p. 104)

solum (p. 105)
spheroidal weathering (p. 99)
talus slope (p. 96)
weathering (p. 94)

Questions for Review

1. Describe the role of external processes in the rock cycle.
2. If two identical rocks were weathered, one mechanically and the other chemically, how would the products of weathering for the two rocks differ?
3. How does mechanical weathering add to the effectiveness of chemical weathering?
4. Describe the formation of an exfoliation dome. Give an example of such a feature.
5. Granite and basalt are exposed at the surface in a hot, wet region.

(a) Which type of weathering will predominate?

(b) Which of the rocks will weather most rapidly? Why?

6. Heat speeds up a chemical reaction. Why then does chemical weathering proceed slowly in a hot desert?

7. How is carbonic acid (H_2CO_3) formed in nature? What results when this acid reacts with potassium feldspar?

8. Relate soil to the Earth system.

9. What factors might cause different soils to develop from the same parent material, or similar soils to form from different parent materials?

10. Which of the controls of soil formation is most important? Explain.

11. How can slope affect the development of soil? What is meant by the term *slope orientation*?

12. List the characteristics associated with each of the horizons in a well-developed soil profile. Which of the horizons constitute the solum? Under what circumstances do soils lack horizons?

13. Distinguish between pedalfers and pedocals.

14. Soils formed in the humid tropics and the Arctic both contain little organic matter. Do both lack humus for the same reasons?

15. What soil type is associated with tropical rain forests? Because this soil supports the growth of lush natural vegetation, is it also excellent for growing crops? Briefly explain. (See Box 5.1.)

16. Before soil particles are carried by water moving in tiny channels called rills, they are moved by an unconfined movement of water across the surface termed _____.

17. List three detrimental effects of soil erosion other than the loss of topsoil from croplands.

18. Name the primary ore of aluminum and describe its formation.

Testing What You Have Learned

To test your knowledge of the material presented in this chapter, answer the following questions:

Multiple-Choice Questions

1. When a rock undergoes _____ weathering, it is broken into smaller and smaller pieces, each retaining the characteristics of the original material.
   a. chemical   **c.** accelerated   **e.** secondary
   b. selective   **d.** mechanical

2. Which of the following sequences correctly lists the basic soil horizons from top to bottom?
   a. A, E, B, C, O   **c.** O, A, E, B, C   **e.** C, B, E, A, O
   b. B, C, O, A, E   **d.** A, O, B, E, C

3. Which factor(s) influence(s) the rate at which a rock weathers?
   a. particle size   **c.** climate   **e.** a., b., and c.
   b. mineral makeup   **d.** only a. and c.

4. The transfer of rock material downslope under the influence of gravity is referred to as _____.
   a. weathering   **c.** sheeting   **e.** mass wasting
   b. erosion   **d.** unloading

5. The concentration of minor amounts of metals that are scattered through unweathered rock into economically valuable deposits by weathering is called _____ enrichment.
   a. auxiliary   **c.** post   **e.** differential
   b. secondary   **d.** alternate

6. When large masses of granite are exposed by erosion, the process called _____ generates concentric, onionlike layers that often break loose.
   a. wedging   **c.** oxidation   **e.** sheeting
   b. cleavage   **d.** laterization

7. Which one of the following is NOT a product you would expect from the complete weathering of granite?
   a. quartz grains   **c.** olivine   **e.** potassium bicarbonate
   b. clay   **d.** silica (soluble)

8. Which one of the following is NOT a control of soil formation?
   a. time   **c.** slope   **e.** abrasion
   b. climate   **d.** parent material

9. The disintegration and decomposition of rock at, or near, Earth's surface is called _____.
   a. weathering   **c.** destruction   **e.** assimilation
   b. mass wasting   **d.** erosion

10. Which one of the following is NOT a process associated with mechanical weathering?
   a. unloading   **c.** organic activity   **e.** thermal expansion
   b. oxidation   **d.** frost wedging

Fill-In Questions

11. The incorporation and transportation of material by a mobile agent, usually water, wind, or ice, is called _____.
12. The layer of rock and mineral fragments produced by weathering and found on Earth's land surface is called _____.
13. The most important agent of chemical weathering is _____.
14. Under most conditions of chemical weathering, the elements aluminum, silicon, and oxygen join with water to produce residual _____ minerals.
15. The depletion of soluble materials from the upper soil is termed _____.

True/False Questions

16. The process of frost wedging is very pronounced in mountainous regions in the middle latitudes. ___
17. Rocks exposed at Earth's surface usually weather at about the same rate. ___
18. Mechanical weathering often decreases the amount of surface area available for chemical weathering. ___
19. The optimum environment for chemical weathering is a combination of cool temperatures and abundant moisture. ___
20. The formation of bauxite is an example of an ore created as a result of enrichment by weathering processes. ___

Answers

1.d; 2.c; 3.e; 4.e; 5.b; 6.e; 7.c; 8.e; 9.a; 10.b; 11. erosion; 12. regolith; 13. water; 14. clay; 15. leaching; 16.T; 17.F; 18.F; 19.F; 20.T

Chapter 6

Sedimentary Rocks

Focus on Learning

To assist you in learning the important concepts in this chapter, you will find it helpful to focus on the following questions:

- What are the two general types of sedimentary rocks and how does each type form?

- What is the primary basis for distinguishing among various types of detrital sedimentary rocks? What criterion is used to subdivide the chemical rocks?

- How is sediment turned into sedimentary rock?

- What is the single most characteristic feature of sedimentary rocks?

- What are the two broad groups of nonmetallic mineral resources?

- Which energy resources are associated with sedimentary rocks?

Eroded sandstone in the Colorado Plateau of Arizona. (Photo by Carr Clifton)

Chapter 5 gave you the background needed to understand the origin of sedimentary rocks. Recall that weathering of existing rocks begins the process. Next, erosional agents such as running water, wind, waves, and ice remove the products of weathering and carry them to a new location where they are deposited. Usually the particles are broken down further during the transport phase. Following deposition, this material, which is now called **sediment**, becomes lithified. In most cases, the sediment is lithified into solid sedimentary rock by the processes of *compaction* and *cementation*.

Thus, the products of mechanical and chemical weathering constitute the raw materials for sedimentary rocks. The word *sedimentary* indicates the nature of these rocks, for it is derived from the Latin *sedimentum*, which means "settling," a reference to solid material settling out of a fluid (water or air). Most, but not all, sediment is deposited in this fashion. Weathered debris is constantly being swept from bedrock, carried away, and eventually deposited in lakes, river valleys, seas, and countless other places. The particles in a desert sand dune, the mud on the floor of a swamp, the gravel in a stream bed, and even household dust are examples of this never-ending process. Because the weathering of bedrock and the transport and deposition of the weathering products are continuous, sediment is found almost everywhere. As piles of sediment accumulate, the materials near the bottom are compacted. Over long periods, these sediments become cemented together by mineral matter deposited in the spaces between particles, forming solid rock.

Geologists estimate that sedimentary rocks account for only about 5 percent (by volume) of Earth's outer 16 kilometers (10 miles). However, the importance of this group of rocks is far greater than this percentage would imply. If we were to sample the rocks exposed at the surface, we would find that the great majority are sedimentary. Indeed, about 75 percent of all rock outcrops on the continents are sedimentary (Figure 6.1). Therefore, we may think of sedimentary rocks as comprising a relatively thin and somewhat discontinuous layer in the uppermost portion of the crust. This fact is readily understood when we consider that sediment accumulates at the surface.

Because sediments are deposited at Earth's surface, the rock layers that they eventually form contain evidence of past events that occurred at the surface. By their very nature, sedimentary rocks contain within them indications of past environments in which their particles were deposited and, in some cases, clues to the mechanisms involved in their transport. Furthermore, it is sedimentary rocks that contain fossils, which are vital tools in the study of the geologic past. Thus, it is largely from this group of rocks that geologists must reconstruct the details of Earth history.

Finally, it should be mentioned that many sedimentary rocks are very important economically. Coal, which is burned to provide a significant portion of U.S. electrical energy, is classified as a sedimentary rock. Our other major energy sources, petroleum and natural gas, are associated with sedimentary rocks. So are major sources of iron, aluminum, manganese, and fertilizer, plus numerous materials essential to the construction industry.

Types of Sedimentary Rocks

Sediment has two principal sources. First, sediment may be an accumulation of material that originates and is transported as solid particles derived from both mechanical and chemical weathering. Deposits of this type are termed *detrital* and the sedimentary rocks that they form are called **detrital sedimentary rocks**. The second major source of sediment is soluble material produced largely by chemical weathering. When these dissolved substances are precipitated by either inorganic or organic processes, the material is known as chemical sediment and the rocks formed from it are called **chemical sedimentary rocks**.

We will now look at each type of sedimentary rock, and some examples of each.

Detrital Sedimentary Rocks

Though a wide variety of minerals and rock fragments may be found in detrital rocks, clay minerals and quartz are the chief constituents of most sedimentary rocks in this category. Recall from Chapter 5 that clay minerals are the most abundant product of the chemical weathering of silicate minerals, especially the feldspars. Clays are fine-grained minerals with sheetlike crystalline structures similar to the micas. The other common mineral, quartz, is abundant because it is extremely durable and very resistant to chemical weathering. Thus, when igneous rocks such as granite are attacked by weathering processes, individual quartz grains are freed.

Other common minerals in detrital rocks are feldspars and micas. Because chemical weathering rapidly transforms these minerals into new substances,

FIGURE 6.1 Sedimentary rocks in Arizona's Monument Valley are very colorful. Sedimentary rocks are exposed at the surface more than igneous and metamorphic rocks. Because they contain fossils and other clues about our geologic past, sedimentary rocks are important in the study of Earth history. (Photo by Carr Clifton)

their presence in sedimentary rocks indicates that erosion and deposition were fast enough to preserve some of the primary minerals from the source rock before they could be decomposed.

Particle size is the primary basis for distinguishing among various detrital sedimentary rocks. Table 6.1 presents the size categories for particles making up detrital rocks. Note that in this context the term *clay* refers only to a particular size and not to the minerals of the same name. Although most clay minerals are of clay size, not all clay-sized sediment consists of clay minerals.

TABLE 6.1 Particle Size Classification for Detrital Rocks.

Size Range (millimeters)	Particle Name	Common Sediment Name	Detrital Rock
>256	Boulder		
64–256	Cobble	Gravel	Conglomerate
4–64	Pebble		or breccia
2–4	Granule		
1/16–2	Sand	Sand	Sandstone
1/256–1/16	Silt	Mud	Shale or
<1/256	Clay		mudstone

Particle size is not only a convenient method of dividing detrital rocks: the sizes of the component grains also provide useful information about environments of deposition. Currents of water or air sort the particles by size; the stronger the current, the larger the particle size carried. Gravels, for example, are moved by swiftly flowing rivers as well as by landslides and glaciers. Less energy is required to transport sand; thus it is common to such features as windblown dunes and some river deposits and beaches. Very little energy is needed to transport clay, so it settles very slowly. Accumulations of these tiny particles are generally associated with the quiet waters of a lake, lagoon, swamp, or certain marine environments.

Common detrital sedimentary rocks, in order of increasing particle size, are shale, sandstone, and conglomerate or breccia. We will now look at each type and how it forms.

Shale

Shale is a sedimentary rock consisting of silt- and clay-sized particles (Figure 6.2). These fine-grained detrital rocks account for well over half of all sedimentary rocks. The particles in these rocks are so small that they cannot be readily identified without

FIGURE 6.2 Shale is a fine-grained detrital rock that is by far the most abundant of all sedimentary rocks. Dark shales containing plant remains are relatively common. (Photo by E. J. Tarbuck)

great magnification and for this reason make shale more difficult to study and analyze than most other sedimentary rocks.

Much of what can be learned is based on particle size. The tiny grains in shale indicate that deposition occurs as the result of gradual settling from relatively quiet, nonturbulent currents. Such environments include lakes, river floodplains, lagoons, and portions of the deep-ocean basins. Even in these "quiet" environments, there is usually enough turbulence to keep clay-sized particles suspended almost indefinitely. Consequently, much of the clay is deposited only after the individual particles coalesce to form larger aggregates.

Sometimes the chemical composition of the rock provides additional information. One example is black shale, which is black because it contains abundant organic matter (carbon). When such a rock is found, it strongly implies that deposition occurred in an oxygen-poor environment such as a swamp, where organic materials do not readily oxidize and decay.

As silt and clay accumulate, they tend to form thin layers which are commonly referred to as *laminae*. Initially the particles in the laminae are oriented randomly. This disordered arrangement leaves a high percentage of open space (called *pore space*) that is filled with water. However, this situation usually changes with time as additional layers of sediment pile up and compact the sediment below. During this phase the clay and silt particles take on a more nearly parallel alignment and become tightly packed. This rearrangement of grains reduces the size of the

pore spaces and forces out much of the water. Once the grains are pressed closely together, the tiny spaces between particles do not readily permit solutions containing cementing material to circulate. Therefore, shales are often described as being weak because they are poorly cemented and therefore not well lithified. The inability of water to penetrate its microscopic pore spaces explains why shale often forms barriers to the subsurface movement of water and petroleum. Indeed, rock layers that contain groundwater are commonly underlain by shale beds that block further downward movement. The opposite is true for underground reservoirs of petroleum. They are often capped by shale beds that effectively prevent oil and gas from escaping to the surface.

It is common to apply the term *shale* to all fine-grained sedimentary rocks, especially in a nontechnical context. However, be aware that there is a more restricted use of the term. In this narrower usage, shale must exhibit the ability to split into thin layers along well-developed, closely spaced planes. If the rock breaks into chunks or blocks, the name *mudstone* is applied. Another fine-grained sedimentary rock which, like mudstone, is often grouped with shale but lacks fissillity is *siltstone*. As its name implies, siltstone is composed largely of silt-sized particles and contains less clay-sized material than shale and mudstone.

Although shale is far more common than other sedimentary rocks, it does not usually attract as much notice as other less abundant members of this group. The reason is that shale does not form prominent outcrops as sandstone and limestone often do. Rather, shale crumbles easily and usually forms a cover of soil that hides the unweathered rock below. This is illustrated nicely in the Grand Canyon, where the gentler slopes of weathered shale are quite inconspicuous and overgrown with vegetation, in sharp contrast with the bold cliffs produced by more durable rocks (Figure 6.3).

Although shale beds may not form striking cliffs and prominent outcrops, some deposits have economic value. Certain shales are quarried to obtain raw material for pottery, brick, tile, and china. Moreover, when mixed with limestone, shale is used to make portland cement. In the future, one type of shale, called oil shale, may become a valuable energy resource.

Sandstone

Sandstone is the name given rocks in which sand-sized grains predominate (Figure 6.4). After shale, sandstone is the most abundant sedimentary rock,

FIGURE 6.3 Sedimentary rock layers exposed in the walls of Arizona's Grand Canyon. Beds of resistant sandstone and limestone produce bold cliffs. By contrast, weaker, poorly cemented shale crumbles easily and produces a gentler slope of weathered debris in which some vegetation is growing. (Photo by Tom Till)

accounting for approximately 20 percent of the entire group. Sandstones form in a variety of environments and often contain significant clues about their origin, including sorting, particle shape, and composition.

Sorting is the degree of similarity in particle size in a sedimentary rock. For example, if all the grains in a sample of sandstone are about the same size, the sand is considered *well sorted*. Conversely, if the rock contains mixed large and small particles, the sand is said to be *poorly sorted*. By studying the degree of sorting, we can learn much about the depositing current. Deposits of windblown sand are usually better sorted than deposits sorted by wave activity. Particles washed by waves are commonly better sorted than materials deposited by streams. Sediment accumulations that exhibit poor sorting usually result when particles are transported for only a relatively short time and then rapidly deposited. For example, when a turbulent stream reaches the gentler slopes at the base of a steep mountain, its velocity is quickly reduced and poorly sorted sands and gravels are deposited.

The shapes of sand grains can also help decipher the history of a sandstone. When streams, winds, or waves move sand and other sedimentary particles, the grains lose their sharp edges and corners and become more rounded as they collide with other particles during transport. Thus, rounded grains likely have been airborne or waterborne. Further, the degree of rounding indicates the distance or time involved in the transportation of sediment by currents of air or water. Highly rounded grains indicate that a great deal of abrasion and hence a great deal of transport has occurred.

Very angular grains, on the other hand, imply two things: that the materials were transported only a short distance before they were deposited, and that some other medium may have transported them. For example, when glaciers move sediment, the particles are usually made more irregular by the crushing and grinding action of the ice.

In addition to affecting the degree of rounding and the amount of sorting that particles undergo, the length of transport by turbulent air and water currents

Close up

FIGURE 6.4 Quartz sandstone. After shale, sandstone is the most abundant sedimentary rock. (Photos by E. J. Tarbuck)

also influences the mineral composition of a sedimentary deposit. Substantial weathering and long transport lead to the gradual destruction of weaker and less stable minerals, including the feldspars and ferromagnesians. Because quartz is very durable, it is usually the mineral that survives the long trip in a turbulent environment.

The preceding discussion has shown that the origin and history of a sandstone can often be deduced by examining the sorting, roundness, and mineral composition of its constituent grains. Knowing this information allows us to infer that a well-sorted, quartz-rich sandstone consisting of highly rounded grains must be the result of a great deal of transport. Such a rock, in fact, may represent several cycles of weathering, transport, and deposition. We may also conclude that a sandstone containing a significant amount of feldspar and angular grains of ferromagnesian minerals underwent little chemical weathering and transport and was probably deposited close to the source area of the particles.

Owing to its durability, quartz is the predominant mineral in most sandstones. When this is the case, the rock may simply be called *quartz sandstone*. When a sandstone contains appreciable quantities of feldspar, the rock is called *arkose*. In addition to feldspar, arkose usually contains quartz and sparkling bits of mica. The mineral composition of arkose indicates that the grains were derived from granitic

source rocks. The particles are generally poorly sorted and angular, which suggests short-distance transport, minimal chemical weathering in a relatively dry climate, and rapid deposition and burial.

A third variety of sandstone is known as *graywacke*. Along with quartz and feldspar, this dark-colored rock contains abundant rock fragments and matrix. *Matrix* refers to the silt- and clay-sized particles found in spaces between larger sand grains. More than 15 percent of graywacke's volume is matrix. The poor sorting and angular grains characteristic of graywacke suggest that the particles were transported only a relatively short distance from their source area and then rapidly deposited. Before the sediment could be reworked and sorted further, it was buried by additional layers of material. Graywacke is frequently associated with submarine deposits made by dense, sediment-choked torrents called turbidity currents.

Conglomerate and Breccia

Conglomerate consists largely of gravels (Figure 6.5). As Table 6.1 indicates, these particles may range in size from large boulders to particles as small as garden peas. The particles are commonly large enough to be identified as distinctive rock types; thus, they can be valuable in identifying the source areas of sediments. More often than not conglomerates are poorly sorted because the openings between the large gravel particles contain sand or mud.

Close up

FIGURE 6.5 Conglomerate is composed primarily of rounded gravel-sized particles. (Photos by E. J. Tarbuck)

Gravels accumulate in a variety of environments and usually indicate the existence of steep slopes or very turbulent currents. The coarse particles in a conglomerate may reflect the action of energetic mountain streams or result from strong wave activity along a rapidly eroding coast. Some glacial and landslide deposits also contain plentiful gravel.

If the large particles are angular rather than rounded, the rock is called *breccia* (Figure 6.6). Because large particles abrade and become rounded very rapidly during transport, the pebbles and cobbles in a breccia indicate that they did not travel far from their source area before they were deposited. Thus, as with many other sedimentary rocks, conglomerates and breccias contain clues to their history. Their particle sizes reveal the strength of the currents that transported them, whereas the degree of rounding indicates how far the particles traveled. The fragments within a sample identify the source rocks that supplied them.

Chemical Sedimentary Rocks

In contrast to detrital rocks, which form from the solid products of weathering, chemical sediments derive from material that is carried *in solution* to lakes and seas. This material does not remain dissolved in the water indefinitely, however. Some of it precipitates to form chemical sediments. These become rocks such as limestone, chert, and rock salt.

This precipitation of material occurs in two ways. *Inorganic* processes such as evaporation and chemical activity can produce chemical sediments. *Organic* (life) processes of water-dwelling organisms also form chemical sediments, said to be of **biochemical** origin (Figure 6.7).

An example of a deposit resulting from inorganic chemical processes is the salt left behind as a body of seawater evaporates. In contrast, many water-dwelling animals and plants extract dissolved mineral matter to form shells and other hard parts. After the organisms die, their skeletons collect by the millions on the floor of a lake or ocean as biochemical sediment.

Limestone

Representing about 10 percent of the total volume of all sedimentary rocks, *limestone* is the most abundant chemical sedimentary rock. It is composed chiefly of the mineral calcite ($CaCO_3$) and forms either by inorganic means or as the result of biochemical processes. Regardless of its origin, the mineral composition of all limestone is similar, yet many different types exist. This is true because limestones are produced under a variety of conditions. Those forms having a marine biochemical origin are by far the most common.

Coral Reefs Corals are one important example of organisms that are capable of creating large quantities of marine limestone. These relatively simple

Close up

FIGURE 6.6 When the gravel-sized particles in a detrital rock are angular, the rock is called breccia. (Photos by E. J. Tarbuck)

Close up

FIGURE 6.7 This rock, called coquina, consists of shell fragments; therefore, it has a biochemical origin. (Photos by E. J. Tarbuck)

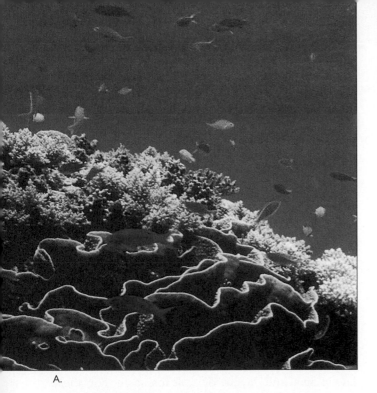

A.

B.

FIGURE 6.8 A. This modern coral reef is at Bora Bora in French Polynesia. (Photo by Nancy Sefton/Photo Researchers) **B.** El Capitan Peak, a massive limestone cliff in Guadalupe Mountains National Park, Texas. The rocks here are an exposed portion of a large reef that formed during the Permian period. (Photo by Steve Elmore/The Stock Market)

invertebrate animals secrete a calcareous (calcite-rich) external skeleton. Although they are small, corals are capable of creating massive structures called *reefs* (Figure 6.8A). Reefs consist of coral colonies made up of great numbers of individuals that live side by side on a calcite structure secreted by the animals. In addition, calcium carbonate-secreting algae live with the corals and help cement the entire structure into a solid mass. A wide variety of other organisms also live in and near reefs.

Certainly the best-known modern reef is Australia's Great Barrier Reef, 2000 kilometers long, but many lesser reefs also exist. They develop in the shallow, warm waters of the tropics and subtropics equatorward of about 30° latitude. Striking examples exist in the Bahamas and Florida Keys.

Of course, not only modern corals build reefs. Corals have been responsible for producing vast quantities of limestone in the geologic past as well. In the United States, reefs of Silurian age are prominent features in Wisconsin, Illinois, and Indiana. In west Texas and adjacent southeastern New Mexico, a massive reef complex formed during the Permian period is strikingly exposed in Guadalupe Mountains National Park (Figure 6.8B).

Coquina and Chalk Although most limestone is the product of biological processes, this origin is not always evident because shells and skeletons may undergo considerable change before becoming lithified into rock. However, one easily identified biochemical limestone is *coquina*, a coarse rock composed of poorly cemented shells and shell fragments (see Figure 6.7). Another less obvious, but nevertheless familiar example, is *chalk*, a soft, porous rock made up almost entirely of the hard parts of microscopic marine organisms no larger than the head of a pin. Among the most famous chalk deposits are those exposed along the southeast coast of England (Figure 6.9).

Inorganic Limestones Limestones having an inorganic origin form when chemical changes or high water temperatures increase the concentration of calcium carbonate to the point that it precipitates. *Travertine*, the type of limestone commonly seen in caves, is an example (see Figure 10.1, p. 193). When travertine is deposited in caves, groundwater is the source of the calcium carbonate. As water droplets become exposed to the air in a cavern, some of the carbon dioxide dissolved in the water escapes, causing calcium carbonate to precipitate.

Another variety of inorganic limestone is *oolitic limestone*. It is a rock composed of small spherical grains called *ooids*. Ooids form in shallow marine waters as tiny "seed" particles (commonly small shell fragments) are moved back and forth by currents. As the grains are rolled about in the warm water, which is supersaturated with calcium carbonate, they become coated with layer upon layer of the precipitate.

FIGURE 6.9 The White Cliffs of Dover. This prominent chalk deposit underlies large portions of southern England as well as parts of northern France. (Photo by Frederica Georgia/Photo Researchers)

Dolostone

Closely related to limestone is *dolostone*, a rock composed of the calcium-magnesium carbonate mineral dolomite. Although dolostone can form by direct precipitation from seawater, it is thought that most originates when magnesium in seawater replaces some of the calcium in limestone. The latter hypothesis is reinforced by the fact that there is practically no young dolostone. Rather, most dolostones are ancient rocks in which there was ample time for magnesium to replace calcium.

Chert

Chert is a name used for a number of very compact and hard rocks made of microcrystalline silica (SiO_2). One well-known form is *flint*, whose dark color results from the organic matter it contains. *Jasper*, a red variety, gets its bright color from iron oxide. The banded form is usually referred to as *agate* (Figure 6.10).

Chert deposits are commonly found in one of two situations: as irregularly shaped nodules in limestone and as layers of rock. The silica composing most chert nodules is believed to have been deposited directly from water. Thus, these nodules have an inorganic origin. However, it is unlikely that a very large percentage of chert layers was precipitated directly from seawater, because seawater is seldom saturated with silica. Hence, beds of chert are thought to have originated largely as biochemical sediment.

Most water-dwelling organisms that produce hard parts make them of calcium carbonate. But some, such as diatoms and radiolarians, produce grasslike silica skeletons. These tiny organisms are able to extract silica even though seawater contains only tiny quantities of the dissolved material. It is from their remains that most chert beds are believed to originate.

Some bedded cherts occur in association with lava flows and layers of volcanic ash. For these occurrences it is probable that the silica was derived from the decomposition of the volcanic ash and not from biochemical sources. Note that when a hand specimen of chert is being examined, there are few reliable criteria by which the mode of origin (inorganic versus biochemical) can be determined.

Like glass, most chert has a conchoidal fracture. Its hardness, ease of chipping, and ability to hold a sharp edge made chert a favorite of Native Americans for fashioning "points" for spears and arrows. Because of its durability and extensive use, "arrowheads" are found in many parts of North America.

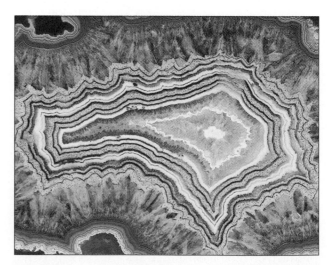

FIGURE 6.10 Agate is the banded form of chert. (Photo by Chip Clark)

Evaporites

Very often evaporation is the mechanism triggering deposition of chemical precipitates. Minerals commonly precipitated in this fashion include halite (sodium chloride, NaCl), the chief component of *rock salt*, and gypsum (hydrous calcium sulfate, $CaSO_4 \cdot 2H_2O$), the main ingredient of *rock gypsum*. Both have significant commercial importance. Halite is familiar to everyone as the common salt used in cooking and seasoning foods. Of course, it has many other uses, from melting ice on roads to making hydrochloric acid, and has been considered important enough that people have sought, traded, and fought over it for much of human history. Gypsum is the basic ingredient of plaster of Paris. This material is used most extensively in the construction industry for wallboard and interior plaster.

In the geologic past, many areas that are now dry land were basins, submerged under shallow arms of a sea that had only narrow connections to the open ocean. Under these conditions, seawater continually moved into the bay to replace water lost by evaporation. Eventually the waters of the bay became saturated and salt deposition began. Such deposits are called **evaporites**.

When a body of seawater evaporates, the minerals that precipitate do so in a sequence which is determined by their solubility. Less-soluble minerals precipitate first and more-soluble minerals precipitate later as salinity increases. For example, gypsum precipitates when about two-thirds to three-quarters of the seawater has evaporated, and halite settles out when nine-tenths of the water has been removed. During the last stages of this process, potassium and magnesium salts precipitate. One of these last-formed salts, the mineral *sylvite*, is mined as a significant source of potassium ("potash") for fertilizer.

On a smaller scale, evaporite deposits may be seen in such places as Death Valley, California. Here, following rains or periods of snowmelt in the mountains, streams flow from the surrounding mountains into an enclosed basin. As the water evaporates, **salt flats** form from dissolved materials left behind as a white crust on the ground (Figure 6.11).

Coal

Coal is quite different from other chemical sedimentary rocks. Unlike other rocks in this category, which are calcite- or silica-rich, coal is made mostly of organic matter. Close examination of a piece of coal under a microscope or magnifying glass often reveals plant structures such as leaves, bark, and wood that have been chemically altered but are still identifiable.

FIGURE 6.11 These salt flats in Death Valley, California, are examples of evaporite deposits and are common in basins located in the arid Southwest. A temporary lake is created when water drains from the adjacent mountains into the basin following periods of rain or snowmelt. When the water evaporates, salts are left behind. (Photo by Tom Till)

This supports the conclusion that coal is the end product of the burial of large amounts of plant material over extended periods.

Coal is commonly called a *fossil fuel*. This is certainly appropriate, for each time we burn coal we are using energy from the sun that was stored by plants many millions of years ago. We are indeed burning a "fossil."

The initial stage in coal formation is the accumulation of large quantities of plant remains. However, special conditions are required for such accumulations, because dead plants normally decompose when exposed to the atmosphere. An ideal environment that allows for the buildup of plant material is a swamp. Because stagnant swamp water is oxygen-deficient, complete decay (oxidation) of the plant material is not possible. At various times during Earth history, such environments have been common. Coal undergoes successive stages of formation. With each successive stage, higher temperatures and pressures drive off impurities and volatiles, as shown in Figure 6.12.

Lignite and bituminous coals are sedimentary rocks, but anthracite is a metamorphic rock. Anthracite forms when sedimentary layers are subjected to the folding and deformation associated with mountain building.

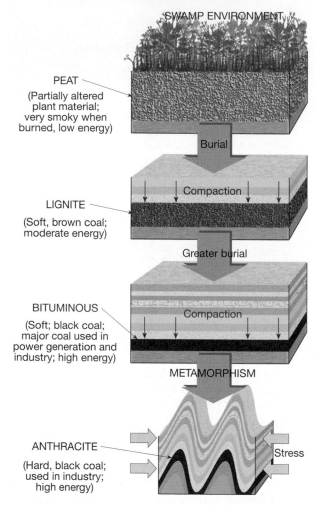

FIGURE 6.12 Successive stages in the formation of coal.

Turning Sediment into Sedimentary Rock

Having examined the general types of sedimentary rocks, let us look at how sediment becomes rock.

Lithification refers to the process by which unconsolidated sediments are transformed into solid sedimentary rocks. One of the most common processes affecting sediments is **compaction**. As sediments accumulate through time, the weight of overlying material compresses the deeper sediments. As the grains are pressed closer and closer, there is a considerable reduction in pore space. For example, when clays are buried beneath several thousand meters of material, the volume of clay may be reduced by as much as 40 percent. Because sands and other coarse sediments are only slightly compressible, compaction is most significant as a lithification process in fine-grained sedimentary rocks such as shale.

Cementation is the most important process by which sediments are converted to sedimentary rocks. The cementing materials are carried in solution by water percolating through the open spaces between particles. Through time, the cement precipitates onto the sediment grains, fills the open spaces, and joins the particles.

Calcite, silica, and iron oxide are the most common cements. It is often a relatively simple matter to identify the cementing material. Calcite cement will effervesce with dilute hydrochloric acid. Silica is the hardest cement and thus produces the hardest sedimentary rocks. An orange or dark red color in a sedimentary rock means that iron oxide is present.

Most sedimentary rocks are lithified by means of compaction and cementation. However, certain chemical sedimentary rocks, such as the evaporates, initially form as solid masses of intergrown crystals, rather than beginning as accumulations of separate particles that later become solid. Other crystalline sedimentary rocks do not begin that way but are transformed into masses of interlocking crystals sometime after the sediment is deposited. For example, with time and burial, loose sediment consisting of delicate calcareous skeletal debris may be recrystallized into a relatively dense crystalline limestone. Because crystals grow until they fill all the available space, pore spaces are frequently lacking in crystalline sedimentary rocks. Unless the rocks later develop joints and fractures, they will be relatively impermeable to fluids like water and oil.

Classification of Sedimentary Rocks

The classification scheme in Table 6.2 divides sedimentary rocks into two major groups: detrital and chemical. Further, we can see that the main criterion for subdividing the detrital rocks is particle size, whereas the primary basis for distinguishing among different rocks in the chemical group is their mineral composition.

As is the case with many (perhaps most) classifications of natural phenomena, the categories presented in Table 6.2 are more rigid than the actual state of nature. In reality, many of the sedimentary rocks classified into the chemical group also contain at least small quantities of detrital sediment. Many limestones, for example, contain varying amounts of mud or sand, giving them a "sandy" or "shaly" quality. Conversely, because practically all detrital rocks are cemented with material that was originally dissolved in water, they too are far from being "pure."

As was the case with the igneous rocks examined in Chapter 3, *texture* is a part of sedimentary rock

TABLE 6.2 Classification of Sedimentary Rocks.

DETRITAL ROCKS			
Texture	Sediment Name and Particle Size	Comments	Rock Name
Clastic	Gravel (>2 mm)	Rounded rock fragments	Conglomerate
		Angular rock fragments	Breccia
	Sand (1/16–2 mm)	Quartz predominates	Quartz sandstone
		Quartz with considerable feldspar	Arkose
		Dark color; quartz with considerable feldspar, clay, and rocky fragments	Graywacke
	Mud (<1/16 mm)	Splits into thin layers	Shale
		Breaks into clumps or blocks	Mudstone

CHEMICAL ROCKS			
Group	Texture	Composition	Rock Name
Inorganic	Clastic or nonclastic	Calcite, $CaCO_3$	Limestone
	Nonclastic	Dolomite, $CaMg(CO_3)_2$	Dolostone
	Nonclastic	Microcrystalline quartz, SiO_2	Chert
	Nonclastic	Halite, $NaCl$	Rock salt
	Nonclastic	Gypsum, $CaSO_4 \cdot 2H_2O$	Rock gypsum
Biochemical	Clastic or nonclastic	Calcite, $CaCO_3$	Limestone
	Nonclastic	Microcrystalline quartz, SiO_2	Chert
	Nonclastic	Altered plant remains	Coal

classification. Two major textures are used in the classification of sedimentary rocks: clastic and nonclastic. The term **clastic** is taken from a Greek word meaning "broken." Rocks that display a clastic texture consist of discrete fragments and particles that are cemented and compacted together. Although cement is present in the spaces between particles, these openings are rarely filled completely. Table 6.2 shows that *all* detrital rocks have a clastic texture. The table also shows that some chemical sedimentary rocks exhibit this texture, too. For example, coquina, the limestone composed of shells and shell fragments, is obviously as clastic as conglomerate or sandstone. The same applies for some varieties of oolitic limestone.

Some chemical sedimentary rocks have a **nonclastic** texture in which the minerals form a pattern of interlocking crystals. The crystals may be microscopically small or large enough to be visible without magnification. Common examples of rocks with nonclastic textures are those deposited when seawater evaporates (Figure 6.13). The materials composing many other nonclastic rocks may actually have originated as detrital deposits. In these instances, the particles probably consisted of shell fragments and other hard parts rich in calcium carbonate or silica. The clastic nature of the grains was subsequently obliterated or obscured because the particles recrystallized when they were consolidated into limestone or chert.

Close up

FIGURE 6.13 Like other evaporites, this sample of rock salt is said to have a nonclastic texture because it is composed of intergrown crystals. (Photos by E. J. Tarbuck)

Nonclastic rocks consist of intergrown crystals, and some may resemble igneous rocks, which are also crystalline. The two rock types are usually easy to distinguish because the minerals that compose

FIGURE 6.14
This outcrop of sedimentary strata illustrates the characteristic layering of this group of rocks. (Photo by Tom Till)

nonclastic sedimentary rocks are quite unlike those found in most igneous rocks. For example, rock salt, rock gypsum, and some forms of limestone consist of intergrown crystals, but the minerals that compose these rocks (halite, gypsum, and calcite) are seldom associated with igneous rocks.

Sedimentary Structures

In addition to variations in grain size, mineral composition, and texture, sediments exhibit a variety of structures. Some, such as graded beds, are created when sediments are accumulating and are a reflection of the transporting medium. Others, such as mud cracks, form after the materials have been deposited and result from processes occurring in the environment. When present, sedimentary structures provide additional information that can be useful in the interpretation of Earth history.

Sedimentary rocks form as layer upon layer of sediment accumulates in various depositional environments. These layers, called **strata**, or **beds**, are probably *the single most common and characteristic feature of sedimentary rocks*. Each stratum is unique. It may be a coarse sandstone, a fossil-rich limestone, a black shale, and so on. When you look at Figure 6.14, or look back through this chapter at Figures 6.1 (p. 116), 6.3 (p. 118), and 6.8B (p. 121), you will see many such layers, each different from the others. The variations in texture, composition, and thickness reflect the different conditions under which each layer was deposited (see Box 6.1).

The thickness of beds ranges from microscopically thin to tens of meters thick. Separating the strata are **bedding planes**, flat surfaces along which rocks tend to separate or break. Changes in the grain size or in the composition of the sediment being deposited can create bedding planes. Pauses in deposition can also lead to layering because chances are slight that newly deposited material will be exactly the same as previously deposited sediment. Generally each bedding plane marks the end of one episode of sedimentation and the beginning of another.

Because sediments usually accumulate as particles that settle from a fluid, most strata are originally deposited as horizontal layers. There are circumstances, however, when sediments do not accumulate

A.

B.

FIGURE 6.15 A. The cut-away section of this sand dune shows cross-bedding. (Photo by John S. Shelton)
B. The cross-bedding of this sandstone indicates it was once a sand dune. (Photo by David Muench)

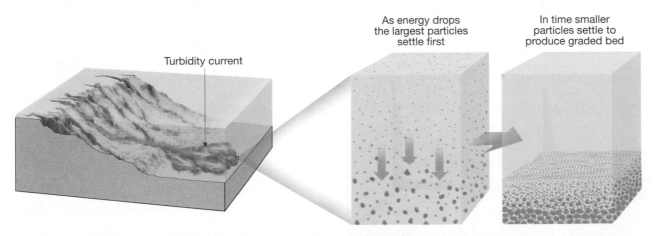

FIGURE 6.16 Graded beds. Each layer grades from coarse at its base to fine at the top.

in horizontal beds. Sometimes when a bed of sedimentary rock is examined, we see layers within it that are inclined to the horizontal. When this occurs, it is called **cross-bedding** and is most characteristic of sand dunes, river deltas, and certain stream channel deposits (Figure 6.15).

Graded beds represent another special type of bedding. In this case the particles within a single sedimentary layer gradually change from coarse at the bottom to fine at the top (Figure 6.16). Graded beds are most characteristic of rapid deposition from water containing sediment of varying sizes. When a current experiences a rapid energy loss, the largest particles settle first, followed by successively smaller grains. The deposition of a graded bed is most often associated with a turbidity current, a

mass of sediment-choked water that is denser than clear water and that moves downslope along the bottom of a lake or ocean.*

As geologists examine sedimentary rocks, much can be deduced. A conglomerate, for example, may indicate a high-energy environment, such as a surf zone or rushing stream, where only coarse materials settle out and finer particles are kept suspended (Figure 6.17). If the rock is arkose, it may signify a dry climate where little chemical alteration of feldspar is possible. Carbonaceous shale is a sign of a low-energy, organic-rich environment, such as a swamp or lagoon.

---

*More on these currents and graded beds may be found in the section on "Submarine Canyons and Turbidity Currents" in Chapter 14.

Box 6.1 Sedimentary Rocks Represent Past Environments

Sedimentary rocks are important in the interpretation of Earth history. By understanding the conditions under which sedimentary rocks form, geologists can often deduce the history of a rock, including information about the origin of its component particles, the method and length of sediment transport, and the nature of the place where the grains eventually came to rest; that is, the environment of deposition. Thus, when a series of sedimentary layers are studied, we can see the successive changes in environmental conditions that occurred at a particular place with the passage of time.

Sediments are deposited at Earth's surface. Thus, they hold many clues about the physical, chemical, and biological conditions that existed in the areas where the materials accumulated. By applying a thorough knowledge of present-day conditions, geologists attempt to reconstruct the ancient environments and geographical relationships of an area at the time a particular set of sedimentary layers were deposited. Such analyses often lead to the creation of maps, which depict the distribution of land and sea, mountains and plains, deserts and glaciers, and other environments of deposition.

Sedimentary environments are commonly placed into one of two broad categories: *terrestrial* (continental) or *marine*. Because the shore zone exhibits characteristics of both, it can be considered transitional between land and sea. Figure 6.A divides the two broad categories of terrestrial and marine into several major sedimentary environments. Chapters 9 through 13, as well as portions of Chapter 14, will describe these environments in detail. Each is an area where sediment accumulates and where organisms live and die. Each produces a characteristic sedimentary rock or assemblage that reflects prevailing conditions.

FIGURE 6.17 In a turbulent stream channel, only large particles settle out. Finer sediments remain suspended and continue their downstream journey. Arrigetch Creek, Brooks Range, Alaska. (Photo by Tom Bean)

Other features found in some sedimentary rocks also give clues to past environments. Ripple marks are such a feature. **Ripple marks** are small waves of sand that develop on the surface of a sediment layer by the action of moving water or air (Figure 6.18A). The ridges form at right angles to the direction of motion. If the ripple marks were formed by air or water moving in essentially one direction, their form will be asymmetrical. These *current ripple marks* will have steeper sides in the downcurrent direction and more gradual slopes on the upcurrent side. Ripple marks produced by a stream flowing across a sandy channel or by wind blowing over a sand dune are two common examples of current ripples. When present in solid rock, they may be used to determine the direction of movement of ancient wind or water currents. Other ripple marks have a symmetrical form. These features, called *oscillation ripple marks*, result from the back-and-forth movement of surface waves in a shallow nearshore environment.

Mud cracks (Figure 6.18B) indicate that the sediment in which they were formed was alternately wet and dry. When exposed to air, wet mud dries out and shrinks, producing cracks. Mud cracks are associated with such environments as shallow lakes and desert basins.

Fossils, the evidence or remains of prehistoric life, are perhaps the most important inclusions found in sedimentary rock (Figure 6.19). Fossils are important tools used to interpret the geologic past. Knowing the nature of the life-forms that existed at

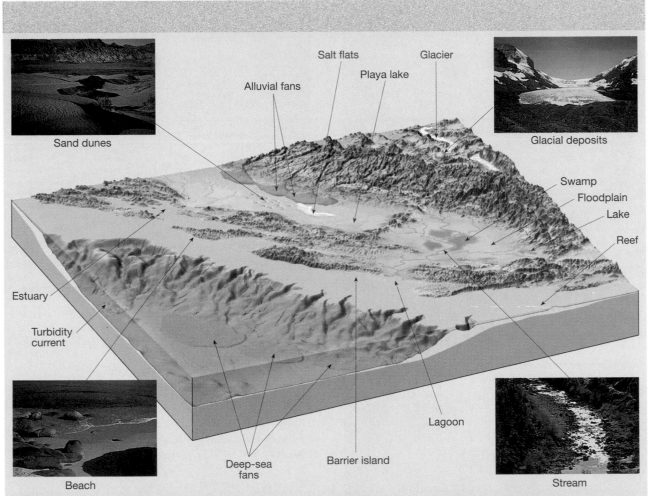

FIGURE 6.A Sedimentary environments are those places where sediment accumulates. Each is characterized by certain physical, chemical, and biological conditions. Because each sediment contains clues about the environment in which it was deposited, sedimentary rocks are important in the interpretation of Earth history. A number of important terrestrial, shoreline (transitional), and marine sedimentary environments are represented in this idealized diagram.

a particular time may help to answer questions about that ancient environment. Further, fossils are important time indicators and play a key role in correlating rocks that are of similar ages but are from different places. Fossils will be examined in more detail in Chapter 18.

Nonmetallic Mineral Resources From Sedimentary Rocks

Earth materials that are not used as fuels or processed for the metals they contain are referred to as **nonmetallic mineral resources**. Realize that use of the word "mineral" is very broad in this economic context, and is quite different from the geologist's strict definition of mineral found in Chapter 2. Nonmetallic mineral resources are extracted and processed either for the nonmetallic elements they contain or for the physical and chemical properties they possess (Table 6.3). Although these resources have diverse origins, many are sediments or sedimentary rocks.

Nonmetallic mineral resources are commonly divided into two broad groups—*building materials* and *industrial minerals*. Because some substances have many different uses, they are found in both categories. For example, limestone, perhaps the most versatile and widely used rock of all, is a building material, used not only as crushed rock and building stone but also in making cement. Moreover, as an industrial mineral, limestone is an ingredient in the manufacture of steel and is used in agriculture to neutralize acidic soils.

Other important building materials include cut stone, aggregate (sand, gravel, and crushed rock), gypsum for plaster and wallboard, clay for tile and bricks, and cement, which is made from limestone

A.

B.

FIGURE 6.18 A. Ripple marks are produced by currents of water or wind. (Photo by Stephen Trimble) **B.** Mud cracks form when wet mud or clay dries out and shrinks. (Photo by Gary Yeowell/Tony Stone Images)

FIGURE 6.19 Fossil of a trilobite, an ancient marine organism, preserved in shale. Fossils are important tools used to determine past environmental conditions. They are also valuable time indicators. (Photo by E. J. Tarbuck)

TABLE 6.3 Uses of nonmetallic minerals.

Mineral	Uses
Apatite	Phosphorus fertilizers
Asbestos (chrysotile)	Incombustible fibers
Calcite	Aggregate; steelmaking; soil conditioning; chemicals; cement; building stone
Clay minerals (kaolinite)	Ceramics; china
Corundum	Gemstones; abrasives
Diamond	Gemstones; abrasives
Fluorite	Steelmaking; aluminum refining; glass; chemicals
Garnet	Abrasives; gemstones
Graphite	Pencil lead; lubricant; refractories
Gypsum	Plaster of Paris
Halite	Table salt; chemicals; ice control
Muscovite	Insulator in electrical applications
Quartz	Primary ingredient in glass
Sulfur	Chemicals; fertilizer manufacture
Sylvite	Potassium fertilizers
Talc	Powder used in paints, cosmetics, etc.

and shale. Cement and aggregate go into the making of concrete, a material that is essential to practically all construction (Figure 6.20).

A wide variety of resources are classified as industrial minerals. People often do not realize the importance of industrial minerals because they see only the products that result from their use and not the minerals themselves. Examples include fluorite and limestone, used in steelmaking; corundum and garnet, used as abrasives to make machinery; and sylvite, used to produce fertilizers for growing food crops.

Energy Resources From Sedimentary Rocks

Coal, petroleum, and natural gas are the primary fuels of our modern industrial economy. Nearly 90 percent of the energy consumed in the United States today comes from these basic fossil fuels. Although

FIGURE 6.20 Crushed stone and sand and gravel are primarily used for aggregate in the construction industry, especially in cement concrete for residential and commercial buildings, bridges, and airports, and as cement concrete or bituminous concrete (asphalt) for highway construction. A large percentage is used without a binder as road base, for road surfacing, and as railroad ballast. The United States produces nearly 2 billion tons of aggregate per year, which represents about one-half of the entire nonenergy mining volume in the country. It is produced commercially in every state and is used in nearly all building construction and in most public works projects. (Photo by Peter Vadnai/The Stock Market)

major shortages of oil and gas may not occur for many years, proven reserves are declining. Despite new exploration, even in very remote regions and severe environments, new sources of oil are not keeping pace with consumption.

Unless large, new petroleum reserves are discovered (which is possible, but not likely), a greater share of our future needs will have to come from coal and from alternative energy sources such as nuclear, solar, wind, tidal, and hydroelectric power. Two fossil fuel alternatives, tar sands and oil shale, are sometimes mentioned as promising new sources of liquid fuels. In the following sections, we will briefly examine the fuels that have traditionally supplied our energy needs.

Coal

Coal has been an important fuel for centuries. In the nineteenth and early twentieth centuries, cheap and plentiful coal powered the industrial revolution. By 1900, coal was providing 90 percent of the energy used in the United States. Although still important,

coal currently provides about 20 percent of the energy needs of the United States (Figure 6.21).

Until the 1950s, coal was an important domestic heating fuel as well as a power source for industry. However, its direct use in the home has been largely replaced by oil, natural gas, and electricity. These fuels are preferred because they are more readily available (delivered via pipes, tanks, or wiring) and cleaner to use. Nevertheless, coal remains the major fuel used in power plants to generate electricity, and it is therefore indirectly an important source of energy for our homes. More than 70 percent of present-day coal usage is for the generation of electricity. As oil reserves gradually diminish in the years to come, the use of coal will probably increase. Expanded coal production is possible because the world has enormous reserves and the technology to mine coal efficiently. In the United States, coal fields are widespread and contain supplies that should last for hundreds of years (Figure 6.22).

Although coal is plentiful, its recovery and use present a number of problems. Surface mining can turn the countryside into a scarred wasteland if careful (and costly) reclamation is not carried out to restore the land. (Today all U.S. surface mines must reclaim the land.) Although underground mining does not scar the landscape to the same degree, it has been costly in terms of human life and health. Underground mining long ago ceased to be a pick-and-shovel operation, and is today a highly mechanized and computerized process. Strong federal safety regulations have made U.S. mining quite safe. The hazards of roof falls, gas explosions, and working with heavy equipment remain, however.

Air pollution is a major problem associated with the burning of coal. Much coal contains significant quantities of sulfur. Despite efforts to remove sulfur before the coal is burned, some remains; when the coal is burned, the sulfur is converted into noxious

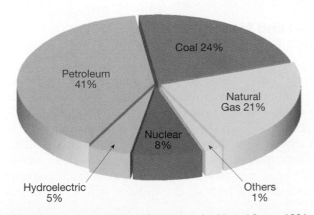

FIGURE 6.21 Consumption of energy in the United States, 1996.

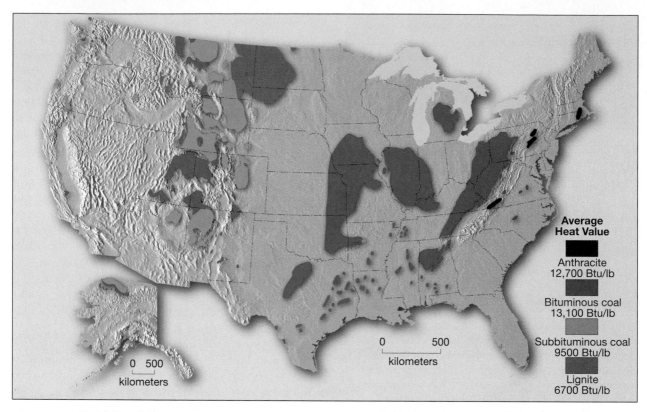

FIGURE 6.22 Coal fields of the United States. (Courtesy of the Bureau of Mines, U.S. Department of the Interior)

Average
Heat Value

Anthracite
12,700 Btu/lb

Bituminous coal
13,100 Btu/lb

Subbituminous coal
9500 Btu/lb

Lignite
6700 Btu/lb

sulfur oxide gases. Through a series of complex chemical reactions in the atmosphere, the sulfur oxides are converted to sulfuric acid, which then falls to Earth's surface as rain or snow. This acid precipitation can have detrimental ecological effects over widespread areas. Although steps have been taken and will continue to be, the air pollution produced by burning coal is still a significant threat to our environment.

Because none of the problems just mentioned are likely to prevent the increased use of this important and abundant fuel, stronger efforts must be made to correct the problems associated with the mining and use of coal.

Oil and Natural Gas

Petroleum and natural gas obviously are not rocks, but we class them as "mineral resources" because they *come from* sedimentary rocks. They are found in similar environments and typically occur together. Both consist of various hydrocarbon compounds (compounds consisting of hydrogen and carbon) mixed together. They may also contain small quantities of other elements such as sulfur, nitrogen, and oxygen. Like coal, petroleum and natural gas are biological products derived from the

remains of organisms. However, the environments in which they form are very different, as are the organisms. Coal is formed mostly from plant material that accumulated in a swampy environment above sea level. Oil and gas, on the other hand, are derived from the remains of both plants and animals having a marine origin.

Petroleum formation is complex and not completely understood. Nevertheless, we know that it begins with the accumulation of sediment in ocean areas that are rich in plant and animal remains. These accumulations must occur where biological activity is high, such as nearshore areas. However, most marine environments are oxygen-rich, which leads to the decay of organic remains before they can be buried by other sediments. Therefore, accumulations of oil and gas are not as widespread as the marine environments that support abundant biological activity. This limiting factor notwithstanding, large quantities of organic matter are buried and protected from oxidation in many offshore sedimentary basins. With increasing burial over millions of years, chemical reactions gradually transform some of the original organic matter into the liquid and gaseous hydrocarbons we call petroleum and natural gas.

Unlike the solid organic matter from which they formed, the newly created petroleum and natural gas are *mobile*. These fluids are gradually squeezed from the compacting, mud-rich layers where they originate into adjacent permeable beds such as sandstone, where openings between sediment grains are larger. Because all of this occurs underwater, the rock layers containing the oil and gas are already saturated with water. But oil and gas are less dense than water, so they migrate upward through the waterfilled pore spaces of the enclosing rocks. Unless something acts to halt this upward migration, the fluids will eventually reach the surface. There the volatile components will evaporate.

Sometimes the upward migration is halted. A geologic environment that allows for economically significant amounts of oil and gas to accumulate underground is termed an **oil trap**. Several geologic structures may act as oil traps. All have two basic conditions in common: a porous, permeable **reservoir rock** that will yield petroleum and natural gas in sufficient quantities to make drilling worthwhile; and a **cap rock**, such as shale, that is virtually impermeable to oil and gas. The cap rock keeps the upwardly mobile oil and gas from escaping at the surface.

The Chapter in Review

The following statements are intended to help you review the primary objectives presented in this chapter.

- *Sedimentary rock* consists of *sediment* that, in most cases, has been *lithified* into solid rock by the processes of *compaction* and *cementation*. Sediment has two principal sources: (1) as *detrital material*, which originates and is transported as solid particles from both mechanical and chemical weathering, which, when lithified, forms detrital sedimentary rocks; and (2) from soluble material produced largely by chemical weathering, which, when precipitated, forms *chemical sedimentary rocks.*

- *Particle size* is the primary basis for distinguishing among various detrital sedimentary rocks. The size of the particles in a detrital rock indicates the energy of the medium that transported them. For example, gravels are moved by swiftly flowing rivers, whereas less energy is required to transport sand. Common detrital sedimentary rocks include *shale* (silt- and clay-sized particles), *sandstone*, and *conglomerate* (rounded gravel-sized particles) or *breccia* (angular gravel-sized particles).

- Precipitation of chemical sediments occurs in two ways: (1) by *inorganic processes*, such as evaporation and chemical activity; or by (2) *organic processes* of water-dwelling organisms that produce sediments of *biochemical origin*. *Limestone*, the most abundant chemical sedimentary rock, consists of the mineral calcite

$(CaCO_3)$ and forms either by inorganic means or as the result of biochemical processes. Inorganic limestones include *travertine*, which is commonly seen in caves, and *oolitic limestone*, consisting of small spherical grains of calcium carbonate. Other common chemical sedimentary rocks include *dolostone* (composed of the calcium-magnesium carbonate mineral dolomite), *chert* (made of microcrystalline quartz), *evaporites* (such as rock salt and rock gypsum), and *coal* (lignite and bituminous).

- *Lithification* refers to the processes by which unconsolidated sediments are transformed into solid sedimentary rock. Most sedimentary rocks are lithified by means of *compaction* and/or *cementation*. Compaction occurs when the weight of overlying materials compresses the deeper sediments. Cementation, the most important process by which sediments are converted to sedimentary rocks, occurs when soluble cementing materials, such as *calcite, silica,* and *iron oxide*, are precipitated onto sediment grains, fill open spaces, and join the particles. Although most sedimentary rocks are lithified by compaction or cementation, certain chemical rocks, such as the evaporites, initially form as solid masses of intergrown crystals.

- Sedimentary rocks can be divided into two main groups: *detrital* and *chemical*. All detrital rocks have a *clastic texture*, which consists of discrete fragments and particles that are cemented and compacted together. The main criterion for subdividing the detrital rocks is particle size. Common detrital rocks include *conglomerate, sandstone*, and *shale*. The primary basis for distinguishing among different rocks in the chemical group is their mineral composition. Some chemical rocks, such as those deposited when seawater evaporates, have a *nonclastic texture* in which the minerals form a pattern of interlocking crystals. However, in reality, many of the sedimentary rocks classified into the chemical group also contain at least small quantities of detrital sediment. Common chemical rocks include *limestone, rock gypsum*, and *coal* (e.g., lignite and bituminous).

- Sedimentary rocks are particularly important in interpreting Earth's history because, as layer upon layer of sediment accumulates, each records the nature of the environment at the time the sediment was deposited. These layers, called *strata*, or *beds*, are probably the single most characteristic feature of sedimentary rocks. Other features found in some sedimentary rocks, such as *ripple marks, mud cracks, cross-bedding*, and *fossils*, also give clues to past environments.

- Earth materials that are not used as fuels or processed for the metals they contain are referred to as *nonmetallic resources*. Many are sediments or sedimentary rocks. The two broad groups of nonmetallic resources are *building materials* and *industrial minerals*. Limestone, perhaps the most versatile and widely used rock of all, is found in both groups.

- *Coal, petroleum*, and *natural gas*, the *fossil fuels* of our modern economy, are all associated with

sedimentary rocks. Coal originates from large quantities of plant remains that accumulate in an oxygen-deficient environment, such as a swamp. More than 70 percent of present-day coal usage is for the generation of electricity. Air pollution from the sulfur oxide gases that form from burning most types of coal is a major environmental problem. Through a series of complex reactions in the atmosphere, the sulfur oxides produced when coal is burned are converted to sulfuric acid, which falls to Earth's surface in rain or snow.

- Oil and natural gas, which commonly occur together in the pore spaces of some sedimentary rocks, consist of various *hydrocarbon compounds* (compounds made of hydrogen and carbon) mixed together. Although petroleum formation is complex and not completely understood, it is associated with the accumulation of sediment in ocean areas that are rich in plant and animal remains that become buried and isolated in an oxygen-deficient environment. As the mobile petroleum and natural gas form, they migrate and accumulate in adjacent permeable beds such as sandstone. If the upward migration is halted by an impermeable rock layer, referred to as a *cap rock*, a geologic environment that allows for economically significant amounts of oil and gas to accumulate underground, called an *oil trap*, develops. The two basic conditions common to all oil traps are (1) a porous, permeable *reservoir rock* that will yield petroleum and/or natural gas in sufficient quantities, and (2) a cap rock.

Key Terms

bedding plane (p. 126)
beds (strata) (p. 126)
biochemical origin (p. 120)
cap rock (p. 133)
cementation (p. 124)
chemical sedimentary rock (p. 115)

clastic (p. 125)
compaction (p. 124)
cross-bedding (p. 127)
detrital sedimentary rock (p. 115)
evaporite (p. 123)
fossil (p. 128)

graded bed (p. 127)
lithification (p. 124)
mud crack (p. 128)
nonclastic (p. 125)
nonmetallic mineral resource (p. 129)
oil trap (p. 133)

reservoir rock (p. 133)
ripple mark (p. 128)
salt flat (p. 123)
sediment (p. 115)
sorting (p. 118)
strata (beds) (p. 126)

Questions for Review

1. How does the volume of sedimentary rocks in Earth's crust compare with the volume of igneous rocks in the crust? Are sedimentary rocks evenly distributed throughout the crust?

2. What minerals are most common in detrital sedimentary rocks? Why are these minerals so abundant?

3. What is the primary basis for distinguishing among various detrital sedimentary rocks?

4. The term *clay* can be used in two different ways. Describe the two meanings of this term.

5. Why does shale usually crumble quite easily?

6. Distinguish between conglomerate and breccia.

7. Distinguish between the two categories of chemical sedimentary rocks.

8. What are evaporite deposits? Name a rock that is an evaporite.

9. How is bituminous coal different from lignite? How is anthracite different from bituminous?

10. Each of the following statements describes one or more characteristics of a particular sedimentary rock. For each statement, name the sedimentary rock that is being described.

    a. An evaporite used to make plaster.
    b. A fine-grained detrital rock that breaks into chunks or blocks.
    c. Dark-colored sandstone containing angular rock particles as well as clay, quartz, and feldspar.
    d. The most abundant chemical sedimentary rock.
    e. A dark-colored, dense, hard rock made of microcrystalline quartz.
    f. A variety of limestone composed of small spherical grains.

11. Compaction is an important lithification process with which sediment size?

12. List three common cements for sedimentary rocks. How might each be identified?

13. What is the primary basis for distinguishing among different chemical sedimentary rocks?

14. Distinguish between clastic and nonclastic textures. What type of texture is common to all detrital sedimentary rocks?

15. What is probably the single most characteristic feature of sedimentary rocks?

16. Distinguish between cross-bedding and graded bedding.

17. Nonmetallic resources are commonly divided into two broad groups. List the two groups and some examples of materials that belong to each.

18. Coal enjoys the advantage of being plentiful. What are some disadvantages associated with the production and use of coal?

19. What is an oil trap? List two conditions common to all oil traps.

Testing What You Have Learned

To test your knowledge of the material presented in this chapter, answer the following questions.

Multiple-Choice Questions

1. The common sedimentary rock consisting of silt- and clay-sized particles is called _____.
   a. limestone c. sandstone e. chert
   b. shale d. arkose

2. Probably the single most characteristic feature of sedimentary rocks is layers called _____, or beds.
   a. horizons c. sheets e. strata
   b. sills d. laminates

3. Sedimentary rocks account for only about _____ percent (by volume) of Earth's outer 16 km; however, about _____ percent of all the rock outcrops on the continents are sedimentary.
   a. 40; 80 c. 30; 70 e. 10; 95
   b. 20; 50 d. 5; 75

4. The _____ rock of an oil trap keeps the upwardly mobile oil and gas from escaping the reservoir rock.
   a. porous   **c.** cap   **e.** sheet
   b. oxidized   **d.** permeable

5. Sedimentary rocks composed of material that originates and is transported as solid particles are called _____ sedimentary rocks.
   a. detrital   **c.** aphanatic   **e.** organic
   b. nonclastic   **d.** chemical

6. Which one of the following sediments would most likely be associated with a low-energy environment, such as a lake, lagoon, or swamp?
   a. sand   **c.** pebbles   **e.** cobbles
   b. clay   **d.** gravel

7. Which one of the following is NOT a chemical sedimentary rock?
   a. chert   **c.** limestone   **e.** coal
   b. conglomerate   **d.** dolostone

8. Which one of the following is NOT a common feature of sedimentary rocks?
   a. ripple marks   **c.** strata   **e.** bedding planes
   b. mud cracks   **d.** phenocrysts

9. The primary basis for distinguishing among various detrital sedimentary rocks is _____ _____.
   a. specific gravity   **c.** particle size   **e.** mineral luster
   b. mineral density   **d.** grain hardness

10. Limestone is composed chiefly of the mineral _____ and forms either by inorganic means or as the result of biochemical processes.
   a. calcite   **c.** halite   **e.** clay
   b. feldspar   **d.** quartz

Fill-In Questions

11. Sedimentary rocks composed of formerly dissolved substances that have been precipitated are called _____ sedimentary rocks.

12. _____, the most abundant chemical sedimentary rock, is made up chiefly of the mineral _____.

13. Chemical sediments formed by the processes of water-dwelling organisms are said to be of _____ origin.

14. Sediment particles that range in size from 1/16 to 2 millimeters are referred to as _____.

15. Perhaps the most important inclusions found in sedimentary rock are the evidence or remains of prehistoric life called _____.

True/False Questions

16. The size of the particles in a detrital rock often indicates the energy of the medium that transported them. ___

17. Bituminous coal and petroleum form in oxygen-rich environments. ___

18. Rock salt is the most abundant chemical sedimentary rock. ___

19. As layer upon layer of sediment accumulates, each records the nature of the environment at the time the sediment was deposited. ___

20. All detrital rocks have a clastic texture. ___

Answers

1.b; 2.e; 3.d; 4.c; 5.a; 6.b; 7.b; 8.d; 9.c; 10.a; 11. chemical; 12. Limestone, calcite; 13. biochemical; 14. sand; 15. fossils; 16.T; 17.F; 18.F; 19.T; 20.T.

Metamorphic Rocks

Deformed metamorphic rocks west of Teakettle Junction, Death Valley, California. (Photo by Michael Collier)

Focus on Learning

To assist you in learning the important concepts in this chapter, you will find it helpful to focus on the following questions:

- What are metamorphic rocks and how do they form?

- In which three geologic settings is metamorphism most likely to occur?

- What are the agents of metamorphism?

- How is the intensity, or degree, of metamorphism reflected in the texture and mineralogy of metamorphic rocks?

- What are the two textural divisions of metamorphic rocks and the conditions associated with the occurrence of each?

- What are the names, textures, and compositions of some common metamorphic rocks?

- What are some mineral and rock resources that are associated with metamorphism?

FIGURE 7.1 Metamorphic rocks outcropping in Shining Rock Wilderness, North Carolina. (Photo by David Muench)

Consider the conditions necessary to fold and distort the rock shown in Figure 7.1. It takes an enormous amount of directed pressure and temperatures hundreds of degrees above surface conditions acting for, typically, thousands to millions of years to produce the deformation shown. Under these extreme conditions rocks respond by folding and flowing, a process called *metamorphism*. This chapter looks at the forces that forge metamorphic rocks, how they change in appearance and mineral makeup. You will see some interesting examples of metamorphic rocks, the environments where they occur, and some economic uses.

Metamorphism

Recall from the section on the rock cycle in Chapter 1 that metamorphism is the transformation of one rock type into another. Metamorphic rocks can be transformed from igneous, sedimentary, or even from other metamorphic rocks. Metamorphism is a very appropriate name for this process because it literally means to "change form." The agents of metamorphism include heat, pressure, and chemically active fluids. The changes that occur are both textural and mineralogical.

Metamorphism occurs incrementally, from slight change (low grade) to dramatic change (high grade). For example, under low-grade metamorphism, the common sedimentary rock *shale* becomes the more compact metamorphic rock called *slate*. Hand samples of these rocks are sometimes difficult to distinguish.

In other cases, high-grade metamorphism causes a transformation so complete that the identity of the original rock cannot be determined. In high-grade metamorphism, such features as bedding planes, fossils, and vesicles that may have existed in the parent rock are obliterated. Further, when rocks at depth are subjected to uneven pressure, they slowly flow and bend into intricate folds (Figure 7.2). In the most extreme metamorphic environments, the temperatures approach those at which rocks melt. However, during metamorphism some material must remain solid, for if complete melting occurs, we have entered the realm of igneous activity.

Where Does Metamorphism Occur?

Metamorphism takes place *where rock is subjected to conditions unlike those in which it formed*. In response to these new conditions, the unstable rock gradually changes until a state of equilibrium with the new environment is reached. Most metamorphic changes occur at the elevated temperatures and pressures that exist in the zone extending from a few kilometers below Earth's surface to the crust–mantle boundary.

FIGURE 7.2 Deformed metamorphic rocks exposed in a road cut in the Eastern Highland of Connecticut. (Photo by Phil Dombrowski)

Metamorphism most often occurs in one of three settings:

1. *When rock is near or touching a mass of magma,* **contact metamorphism** takes place. Here the changes are caused primarily by the high temperatures of the molten material, which in effect "bake" the surrounding rock.
2. *During mountain-building,* great quantities of rock are subjected to the intense pressures and temperatures associated with large-scale deformation called **regional metamorphism**. The end result may be extensive areas of metamorphic rocks. The greatest volume of metamorphic rock is produced in this fashion.
3. The least common type of metamorphism occurs along *fault zones.* Here rock is broken and distorted as crustal blocks on opposite sides of a fault grind past one another.

Extensive areas of metamorphic rocks are exposed on every continent in the relatively flat regions known as *shields.* These areas include eastern Canada, Brazil, much of Africa, India, and half of Australia. Moreover, metamorphic rocks are an important component of many mountain belts, where they make up a large portion of a mountain's crystalline core. Even those portions of the stable continental interiors that are covered by sedimentary rocks are underlain by metamorphic basement rocks. In all of these settings the metamorphic rocks are usually highly deformed and intruded by igneous

masses. Indeed, significant parts of Earth's continental crust are composed of metamorphic and associated igneous rocks.

Unlike some igneous and sedimentary processes that take place in surface or near-surface environments, metamorphism usually occurs deep within Earth beyond our direct observation. Notwithstanding this significant obstacle, geologists have developed techniques that have allowed them to learn a great deal about the conditions under which metamorphic rocks form. Thus, metamorphic rocks provide important clues about the geologic processes that operate within Earth's crust.

Agents of Metamorphism

As stated earlier, the agents of metamorphism include *heat, pressure,* and *chemically active fluids.* During metamorphism, rocks are often subjected to all three metamorphic agents simultaneously. However, the degree of metamorphism and the contribution of each agent vary greatly from one environment to another. In low-grade metamorphism, rocks are subjected to temperatures and pressures only slightly greater than those associated with the lithification of sediments. High-grade metamorphism, on the other hand, involves extreme conditions closer to those at which rocks melt.

In addition, the mineral makeup of the parent rock determines, to a large extent, the degree to which each metamorphic agent will cause change. For example, when intruding magma forces its way into existing rock, hot, ion-rich fluids (mostly water) circulate through the host rock. If the host rock is quartz sandstone, very little alteration may take place. On the other hand, if the host rock is limestone, the impact of these fluids can be dramatic and the effects of metamorphism may extend for several kilometers from the magma body.

Heat as a Metamorphic Agent

Perhaps the most important agent of metamorphism is *heat* because it provides the energy to drive chemical reactions that result in the recrystallization of minerals. Rocks formed near Earth's surface may be subjected to intense heat when they are intruded by molten material rising from below. The effects of this contact metamorphism are most apparent when it occurs at or near the surface where the temperature contrast between the magma and the host rock is most pronounced. Here the adjacent host rock is "baked" by the emplaced magma. In this high-temperature and low-pressure environment, the boundary that forms between the intruding magma and the altered rocks is usually quite distinct.

In addition to magma rising and metamorphosing near-surface rocks, rocks near the surface may be slowly thrust downward to become metamorphosed at depth. As we discussed earlier, Earth materials are continually being transported to great depths at convergent plate boundaries. Recall that temperatures increase with depth at a rate known as the *geothermal gradient*. In the upper crust, this increase in temperature averages between 20°C and 30°C per kilometer. When buried to a depth of only a few kilometers, certain minerals, such as clay, become unstable and begin to recrystallize into minerals, such as muscovite (mica), that are stable in this environment. Other minerals, particularly those found in crystalline igneous rocks, are stable at relatively high temperatures and pressures and therefore require burial to 20 kilometers or more before metamorphism will occur.

Pressure as a Metamorphic Agent

Pressure, like temperature, also increases with depth. Buried rocks are subjected to the force, or **stress**, exerted by the load above (Figure 7.3A). This con-

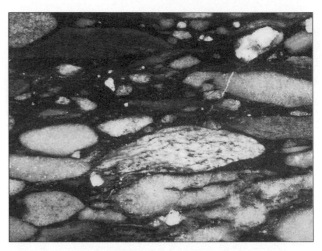

FIGURE 7.4 Metaconglomerate. These once nearly spherical pebbles have been heated and stretched into elongated structures. (Photo by E. J. Tarbuck)

fining pressure is analogous to water pressure where the force is applied equally in all directions. The deeper you go in the ocean, the greater the pressure on all sides. The same is true for rocks at depth.

In addition to the confining pressure exerted by the load of material above, rocks are also subjected to stresses during mountain building (Figure 7.3B). These forces are directional, not equal from all directions. Most often these directional forces are *compressional* and act to shorten a rock body. In some environments, however, the stresses are *extensional* and tend to elongate, or pull apart, the rock mass, with results visible in Figure 7.4.

Directional stress can also cause a rock to **shear**. Shearing is similar to the slippage that occurs between individual cards when you hold a deck flat between your hands and slide your hands in opposite directions, shearing the deck. In near-surface environments shearing results when relatively brittle rock breaks into thin slabs that are forced to slide past one another. This deformation grinds and pulverizes the original mineral grains into small fragments. By contrast, rock located at great depths is warmer and behaves plastically during deformation. This accounts for its ability to flow and bend into intricate folds when subjected to shearing (see Figure 7.2).

Chemical Activity as a Metamorphic Agent

Chemically active fluids also enhance the metamorphic process. Most commonly, the fluid is water containing ions in solution. Water is plentiful, because some water is contained in the pore spaces of virtually every rock. In addition, many minerals are *hydrated* (have water bound chemically) and thus contain water within their crystalline structures.

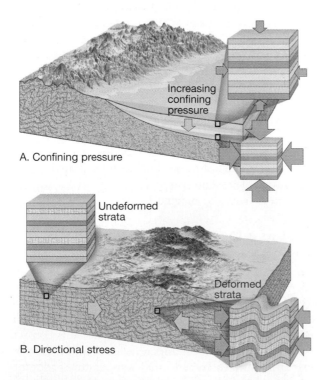

A. Confining pressure

Undeformed strata

Deformed strata

B. Directional stress

FIGURE 7.3 Pressure (stress) as a metamorphic agent. **A.** In a depositional environment, as confining pressure increases, rocks deform by decreasing in volume. **B.** During mountain building, directional stress shortens and deforms rock strata.

When deep burial occurs, water is forced out of the mineral structures and thus becomes available to aid in chemical reactions. Further, as heat is applied, the dehydration of minerals releases water. Water that surrounds the crystals acts as a catalyst by aiding ion migration. In some instances the minerals recrystallize to form more stable configurations. In other cases, ion exchange among minerals results in the formation of completely new minerals.

Complete alteration of rock by hot, mineral-rich water has been observed in the near-surface environment of Yellowstone National Park. On a much larger scale similar activity occurs along the mid-ocean ridge system. Here, seawater circulates through the still-hot basaltic rocks, transforming iron- and magnesium-rich minerals into metamorphic minerals such as serpentine and talc.

How Metamorphism Alters Rocks

The metamorphic process causes many changes in rocks, including increased density, growth of larger crystals, reorientation of the mineral grains into a layered or banded texture, and the transformation of low-temperature minerals into high-temperature minerals. Further, the introduction of ions generates new minerals, some of which are economically important. Thus, the grade of metamorphism is reflected in the *texture* and *mineralogy* (mineral makeup) of metamorphic rocks.

Textural Changes

When rocks are subjected to low-grade metamorphism, they become more compact and thus more dense. A common example is conversion of the sedimentary rock *shale* into the metamorphic rock *slate*. When shale is subjected to temperatures and pressures only slightly greater than those of the sedimentary processes that formed it, slate results. In this case, the directed pressure causes the microscopic clay minerals in shale to align into the more compact arrangement found in slate. This realignment of particles that occurs when shale is converted to slate gives slate a distinctive **texture**. (Recall that texture is the size, shape, and distribution of particles that constitute a rock.)

Foliated Textures Under more extreme conditions, pressure causes mineral grains in a rock to do much more than just realign. Pressure can cause certain minerals to *recrystallize*. In general, recrystallization encourages the growth of larger crystals. Consequently, many metamorphic rocks consist not

of microscopic crystals, but of visible crystals, much like coarse-grained igneous rocks.

The crystals of some minerals, such as micas (platy minerals) and hornblende (needlelike minerals), will recrystallize with a *preferred orientation*. The new orientations will be essentially perpendicular to the direction of the compressional force as shown on the right side in Figure 7.5. The resulting mineral alignment usually gives the rock a layered or banded texture termed **foliation** (Figure 7.5). Simply put, *a foliated texture results whenever the minerals and structural features of a metamorphic rock are forced into parallel alignment.*

Various types of foliation exist, depending largely upon the grade of metamorphism and the mineralogy of the parent rock. We will look at three: rock or slaty cleavage; schistosity; and gneissic texture.

Rock or Slaty Cleavage During the transformation of shale to slate, clay minerals (stable at the surface) recrystallize into minute mica flakes (stable at much higher temperatures and pressures). Further, these platy mica crystals become aligned so that their flat surfaces are nearly parallel. Consequently, slate can be split easily along these layers of mica grains into rather flat slabs. This property is called **rock cleavage** or **slaty cleavage**, to differentiate it from the type of cleavage exhibited by minerals (Figure 7.6). Because the mica flakes composing slate are minute, slate is not visibly foliated. But slate is considered foliated because it can be split easily into slabs, evidence that its minerals are aligned.

Schistosity Under more extreme temperature–pressure regimes, the very fine mica grains in slate will grow many times larger. These mica crystals, which are up to a centimeter in diameter, give the rock a scaly appearance. This type of foliation is called **schistosity**, and a rock having this texture is called *schist*. Many types of schist exist depending on the original parent rock, and they are named according to their mineral constituents—mica schist, talc schist, and so on. By far the most abundant are the mica schists.

Gneissic Texture During high-grade metamorphism, ion migrations can be extreme enough to cause minerals to segregate. An example is Figure 7.5, lower right. Notice that the dark and light silicate minerals have separated, giving the rock a banded appearance called **gneissic texture**. Metamorphic rocks with this texture are called *gneiss* (pronounced "nice") and are quite common. Gneiss often forms from the metamorphism of granite or diorite but can form from

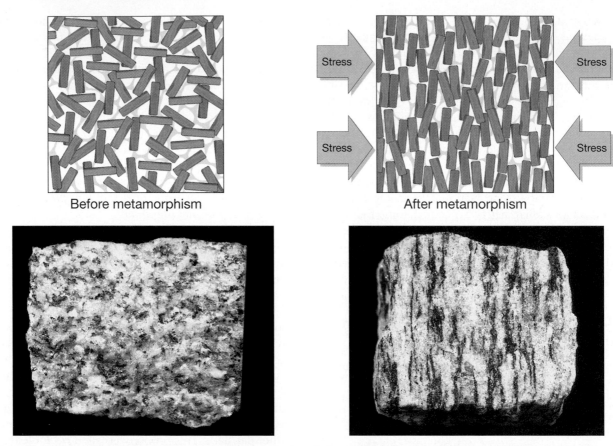

FIGURE 7.5 Under directed pressure, flat or needle-shaped minerals (top left) become reoriented or recrystallized so that they are aligned at right angles to the stress (top right). The resulting parallel orientation of mineral grains gives the rock a foliated texture. If the coarse-grained igneous rock (granite, bottom left) underwent intense metamorphism, it could end up closely resembling the metamorphic rock on the bottom right (gneiss). (Photos by E. J. Tarbuck)

gabbro or even by the high-grade metamorphism of shale. Although foliated, gneiss will not usually split parallel to the crystals as easily as will slate.

Nonfoliated Texture Not all metamorphic rocks have a foliated texture. Those that do not are **nonfoliated**. Metamorphic rocks composed of only one mineral that exhibits equidimensional crystals usually are not visibly foliated. For example, when a fine-grained limestone (made of a single mineral, calcite) is metamorphosed, the small calcite crystals combine to form relatively large interlocking crystals. The resulting rock, *marble*, has a texture similar to that of a coarse-grained igneous rock. Although most marbles are nonfoliated, microscopic investigation of marble may reveal some flattening and parallelism of the grains.

Further, some limestones contain thin layers of clay minerals that may become distorted during metamorphism. These "impurities" will often appear as curved bands of dark material flowing through the marble, a clear indication of metamorphism.

Mineralogical Changes

In the metamorphism of shale to slate, you saw that clay minerals recrystallize to form mica crystals. During most recrystallization, including this example, the chemical composition of the rock does not change (except for the loss of water and carbon dioxide). Rather, the existing minerals and available ions in the water will recombine to form minerals that are stable in the new environment. A common example is when limestone ($CaCO_3$), containing abundant quartz (SiO_2), is heated during contact metamorphism. The calcite and quartz crystals chemically react to form wollastonite ($CaSiO_3$), and carbon dioxide is liberated.

In some environments, however, new materials are actually introduced during the metamorphic process. For example, rock adjacent to a large magma body would acquire new elements from **hydrothermal solutions** (hot water). Many metallic ore deposits are formed by the deposition of minerals from hydrothermal solutions.

as particle-filled clouds called *black smokers*. Upon mixing with the cold seawater, the sulfides precipitate to form massive metallic deposits. This is the origin of the copper ores mined today on the island of Cyprus. Some of Earth's richest copper deposits have formed in this manner.

Common Metamorphic Rocks

Foliated Rocks

What follows is an overview of common metamorphic rocks. Please refer to Table 7.1 throughout this discussion.

Slate To review, *slate* is a very fine-grained foliated rock composed of minute mica flakes. The most noteworthy characteristic of slate is its excellent rock cleavage, or tendency to break into flat slabs. This property traditionally made slate a most useful rock for roof and floor tile, blackboards, and billiard tables (Figure 7.7).

Slate is most often generated by the low-grade metamorphism of shale, although less frequently it is metamorphosed from volcanic ash. Slate's color depends on its mineral constituents. Black (carbonaceous) slate contains organic material (carbon-bearing), red slate gets its color from iron oxide, and green slate usually contains chlorite, a mica-like mineral formed by the metamorphism of iron-rich silicates.

Because slate forms during low-grade metamorphism, evidence of shale's original bedding planes is often preserved. However, the orientation of slate's rock cleavage is at a pronounced angle to the original sedimentary layering (Figure 7.8). Thus, unlike shale, which splits along bedding planes, slate splits across them.

Phyllite *Phyllite* represents a gradation in metamorphism between slate and schist. Its

FIGURE 7.6 Rock cleavage or slaty cleavage in metamorphic rock, California. The parallel mineral alignment in this rock allows it to split easily into the flat plates visible in the photo. (Photo by E. J. Tarbuck)

With the development of plate tectonics, it became clear that some metal-rich hydrothermal deposits originate along ancient spreading centers (mid-ocean ridges). As seawater percolates through newly formed oceanic crust, it dissolves metallic sulfides from the basaltic rocks. The hot, metal-rich fluids rise along fractures and gush from the seafloor

TABLE 7.1 Common metamorphic rocks.

Metamorphic Rock	Texture	Parent Rock	Comments
Slate	Foliated	Shale	Very fine grained
Phyllite	Foliated	Shale	Fine- to medium-grained
Schist	Foliated	Shale, granitic and volcanic rocks	Coarse-grained micaceous minerals
Gneiss	Foliated	Shale, granitic and volcanic rocks	Coarse-grained (non-micaceous)
Marble	Nonfoliated	Limestone, dolostone	Composed of interlocking calcite grains
Quartzite	Nonfoliated	Quartz sandstone	Composed of interlocking quartz grains
Hornfels	Nonfoliated	Any fine-grained material	Fine-grained
Migmatite	Weakly foliated	Mixture of granitic and mafic rocks	Composed of contorted layers
Mylonite	Weakly foliated	Any material	Hard, fine-grained rock
Metaconglomerate	Weakly foliated	Quartz-rich conglomerate	Strongly stretched pebbles
Amphibolite	Weakly foliated	Mafic volcanic rocks	Coarse-grained

FIGURE 7.7 Slate used for roofing material on a house in Switzerland. (Photo by E. J. Tarbuck)

constituent platy minerals are larger than those in slate, but not yet large enough to be clearly identifiable. Although phyllite appears similar to slate, it can be easily distinguished from slate by its glossy sheen (Figure 7.9). Phyllite usually exhibits rock cleavage and is composed mainly of very fine crystals of either muscovite or chlorite.

Schist *Schists* are strongly foliated rocks that can be readily split into thin flakes or slabs. By definition, schists contain more than 50 percent platy and elongated minerals that commonly include mica (muscovite, biotite) and amphibole. Like slate, the

parent material from which many schists originate is shale, but to form schist, the metamorphism is more intense. In addition, most schists are products of major mountain building episodes.

The term *schist* describes the texture of a rock regardless of composition. For example, schists composed primarily of the micas muscovite and biotite are called *mica schists*. Depending upon the degree of metamorphism and composition of the parent rock, mica schists often contain *accessory minerals* unique to metamorphic rocks. A common accessory mineral is the gemstone *garnet*, in which case the rock is called *garnet-mica schist* (Figure 7.10). Some schists contain another accessory mineral, *graphite*, which is used as pencil "lead," graphite fibers (used in fishing rods), and lubricant (commonly for locks).

Gneiss *Gneiss* is the term applied to banded metamorphic rocks that contain mostly elongated and granular (as opposed to platy) minerals. The most common minerals in gneiss are quartz, potassium feldspar, and sodium feldspar. Lesser amounts of muscovite, biotite, and hornblende are common. The segregation of light and dark silicates is developed in gneisses, giving them a characteristic banded appearance. Thus, most gneisses consist of alternating bands of white or reddish feldspar-rich zones and layers of dark ferromagnesian minerals (see Figure 7.5, lower right). These banded gneisses are often deformed while in a plastic state into rather intricate folds. Some gneisses will split along the layers of platy minerals, but most break in an irregular fashion.

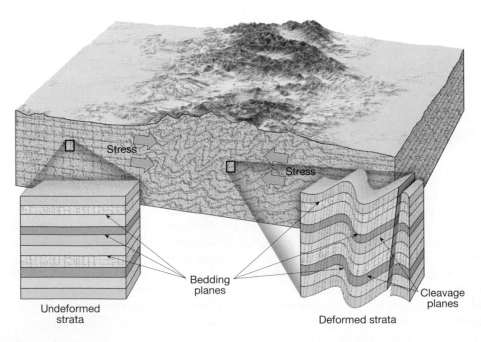

FIGURE 7.8 Illustration showing the relationship between slaty cleavage and bedding planes.

FIGURE 7.9 Phyllite (left) can be distinguished from slate (right) by its glossy sheen. (Photo by E. J. Tarbuck)

FIGURE 7.10 Garnet-mica schist. The dark-red garnet crystals and the light-colored mica matrix formed during the metamorphism of shale. (Photo by E. J. Tarbuck)

FIGURE 7.11 Marble, a crystalline rock formed by the metamorphism of limestone. (Photo by E. J. Tarbuck)

Common Nonfoliated Rocks

Marble *Marble* is a coarse, crystalline rock whose parent rock was limestone or dolostone (Figure 7.11). Pure marble is white and composed essentially of the mineral calcite. Because of its attractive color and relative softness (hardness of 3), marble is a popular building stone in banks and government buildings. White marble is particularly prized as a stone from which to carve monuments and statues, such as the famous statue of David by Michelangelo.

Unfortunately, because marble is basically calcium carbonate, it is readily attacked by acid rain. Some historic monuments and tombstones already show severe chemical weathering.

Often the limestone from which marble forms contains impurities that color the marble. Thus, marble can be pink, gray, green, or even black. Also, when impure limestone is metamorphosed, the resulting marble may contain a variety of accessory minerals (chlorite, mica, garnet, and commonly wollastonite). When marble forms from limestone interbedded with shales, it will appear banded. Under extreme deformation, such banded marble may become highly contorted and give the rock a rather artistic design.

Quartzite *Quartzite* is a very hard metamorphic rock most often formed from quartz sandstone (Figure 7.12). Under moderate-to-high-grade metamorphism, the quartz grains in sandstone fuse like chips of glass melting together (inset in Figure 7.12). The recrystallization is so complete that when broken, quartzite will not split between the original quartz grains, but through them. In some instances, such sedimentary features as crossbedding are preserved and give the rock a banded appearance. Quartzite is typically white, but iron oxide may produce reddish or pinkish stains. Dark mineral grains may impart a gray color.

|←————— 12 cm —————→|

Quartz sand grains

Photomicrograph (× 26.6)
Sample width is 1.23 mm

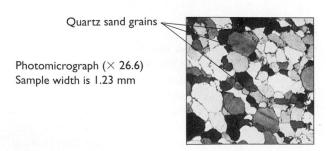

FIGURE 7.12 Quartzite is a nonfoliated metamorphic rock formed from quartz sandstone. The photomicrograph shows the interlocking quartz grains typical of quartzite. (Photos by E. J. Tarbuck)

Environments That Metamorphose Rocks

Recall that metamorphic rocks commonly form in three environments: in contact with magma, during mountain building, and along fault zones. Here is a look at what goes on in each environment.

Contact Metamorphism

Contact metamorphism occurs when magma invades cooler rock. Here, a zone of alteration called an **aureole** (or halo) forms around the emplaced magma (Figure 7.13). Small intrusive magma bodies that form dikes and sills have aureoles only a few centimeters thick. On the other hand, large magma bodies that form batholiths and laccoliths may form zones of metamorphic rocks a few kilometers or more in thickness (Figure 7.13).

These large aureoles often consist of distinct zones of metamorphism. Near the magma body, high-temperature minerals such as garnet may form, whereas farther away such low-grade minerals as chlorite are produced. In addition to the size of the intrusive magma body, the mineral composition of the host rock and the availability of water greatly affect the size of the aureole produced. In chemically active rock such as limestone, the zone of alteration can extend 10 kilometers or more from the magma body. Here the occurrence of minerals such as garnet and wollastonite marks the boundary of metamorphism.

Most contact metamorphic rocks are fine-grained, dense, tough, and of various chemical compositions. For example, during contact metamorphism, clay minerals are baked, as if placed in a kiln, and can generate a very hard, fine-grained rock resembling porcelain. Because directional pressure is not a major factor in forming these rocks, they generally are not foliated.

When larger igneous masses are involved in contact metamorphism, *hydrothermal solutions* that originate within the magma can migrate great distances. These solutions percolate through the host rock, chemically reacting with it and greatly enhancing the metamorphic process. Further, these hydrothermal solutions are thought to be the source of metal ore deposits in metamorphic rocks. These deposits include ores of copper, zinc, lead, iron, and gold.

Contact metamorphism is clearly distinguishable only when it occurs at the surface or in a near-surface environment where the temperature contrast between the magma and host rock is great.

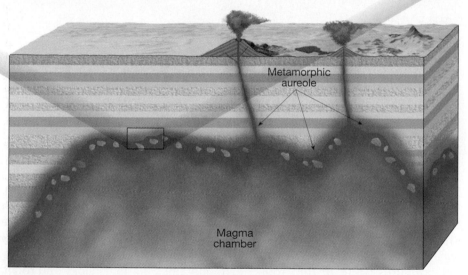

FIGURE 7.13 Contact metamorphism produces a zone of alteration called an *aureole* around an intrusive igneous body. In the photo, the dark layer, called a *roof pendant*, consists of metamorphosed host rock adjacent to the upper part of the light-colored igneous pluton. The term *roof pendant* implies that the rock was once the roof of a magma chamber. Sierra Nevada, near Bishop, California. (Photo by Michael Collier)

Undoubtedly, contact metamorphism is also an active process at great depth. However, its effect is blurred owing to the general alteration caused by regional metamorphism.

Regional Metamorphism

By far the greatest quantity of metamorphic rock is produced during regional metamorphism. As stated earlier, regional metamorphism takes place at considerable depth over an extensive area and is associated with mountain building. During mountain building, a large segment of Earth's crust is intensely squeezed into a highly deformed mass (Figure 7.14). As the rocks are folded and faulted, the crust is shortened and thickened, like a rumpled carpet. This general thickening of the crust results in mountain terrains that are heaved high above sea level.

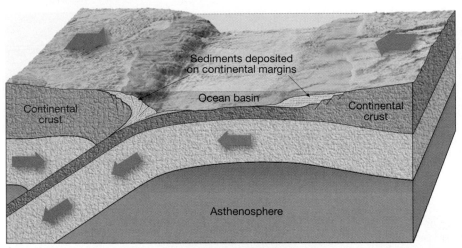

A.

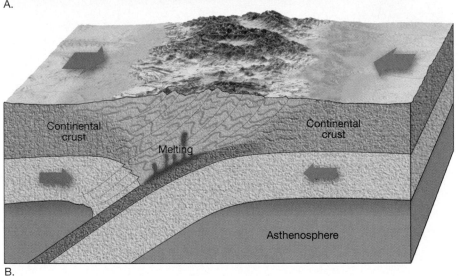

B.

FIGURE 7.14 Regional metamorphism occurs where rocks are squeezed between two converging plates during mountain building.

Although material is obviously elevated to great heights during mountain building, an equally large quantity of rock is forced downward, where it is exposed to high temperatures and pressures. Here in the "roots" of mountains, the most intense metamorphic activity occurs. Some of the deformed rock is thought to be heated enough to melt and thereby generate magma. This magma, being less dense than the surrounding rock, migrates upward. Magmas emplaced in a near-surface environment will cause contact metamorphism within the zone of regional metamorphism.

Consequently, the cores of many mountain ranges consist of intrusive igneous bodies surrounded by high-grade metamorphic rocks. As these deformed rock masses are uplifted, erosion removes the overlying material to expose the igneous and metamorphic rocks composing the central core of the mountain range.

Zones of Metamorphism In regional metamorphism, there usually exists a gradation in intensity. As we progress from areas of low-grade metamorphism to areas of high-grade metamorphism, changes in mineralogy and rock texture can be observed.

For a simplified example, we will use the sedimentary rock shale. Under low-grade metamorphism it yields the metamorphic rock slate (Figure 7.15). In high-temperature, high-pressure environments, slate will turn into phyllite and then into mica schist. Under more extreme conditions, the micas in schist will recrystallize into minerals such as feldspar and hornblende, and eventually generate a gneiss.

You can see this transition by approaching the Appalachian Mountains from the west. Beds of shale, which once extended over large areas of the eastern United States, still occur as nearly flat-lying strata in Ohio. However, in the broadly folded Appalachians of central Pennsylvania, these beds are inclined and composed of low-grade slate.

FIGURE 7.15 Idealized illustration of progressive regional metamorphism. From left to right, we progress from low-grade metamorphism (slate) to high-grade metamorphism (gneiss). (Photos by E. J. Tarbuck)

As we progress farther eastward to the intensely deformed crystalline Appalachians, we find large outcrops of schist and gneiss, some of which are perhaps remnants of once flat-lying shale beds. The most intense zones of metamorphism are found in Vermont and New Hampshire, often near igneous intrusions.

Changes in Mineralogy In addition to the textural changes just described, we encounter changes in mineralogy as we progress from low-grade to high-grade metamorphism. The typical transition in mineralogy that results from the regional metamorphism of shale is shown in Figure 7.16. The first new mineral to be produced in the formation of slate is chlorite. As we move toward the region of high-grade metamorphism, chlorite is replaced by

ever-greater amounts of muscovite and biotite. Mica schists form under more extreme conditions and may also contain garnet and staurolite crystals. At temperatures and pressures approaching the melting point of rock, sillimanite forms. Sillimanite is a high-temperature metamorphic mineral used to make refractory porcelains such as those used in spark plugs.

Migmatites In the most extreme environments, even the highest-grade metamorphic rocks are subjected to change. For example, in a near-surface environment where temperatures approach 750°C, a schist or gneiss having a chemical composition similar to granite will begin to melt. However, recall from our discussion of igneous rocks that not all minerals melt at the same temperature. The light-colored silicates, usually quartz

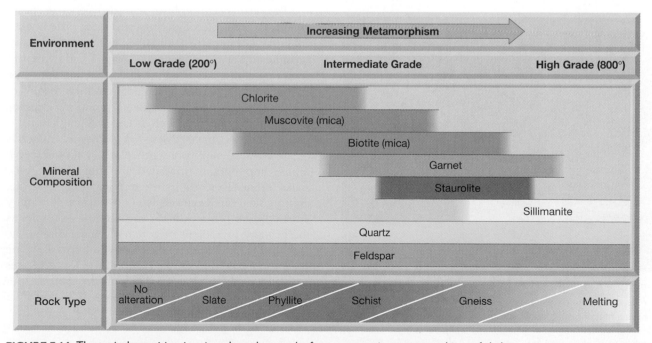

FIGURE 7.16 The typical transition in mineralogy that results from progressive metamorphism of shale.

and potassium feldspar, will melt first, whereas the mafic silicates, such as amphibole and biotite, will remain solid. If this partially melted rock cools, the light bands will be made of crystalline igneous rock while the dark bands will consist of unmelted metamorphic material. Rocks of this type fall into a transitional zone somewhere between "true" igneous rocks and "true" metamorphic rocks. They are called **migmatites** (Figure 7.17).

Metamorphism Along Fault Zones

Near the surface, rock behaves like a brittle solid. Consequently, movement along a fault zone crushes, pulverizes, and generally mills the rock into fine components. The result is a loosely coherent rock called *fault breccia* that is composed of broken and distorted rock fragments. However, most of the intense deformation associated with fault zones occurs at great depths. In this environment the rocks deform by ductile flow, which generates elongated grains that often give the rock a foliated or lineated appearance. Rocks formed in this manner are termed *mylonites*.

Worldwide, the quantity of metamorphic rock generated solely by faulting is very small when compared to the other two processes. Nevertheless, in some areas these granulated rocks are quite abundant. For example, movements along California's San Andreas fault have created a zone of fault breccia and related rock types nearly 1000 kilometers long and up to 3 kilometers wide.

FIGURE 7.17 Migmatite. The lightest colored layers are igneous rock composed of quartz and feldspar, while the darker layers have a metamorphic origin. (Photo by Stephen Trimble)

Metamorphism and Mineral Resources

The role of metamorphism in producing mineral deposits is frequently tied to igneous processes. For example, many of the most important metamorphic ore deposits are produced by contact metamorphism. Here the host rock is recrystallized and chemically altered by heat, pressure, and hydrothermal solutions from an intruding igneous body.

The extent to which the host rock is altered depends on both the nature of the intruding magma and the nature of the host rock. Some resistant materials, such as quartz sandstone, may show very little alteration. Others, including limestone, may exhibit the effects of metamorphism for several kilometers from the igneous body. In limestone, hot, ion-rich fluids moving through the rock cause chemical reactions that produce useful minerals such as garnet and corundum (both used as abrasives).

Further, these reactions release carbon dioxide, which greatly facilitates the outward migration of metallic ions. Thus, extensive aureoles of metal-rich deposits commonly surround igneous bodies that have invaded limestone strata. The most common metallic minerals associated with contact metamorphism are sphalerite (zinc), galena (lead), chalcopyrite (copper), magnetite (iron), and bornite (copper) (Figure 7.18). The hydrothermal ore deposits might be disseminated throughout the altered zone or exist as concentrated masses next to the intrusive body or at the periphery of the metamorphic zone.

Regional metamorphism can also generate useful mineral deposits. Recall that at convergent plate boundaries the oceanic crust and the sediments that have accumulated at the continental margins are carried to great depths. In these high-temperature, high-pressure environments, both the mineralogy and the texture of the subducted materials are altered, producing deposits of nonmetallic minerals such as talc and graphite.

So far, we have discussed accessory minerals created by metamorphic processes. Of equal economic importance, however, are the metamorphic rocks themselves. In many regions, slate, marble, and quartzite are quarried for construction. Ordinary crushed rock has a number of uses, for example, as the primary ingredient of concrete. According to the U.S. Bureau of Mines, the total amount of stone (including sand and gravel) used in the United States annually is 8000 kilograms per citizen. This is four times the combined amount required of all the other mineral resources.

Box 7.1   Impact Metamorphism and Tektites

It now is clear that comets and asteroids have collided with Earth far more frequently than once thought. The evidence: More than a hundred giant impact structures called *astroblemes* have been identified to date. Many had been thought to result from some poorly understood volcanic process. Most astroblemes are too old to look like an impact crater, but a notable exception is the very fresh-looking Meteor Crater in Arizona.

One earmark of astroblemes is impact or shock metamorphism. When high-velocity projectiles (comets, asteroids) impact Earth's surface, pressures reach millions of atmospheres and temperatures exceed 2000°C momentarily. The result is pulverized, shattered, and melted rock. The products of these impacts, called *impactiles*, include fused mixtures of fragmented rock and melted material, plus glass-rich ejecta that resemble volcanic bombs. In some cases, a very dense form of quartz and minute diamonds are found. These high-pressure minerals indicate shock metamorphism. The best-known impactite occurrences in North America are at Meteor Crater and at astroblemes in Indiana and Ohio.

FIGURE 7.A Tektites recovered from Nullarbor Plain, Australia. (Photo by Bill Mason/Smithsonian Institution)

Where impact craters are relatively fresh, shock-melted ejecta and rock fragments ring the impact site. Although most material is deposited close to its source, some ejecta can travel great distances. One example is *tektites*, beads of silica-rich glass, some of which have been aerodynamically shaped like teardrops during flight (Figure 7.A). Most tektites are no more than a few centimeters across and are jet-black to dark green or yellowish. In Australia, millions of tektites are strewn over an area seven times the size of Texas. Several such tektite groupings have been identified worldwide, one stretching nearly halfway around the globe.

No tektite falls have been observed, so their origin is not known with certainty. Because tektites are much higher in silica than volcanic glass (obsidian), a volcanic origin is unlikely. Most researchers agree that tektites are the result of impacts of large projectiles.

One hypothesis suggests an extraterrestrial origin for tektites. Asteroids may have struck the Moon with such force that ejecta "splashed" outward fast enough to escape the Moon's gravity. Others argue that tektites are terrestrial, but an objection is that some groupings, such as Australia's, lack an identifiable impact crater. However, the object that produced the Australian tektites might have struck the continental shelf, leaving the remnant crater out of sight below sea level. Evidence supporting a terrestrial origin includes tektite falls in western Africa that appear to be the same age as a crater in the same region.

FIGURE 7.18 Bornite, an ore of copper often associated with contact metamorphism. (Photo by E. J. Tarbuck)

The Chapter in Review

The following statements are intended to help you review the primary objectives presented in this chapter.

- *Metamorphism* is the transformation of one rock type into another. *Metamorphic rocks* form from preexisting rocks (either igneous, sedimentary, or other metamorphic rocks) that have been altered by the agents of metamorphism, which include *heat, pressure*, and *chemically active fluids*. During metamorphism some of the material must remain solid. The changes that occur in the rocks are textural as well as mineralogical.

- Metamorphism most often occurs in one of three settings: (1) when rock is in contact with or near a mass of magma, *contact metamorphism* occurs; (2) during mountain building, where extensive areas of rock undergo *regional metamorphism*; or (3) *along fault zones*. The greatest volume of metamorphic rock is produced during regional metamorphism.

- The three agents of metamorphism include heat, pressure, and chemically active fluids. The mineral makeup of the parent rock determines, to a large extent, the degree to which each metamorphic agent will cause change. Heat is perhaps the most important agent because it provides the energy to drive chemical reactions that result in the recrystallization of minerals. Pressure, like temperature, also increases with depth. When subjected to *stress* exerted by the load above or during mountain building, rock located at great depth is warmed and behaves plastically. Chemically active fluids, most commonly water containing ions in solution, also enhance the metamorphic process by acting as a catalyst and aiding ion migration.

- *The grade of metamorphism is reflected in the texture and mineralogy of metamorphic rocks.* During the metamorphic process, rocks become more dense and, as the degree of metamorphism intensifies, recrystallization can cause mineral crystals to grow larger. Minerals with a sheet or elongated structure often become oriented essentially perpendicular to the direction of compressional force, giving the rock a layered or banded appearance termed *foliation*. Foliation can result in a fine crystalline rock having *slaty cleavage*, such as slate, or in rocks with larger crystals, a scaly appearance called *schistosity*. During high-grade metamorphism, ion migrations can cause

minerals to segregate into bands. Metamorphic rocks with a banded texture are called *gneiss*. Metamorphic rocks composed of only one mineral forming equidimensional crystals are often *nonfoliated*. Most *marble*, the metamorphic equivalent of limestone, is nonfoliated. Further, metamorphism can cause the transformation of low-temperature minerals into high-temperature minerals and, through the introduction of ions from *hydrothermal solutions*, generate new minerals, some of which form economically important metallic ore deposits.

- Common foliated metamorphic rocks include *slate, phyllite*, various types of *schists* (e.g., garnet-mica schist), and *gneiss*. Nonfoliated rocks include *marble* (parent rock—limestone) and *quartzite* (most often formed from quartz sandstone).

- The three geologic environments that metamorphic rocks commonly occur in are (1) in contact with igneous rocks (contact metamorphism), (2) during dynamic episodes associated with mountain building called *regional metamorphism*, or (3) along fault zones. Contact metamorphism occurs when rocks are in contact with igneous bodies and a zone of alteration called an *aureole* forms around the emplaced magma. Most contact metamorphic rocks are fine-grained, dense, tough rocks of various chemical compositions and, since directional pressure is not a major factor, are not generally foliated. Regional metamorphism takes place at considerable depths over an extensive area and is associated with the process of mountain building. A gradation in the intensity of metamorphism usually exists in regional metamorphism, with the intensity of metamorphism (low- to high-grade) reflected in the texture and mineralogy of the rock. In the most extreme metamorphic environments, rocks, called *migmatites*, that fall into a transition zone somewhere between "true" igneous rocks and "true" metamorphic rocks may form. During metamorphism along fault zones, rocks are deformed by ductile flow that generates elongated grains that often give the rock a foliated or lineated appearance. Compared to the other two processes, the quantity of metamorphic rock generated solely by faulting is very small.

- Many of the most important metamorphic ore

deposits are produced by contact metamorphism. Extensive aureoles of metal-rich deposits commonly surround igneous bodies where ions have invaded limestone strata. The most common metallic minerals associated with contact metamorphism are sphalerite (zinc), galena (lead), chalcopyrite (copper), magnetite (iron), and bornite (copper). Of equal economic importance are the metamorphic rocks themselves. In many regions, slate, marble, and quartzite are quarried for a variety of construction purposes.

Key Terms

aureole (p. 146)
contact metamorphism
 (p. 139)
foliation (p. 141)
gneissic texture (p. 141)

hydrothermal solution
 (p. 142)
migmatite (p. 150)
nonfoliated (p. 142)

regional metamorphism
 (p. 139)
rock cleavage (p. 141)
schistosity (p. 141)
shear (p. 140)

slaty cleavage (p. 141)
stress (p. 140)
texture (p. 141)

Questions for Review

1. What is metamorphism? What are the agents that change rocks?

2. What is foliation? Distinguish between *rock cleavage* and *schistosity.*

3. List some changes that might occur to a rock in response to metamorphic processes.

4. Slate and phyllite resemble each other. How might you distinguish one from the other?

5. Each of the following statements describes one or more characteristics of a particular metamorphic rock. For each statement, name the metamorphic rock that is being described.

 a. Calcite-rich and nonfoliated.
 b. Foliated and composed mainly of granular materials.
 c. Represents a grade of metamorphism between slate and schist.
 d. Very fine-grained and foliated; excellent rock cleavage.

 e. Foliated and composed of more than 50 percent platy minerals.
 f. Often composed of alternating bands of light and dark silicate minerals.
 g. Hard, nonfoliated rock resulting from contact metamorphism.

6. Distinguish between contact metamorphism and regional metamorphism. Which creates the greatest quantity of metamorphic rock?

7. What feature would make schist and gneiss easily distinguishable from quartzite and marble?

8. Briefly describe the textural and mineralogical differences among slate, mica schist, and gneiss. Which one of these rocks represents the highest degree of metamorphism?

9. Are migmatites associated with high-grade or low-grade metamorphism?

10. Metamorphic ore deposits are often related to igneous processes. Provide an example.

Testing What You Have Learned

To test your knowledge of the material presented in this chapter, answer the following questions:

Multiple-Choice Questions

1. Which of the following is NOT an agent of metamorphism?
 a. pressure
 b. age
 c. heat
 d. chemically-active fluids
 e. both b. and d.

2. Which one of the following rocks represents the most extreme metamorphic environment?
 a. schist
 b. slate
 c. shale
 d. gneiss
 e. phyllite

3. Rock adjacent to a large magma body can be altered by hot water solutions, called _____ solutions, released during the later stages of crystallization.
   a. temperate   **c.** mineraloid   **e.** hydrothermal
   b. aqua   **d.** somatic

4. The alignment of minerals in a metamorphic rock that gives the rock a layered or banded appearance is termed _____.
   a. foliation   **c.** migration   **e.** hydration
   b. elongation   **d.** orientation

5. During contact metamorphism, a zone of alteration called a(n) _____ (or halo) forms around the emplaced magma.
   a. aureole   **c.** breccia   **e.** zonal
   b. tektite   **d.** folite

6. A metamorphic rock that exhibits mineral segregation due to an extreme temperature-pressure regime is called _____.
   a. nonfoliated   **c.** schist   **e.** phyllite
   b. gneiss   **d.** marble

7. The greatest volume of metamorphic rock is produced by _____ metamorphism.
   a. fault zone   **c.** transform   **e.** contact
   b. regional   **d.** secondary

8. The metamorphic equivalent of limestone is _____.
   a. slate   **c.** marble   **e.** phyllite
   b. gneiss   **d.** quartzite

9. Which one of the following is NOT a mineral produced during metamorphism?
   a. galena   **c.** sphalerite   **e.** staurolite
   b. calcite   **d.** garnet

10. Perhaps the most important agent of metamorphism is _____.
   a. chemical fluids   **c.** lithification   **e.** heat
   b. pressure   **d.** foliation

Fill-In Questions

11. The metamorphic rock _____ is a very fine-grained foliated rock composed of minute mica flakes.
12. When a rock is near a mass of magma, _____ metamorphism takes place.
13. Rocks that fall into a transition between "true" igneous rocks and "true" metamorphic rocks are called _____.
14. A coarse-grained metamorphic rock with a scaly appearance is called _____.
15. With increasing metamorphism, the size of the mineral crystals in a rock often becomes _____.

True/False Questions

16. Metamorphic rocks composed of only one mineral, which forms equidimensional crystals, are as a rule visibly foliated. ___
17. The force exerted on buried rock by the load above is referred to as stress. ___
18. Several metamorphic rocks are economically important. ___
19. The degree of metamorphism is reflected in the texture and mineralogy of metamorphic rocks. ___
20. Marble is a very hard metamorphic rock most often formed from sandstone. ___

Answers

1.b; 2.d; 3.e; 4.a; 5.a; 6.b; 7.b; 8.c; 9.b; 10.e; 11. slate; 12. contact; 13. migmatites; 14. schist; 15. larger; 16.F; 17.T; 18.T; 19.T; 20.F

Mass Wasting

C h a p t e r **8**

Focus on Learning

To assist you in learning the important concepts in this chapter, you will find it helpful to focus on the following questions:

- What is the process of mass wasting?

- What role does mass wasting play in the development of valleys?

- What are the factors that control mass wasting?

- What criteria are used to divide and describe the various types of mass wasting?

- What are the general characteristics of slump, rockslide, mudflow, earthflow, and creep?

This southern California hillside broke loose and destroyed nine homes following heavy rains in March, 1995. (Photo by Rod Rolle/Gamma-Liaison)

arth's surface is never perfectly flat but instead consists of slopes. Some are steep and precipitous; others are moderate or gentle. Some are long and gradual; others are short and abrupt. Some slopes are mantled with soil and covered by vegetation; others consist of barren rock and rubble. Their form and variety are great. Taken together, slopes are the most common elements in our physical landscape. Although most slopes appear to be stable and unchanging, they are not static features because the force of gravity causes material to move downslope. At one extreme, the movement may be gradual and practically imperceptible. At the other extreme, it may consist of a thundering rockfall or avalanche.

Occasionally, news media report the terrifying and often grim details of landslides. For example, on May 31, 1970, a gigantic rock avalanche buried more than 20,000 people in Yungay and Ranrahirca, Peru. There was little warning of the impending disaster; it began and ended in just a matter of a few minutes. The avalanche started 14 kilometers from Yungay, near the summit of the 6700-meter (22,000-foot) Nevados Huascaran, the loftiest peak in the Peruvian Andes. Triggered by the ground motion from a strong offshore earthquake, a huge mass of rock and ice broke free from the precipitous north face of the mountain. After plunging nearly a kilometer, the material pulverized on impact and immediately began rushing down the mountainside, made fluid by trapped air and melted ice.

The initial mass ripped loose additional millions of tons of debris as it roared downhill. Although the material followed a previously eroded gorge, a portion of the debris jumped a 200–300-meter bedrock ridge that had protected Yungay from similar events in the past and buried the entire city. After inundating another town in its path, Ranrahirca, the mass of debris finally reached the bottom of the valley where its momentum carried it across the Rio Santa and tens of meters up the valley wall on the opposite side.

Fortunately, landslide disasters that kill hundreds or thousands of people are rare. However, property damages each year can be substantial. In the United States alone, losses related to mass wasting events exceed a billion dollars annually.

As with many geologic hazards, the tragic rock avalanche in Peru was triggered by a natural event—in this case, an earthquake. In fact, most mass wasting events, whether spectacular or subtle, are the result of circumstances that are completely independent of human activities. In places where mass wasting is a recognized threat, steps can often be taken to control downslope movements or limit the damages that such movements can cause. If the potential for mass wasting goes unrecognized or is ignored, the results can be costly and dangerous. It should also be pointed out that, although most downslope movements occur whether people are present or not, many occurrences each year are aggravated or even triggered by human actions.

Mass Wasting and Landform Development

Landslides are spectacular examples of a common geologic process called mass wasting. **Mass wasting** refers to the downslope movement of rock, regolith, and soil under the direct influence of gravity. It is distinct from the erosional processes that are examined in subsequent chapters because mass wasting does not require a transporting medium.

In the evolution of most landforms, mass wasting is the step that follows weathering. By itself weathering does not produce significant landforms. Rather, landforms develop as the products of weathering are removed from the places where they originate. Once weathering weakens and breaks rock apart, mass wasting transfers the debris downslope, where a stream, acting as a conveyor belt, usually carries it away. Although there may be many intermediate stops along the way, the sediment is eventually transported to its ultimate destination, the sea.

The combined effects of mass wasting and running water produce stream valleys, which are the most common and conspicuous of Earth's landforms. If streams alone were responsible for creating the valleys in which they flow, valleys would be very narrow features. However, the fact that most river valleys are much wider than they are deep is a strong indication of the significance of mass wasting processes in supplying material to streams. This is illustrated by the Grand Canyon (Figure 8.1). The walls of the canyon extend far from the Colorado River owing to the transfer of weathered debris downslope to the river and its tributaries by mass-wasting processes. In this manner, streams and mass wasting combine to modify and sculpture the surface. Of course, glaciers, groundwater, waves, and wind are also important agents in shaping landforms and developing landscapes.

Controls and Triggers of Mass Wasting

Gravity is the controlling force of mass wasting, but several factors play an important role in overcoming inertia and triggering downslope move-

FIGURE 8.1 The walls of the Grand Canyon extend far from the channel of the Colorado River. This results primarily from the transfer of weathered debris downslope to the river and its tributaries by mass-wasting processes. (Photo by Tom Till)

ments. Among these factors are saturation of material with water, oversteepening of slopes, removal of anchoring vegetation, and ground vibrations from earthquakes.

The Role of Water

When the pores in sediment become filled with water, the cohesion among particles is destroyed, allowing them to slide past one another with relative ease. For example, when sand is slightly moist, it sticks together quite well. However, if enough water is added to fill the openings between the grains, the sand will ooze out in all directions. Thus, saturation reduces the internal resistance of materials, which are then easily set in motion by the force of gravity. When clay is wetted, it becomes very slick—another example of the "lubricating" effect of water. Water also adds considerable weight to a mass of material. The added weight in itself may be enough to cause the material to slide or flow downslope.

Oversteepened Slopes

Oversteepening of slopes is another cause of many mass movements. Unconsolidated, granular sand-sized or coarser particles assume a stable slope called the **angle of repose**, the steepest angle at which material remains stable. Depending on the size and shape of the particles, the angle varies from 25 to 40 degrees. The larger, more angular particles maintain the steepest slopes. If the angle is increased, the rock debris will adjust by moving downslope.

Oversteepening is not just important because it triggers movements of unconsolidated granular materials. Oversteepening also produces unstable slopes and mass movements in cohesive soils, regolith, and bedrock. The response will not be immediate, as with loose, granular material, but sooner or later, one or more mass-wasting processes will eliminate the oversteepening and restore stability to the slope.

There are many situations in nature where this takes place. A stream undercutting a valley wall and waves pounding against the base of a cliff are but two familiar examples. Furthermore, through their activities, people often create oversteepened and unstable slopes that become prime sites for mass wasting.

Vegetation

Plants protect against erosion and contribute to the stability of slopes because their root systems bind soil and regolith together. Where plants are lacking, mass wasting is enhanced, especially if slopes are steep and water is plentiful. When anchoring vegetation is removed by forest fires or by people (for timber, farming, or development), surface materials frequently move downslope.

An unusual but well-known example occurred several decades ago on steep slopes near Menton, France. Farmers replaced olive trees, which have deep roots, with a more profitable, but shallow-rooted crop, carnations. When the less stable slope failed, the landslide took eleven lives.

Earthquakes as Triggers

Conditions favoring mass wasting may exist in an area for a long time without movement occurring. An additional factor is sometimes necessary to trigger the movement. Among the more important and dramatic triggers are earthquakes. An earthquake and its aftershocks can dislodge enormous volumes of rock and unconsolidated material. The event in the Peruvian Andes described at the beginning of the chapter is one tragic example. In many areas that are jolted by earthquakes, it is not ground vibrations directly, but landslides and ground subsidence triggered by the vibrations that cause the greatest damage. Figure 8.2 illustrates this effect.

Classification of Mass-Wasting Processes

There is a broad array of different processes that geologists call mass wasting. Four are illustrated in Figure 8.3. Generally, the different types are divided

and described on the basis of the type of material involved, the kind of motion displayed, and the velocity of the movement.

Type of Material

The classification of mass-wasting processes on the basis of material involved in the movement depends upon whether the descending mass began as unconsolidated material or as bedrock. If soil and regolith dominate, we typically see terms such as "debris," "mud," or "earth" used in the description. On the other hand, when a mass of bedrock breaks loose and moves downslope, the term "rock" may be part of the description.

Type of Motion

In addition to characterizing the type of material involved in a mass-wasting event, the way in which the material moves may also be important. Generally, the kind of motion is described as either a fall, a slide, or a flow.

When the movement involves the free-fall of detached individual pieces of any size, it is termed a **fall**. Fall is a common form of movement on slopes that are so steep that loose material cannot remain on the surface. The rock may fall directly to the base of the slope or move in a series of leaps and bounds over other rocks along the way. Many falls result when freeze and thaw cycles or the action of plant roots loosen rock to the point that gravity takes over. Although signs along bedrock cuts on highways warn of falling rock few of us have actually witnessed such an event. However, as Figure 8.4 illustrates, they do indeed occur. In fact, this is the primary way in which **talus slopes** are built and maintained (Figure 8.5). Sometimes falls may trigger

FIGURE 8.2 Various forms of mass wasting can be triggered by earthquakes. This home in Pacific Palisades, California, was destroyed by a landslide triggered by the January 1994 Northridge earthquake. In some cases, damages from earthquake-induced mass wasting are greater than damages caused directly by an earthquake's ground vibrations. (Photo by Chromo Sohm/The Stock Market)

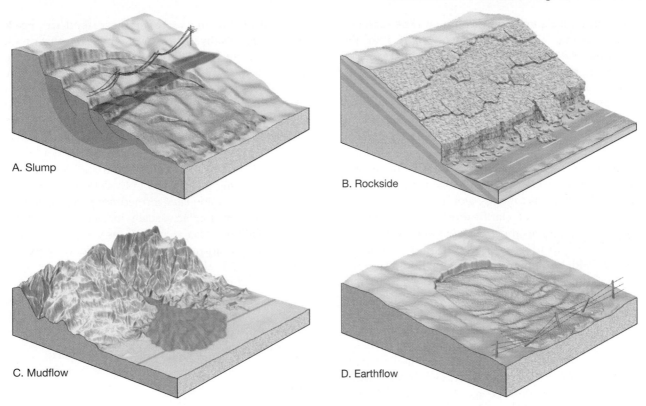

A. Slump

B. Rockside

C. Mudflow

D. Earthflow

FIGURE 8.3 The four processes illustrated here are all considered to be relatively rapid forms of mass wasting. Because material in slumps **A.** and rockslides **B.** move along well-defined surfaces, they are said to move by sliding. By contrast, when material moves downslope as a viscous fluid, the movement is described as a flow. Mudflow **C.** and earthflow **D.** advance downslope in this manner.

FIGURE 8.5 Talus is a slope built of angular rock fragments. Mechanical weathering, especially frost wedging, loosens pieces of bedrock, which then fall to the base of the cliff. With time, a series of steep, cone-shaped accumulations build up at the base of the mountain. (Photo by E. J. Tarbuck)

FIGURE 8.4 Rockfall blocking a highway near Bozeman, Montana. (Photo by Phil Farnes/Photo Researchers)

other forms of downslope movement. For example, recall that the Yungay disaster described at the beginning of the chapter was initiated by a mass of free-falling material that broke from the nearly vertical summit of Nevado Huascaran.

Many mass-wasting processes are described as **slides**. Slides occur whenever material remains fairly coherent and moves along a well-defined surface. Sometimes the surface is a joint, a fault, or a bedding plane that is approximately parallel to the slope. However, in the case of the movement called slump, the descending mass moves along a curved surface of rupture. A note of clarification is appropriate at this point. Sometimes the word *slide* is used as a synonym for the word *landslide*. It should be pointed out that although many people, including geologists, use the term, the word *landslide* has no specific definition in geology. Rather, it should be considered as a popular nontechnical term used to describe all perceptible forms of mass wasting, including those in which sliding does not occur.

The third type of movement common to mass wasting processes is termed **flow**. Flow occurs when material moves downslope as a viscous fluid. Most flows are saturated with water and typically move as lobes or tongues.

Rate of Movement

The event described at the beginning of this chapter clearly involved rapid movement. The rock and debris moved downslope at speeds well in excess of 200 kilometers (125 miles) per hour. This most rapid type of mass movement is termed a **rock avalanche**. Many researchers believe that rock avalanches, such as the one that produced the scene in Figure 8.6, must literally "float on air" as they move downslope. That is, high velocities result when air becomes trapped and compressed beneath the falling mass of debris, allowing it to move as a buoyant, flexible sheet across the surface.

Most mass movements, however, do not move with the speed of a rock avalanche. In fact, a great deal of mass wasting is imperceptibly slow. One process that we will examine later, termed *creep*, results in particle movements that are usually measured in millimeters or centimeters per year. Thus, as you can see, rates of movement can be spectacularly sudden or exceptionally gradual. Although various types of mass wasting are often classified as either rapid or slow, such a distinction is highly subjective because there is a wide range of rates between the two extremes. Even the velocity of a single process at a particular site can vary considerably from one time to another.

Slump

Slump refers to the downward sliding of a mass of rock or unconsolidated material moving as a unit along a curved surface (Figure 8.3A). Usually the slumped material does not travel spectacularly fast nor very far. This is a common form of mass wasting, especially in thick accumulations of cohesive materials such as clay. The rupture surface is characteristically spoon-shaped and concave upward or

FIGURE 8.6 Debris deposited atop Sherman Glacier in Alaska by a rock avalanche. The event was triggered by a tremendous earthquake in March 1964. (Photo by Austin Post, U.S. Geological Survey)

outward. As the movement occurs, a crescent-shaped scarp is created at the head and the block's upper surface is sometimes tilted backwards. Although slump may involve a single mass, it often consists of multiple blocks. Sometimes water is impounded between the base of the scarp and the top of the tilted block. As this water percolates downward along the surface of rupture, it may promote further instability and additional movement.

Slump commonly occurs because a slope has been oversteepened. The material on the upper portion of a slope is held in place by the material at the bottom of the slope. As this anchoring material at the base is removed, the material above is made unstable and reacts to the pull of gravity. One relatively common example is a valley wall that becomes oversteepened by a meandering river. Another is a coastal cliff that has been undercut by wave action at its base. Slumping may also occur when a slope is overloaded, causing internal stress on the material below. This type of slump often occurs where weak, clay-rich material underlies layers of stronger, more resistant rock such as sandstone. The seepage of water through the upper layers reduces the strength of the clay below and slope failure results.

Rockslide

Rockslides occur when blocks of bedrock break loose and slide down a slope (see Figure 8.3B, p. 159). If the material involved is largely unconsolidated, the term **debris slide** is used instead. Such events are among the fastest and most destructive mass movements. Usually rockslides take place in a geologic setting where the rock strata are inclined, or where joints and fractures exist parallel to the slope. When such a rock unit is undercut at the base of the slope, it loses support and the rock eventually gives way.

Sometimes the rockslide is triggered when rain or melting snow lubricates the underlying surface to the point that friction is no longer sufficient to hold the rock unit in place. As a result, rockslides tend to be more common during the spring, when heavy rains and melting snow are most prevalent.

Earthquakes may trigger rockslides and other mass movements. The 1811 earthquake at New Madrid, Missouri, for example, caused slides in an area of more than 13,000 square kilometers (5000 square miles) along the Mississippi River valley. A more recent example occurred on August 17, 1959, when a severe earthquake west of Yellowstone

National Park triggered a massive slide in the canyon of the Madison River in southwestern Montana. In a matter of moments an estimated 27 million cubic meters of rock, soil, and trees slid into the canyon. The debris dammed the river and buried a campground and highway. More than twenty unsuspecting campers perished.

The Gros Ventre River flows west from the northernmost part of the Wind River Range in northwestern Wyoming, through the Grand Teton National Park, and eventually empties into the Snake River. On June 23, 1925, a classic rockslide took place in its valley, just east of the small town of Kelly. In the span of just a few minutes a great mass of sandstone, shale, and soil crashed down the south side of the valley, carrying with it a dense pine forest. The volume of debris, estimated at 38 million cubic meters (50 million cubic yards), created a 70-meter-high dam on the Gros Ventre River (Figure 8.7). Because the river was completely blocked, a lake was created. It filled so quickly that a house that had been 18 meters (60 feet) above the river was floated off its foundation 18 hours after the slide. In 1927, the lake overflowed the dam, partially draining the lake and resulting in a devastating flood downstream.

Why did the Gros Ventre rockslide take place? Figure 8.8 is a diagrammatic cross-sectional view of the geology of the valley. Notice the following points: (1) the sedimentary strata in this area dip (tilt) 15–21 degrees; (2) underlying the bed of sandstone is a relatively thin layer of clay; and (3) at the bottom of the valley the river had cut through much of the sandstone layer. During the spring of 1925, water from heavy rains and melting snow seeped through the sandstone, saturating the clay below. Because much of the sandstone layer had been cut through by the Gros Ventre River, the layer had virtually no support at the bottom of the slope. Eventually the sandstone could no longer hold its position on the wetted clay, and gravity pulled the mass down the side of the valley. The circumstances at this location were such that the event was inevitable.

Mudflow

Mudflow is a relatively rapid type of mass wasting that involves a flowage of debris containing a large amount of water (Figure 8.3C, p. 159). Mudflows are most characteristic of semiarid mountainous regions and are also common on the slopes of some volcanoes. Because of their fluid properties, mudflows follow canyons and stream channels.

FIGURE 8.7 Although the Gros Ventre rockslide occurred in 1925, the scar left on the side of Sheep Mountain is still a prominent feature. (Photo by Stephen Trimble)

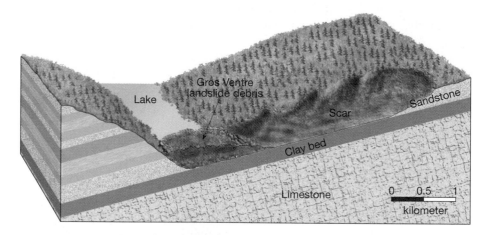

FIGURE 8.8 Cross-sectional view of the Gros Ventre rockslide. The slide occurred when the tilted and undercut sandstone bed could no longer maintain its position atop the saturated bed of clay. (After W. C. Alden, "Landslide and Flood at Gros Ventre, Wyoming," *Transactions* (AIME) 76 (1928): 348)

Mudflows in Semiarid Regions

Although rains in semiarid regions are infrequent, they are typically heavy when they occur. When a cloudburst or rapidly melting mountain snows create a sudden flood, large quantities of soil and regolith are washed into nearby stream channels because there is usually little or no vegetation to anchor the surface material. The end product is a flowing tongue of well-mixed mud, soil, rock, and water. Its consistency may range from that of wet concrete to a soupy mixture not much thicker than muddy water. The rate of flow therefore depends not only on the slope but on the water content as well. When dense, mudflows are capable of carrying or pushing large boulders, trees, and even houses with relative ease.

Mudflows pose a serious hazard to development in dry mountainous areas such as southern California. Here construction of homes on canyon hillsides and the removal of native vegetation by brush fires and other means have increased the frequency of these destructive events (Figure 8.9). Moreover, when a mudflow reaches the end of a steep, narrow canyon, it spreads out, covering the area beyond the mouth of the canyon with a mixture of wet debris. This material contributes to the buildup of fanlike deposits at canyon mouths.* The fans are relatively

---

*These structures are called *alluvial fans* and will be discussed in greater detail in Chapters 9 and 12.

FIGURE 8.9 Severe damage resulted when a mudflow buried the lower portion of this house located near the mouth of a canyon in southern California. (Photo by James E. Patterson)

easy to build on, often have nice views, and are close to the mountains; thus, like the nearby canyons, many have become preferred sites for development. Because mudflows occur only sporadically, the public is often unaware of the potential hazard of such sites. This ignorance has led to many serious problems.

Lahars

Mudflows are also common on the slopes of some volcanoes, in which case they are termed **lahars**. The word originated in Indonesia, a volcanic region that has experienced many of these often destructive events. Lahars result when highly unstable layers of ash and debris become saturated with water and flow down steep volcanic slopes, generally following existing stream channels. Some are initiated when heavy rainfalls erode volcanic deposits. Others are triggered when large volumes of ice and snow are suddenly melted by heat flowing to the surface from within the volcano or by the hot gases and near molten debris emitted during a violent eruption.

When Mount St. Helens erupted in May 1980, several lahars were created. The flows and accompanying floods raced down the valleys of the north and south forks of the Toutle River at speeds that were often in excess of 30 kilometers per hour. Fortunately, the affected area was not densely settled. Nevertheless, more than 200 homes were destroyed or severely damaged. Most bridges met a similar fate.

In November 1985, lahars were produced when Navado del Ruiz, a 5300-meter (17,400-foot) volcano in the Andes Mountains of Colombia erupted. The eruption melted much of the snow and ice that capped the uppermost 600 meters of the peak, producing torrents of hot viscous mud, ash, and debris. The lahars moved outward from the volcano, following the valleys of three rain-swollen rivers that radiate from the peak. The flow that moved down the valley of the Lagunilla River was the most destructive, devastating the town of Armero, 48 kilometers from the mountain. Most of the more than 25,000 deaths caused by the event occurred in this once-thriving agricultural community. Death and property damage also occurred in thirteen other villages within the 180-square-kilometer disaster area. Although a great deal of pyroclastic material was explosively ejected from Nevado del Ruiz, it was the lahars triggered by this eruption that made this such a devastating natural disaster. In fact, it was the worst volcanic disaster since 28,000 people died following the 1902 eruption of Mount Pelée on the Caribbean island of Martinique.*

Earthflow

Unlike mudflows, which are usually confined to channels in semiarid regions, **earthflows** most often form on hillsides in humid areas during times of heavy precipitation or snowmelt (see Figure 8.3D, p. 159). When water saturates the soil and regolith on a hillside, the material may break away, leaving a scar on the slope and forming a tongue- or teardrop-shaped mass that flows downslope (Figure 8.10). The materials most commonly involved are rich in clay and silt and contain only small proportions of sand and coarser particles. Earthflows range in size from bodies a few meters long, a few meters wide, and less than a meter deep to masses more than a kilometer long, several hundred meters wide, and more than 10 meters deep. Because earthflows are quite viscous, they generally move at slower rates than the more fluid mudflows described in the preceding section. They are characterized by a slow and persistent movement and may remain active for periods ranging from days to years. Depending on the steepness of the slope and the material's consistency, measured velocities range from less than 1 millimeter per day up to several meters per day. Over the time span that

*A discussion of the Mount Pelée eruption can be found in the section on composite cones in Chapter 4.

FIGURE 8.10 This small, tongue-shaped earthflow occurred on a newly formed slope along a recently constructed highway. It formed in clay-rich material following a period of heavy rain. Notice the small slump at the head of the earthflow. (Photo by E. J. Tarbuck)

earthflows are active, movement is typically faster during wet periods than during drier times. In addition to occurring as isolated hillside phenomena, earthflows commonly take place in association with large slumps. In this situation, they may be seen as tonguelike flows at the base of the slump block.

A special type of earthflow, known as *liquefaction*, sometimes occurs in association with earthquakes. Porous, clay- to sand-sized sediments that are saturated with water are most vulnerable. When shaken suddenly, the grains lose cohesion and the ground flows. Liquefaction can cause buildings to sink or tip on their sides and underground storage tanks and sewer lines to float upward. To say the least, damage can be substantial.

Slow Movements

Movements such as rockslides, rock avalanches, and lahars are certainly the most spectacular and catastrophic forms of mass wasting. As these events have been known to kill thousands, they deserve intensive study so that, through more effective prediction, timely warnings and better controls can help save lives. However, because of their large size and spectacular nature, they give us a false impression of their importance as a mass-wasting process. Indeed, sudden movements are responsible for moving less material than the slower and far more subtle action of creep. Whereas rapid types of mass wasting are characteristic of mountains and steep hillsides, creep can take place on gentle slopes and is thus much more widespread.

Creep

Creep is a type of mass wasting that involves the gradual downhill movement of soil and regolith. One of the primary causes of creep is the alternate expansion and contraction of surface material caused by freezing and thawing or wetting and drying. As shown in Figure 8.11, freezing or wetting lifts particles at right angles to the slope, and thawing or drying allows the particles to fall back to a slightly lower level. Each cycle therefore moves the material a short distance downhill. Creep may also be initiated if the ground becomes saturated with water. Following a heavy rain or snowmelt, a waterlogged soil may lose its internal cohesion, allowing gravity to pull the material downslope. Because creep is imperceptibly slow, the process cannot be observed in action. What can be observed, however, are the effects of creep. Creep causes fences and utility poles to tilt and retaining walls to be displaced.

Solifluction

Solifluction is a form of mass wasting that is common in regions underlain by **permafrost**. Permafrost refers to the permanently frozen ground that occurs in association with Earth's harsh tundra and ice cap climates (see Box 8.1). Solifluction may be regarded as a form of creep in which unconsolidated, water-saturated material gradually moves downslope. Solifluction occurs in a zone above the permafrost called the *active layer*, which thaws in summer and refreezes in winter. During the summer season, water is unable to percolate into the impervious permafrost layer below. As a result, the active layer

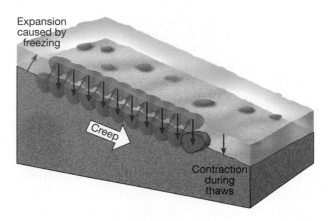

FIGURE 8.11 The repeated expansion and contraction of the surface material causes a net downslope migration of rock particles—a process called *creep.*

Box 8.1   The Sensitive Permafrost Landscape

Many of the mass-wasting events described in Chapter 8 had sudden and disastrous impacts on people. When the activities of people cause ice contained in permanently frozen ground to melt, the impact is more gradual and less deadly. Nevertheless, because permafrost regions are sensitive and fragile landscapes, the scars resulting from poorly planned actions can remain for generations.

Permanently frozen ground, known as *permafrost*, occurs in regions where summertime temperatures do not get sufficiently high for long enough periods to melt more than a shallow surface layer. Deeper ground remains frozen throughout the year. Permafrost covers more than 80 percent of Alaska and about 50 percent of Canada as well as a substantial portion of northern Siberia.

When changes occur in the surface environment, such as the clearing of the insulating vegetation mat or the building of roads and other structures, the delicate thermal balance is disturbed, and thawing of the permafrost can result (Figure 8.A). This thawing, in turn, produces unstable ground that may be susceptible to slides, slumps, subsidence, and severe frost heaving. The many environmental problems stemming from human activities in the Arctic during and following World War II demonstrated the need for a thorough understanding of the nature of permafrost.

As Figure 8.B illustrates, when a heated structure is built directly on permafrost that contains a high proportion of ice, thawing can cause the

FIGURE 8.A When a rail line was built across this permafrost landscape in Alaska, the ground subsided. (Photo by Lynn A. Yehle, U.S. Geological Survey)

building to sink into the soggy material beneath the foundation. One solution is to place buildings and other structures on piles. Such piles allow subfreezing air to circulate between the floor of the building and the soil and therefore cause minimal disturbance to the frozen ground.

When oil was discovered on Alaska's North Slope, many people were concerned about the building of a pipeline linking the oil fields of Prudhoe Bay to the ice-free port of Valdez 1300 kilometers to the south. There was serious concern about the impact of such a massive project on the sensitive permafrost environment. Many were also worried about the effects of possible oil spills after the pipeline was completed.

Because the oil that was to move through the pipeline had to be hot (about 60°C) in order to flow properly, special engineering procedures had to be developed to minimize thawing. The procedures that were used included insulating the pipe to reduce heat flow, elevating portions of the pipeline above ground level in areas of ice-rich permafrost, and, in some instances, placing cooling devices in the ground. The Alaska pipeline is clearly one of the most complex and costly projects ever built in the Arctic tundra. Detailed studies and careful engineering helped minimize adverse effects resulting from the disturbance of frozen ground.

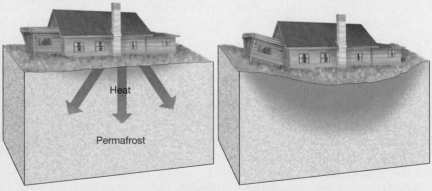

FIGURE 8.B This building, located south of Fairbanks, Alaska, subsided because of thawing permafrost. Notice that the right side, which was heated, settled much more than the unheated porch on the left.

becomes saturated and slowly flows. The process can occur on slopes as gentle as 2–3 degrees. Where there is a well-developed mat of vegetation, a solifluction sheet may move in a series of well-defined lobes or as a series of partially overriding folds (Figure 8.12).

FIGURE 8.12 Solifluction lobes northeast of Fairbanks, Alaska. Solifluction occurs when the active layer thaws in summer. (Photo by James E. Patterson)

The Chapter in Review

The following statements are intended to help you review the primary objectives presented in this chapter.

- *Mass wasting* refers to the downslope movement of rock, regolith, and soil under the direct influence of gravity. In the evolution of most landforms, mass wasting is the step that follows weathering. The combined effects of mass wasting and erosion by running water produce stream valleys.

- *Gravity is the controlling force of mass wasting.* Other factors that play an important role in overcoming inertia and triggering downslope movements are saturation of the material with water and oversteepening of slopes beyond the *angle of repose.*

- The various processes included under the name of mass wasting are divided and described on the basis of (1) the type of material involved (debris, mud, earth, or rock); (2) the type of motion (fall, slide, or flow); and (3) the rate of movement (rapid or slow).

- The various kinds of mass wasting include the fast forms called *slump*, the downward sliding of a mass of rock or unconsolidated material moving as a unit along a curved surface; *rockslide*, blocks of bedrock breaking loose and sliding downslope; *mudflow*, a relatively rapid flow of debris containing a large amount of water; and *earthflow*, an unconfined flow of saturated, clay-rich soil that most often occurs on a hillside in a humid area following heavy precipitation or snowmelt. Slow movements include *creep*, the gradual downhill movement of soil and regolith, and *solifluction*, a form of mass wasting that is common in regions underlain by *permafrost* (permanently frozen ground associated with tundra and ice cap climates).

Key Terms

angle of repose (p. 157)
creep (p. 164)
debris slide (p. 161)
earthflow (p. 163)

fall (p. 158)
flow (p. 160)
lahar (p. 163)
mass wasting (p. 156)

mudflow (p. 161)
permafrost (p. 164)
rock avalanche (p. 160)
rockslide (p. 161)

slide (p. 160)
slump (p. 160)
solifluction (p. 164)
talus slope (p. 158)

Questions for Review

1. Describe how mass-wasting processes contribute to the development of stream valleys.
2. What is the controlling force of mass wasting?
3. How does water affect mass wasting?
4. Describe the significance of the angle of repose.
5. Distinguish among fall, slide, and flow.
6. Why can rock avalanches move at such great speeds?
7. Slump and rockslide both move by sliding. In what ways do these processes differ?
8. What factors led to the massive rockslide at Gros Ventre, Wyoming?

9. Compare and contrast mudflow and earthflow.
10. Describe the mass wasting that occurred at Mount St. Helens during its active period in 1980 and at Nevado del Ruiz in 1985.
11. Describe the mechanism that leads to the slow downslope movement called creep.
12. Why is solifluction only a summertime phenomenon?
13. What is permafrost? What areas in the Northern Hemisphere are affected?

Testing What You Have Learned

To test your knowledge of the material presented in this chapter, answer the following questions:

Multiple-Choice Questions

1. Which one of the following is NOT a kind of motion used to describe mass wasting processes?
   a. flow   **c.** fall
   b. roll   **d.** slide

2. The permanently frozen ground that occurs in association with Earth's harsh tundra and ice caps is referred to as _____.
   a. solifluction   **c.** permafrost   **e.** a glacier
   b. lahar   **d.** slump

3. The type of mass wasting that involves the gradual downhill movement of soil and regolith is _____.
   a. earthflow   **c.** rockslide   **e.** slump
   b. mudslide   **d.** creep

4. Which one of the following is the controlling force of mass wasting?
   a. rock type   **c.** temperature   **e.** gravity
   b. climate   **d.** clay content

5. The most rapid of mass wasting events are _____.
   a. slumps   **c.** earthflows   **e.** rock avalanches
   b. mudflows   **d.** rockslides

6. The combined effects of _____ _____ and running water produce stream valleys.
   a. contact metamorphism   **c.** soil formation   **e.** chemical weathering
   b. normal faulting   **d.** mass wasting

7. When a mass of material moves downslope along a curved surface the process is called _____.
   a. slump   **c.** mudflow   **e.** earthflow
   b. rockslide   **d.** creep

8. The steepest angle at which material remains stable is called the angle of _____.
   a. repose   **c.** stability   **e.** slope
   b. rest   **d.** talus

9. The most widespread form of mass wasting is _____.
   a. rockslide   **c.** earthflow   **e.** solifluction
   b. mudflow   **d.** creep

10. The imperceptibly slow mass wasting process that causes fences and utility poles to tilt is called
   _____.
   a. creep   **c.** rockslide   **e.** debris slide
   b. mudflow   **d.** slump

Fill-In Questions

11. The transfer of rock material downslope under the influence of gravity is referred to as _____
   _____.

12. A(n) _____ occurs when blocks of bedrock break loose and slide down a slope.

13. A(n) _____ is a mudflow that occurs on the slope of a volcano when unstable layers of ash and debris become saturated and flow.

14. Permanently frozen ground, called _____, occurs in Earth's harsh tundra and ice cap climates.

15. The downward slipping of a mass of rock or unconsolidated material moving as a unit along a curved surface is called _____.

True/False Questions

16. A large pile of angular rock fragments that forms at the base of a steep slope is called a talus slope. ___

17. One of the primary causes of creep is the alternate expansion and contraction of surface material caused by freezing and thawing or wetting and drying. ___

18. The downward movement of incoherent material as a viscous fluid is termed slide. ___

19. Saturating the pore spaces of soil and regolith with water will usually decrease the likelihood of downslope movement. ___

20. Solifluction is usually associated with humid, tropical climates. ___

Answers

1.b; 2.c; 3.d; 4.e; 5.e; 6.d; 7.a; 8.a; 9.d; 10.a; 11. mass wasting; 12. rockslide; 13. lahar; 14. permafrost; 15. slump; 16.T; 17.T; 18.F; 19.F; 20.F

Running Water

Focus on Learning

To assist you in learning the important concepts in this chapter, you will find it helpful to focus on the following questions:

- What is the hydrologic cycle? What is the source of energy that powers the cycle?

- What are the factors that determine the velocity of water in a stream?

- How does urbanization affect the discharge of a stream?

- In what way is base level related to a stream's ability to erode?

- The work of a stream includes what three processes?

- What are the two general types of stream valleys and some features associated with each?

- What are some common drainage patterns produced by streams?

- What are the three sequential stages into which the evolution of a valley is often divided?

Kepler Cascades on the Firehole River, Yellowstone National Park, Wyoming. (Photo by Carr Clifton)

The amount of water on Earth is immense, an estimated 1.36 billion cubic kilometers (326 million cubic miles). Of this total, the vast bulk—97.2 percent—is part of the world ocean. Ice sheets and glaciers account for another 2.15 percent, leaving only 0.65 percent to be divided among lakes, streams, subsurface water, and the atmosphere (Figure 9.1). Although the percentages of Earth's total water found in each of the latter sources is but a small fraction of the total inventory, the absolute quantities are great.

Earth as a System: The Hydrologic Cycle

All the rivers run into the sea; yet the sea is not full; unto the place from whence the rivers come, thither they return again.

(Ecclesiastes 1:7)

As the perceptive writer of Ecclesiastes indicated, water is continually on the move, from the ocean to the land and back again in an endless cycle.

The water found in each of the reservoirs depicted in Figure 9.1 does not remain in these places indefinitely. Water can readily change from one state of matter (solid, liquid, or gas) to another at the temperatures and pressures that occur at Earth's surface. Therefore, water is constantly moving among the hydrosphere, the atmosphere, the solid Earth, and the biosphere. This unending circulation of Earth's water supply is called the **hydrologic**

cycle. The cycle shows us many critical interrelationships among different parts of the Earth system.

The hydrologic cycle is a gigantic worldwide system powered by energy from the Sun in which the atmosphere provides the vital link between the oceans and continents (Figure 9.2). Water evaporates into the atmosphere from the ocean and to a much lesser extent from the continents. Winds transport this moisture-laden air, often great distances, until conditions cause the moisture to condense into clouds, and precipitation to fall. The precipitation that falls into the ocean has completed its cycle and is ready to begin another. The water that falls on the continents, however, must make its way back to the ocean.

What happens to precipitation once it has fallen on land? A portion of the water soaks into the ground (called **infiltration**), slowly moving downward, then laterally, finally seeping into lakes, streams, or directly into the ocean. When the rate of rainfall exceeds Earth's ability to absorb it, the surplus water flows over the surface into lakes and streams, a process called **runoff**. Much of the water that infiltrates or runs off eventually returns to the atmosphere because of evaporation from the soil, lakes, and streams. Also, some of the water that infiltrates the ground surface is absorbed by plants, which then release it into the atmosphere. This process is called **transpiration**. Each year a field of crops may transpire the equivalent of a water layer 60 centimeters (2 feet) deep over the entire field. The same area of trees may pump twice this amount into the atmosphere. Because we cannot clearly distinguish between the amount of water that is evaporated and the amount that is transpired by plants, the term *evapotranspiration* is often used for the combined effect.

When precipitation falls in very cold areas—at high elevations or high latitudes—the water may not immediately soak in, run off, or evaporate. Instead, it may become part of a snowfield or a glacier. In this way, glaciers store large quantities of water on land. If present-day glaciers were to melt and release all their water, sea level would rise by several tens of meters. This would submerge many heavily populated coastal areas. As we shall see in Chapter 11, over the past two million years, huge ice sheets have formed and melted on several occasions, each time changing the balance of the hydrologic cycle.

Figure 9.2 also shows Earth's overall *water balance*, or the volume of water that passes through each part of the cycle annually. The amount of water vapor in the air at any one time is just a tiny fraction of Earth's total water supply. But the *absolute* quantities that are cycled through the atmosphere over a 1-year

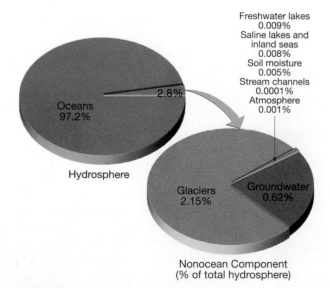

FIGURE 9.1 Distribution of Earth's water.

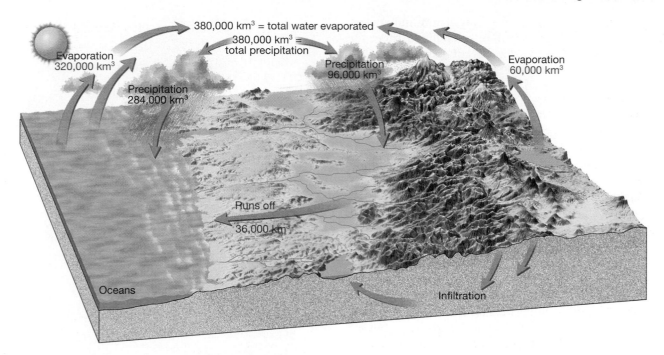

380,000 km³ = total water evaporated

380,000 km³ = total precipitation

Evaporation 320,000 km³

Precipitation 284,000 km³

Precipitation 96,000 km³

Evaporation 60,000 km³

Runs off 36,000 km³

Oceans

Infiltration

FIGURE 9.2 Earth's water balance. About 320,000 cubic kilometers of water are evaporated each year from the oceans, while evaporation from the land (including lakes and streams) contributes 60,000 cubic kilometers of water. Of this total of 380,000 cubic kilometers of water, about 284,000 cubic kilometers fall back to the ocean, and the remaining 96,000 cubic kilometers fall on Earth's land surface. Because 60,000 cubic kilometers of water evaporate from the land, 36,000 cubic kilometers of water remain to erode the land during the journey back to the oceans.

period are immense—some 380,000 cubic kilometers—enough to cover Earth's entire surface to a depth of about 1 meter (39 inches).

It is important to know that the water cycle is *balanced*. Because the total amount of water vapor in the atmosphere remains about the same, the average annual precipitation worldwide must be equal to the quantity of water evaporated. However, for all of the continents taken together, precipitation exceeds evaporation. Conversely, over the oceans, evaporation exceeds precipitation. As the level of the world ocean is not dropping, the system must be in balance. In Figure 9.2, the 36,000 cubic kilometers of water that annually runs off from the land to the ocean causes enormous erosion. In fact, this immense volume of moving water is *the single most important agent sculpting Earth's land surface*.

To summarize, the hydrologic cycle represents the continuous movement of water from the oceans to the atmosphere, from the atmosphere to the land, and from the land back to the sea. The wearing down of Earth's land surface is largely attributable to the last of these steps and is the primary focus of the remainder of this chapter.

Running Water

Of all the geological processes, running water may have the greatest impact on people. We depend on rivers for energy, travel, and irrigation. Their fertile floodplains have fostered human progress since the dawn of civilization. As the dominant agent of landscape alteration, streams have shaped much of our physical environment.

Although we have always depended to a great extent on running water, its source eluded us for centuries. Not until the sixteenth century did we realize that streams were supplied by surface runoff and underground water, which ultimately had their sources as rain and snow.

Runoff initially flows in broad, thin sheets across the ground, appropriately termed **sheet flow**. The amount of water that runs off in this manner rather than sinking into the ground depends upon the **infiltration capacity** of the soil. Infiltration capacity is controlled by many factors, including: (1) the intensity and duration of the rainfall, (2) the prior wetted condition of the soil, (3) the soil texture, (4) the slope of the land, and (5) the nature of the vegetative cover. When the soil becomes saturated, sheet flow commences as a layer only a few millimeters thick. After flowing as a thin, unconfined sheet for only a short distance, threads of current typically develop and tiny channels called **rills** begin to form and carry the water to a stream.

To some, the term *stream* implies relative size. That is to say, streams are thought of as being larger than creeks or brooks but smaller than rivers. In geology, however, this is not the case. Here the word **stream** is used to denote channelized flow of any size, from the smallest trickle to the mightiest river. It should be pointed out, however, that although the terms *river* and *stream* are used interchangeably, the term *river* is often preferred when describing a main stream into which several tributaries flow.

The remainder of this chapter will concentrate on that part of the hydrologic cycle in which the water moves in stream channels. The discussion will deal primarily with streams in humid regions. Streams are also important in arid landscapes, but we will examine that in Chapter 12, "Deserts and Wind."

Streamflow

Water makes its way to the sea under the influence of gravity. The time required for the journey depends on the velocity of the stream. Velocity is the distance that water travels in a unit of time. Water in some sluggish streams travels at less than 0.8 kilometer (0.5 mile) per hour, whereas water in a few rapid streams reaches speeds as high as 32 kilometers (20 miles) per hour. Velocities are measured at gauging stations. Along straight stretches, the highest velocities are near the center of the channel just below the surface, where friction is lowest (Figure 9.3). But when a stream curves, its zone of maximum speed shifts toward its outer bank.

The ability of a stream to erode and transport materials depends on its velocity. Even slight variations in velocity can lead to significant changes in how much sediment can be transported by the water. Several factors determine the velocity of a stream, including (1) gradient; (2) shape, size, and roughness of the channel; and (3) discharge.

Gradient Certainly one of the most obvious factors controlling stream velocity is the **gradient**, or slope, of a stream channel. Gradient is typically expressed as the vertical drop of a stream over a fixed distance. Gradients vary considerably from one stream to another as well as along the course of a given stream.

Portions of the lower Mississippi River, for example, have gradients of 10 centimeters per kilometer and less. By way of contrast, some steep mountain stream channels decrease in elevation at a rate of

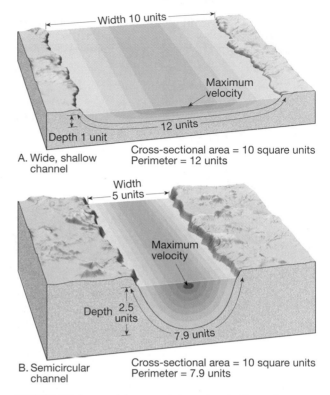

FIGURE 9.3 In a straight stream channel water is moving fastest in the center, just below the surface. This figure also illustrates the influence of channel shape on velocity. Although the cross-sectional area of these channels is the same, the semicircular channel has less water in contact with the channel, and hence less frictional drag. As a result, the water will flow more rapidly in this channel, all other factors being equal.

more than 40 meters per kilometer, 400 times more abruptly than the lower Mississippi. The steeper the gradient, the more energy available for streamflow. If two streams were identical in every respect except gradient, the stream with the higher gradient would obviously have the greater velocity.

Channel Characteristics The *cross-sectional shape* of a channel determines the amount of water in contact with the channel and hence affects the frictional drag. The most efficient channel is one with the least perimeter for its cross-sectional area. Figure 9.3 compares two channel shapes. Although the cross-sectional area of both is identical, the semicircular shape has less water in contact with the channel and therefore less frictional drag. As a result, if all other factors are equal, the water will flow more rapidly in the semicircular channel.

The size and roughness of the channel also affect the amount of friction. An increase in the size of a channel reduces the ratio of perimeter to cross-sectional area and therefore increases the efficiency of flow. The effect of roughness is obvious. A smooth channel promotes a more uniform flow, whereas an irregular channel filled with boulders creates enough turbulence to significantly retard the stream's forward motion.

Discharge The **discharge** of a stream is the amount of water flowing past a certain point in a given unit of time (see Box 9.1). This is usually measured in cubic meters per second or cubic feet per second. Discharge is determined by multiplying a stream's cross-sectional area by its velocity:

Discharge (m³/second) =
Channel width (meters) × Channel depth (meters)
× Velocity (meters/second)

The largest river in North America, the Mississippi, discharges an average of 17,300 cubic meters per second. Although this is a huge quantity of water, it is nevertheless dwarfed by the mighty Amazon, the world's largest river. Draining an area that is nearly three-quarters the size of the conterminous United States and that averages about 200 centimeters of rain per year, the Amazon discharges 12 times more water than the Mississippi. In fact, it has been estimated that the flow of the Amazon accounts for about 15 percent of all the fresh water discharged into the ocean by all of the world's rivers. Just one day's discharge would supply the water needs of New York City for 9 years!

The discharges of most rivers are far from constant. This is true because of such variables as rainfall and snowmelt. When discharge changes, then the factors noted earlier must also change. When discharge increases, the width or depth of the channel must increase or the water must flow faster, or some combination of these factors must change. Indeed, measurements show that when the amount of water in a stream increases, the width, depth, and velocity all increase in an orderly fashion (Figure 9.4). To handle the additional water, the stream will increase the size of its channel by widening and deepening it. As we saw earlier, when the size of the channel increases, proportionally less of the water is in contact with the bed and banks of the channel. This means that friction, which acts to retard the flow, is reduced. The less friction, the more swiftly the water will flow.

Changes Downstream

One useful way of studying a stream is to examine its **longitudinal profile**. Such a profile is simply a cross-sectional view of a stream from its source area (called the **head** or **headwaters**) to its **mouth**, the point downstream where the river empties into another water body. By examining Figure 9.5, you

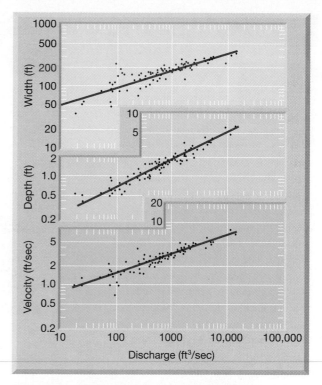

FIGURE 9.4 Relationship of width, depth, and velocity to discharge of the Powder River at Locate, Montana. As discharge increases, width, depth, and velocity all increase in an orderly fashion. (From L. B. Leopold and Thomas Maddock, Jr., *U.S. Geological Survey Professional Paper 252*, 1953)

can see that the most obvious feature of a typical longitudinal profile is a constantly decreasing gradient from the head to the mouth. Although many local irregularities may exist, the overall profile is a smooth concave-upward curve.

The longitudinal profile shows that the gradient decreases downstream. To see how other factors change in a downstream direction, observations and measurements must be made. When data are collected from successive gauging stations along a river, they show that discharge increases toward the mouth. This should come as no surprise because, as we move downstream, more and more tributaries contribute water to the main channel. In the case of the Amazon, for example, about a thousand tributaries join the main river along its 6500-kilometer course across South America. Furthermore, in most humid regions, additional water is continually being added from the groundwater supply. Because this is the case, the width, depth, and velocity all must change in response to the increased volume of water carried by the stream. Indeed, the downstream changes in these variables have been shown to vary in a manner similar to what occurs when discharge increases at one place; that is, width, depth, and velocity all increase systematically.

Box 9.1 The Effect of Urbanization on Discharge

When rains occur, stream discharge increases. If the rains are sufficiently heavy, the ability of the channel to contain the discharge is exceeded, and water spills over the banks as a flood. Floods are natural events that should be expected. However, when cities are built, the magnitude and frequency of flooding increases. The top portion of Figure 9.A is a hypothetical hydrograph that shows the time relationship between a rainstorm and the occurrence of flooding. Notice that the water level in the stream does not rise at the onset of precipitation because time is needed for water to move from the place where it fell to the stream. This time difference is called the *lag time*.

When an area changes from being predominantly rural to largely urban, streamflow is affected. The effect of urbanization on streamflow is illustrated by the bottom hydrograph in Figure 9.A. Notice that after urbanization the peak discharge during a flood is greater, and that the lag time between precipitation and flood peak is shorter than before urbanization. The explanation for this effect is relatively simple. The construction of streets, parking lots, and buildings covers over the ground that once soaked up water. Thus, less water infiltrates the ground, and the rate and amount of runoff increase. Further, because much less water soaks into the ground, the low-water (dry-season) flow in urban streams, which is maintained by the seepage of

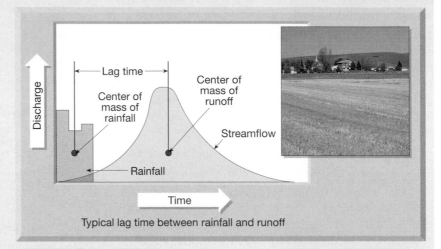

Typical lag time between rainfall and runoff

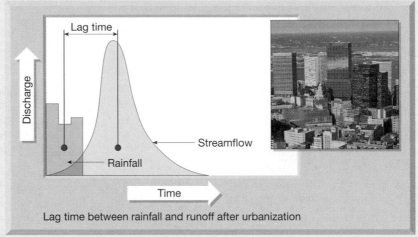

Lag time between rainfall and runoff after urbanization

FIGURE 9.A When an area changes from rural to urban, the lag time between rainfall and flood peak is shortened. The flood peak is also higher following urbanization. (After L. B. Leopold, U.S. Geological Survey)

groundwater into the channel, is greatly reduced. As one might expect, the magnitude of these effects is a function of the percentage of land that is covered by impermeable surfaces.

Urbanization is just one example of human interference with streams.

There are many other ways that land use inadvertently influences the flow of streams and the work they carry out. Moreover, there are also many ways by which people intentionally attempt to manipulate and control streams. Some of these are discussed at appropriate points in this chapter.

The observed increase in average velocity that occurs downstream contradicts our intuitive impressions concerning wild, turbulent, mountain streams and wide, placid, rivers in the lowlands. A mental picture that we may have of "old man river just rollin' along" is really correct. The mountain stream has much higher instantaneous, turbulent velocities,

but the water moves vertically, laterally, and actually upstream in some cases (Figure 9.6). Thus, the average flow velocity may be lower than in a wide, placid river just "rollin' along" very efficiently with far less turbulence.

In the headwaters region where the gradient is steepest, the water must flow in a relatively small

FIGURE 9.5 A longitudinal profile is a cross section along the length of a stream. Note the concave-upward curve of the profile, with a steeper gradient upstream and a gentler gradient downstream.

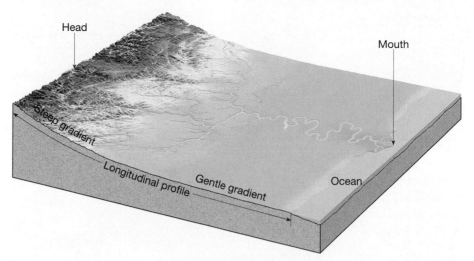

and often boulder-strewn channel. The small channel and rough bed create great drag and inhibit movement by sending water in all directions with almost as much backward motion as forward motion. However, as one progresses downstream, the material on the bed of the stream becomes much smaller, offering less resistance to flow, and the width and depth of the channel increase to accommodate the greater discharge. These factors, especially the wider and deeper channel, permit the water to flow more freely and hence more rapidly.

In summary, we have seen that an inverse relationship exists between gradient and discharge. Where the gradient is high, the discharge is small, and where the discharge is great, the gradient is small. Stated another way, a stream can maintain a higher velocity near its mouth even though it has a lower gradient than upstream because of the greater discharge, larger channel, and smoother bed.

Base Level and Graded Streams

In 1875, John Wesley Powell, the pioneering geologist who first explored the Grand Canyon and later headed the U.S. Geological Survey, introduced the concept that there is a downward limit to stream erosion, which he called *base level*. Although the idea is relatively straightforward, it is nevertheless a key concept in the study of stream activity. **Base level** is defined as the lowest elevation to which a stream can erode its channel. Essentially this is the level at which the mouth of a stream enters the ocean, a lake, or another stream. Base level accounts for the fact that most stream profiles have low gradients near their mouths, because the streams are approaching the elevation below which they cannot erode their beds.

Two general types of base level are recognized. Sea level is considered the **ultimate base level**, because it usually represents the lowest level to which

a stream can erode the land. **Local** or **temporary base levels** include lakes, resistant layers of rock, and main streams that act as base levels for their tributaries. All have the capacity to limit a stream at a certain level.

For example, when a stream enters a lake, its velocity quickly approaches zero and its ability to erode ceases. Thus, the lake prevents the stream from eroding below its level at any point upstream from the lake. However, because the outlet of the lake can cut downward and drain the lake, the lake is only a temporary hindrance to the stream's ability to downcut its channel. In a similar manner, the layer of resistant rock at the lip of the waterfall in Figure 9.7 acts as temporary base level. Until the ledge of

FIGURE 9.6 Rapids are common in mountain streams where the gradient is steep and the channel is rough and irregular. (Photo by Bruce Gaylord/Visuals Unlimited)

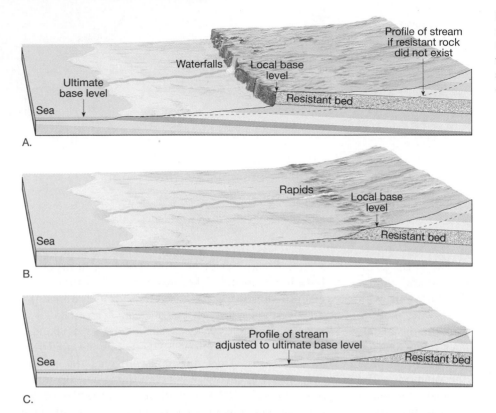

FIGURE 9.7 A resistant layer of rock can act as a local (temporary) base level. Because the durable layer is eroded more slowly, it limits the amount of downcutting upstream.

Within figure A: Ultimate base level, Sea, Waterfalls, Local base level, Profile of stream if resistant rock did not exist, Resistant bed

Within figure B: Sea, Rapids, Local base level, Resistant bed

Within figure C: Sea, Profile of stream adjusted to ultimate base level, Resistant bed

hard rock is eliminated, it will limit the amount of downcutting upstream.

Any change in base level will cause a corresponding readjustment of stream activities. When a dam is built along a stream course, the reservoir that forms behind it raises the base level of the stream (Figure 9.8). Upstream from the dam the stream gradient is reduced, lowering its velocity and, hence, its sediment-transporting ability. The stream, now unable to transport all of its load, will deposit material, thereby building up its channel. This process continues until the stream again has a gradient sufficient to carry its load. The profile of the new channel would be similar to the old, except that it would be somewhat higher.

If, on the other hand, the base level should be lowered, either by uplifting of the land or by a drop in sea level, the stream would again readjust. The stream, now above base level, would have excess energy and downcut its channel to establish a balance with its new base level. Erosion would first progress near the mouth, then work upstream until the stream profile was adjusted along its full length.

The observation that streams adjust their profiles to changes in base level led to the concept of a graded stream. A **graded stream** has the correct slope and other channel characteristics necessary to maintain just the velocity required to transport the material supplied to it. On the average, a graded

system is neither eroding nor depositing material but is simply transporting it. Once a stream has reached this state of equilibrium, it becomes a self-regulating system in which a change in one characteristic causes an adjustment in the others to counteract the effect. Referring again to our example of a stream adjusting to a lowering of its base level, the stream would not be graded while it was cutting its new channel but would achieve this state after downcutting had ceased.

Work of Streams

The work of streams includes erosion, transportation, and deposition. These activities go on simultaneously in all stream channels, even though they are presented individually here.

Erosion

Erosion is the removal of rock and soil. Much of the material carried by streams reaches them through underground water, overland flow, and mass wasting. Streams also create some of their load by eroding their own channels. If a channel is composed of bedrock, most of the erosion is accomplished by the abrasive action of water armed with sediment, a process analogous to sandblasting. Pebbles caught in swirling eddies act like cutting tools and bore circular "potholes" into the channel floor. In channels composed

Work of Streams **177**

FIGURE 9.8 When a dam is built and a reservoir forms, the stream's base level is raised. This reduces the stream's velocity and leads to deposition and a reduction of the gradient upstream from the reservoir.

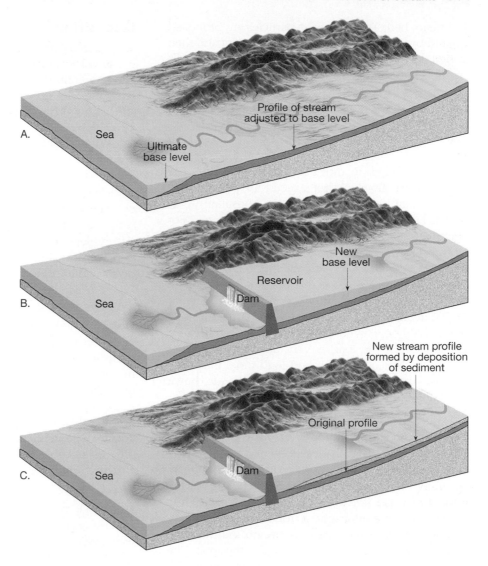

of loose material, considerable lifting and removal can be accomplished by the impact of water alone.

Transportation

Streams transport their load of sediment in three ways: (1) in solution (**dissolved load**); (2) in suspension (**suspended load**); and (3) scooting or rolling along the bottom (**bed load**).

Dissolved Load Most of the *dissolved load* is brought to the stream by groundwater and to a lesser degree it is acquired by dissolving rock along the stream's course. The quantity of material carried in solution is highly variable and depends on climate and the geologic setting. Usually the dissolved load is expressed as parts of dissolved material per million parts of water (parts per million or ppm). Although some rivers may have a dissolved load of 1000 ppm or more, the average figure for the world's rivers is estimated at 115 to 120 ppm. Almost 4 billion metric tons of dissolved mineral matter are supplied to the oceans each year by streams.

Suspended Load Most streams (but not all) carry the largest part of their load in *suspension*. Indeed, the visible cloud of sediment suspended in the water is the most obvious portion of a stream's load. Usually only sand, silt, and clay can be carried this way, but during a flood larger particles are transported as well. Also, during a flood, the total quantity of material carried in suspension increases dramatically, as can be verified by anyone whose home has been a site for the deposition of this material.

Bed Load A portion of a stream's load of solid material consists of sediment too large to be carried in suspension. These coarser particles move along the bottom of the stream and constitute the *bed load*. In terms of the erosional work accomplished by a downcutting stream, the grinding action of the bed load is of great importance.

The particles making up the bed load move along the bottom by rolling, sliding, and saltation. Sediment moving by **saltation** appears to jump or skip along the stream bed. This occurs as particles are propelled upward by collisions or lifted by the current and then carried downstream a short distance until gravity pulls them back to the bed of the stream. Particles that are too large or heavy to move by saltation either roll or slide along the bottom, depending on their shapes.

Unlike the suspended and dissolved loads, which are constantly in motion, the bed load is in motion only intermittently, when the force of the water is sufficient to move the larger particles. The bed load usually does not exceed 10 percent of a stream's total load, although it may constitute up to 50 percent of the total load of a few streams. For example, consider the distribution of the 750 million tons of material carried to the Gulf of Mexico by the Mississippi River each year. Of this total, it is estimated that approximately 67 percent is carried in suspension, 26 percent in solution, and the remaining 7 percent as bed load. Estimates of a stream's bed load should be viewed cautiously, however, because this fraction of the load is very difficult to measure accurately. Not only is the bed load more inaccessible than the suspended and dissolved loads, but it moves primarily during periods of flooding when the bottom of a stream channel is most difficult to study.

Capacity and Competence A stream's ability to carry solid particles is typically described using two criteria. First, the maximum load of solid particles that a stream can transport is termed its **capacity**. The greater the amount of water flowing in a stream (discharge), the greater the stream's capacity for hauling sediment. Second, the **competence** of a stream indicates the maximum particle size that a stream can transport. The stream's velocity determines its competence; the stronger the flow, the larger the particles it can carry in suspension and as bed load. It is a general rule that the competence of a stream increases as the square of its velocity. Thus, if the velocity of a stream doubles, the impact force of the water increases four times; if the velocity triples, the force increases nine times, and so forth. Hence, the large boulders that are often visible during a low-water stage and that seem immovable can, in fact, be transported during flood stage because of the stream's increased competence.

By now it should be clear why the greatest erosion and transportation of sediment occur during floods. The increase in discharge results in greater capacity; the increased velocity produces greater competence. With rising velocity the water becomes more turbulent, and larger and larger particles are set in motion. In the course of just a few days, or perhaps just a few hours, a stream in flood stage can erode and transport more sediment than it does during months of normal flow (Figure 9.9).

Deposition

Whenever a stream slows down, the situation reverses. As its velocity decreases, its competence is reduced and sediment begins to drop out, largest particles first. Each particle size has a *critical settling velocity*. As streamflow drops below the critical settling velocity of a certain particle size, sediment in that category begins to settle out. Thus, stream transport provides a mechanism by which solid particles of various sizes are separated. This process, called **sorting**, explains why particles of similar size are deposited together.

The well-sorted material typically deposited by a stream is called **alluvium**, the general term for any stream-deposited sediment. Many different depositional features are composed of alluvium. Some occur within stream channels, some occur on the valley floor adjacent to the channel, and some exist at the mouth of the stream.

Deltas When a stream enters the relatively still waters of an ocean or lake, its velocity drops abruptly, and the resulting deposits form a **delta** (Figure 9.10). As the delta grows outward, effectively lengthening the river, the gradient of the river continually lessens. This causes the stream to seek a shorter route to base level. Frequently the main channel divides into several smaller ones called **distributaries**. These shifting channels act in an opposite way from tributaries, *distributing* water instead of contributing it. Rather

FIGURE 9.9 The suspended load is clearly visible because it gives this flooding river a brown "muddy" appearance. During floods, both capacity and competence increase. Therefore, the greatest erosion and sediment transport occur during these high-water periods. This muddy torrent washed out a section of Interstate 5 near Coalinga, California, in March 1995. (Photo by AP/Wide World Photos)

than carrying water into the main channel, distributaries carry water away from the main channel. After numerous shifts of the channel, a delta may grow into a rough triangular shape like the Greek letter delta (Δ), for which it is named. Note, however, that many deltas do not exhibit the idealized shape. Differences in the configurations of shorelines and variations in the nature and strength of wave activity result in many shapes.

Many large rivers have deltas extending over thousands of square kilometers. The delta of the Mississippi River is one example. It resulted from the accumulation of huge quantities of sediment derived from the vast region drained by the river and its tributaries. Today, New Orleans rests where there was ocean less than 5000 years ago. Figure 9.11 shows that portion of the Mississippi delta that has been built over the past 5000 to 6000 years. As shown, the delta is actually a series of seven coalescing subdeltas. Each formed when the river left its existing channel in favor of a shorter, more direct path to the Gulf of Mexico. The individual subdeltas interfinger and partially cover one another to produce a very complex structure. The present subdelta, called a *bird-foot* delta because of the configuration of its distributaries, has been built by the Mississippi in the last 500 years.

Natural Levees Some rivers occupy valleys with broad, flat floors and build **natural levees** that parallel their channels on both banks (Figure 9.12). Natural levees are built by successive floods over many years. When a stream overflows its banks, its velocity immediately diminishes, leaving coarse sediment deposited in strips bordering the channel. As the water spreads out over the valley, a lesser amount of fine sediment is deposited over the valley floor. This uneven distribution of material produces the very gentle slope of the natural levee.

The natural levees of the lower Mississippi rise 6 meters (20 feet) above the valley floor. The area behind the levee is characteristically poorly drained for the obvious reason that water cannot flow up the levee and into the river. Marshes called **back swamps** result.

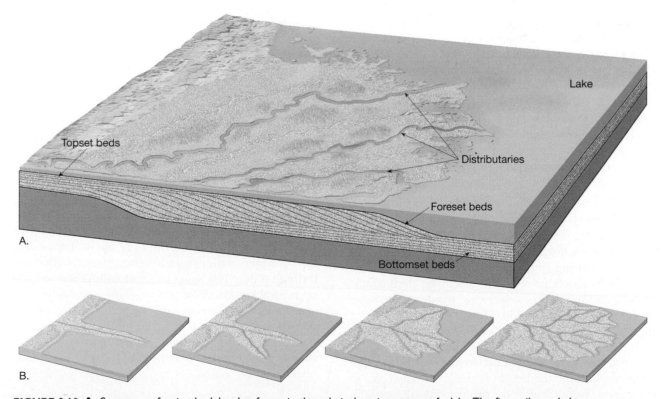

FIGURE 9.10 A. Structure of a simple delta that forms in the relatively quiet waters of a lake. The finer silts and clays will settle out some distance from the mouth into nearly horizontal layers called *bottomset beds*. Prior to the accumulation of bottomset beds, *foreset beds* begin to form. These beds are composed of coarse sediment, which is dropped almost immediately upon entering a lake or ocean, forming sloping layers. The foreset beds are usually covered by thin, horizontal *topset beds* deposited during floodstage. **B.** Growth of a simple delta. As a stream extends its channel, the gradient is reduced. Frequently, during flood stage the river is diverted to a higher-gradient route, forming a new distributary. Old abandoned distributaries gradually fill with sediment and vegetation. (After Ward's Natural Science Establishment, Inc., Rochester, N.Y.)

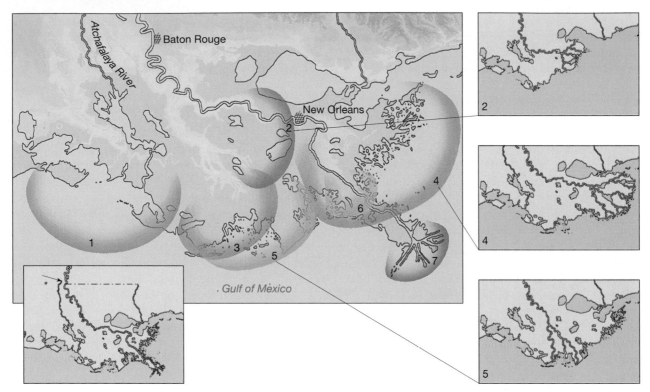

FIGURE 9.11 During the past 5000 to 6000 years, the Mississippi River has built a series of seven coalescing subdeltas. The numbers indicate the order in which the subdeltas were deposited. The present bird-foot delta (number 7) represents the activity of the past 500 years. Without ongoing human efforts, the present course of the Mississippi will shift and follow the path of the Atchafalaya River (see arrows in inset). (After C. R. Kolb and J. R. Van Lopik, *Depositional Environments of the Mississippi River Deltaic Plain*, p. 22. Copyright © 1966 by the Houston Geological Society.)

A tributary stream that cannot enter a river because levees block the way often has to flow parallel to the river until it can breach the levee. Such streams are called **yazoo tributaries** after the Yazoo River, which parallels the Mississippi for over 300 kilometers.

Artificial Levees Sometimes *artificial levees* are built along rivers to control flooding. Artificial levees are usually easy to distinguish from natural levees because their slopes are much steeper. When a river is confined by levees during periods of high water, it deposits material in its channel as the discharge diminishes. This is sediment that otherwise would have been dropped on the floodplain. Thus, each time there is a high flow, deposits are left on the river bed and the bottom of the channel is built up. With the buildup of the bed, less water is required to overflow the original levee. As a result, the height of the levee may have to be raised periodically to protect the floodplain. Moreover, many artificial levees are not built to withstand periods of extreme flooding. For example, levee failures were numerous in the Midwest during the summer of 1993, when the upper Mississippi and many of its tributaries experienced record floods (Figure 9.13).

Stream Valleys

Stream valleys can be divided into two general types. Narrow V-shaped valleys and wide valleys with flat floors exist as the ideal forms, with many gradations between.

Most stream valleys are much broader at the top than is the width of their channel at the bottom. This would not be the case if the only agent responsible for eroding valleys were the streams flowing through them. The sides of most valleys are shaped by a combination of weathering, overland flow, and mass wasting. In some arid regions, where downcutting is rapid and weathering is slow, and in places where rock is particularly resistant, narrow valleys may not be V-shaped but rather may have nearly vertical walls.

Narrow Valleys

The Yellowstone River provides an excellent example of a narrow valley (Figure 9.14). A narrow V-shaped valley indicates that the primary work of the stream has been downcutting toward base level. The most prominent features of a narrow valley are *rapids* and

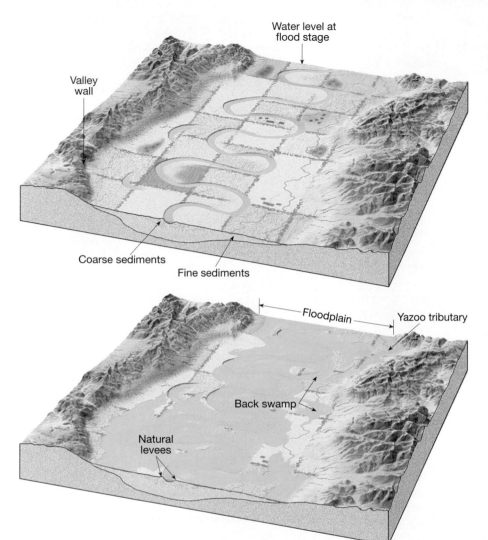

Water level at
flood stage

Valley
wall

Coarse sediments

Fine sediments

Floodplain

Yazoo tributary

Back swamp

Natural
levees

FIGURE 9.12 Natural levees are gently sloping structures that are created by repeated floods. Because the ground next to the stream channel is higher than the adjacent floodplain, back swamps and yazoo tributaries may develop.

FIGURE 9.13 Water rushes through a break in an artificial levee in Monroe County, Illinois. During the record-breaking 1993 Midwest floods, many artificial levees could not withstand the force of the floodwaters. Sections of many weakened structures were overtopped or simply collapsed. (Photo by James A. Finley/AP/Wide World Photos)

FIGURE 9.14 V-shaped valley of the Yellowstone River. The rapids and waterfalls indicate that the river is vigorously downcutting. (Photo by Art Wolfe)

waterfalls. Both occur where the stream profile drops rapidly, a situation usually caused by variations in the erodibility of the bedrock into which a stream channel is cutting. Resistant beds create rapids by acting as a temporary base level upstream while allowing downcutting to continue downstream. Once erosion has eliminated the resistant rock, the stream profile smooths out again. Waterfalls are places where the stream profile makes a vertical drop.

Wide Valleys

Once a stream has cut its channel closer to base level, downward erosion becomes less dominant. At this point more of the stream's energy is directed from side to side. The result is a widening of the valley as the river cuts away at one bank and then at the other (Figure 9.15).

Floodplains The side-to-side cutting of a stream eventually produces a flat valley floor, or **floodplain**. It is appropriately named because the river is confined to its channel, except during flood stage, when it overflows its banks and inundates the floodplain.

When a river erodes laterally, creating a floodplain as just described, it is called an *erosional floodplain.* Floodplains can be depositional in nature as well. *Depositional floodplains* are produced by a

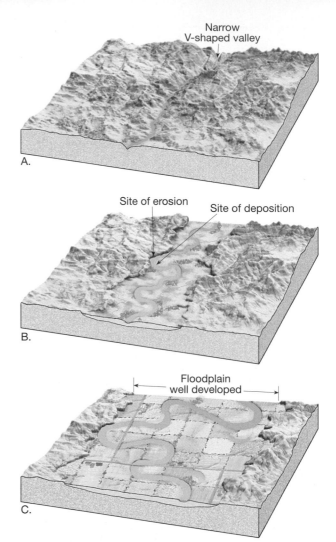

FIGURE 9.15 Stream eroding its floodplain.

major fluctuation in conditions, such as a change in base level. The floodplain in California's Yosemite Valley is one such feature, and was produced when a glacier gouged the former stream valley deeper by about 300 meters (1000 feet). After the glacial ice melted, the stream readjusted to its former base level by refilling the valley with alluvium.

Meanders Streams that flow on floodplains move in sweeping bends called **meanders** (Figure 9.15). Meanders continually change position by eroding sideways and slightly downstream. The sideways movement occurs because the maximum velocity of the stream shifts toward the outside of the bend, causing erosion of the outer bank (Figure 9.16). At the same time, the reduced current at the inside of the meander results in the deposition of coarse sediment, especially sand, called a *point bar.* Thus, by eroding its outer bank and depositing material along its inner bank, a stream moves sideways without changing its channel size.

FIGURE 9.16 Lateral movement of meanders. By eroding its outer bank and depositing material on the inside of the bend, a stream is able to shift its channel.

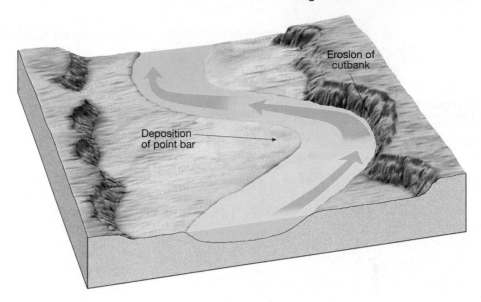

Erosion of cutbank

Deposition of point bar

Because of the slope of the channel, erosion is more effective on the downstream side of a meander. Therefore, in addition to migrating laterally, the bends also gradually migrate down the valley. Sometimes the downstream migration of a meander is slowed when it reaches a more resistant portion of the floodplain. This allows the next meander upstream to overtake it. Gradually the neck of land between the meanders is narrowed. When they get close enough, the river erodes through the narrow neck of land to the next loop (Figure 9.17). The new, shorter channel segment is called a **cutoff** and, because of its shape, the abandoned bend is called an **oxbow lake**.

Artificial Cutoffs One method of flood control involves straightening a channel by creating *artificial cutoffs*. The idea is that by shortening the stream, the gradient and, hence, the velocity are increased. By increasing velocity, the larger discharge associated with flooding can be dispersed more rapidly. Since the early 1930s, the Army Corps of Engineers has created many artificial cutoffs on the Mississippi for the purpose of increasing the channel efficiency and reducing the threat of flooding. In all, the river has been shortened more than 240 kilometers (150 miles). The program has been somewhat successful in reducing the height of the river in flood. However, because the river's tendency to meander still exists, preventing the river from returning to its previous condition has been difficult.

Drainage Basins and Patterns

Every stream, no matter how large or small, has a **drainage basin**. This is the land area that contributes water to the stream. The drainage basin of one stream is separated from the drainage basin of another by an imaginary line called a **divide** (Figure 9.18). Divides range in scale from a ridge separating two small gullies on a hillside to a *continental divide* that splits continents into enormous drainage basins. For example, the continental divide that runs somewhat north–south through the Rocky Mountains separates the drainage which flows west to the Pacific Ocean from that which flows to the Gulf of Mexico. Although divides separate the drainage of two streams, if both streams are tributaries of the same river, they are both a part of the river's drainage system.

Drainage systems are networks of streams that together form distinctive patterns. The nature of a drainage pattern can vary greatly from one type of terrain to another, primarily in response to the kinds of rock on which the streams developed or the structural pattern of faults and folds.

Certainly the most commonly encountered drainage pattern is the **dendritic** pattern (Figure 9.19A). This pattern of irregularly branching tributary streams resembles the branching pattern of a deciduous tree. In fact, the word *dendritic* means "treelike." The dendritic pattern forms where the underlying material is relatively uniform. Because the surface material is essentially uniform in its resistance to erosion, it does not control the pattern of streamflow. Rather, the pattern is determined chiefly by the direction of slope of the land.

When streams diverge from a central area like spokes from the hub of a wheel, the pattern is said to be **radial** (Figure 9.19B). This pattern typically develops on isolated volcanic cones and domal uplifts.

Figure 9.19C illustrates a **rectangular** pattern, in which many right-angle bends can be seen. This

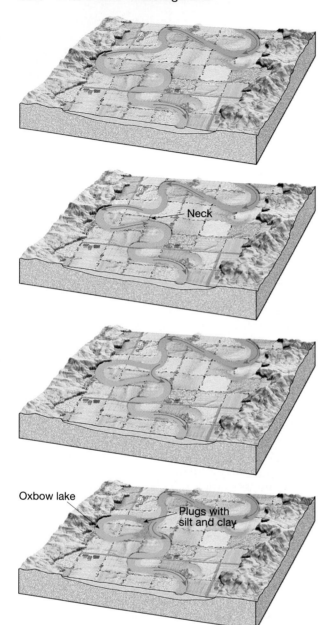

Neck

Oxbow lake

Plugs with silt and clay

FIGURE 9.17 Formation of a cutoff and oxbow lake.

pattern develops when the bedrock is crisscrossed by a series of joints and/or faults. Because these structures are eroded more easily than unbroken rock, their geometric pattern guides the directions of valleys.

Figure 9.19D illustrates a **trellis** drainage pattern, a rectangular pattern in which tributary streams are nearly parallel to one another and have the appearance of a garden trellis. This pattern forms in areas underlain by alternating bands of resistant and less resistant rock.

Stages of Valley Development

Two centuries ago, many people believed Earth to be only a few thousand years old, and that rainwater flowed down valleys that had been there since the beginning of time. An early geologist named James Hutton believed otherwise and proposed that streams were responsible for cutting the valleys in which they flowed. Later geologic work substantiated Hutton's proposal and further revealed that the development of stream valleys progresses in a somewhat predictable fashion. To learn about the evolution of a valley, it is helpful to divide its development into three stages: youth, maturity, and old age.

As long as the stream is downcutting, it is considered youthful. Rapids, an occasional waterfall, and a narrow V-shaped valley are all visible signs of vigorous downcutting. Other characteristics of youth include a steep gradient, little or no floodplain, and a relatively straight course without meanders (Figure 9.20A). The valley of the Yellowstone River pictured in Figure 9.14 provides an excellent example of the youthful stage of valley development.

When a stream reaches maturity, downward erosion diminishes and lateral erosion dominates. Thus, the mature stream begins to create a floodplain and meander upon it (Figures 9.20B and C). In contrast to the gradient of a youthful stream, the gradient of a mature stream is much lower and the profile is much smoother because all rapids and waterfalls have been eliminated.

A stream enters old age after it has cut its floodplain several times wider than its *meander belt*, which is the width of the meander (Figure 9.20D). When this stage is reached, the stream is rarely near the valley walls; hence, it ceases to significantly enlarge the floodplain. Thus, the primary work of a river in an old-age valley is the reworking of unconsolidated floodplain deposits. Because this task is easier than cutting bedrock, a stream in an old-age valley shifts more rapidly than a stream in a mature valley. For example, some meanders of the lower Mississippi move about 20 meters (over 60 feet) a year, and its large floodplain is dotted with oxbow lakes and old cutoffs. Natural levees are also common features of old-age valleys and, when present, are accompanied by back swamps and yazoo tributaries.

Thus far we have assumed that the base level of a stream remains constant as a river progresses from youth to old age. On many occasions, however, the land is uplifted or the base level is lowered. The effect of uplifting on a youthful stream is to increase

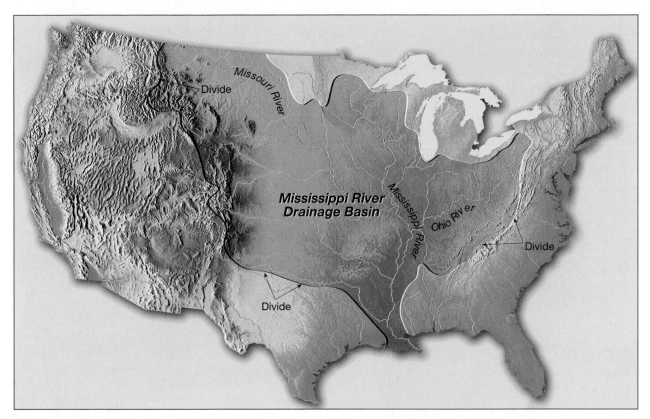

FIGURE 9.18 A *drainage basin* is the land area drained by a stream and its tributaries. The drainage basin of the Mississippi River, North America's largest river, covers about 3 million square kilometers. *Divides* are the boundaries that separate drainage basins from each other. Drainage basins and divides exist for all streams.

its gradient and accelerate its rate of downcutting. However, uplifting of a mature stream would cause it to abandon lateral erosion and revert to downcutting. Rivers of this type are said to be **rejuvenated** ("made young again"). The meanders stay in the same place but become deeper, and are called **entrenched meanders** (Figure 9.21). Mature streams may eventually readjust to uplift by cutting a new floodplain at a level below the old one. The remnants of the old higher floodplain often remain as flat surfaces called *terraces* (Figure 9.22).

Two additional points concerning valley development should be made. First, the time required for a stream to reach any given stage depends on several factors, including the erosive ability of the stream, the nature of the material through which the stream must cut, and the stream's height above base level. Consequently, a stream that starts out very near base level and has to cut only through unconsolidated sediments may reach maturity in a matter of a few hundred years. On the other hand, the Colorado River, where it is actively cutting the Grand Canyon, has retained its youthful nature for an estimated five million to six million years. Second, individual portions of a stream reach each stage at different times. Often the lower reaches of a stream attain old age while the headwaters are still youthful in character.

The Chapter in Review

The following statements are intended to help you review the primary objectives presented in this chapter.

• The *hydrologic cycle* describes the continuous interchange of water among the oceans, atmosphere, and continents. Powered by energy from the Sun, it is a global system in which the atmosphere provides the link between the oceans and continents. The processes involved in the hydrologic cycle include *precipitation, evaporation, infiltration* (the movement of water into rocks or soil through cracks and pore spaces), *runoff* (water that flows over the

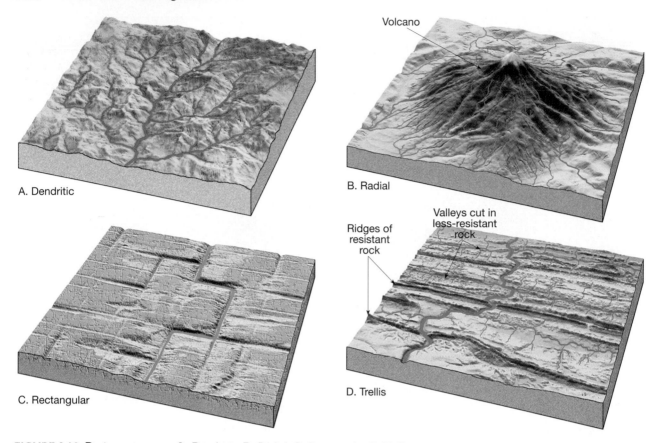

FIGURE 9.19 Drainage patterns. **A.** Dendritic. **B.** Radial. **C.** Rectangular. **D.** Trellis.

land), and *transpiration* (the release of water vapor to the atmosphere by plants). *Running water is the single most important agent sculpturing Earth's land surface.*

- The amount of water that runs off the land rather than sinking into the ground depends upon the *infiltration capacity* of the soil. Initially, runoff flows as broad, thin sheets across the ground, appropriately termed *sheet flow.* After a short distance, threads of current typically develop and tiny channels called *rills* begin to form.

- The factors that determine a stream's *velocity* are *gradient* (slope of the stream channel), *cross-sectional shape, size* and *roughness* of the channel, and the stream's *discharge* (amount of water passing a given point per unit of time, frequently measured in cubic feet per second). Most often, the gradient and roughness of a stream decrease downstream, while width, depth, discharge, and velocity increase.

- The two general types of *base level* (the lowest point to which a stream may erode its channel) are (1) *ultimate base level* (sea level) and (2) *temporary,* or *local base level.* Any change in base

level will cause a stream to adjust and establish a new balance. Lowering base level will cause a stream to erode, whereas raising base level results in deposition of material in the channel.

- The work of a stream includes *erosion* (the incorporation of material), *transportation* (as *dissolved load, suspended load*, and *bed load*), and, whenever a stream's velocity decreases, *deposition.* A stream's ability to carry solid particles is typically described using two criteria: *capacity* (the maximum load of solid particles a stream can transport) and *competence* (the maximum particle size a stream can transport). The depositional features of a stream include *deltas* and *natural levees.*

- Although many gradations exist, the two general types of stream valleys are (1) *narrow V-shaped valleys* and (2) *wide valleys with flat floors.* Because the dominant activity is downcutting toward base level, narrow valleys often contain *waterfalls* and *rapids.* When a stream has cut its channel closer to base level, its energy is directed from side-to-side, and erosion produces a flat valley floor, or *floodplain.*

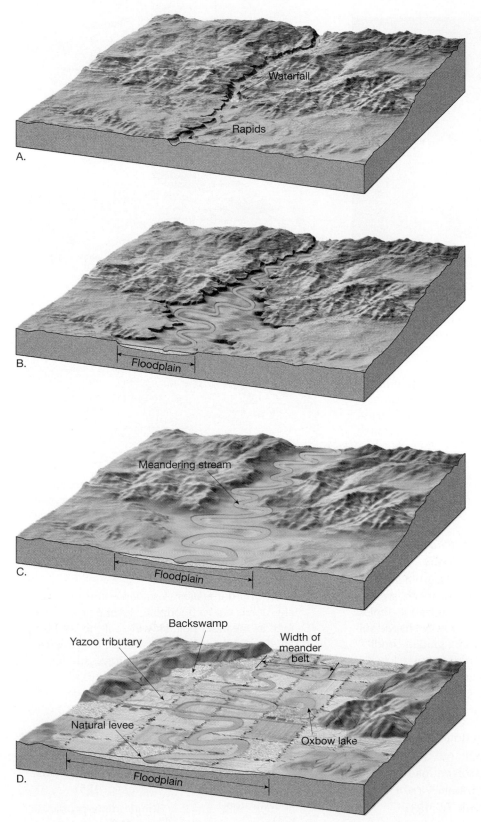

A.

Waterfall

Rapids

B.

Floodplain

C.

Meandering stream

Floodplain

D.

Backswamp

Yazoo tributary

Width of meander belt

Natural levee

Oxbow lake

Floodplain

FIGURE 9.20 Stages of valley development. **A.** Youth. The youthful stage is characterized by downcutting and a V-shaped valley. **B.** and **C.** Maturity. Once a stream has sufficiently lowered its gradient, it begins to erode laterally, producing a wide valley. **D.** Old age. After the valley has been cut several times wider than the width of the meander belt, it has entered old age.

FIGURE 9.21 A close-up view of entrenched meanders in Canyonlands National Park, Utah. The meandering river became entrenched because of the uplift of the Colorado Plateau. (Photo by Michael Collier)

Streams that flow upon floodplains often move in sweeping bends called *meanders*. Widespread meandering may result in shorter channel segments, called *cutoffs*, and/or abandoned bends, called *oxbow lakes*.

- The land area that contributes water to a stream is called a *drainage basin*. Drainage basins are separated by an imaginary line called a *divide*. Common *drainage patterns* (the form of a network of streams) produced by a main channel and its tributaries include (1) *dendritic*, (2) *radial*, (3) *rectangular*, and (4) *trellis*.

- The evolution of a valley has been arbitrarily divided into three sequential stages: *youth, maturity,* and *old age*. A youthful stream, with its rapids, occasional waterfalls, and a narrow V-shaped valley, is downcutting to establish a graded condition with its base level. When a stream reaches maturity, downward erosion diminishes and later-

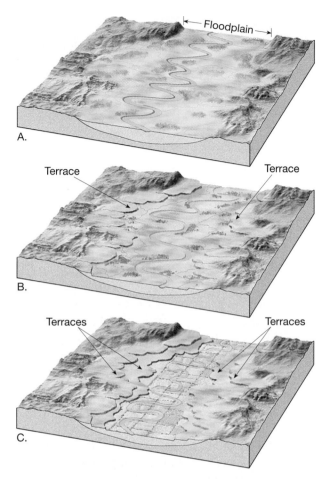

FIGURE 9.22 A. A stream near base level has developed a floodplain. **B.** Because of a lowering of base level or regional uplift, the river downcuts through the former floodplain and creates a new floodplain. The remnants of the former floodplain now exist as *terraces*. **C.** A second set of terraces form in response to another change in base level.

al erosion dominates. A stream enters old age after it has cut its floodplain several times wider than its meander belt. However, if the land is uplifted or the base level is lowered, the progression from youth to old age is interrupted and the stream is said to be *rejuvenated*.

Key Terms

alluvium (p. 178)
back swamp (p. 179)
base level (p. 175)
bed load (p. 177)
capacity (p. 178)
competence (p. 178)
cutoff (p. 183)
delta (p. 178)

dendritic pattern (p. 183)
discharge (p. 173)
dissolved load (p. 177)
distributary (p. 178)
divide (p. 183)
drainage basin (p. 183)
entrenched meander (p. 185)

floodplain (p. 182)
graded stream (p. 176)
gradient (p. 172)
head (headwaters) (p. 173)
hydrologic cycle (p. 170)
infiltration (p. 170)

infiltration capacity (p. 171)
local (temporary) base level (p. 175)
longitudinal profile (p. 173)
meander (p. 182)
mouth (p. 178)

natural levee (p. 179)
oxbow lake (p. 183)
radial pattern (p. 183)
rectangular pattern
  (p. 183)
rejuvenated stream
  (p. 185)

rills (p. 172)
runoff (p. 170)
saltation (p. 177)
Sheet flow (p. 171)
sorting (p. 178)
stream (p. 172)

suspended load (p. 177)
temporary (local) base
  level (p. 175)
transpiration (p. 170)
trellis pattern (p. 184)

ultimate base level
  (p. 175)
yazoo tributary (p. 180)

Questions for Review

1. Describe the movement of water through the hydrologic cycle. Once precipitation has fallen on land, what paths are available to it?

2. Over the oceans the quantity of water lost through evaporation is greater than the amount gained from precipitation. Why then does sea level not drop?

3. A stream starts out 2000 meters above sea level and travels 250 kilometers to the ocean. What is its average gradient in meters per kilometer?

4. Suppose that the stream mentioned in Question 3 developed extensive meanders so that its course was lengthened to 500 kilometers. Calculate its new gradient. How does meandering affect gradient?

5. When the discharge of a stream increases, what happens to the stream's velocity?

6. What typically happens to channel width, channel depth, velocity, and discharge from the point where a stream begins to the point where it ends? Briefly explain why these changes take place.

7. Why do most streams have low gradients near their mouths?

8. When an area changes from predominantly rural to largely urban, how is streamflow affected?

9. Define *base level*. Name the main river in your area. For what streams does it act as base level?

10. List three ways in which a stream transports its load.

11. If you collect a jar of water from a stream, what part of its load will settle to the bottom of the jar? What portion will remain in the water?

12. Distinguish between competence and capacity.

13. Why does a river flowing across a delta eventually change course?

14. Briefly describe the formation of a natural levee. How is this feature related to back swamps and yazoo tributaries?

15. What is an artificial cutoff? What is its purpose?

16. What is a divide?

17. Each of the following statements refers to a particular drainage pattern. Identify the pattern.
   a. Streams diverging from a central high area such as a dome.
   b. Branching "treelike" pattern.
   c. A pattern that develops when bedrock is criss-crossed by joints and faults.

18. Why is it possible for a youthful valley to be older (in years) than a mature valley?

19. Do streams flowing in mature and old-age valleys make good political boundaries? Explain.

Testing What You Have Learned

To test your knowledge of the material presented in this chapter, answer the following questions:

Multiple-Choice Questions

1. The ability of a stream to erode is directly related to its _____.
   a. depth   c. load   e. temperature
   b. velocity   d. width

2. Which of the following is NOT a process included in the hydrologic cycle?
   a. precipitation   c. infiltration   e. runoff
   b. evaporation   d. erosion

3. The finer silts and clays will settle out of a stream some distance from the mouth and may form the horizontal layers of a delta called _____ beds.
   a. topset   c. intermediate   e. foreset
   b. offshore   d. bottomset

4. The single most important agent sculpturing Earth's land surface is _____.
   a. running water   c. ice sheets   e. volcanism
   b. wind   d. ocean waves

5. Which one of the following is NOT one of the four basic drainage patterns?
   a. radial   c. divisional   e. dendritic
   b. trellis   d. rectangular

6. Which one of the following is NOT a way a stream transports its load of sediment?
   a. abrasion   c. solution   e. rolling
   b. suspension   d. saltation

7. The time required for water to reach a stream from where it fell is called the _____ time.
   a. delay   c. flow   e. lag
   b. discharge   d. peak

8. Which one of the following features is least likely be found in a wide valley?
   a. floodplain   c. waterfall   e. cutoff
   b. oxbow lake   d. meander

9. The lowest level to which a stream can erode its channel is the stream's _____.
   a. meander   c. gradient   e. base level
   b. discharge   d. divide

10. When a stream slows down, its competence is reduced and sediment begins to be _____.
    a. eroded   c. abraded   e. deposited
    b. suspended   d. transported

Fill-In Questions

11. A stream's _____ is the slope of its channel expressed as the vertical drop of the stream over a specified distance.

12. The unending circulation of Earth's water supply is called the _____ _____.

13. An imaginary line called a _____ separates the drainage basin of one stream from another.

14. Streams that flow upon floodplains often move in sweeping bends called _____.

15. The _____ of a stream, usually expressed in cubic meters per second or cubic feet per second, is the amount of water flowing past a certain point in a given period of time.

True/False Questions

16. Most streams, but not all, carry the largest part of their load in solution. ___

17. Floodplains can be either erosional or depositional in nature. ___

18. The most commonly encountered drainage pattern is the trellis pattern. ___

19. The rejuvenation of a mature stream often produces entrenched meanders. ___

20. A lake prevents a stream from eroding below its level at any point upstream from the lake. ___

Answers

1.b; 2.d; 3.d; 4.a; 5.c; 6.a; 7.e; 8.c; 9.e; 10.e; 11. gradient; 12. hydrologic cycle; 13. divide; 14. meanders; 15. discharge; 16.F; 17.T; 18.F; 19.T; 20.T.

Groundwater

Focus on Learning

To assist you in learning the important concepts in this chapter, you will find it helpful to focus on the following questions:

- What is the importance of groundwater as a resource and as a geological agent?

- What is groundwater and how does it move?

- What are some common natural phenomena associated with groundwater?

- What are some environmental problems associated with groundwater?

- How is karst topography produced?

These cave decorations in Leman Caves are called the "Pearly Gates." Great Basin National Park, Nevada. (Photo by Tom Bean)

Worldwide, wells and springs provide water for cities, crops, livestock, and industry. In the United States, groundwater is the source of about 40 percent of the water used for all purposes (except hydroelectric power generation and powerplant cooling). Groundwater is the drinking water for more than 50 percent of the population, is 40 percent of the water used for irrigation, and provides more than 25 percent of industry's needs. In some areas, however, overuse of this basic resource has resulted in water shortage, streamflow depletion, land subsidence, contamination by saltwater, increased pumping cost, and groundwater pollution.

Importance of Underground Water

Groundwater is one of our most important and widely available resources, yet people's perceptions of the subsurface environment from which it comes are often unclear and incorrect (see Box 10.1). The reason is that the groundwater environment is largely hidden from view except in caves and mines, and the impressions people gain from these subsurface openings are misleading. Observations on the land surface give an impression that Earth is "solid." This view remains when we enter a cave and see water flowing in a channel that appears to have been cut into solid rock.

Because of such observations, many people believe that groundwater occurs only in underground "rivers." In reality, most of the subsurface environment is not "solid" at all. It includes countless tiny *pore spaces* between grains of soil and sediment, plus narrow joints and fractures in bedrock. Together, these spaces add up to an immense volume. It is in these small openings that groundwater collects and moves.

Considering the entire hydrosphere, or all of Earth's water, only about six-tenths of one percent occurs underground. Nevertheless, this small percentage, stored in the rocks and sediments beneath Earth's surface, is a vast quantity. When the oceans are excluded and only sources of freshwater are considered, the significance of groundwater becomes more apparent.

Table 10.1 contains estimates of the distribution of freshwater in the hydrosphere. Clearly the largest volume occurs as glacial ice. Second in rank is groundwater, with slightly more than 14 percent of the total. However, when ice is excluded and just liquid water is considered, more than 94 percent of all freshwater is groundwater. Without question, *groundwater represents the largest reservoir of freshwater that is readily*

TABLE 10.1 Fresh water of the hydrosphere.

Parts of the Hydrosphere	Volume of Freshwater (km³)	Share of Total Volume of Freshwater (percent)	Rate of Water Exchange
Ice sheets and glaciers	24,000,000	84.945	8000 years
Groundwater	4,000,000	14.158	280 years
Lakes and reservoirs	155,000	0.549	7 years
Soil moisture	83,000	0.294	1 year
Water vapor in the atmosphere	14,000	0.049	9.9 days
River water	1,200	0.004	11.3 days
Total	28,253,200	100.00	

(*Source*: U.S. Geological Survey Water Supply Paper 2220, 1987)

available to humans. Its value in terms of economics and human well-being is incalculable.

Geologically, groundwater is important as an erosional agent. The dissolving action of groundwater slowly removes rock, allowing surface depressions known as sinkholes to form as well as creating subterranean caverns (Figure 10.1). Groundwater is also an equalizer of streamflow. Much of the water that flows in rivers is not direct runoff from rain and snowmelt. Rather, a large percentage of precipitation soaks in and then moves slowly underground to stream channels. Groundwater is thus a form of storage that sustains streams during periods when rain does not fall. When we see water flowing in a river during a dry period, it is water from rain that fell at some earlier time and was stored underground.

Distribution of Underground Water

When rain falls, some of the water runs off, some evaporates, and the remainder soaks into the ground. This last path is the primary source of practically all underground water. The amount of water that takes each of these paths, however, varies greatly both in time and space. Influential factors include steepness of slope, nature of surface material, intensity of rainfall, and type and amount of vegetation. Heavy rains falling on steep slopes underlain by impervious materials will obviously result in a high percentage of the water running off. On the other hand, if rain falls steadily and gently on more gradual slopes composed of materials that are easily penetrated by the water, a much larger percentage of water soaks into the ground.

FIGURE 10.1 The dissolving action of groundwater creates caverns. Carlsbad Caverns National Park, New Mexico. (Photo by Harold Hoffman/Photo Researchers)

Some of the water that soaks in does not travel far, because it is held by molecular attraction as a surface film on soil particles. A portion of this moisture evaporates back into the atmosphere. Much of the remainder is used by plants between rains. But water that is not held near the surface penetrates downward until it reaches a zone where all of the open spaces in sediment and rock are completely filled with water. This is called the **zone of saturation**. Water within it is called **groundwater**. The upper limit of this zone is known as the **water table**. The area above the water table where the soil, sediment, and rock are not saturated is called the **zone of aeration** (Figure 10.2). The open spaces here are filled mainly with air.

The Water Table

The water table, the upper limit of the zone of saturation, is a very significant feature of the groundwater system. The water table level is important in predicting the productivity of wells, explaining the changes in the flow of springs and streams, and accounting for fluctuations in the levels of lakes.

Although we cannot observe the water table directly, its elevation can be mapped and studied in detail where wells are numerous, because the water level in wells coincides with the water table. Such maps reveal that the water table is rarely level, as we might expect a table to be. Instead, its shape is usually a subdued replica of the surface topography, reaching its highest elevations beneath hills and then descending toward valleys (Figure 10.2). Where a wetland (swamp) is encountered, the water table is right at the surface. Lakes and streams generally occupy areas low enough that the water table is above the land surface.

A number of factors contribute to the irregular surface of the water table. The most important cause is that groundwater moves very slowly and at varying rates under different conditions. Because of this, water tends to "pile up" beneath high areas between stream valleys. If rainfall were to cease completely, these water table "hills" would slowly subside and gradually approach the level of the valleys. However, new supplies of rainwater are usually added frequently enough to prevent this. Nevertheless, in times of extended drought, the water table may drop enough to dry up shallow wells. Other causes for the uneven water table are variations in rainfall and permeability from place to place.

The relationship between the water table and a stream in a humid region is illustrated in Figure 10.3A. Even during dry periods, the movement of groundwater into the channel maintains a flow in the stream. In situations such as this, streams are said to be **effluent**.

Box 10.1   Dowsing for Water*

"Water dowsing" refers in general to the practice of using a forked stick, rod, pendulum, or similar device to locate underground water, minerals, or other hidden or lost substances, and it has been a subject of discussion and controversy for hundreds of years.

Although tools and methods vary widely, most dowsers (also called diviners or water witches) probably still use the traditional forked stick, which may come from a variety of trees, including the willow, peach, and witch hazel (Figure 10.A). Other dowsers may use keys, wire coat hangers, pliers, wire rods, pendulums, or various kinds of elaborate boxes and electrical instruments.

In the classic method of using a forked stick, one fork is held in each hand with the palms upward and the bottom of the "Y" pointed skyward. The dowser then walks back and forth over the area to be tested. When a source of water is detected, the butt end of the stick is supposed to rotate or be attracted downward.

According to dowsers, the attraction of the water may be so great that the bark peels off as the rod twists in the hands. Some dowsers are said to have suffered blistered or bloody hands from the twisting.

The exact origin of the divining rod in Europe is not known. The first detailed description of it is in Johannes Agricola's *De Re Metallica* (1556), a description of German mines and mining methods. The device was introduced into England during the reign of Elizabeth I (1558–1603) to locate mineral deposits, and soon afterward it was adopted as a water finder throughout Europe.

Water dowsing seems to be a mainly European cultural phenomenon, completely unknown to New World Indians and Eskimos. It was carried to America by some of the earliest settlers from England and Germany. Although the published record was very slight at first, water dowsing or witching began to be mentioned after 1675 in connection with witches and witchcraft. Two articles condemning it appeared in the 1821 and 1826 issues of the *American Journal of Science* and were among the first in a long line of treatises on water witching.

Despite almost unanimous condemnation by geologists and technicians, the practice of water dowsing has spread throughout America. Thousands of dowsers are currently active in the United States.

Case histories and demonstrations of dowsers may seem convincing, but when dowsing is exposed to scientific examination, it presents a very different picture. A study in Australia compared geologist's successes at locating groundwater with those of dowsers. The dowsers caused twice as many dry holes to be dug as did the geologists. At Iowa State University, dowsers were given the task of finding water along a prescribed path on campus. They were not able to "discover" the water mains directly beneath them.

What does it mean to say that a dowser is successful? The dowser may find water, but how much? And of what quality? At what rate can it be withdrawn? For how long and with what impact on other wells and on nearby streams?

The natural explanation of "successful" water dowsing is that in many areas water would be hard to miss. The dowser commonly implies that

FIGURE 10.A Most dowsers still use the traditional forked stick.

the spot indicated by the rod is the *only* one where water could be found, but this is not necessarily true. In a region of adequate rainfall and favorable geology, it is difficult *not* to drill and find water!

Some water exists under the surface almost everywhere. This explains why many dowsers appear to be successful. To locate groundwater accurately, however, as to depth, quantity, and quality, a number of techniques must be used. Hydrologic, geologic, and geophysical knowledge is needed to determine the depths and extent of the different water-bearing strata and the quantity and quality of water found in each. The area must be thoroughly tested and studied to determine these facts.

*Much of this information is based on material prepared by the U.S. Geological Survey.

By contrast, in arid regions where the water table is far below the surface, groundwater cannot contribute to streamflow. Therefore, the only permanent streams in such areas are those that originate in wet regions and then happen to traverse the desert. (Examples are the Nile River in Egypt and the Colorado River in the American Southwest.) Under these conditions the zone of saturation below the valley floor is supplied by downward

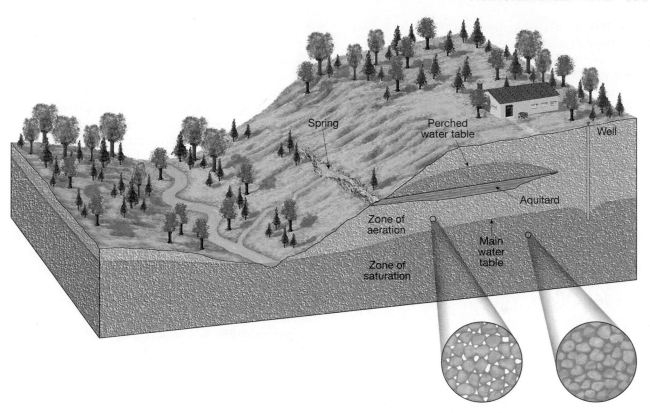

FIGURE 10.2 This diagram illustrates the relative positions of many features associated with subsurface water.

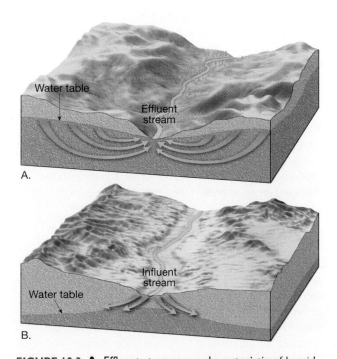

FIGURE 10.3 A. Effluent streams are characteristic of humid areas and are supplied by water from the zone of saturation. **B.** Influent streams are found in deserts. Seepage from such streams produces an upward bulge in the water table.

seepage from the stream channel which, in turn, produces an upward bulge in the water table. Streams that provide water to the water table in this manner are called **influent streams** (Figure 10.3B).

How Groundwater Moves

We noted the common misconception that groundwater occurs in underground rivers that resemble surface streams. Although subsurface streams do exist, they are not common. Rather, as you learned in the preceding sections, groundwater exists in the pore spaces and fractures in rock and sediment. Thus, contrary to any impressions of rapid flow that an underground river might evoke, the movement of most groundwater is exceedingly slow, from pore to pore. By exceedingly slow, we mean anywhere from millimeters per year to perhaps a kilometer per year, depending on conditions.

The energy that makes groundwater move is provided by the force of gravity. In response to gravity, water moves from areas where the water table is high to zones where the water table is lower. This means that water gravitates toward a stream channel, lake, or spring as shown in Figure 10.2. Although some water takes the most direct path down the slope of the water table, much of the water follows long curving paths toward the zone of discharge.

Figure 10.4 shows how water percolates into a stream from all possible directions. Some paths actually turn upward, apparently against the force of gravity, and enter through the bottom of the channel. This is easily explained: The deeper you go into the zone of saturation, the greater is the water pressure. Thus, the looping curves followed by water in the saturated zone may be thought of as a compromise between the downward pull of gravity and the tendency of water to move toward areas of reduced pressure. As a result, water at any given height is under greater pressure beneath a hill than beneath a stream channel, and the water tends to migrate toward points of lower pressure.

Porosity

Depending upon the nature of the subsurface material, the groundwater flow and the amount of water stored are highly variable. Water soaks into the ground because bedrock, sediment, and soil contain countless voids, or openings. These openings are similar to those of a sponge and are often called pore spaces. The quantity of groundwater that can be stored depends on the **porosity** of the material, which is the percentage of the total volume of rock or sediment that consists of pore spaces. Voids most often are spaces between sedimentary particles, but also common are joints, faults, cavities formed by the dissolving of soluble rocks such as limestone, and vesicles (voids left by gases escaping from lava).

Sediment is commonly quite porous, and open spaces may occupy 10 to 50 percent of the sediment's total volume. Pore space depends on the size and

shape of the grains, how they are packed together, the degree of sorting, and in sedimentary rocks, the amount of cementing material. For example, clay may have a porosity as high as 50 percent, whereas some gravels may have only 20 percent voids.

Where sediments of various sizes are mixed, the porosity is reduced because the finer particles tend to fill the openings among the larger grains. Most igneous and metamorphic rocks, as well as some sedimentary rocks, are composed of tightly interlocking crystals so the voids between the grains may be negligible. In these rocks, fractures must provide the voids.

Permeability, Aquitards and Aquifers

Porosity alone cannot measure a material's capacity to yield groundwater. Rock or sediment may be very porous, yet still not allow water to move through it. The pores must be *connected* to allow water flow, and they must be *large enough* to allow flow. Thus, the **permeability** of a material, its ability to *transmit* a fluid, is also very important.

Groundwater moves by twisting and turning through interconnected small openings. The smaller the pore spaces, the slower the water moves. Clay's ability to store water is great owing to its high porosity, but its pore spaces are so small that water is unable to move through it. Thus, clay's porosity is high but its permeability is poor.

Impermeable layers that hinder or prevent water movement are termed **aquitards**. Clay is a good example. On the other hand, larger particles, such as sand or gravel, have larger pore spaces. Therefore, the water moves with relative ease. Permeable rock strata or sediment that transmit groundwater freely are called **aquifers**. Sands and gravels are common examples.

In summary, you have seen that porosity is not always a reliable guide to the amount of groundwater that can be produced, and permeability is significant in determining the rate of groundwater movement and the quantity of water that might be pumped from a well.

Springs

Springs have aroused the curiosity and wonder of people for thousands of years. The fact that springs were, and to some people still are, rather mysterious phenomena is not difficult to understand, for here is water flowing freely from the ground in all kinds of weather in seemingly inexhaustible supply but with no obvious source.

Not until the middle of the 1600s did the French physicist Pierre Perrault invalidate the age-old

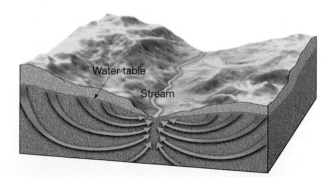

FIGURE 10.4 Arrows indicate groundwater movement through uniformly permeable material. The looping curves may be thought of as a compromise between the downward pull of gravity and the tendency of water to move toward areas of reduced pressure.

FIGURE 10.5 Minnie Miller Springs, Thousand Springs Preserve, along the Snake River near Hagerman, Idaho. (Photo by William H. Mullins/Photo Researchers)

assumption that precipitation could not adequately account for the amount of water emanating from springs and flowing in rivers. Over several years, Perrault computed the quantity of water that fell on France's Seine River basin. He then calculated the mean annual runoff by measuring the river's discharge. After allowing for the loss of water by evaporation, he showed that there *was* sufficient water remaining to feed the springs. Thanks to Perrault's pioneering efforts and the measurements by many afterward, we now know that the source of springs is water from the zone of saturation and that the ultimate source of this water is precipitation.

Whenever the water table intersects Earth's surface, a natural outflow of groundwater results, which we call a **spring**. Springs form when an aquitard blocks the downward movement of groundwater and forces it to move laterally. Where the permeable bed outcrops in a valley, a line of springs results. Figure 10.5 shows an example of this. Here, the aquifer consists of basaltic lava flows containing many joints and countless vesicles. Another situation

leading to the formation of a spring is illustrated in Figure 10.2. Here an aquitard is situated above the main water table. As water percolates downward, a portion of it is intercepted by the aquitard, thereby creating a localized zone of saturation called a **perched water table**.

Many geological situations lead to the formation of springs because subsurface conditions vary greatly from place to place. Even in areas underlain by impermeable crystalline rocks, permeable zones may exist in the form of fractures or solution channels. If these openings fill with water and intersect the ground surface along a slope, a spring will result.

Wells

The most common device used by people for removing groundwater is the **well**, a hole bored into the zone of saturation (see Figure 10.2). Wells serve as small reservoirs into which groundwater migrates and from which it can be pumped to the surface. The use of wells dates back many centuries and continues to be an important method of obtaining water today. By far the single greatest use of this water in the United States is irrigation for agriculture. More than 65 percent of the groundwater used each year is for this purpose. Industrial uses rank a distant second, followed by the amount used by homes in cities and rural areas.

The level of the water table may fluctuate considerably during the course of a year, dropping during dry seasons and rising following periods of rain. Therefore, to ensure a continuous supply of water, a well must penetrate below the water table. Whenever water is withdrawn from a well, the water table around the well is lowered. This effect, termed **drawdown**, decreases with increasing distance from the well. The result is a depression in the water table, roughly conical in shape, known as a **cone of depression** (Figure 10.6). Because the cone of

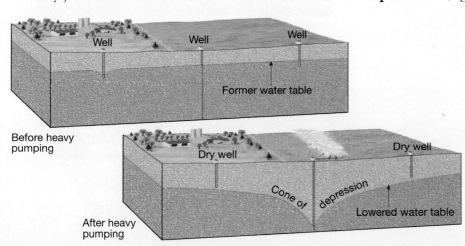

FIGURE 10.6 A cone of depression in the water table often forms around a pumping well. If heavy pumping lowers the water table, some wells may be left dry.

depression increases the slope of the water table near the well, groundwater will flow more rapidly toward the opening. For most small domestic wells, the cone of depression is negligible. However, when wells are heavily pumped for irrigation or industrial purposes, the withdrawal of water can be great enough to create a very wide and steep cone of depression. This may substantially lower the water table in an area and cause nearby shallow wells to become "high and dry" (Figure 10.6).

Artesian Wells

In most wells, water cannot rise on its own. If water is first encountered at 30 meters' depth, it remains at that level, fluctuating perhaps a meter or two with seasonal wet and dry periods. However, in some wells, water rises, sometimes overflowing at the surface. Such wells are abundant in the *Artois* region of northern France, and so we call these self-rising wells *artesian.*

The term **artesian** is applied to any situation in which groundwater rises in a well above the level where it was initially encountered. For such a situation to occur, two conditions must exist (Figure 10.7): (1) water must be confined to an aquifer that is inclined so that one end is exposed at the surface, where it can receive water; and (2) impermeable layers (aquitards), both above and below the aquifer, must be present to prevent the water from escaping. When such a layer is tapped, the pressure created by the weight of the water above will force the water to rise. If there were no friction the water in the well would rise to the level of the water at the top of the aquifer. However, friction reduces the height of this pressure surface. The greater the distance from the recharge area (area where water enters the inclined aquifer), the greater the friction and the less the rise of water.

In Figure 10.7, Well 1 is a *nonflowing artesian well*, because at this location the pressure surface is below ground level. When the pressure surface is above the ground and a well is drilled into the aquifer, a *flowing artesian well* is created (Well 2, Figure 10.7).

Artesian systems act as conduits, transmitting water from remote areas of recharge great distances to the points of discharge. In this manner, water that fell in central Wisconsin years ago is now taken from the ground and used by communities many kilometers away in Illinois. In South Dakota, such a system brings water from the Black Hills in the west, eastward across the state.

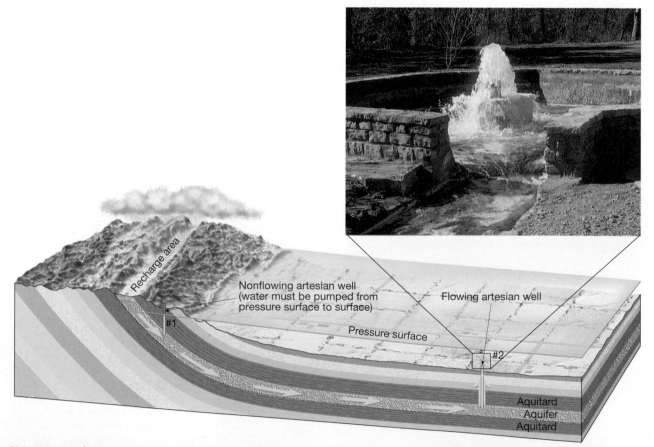

FIGURE 10.7 Artesian systems occur when an inclined aquifer is surrounded by impermeable beds. (Photo by E. J. Tarbuck)

On a different scale, city water systems may be considered as examples of artificial artesian systems. The water tower, into which water is pumped, may be considered the area of recharge, the pipes the confined aquifer, and the faucets in homes the flowing artesian wells.

Environmental Problems of Groundwater

As with many of our valuable natural resources, groundwater is being exploited at an increasing rate. In some areas, overuse threatens the groundwater supply. In other places, groundwater withdrawal has caused the ground and everything resting upon it to sink. Still other localities are concerned with the possible contamination of their groundwater supply.

Groundwater as a Nonrenewable Resource

Many natural systems tend to establish a condition of equilibrium. The groundwater system is no exception. The water table's height reflects a balance between the rate of water added by precipitation and the rate of water removed by discharge and withdrawal. Any imbalance will either raise or lower the water table. A long-term drop in the water table can occur if there is either a decrease in recharge due to a prolonged drought, or an increase in groundwater discharge or withdrawal.

In some regions groundwater has been and continues to be treated as a *nonrenewable resource*. Where this occurs, the water available to recharge the aquifer falls significantly short of the amount being withdrawn. The High Plains, a relatively dry region that extends from South Dakota to western Texas, provides an example. Here an extensive agricultural economy is largely dependent on irrigation (Figure 10.8). Widespread permeable layers of sand and gravel have permitted the construction of high-yield wells almost anywhere in the region. As a result an estimated 168,000 wells are irrigating more than 65,000 square kilometers (16 million acres) of land. In the southern part of this region, which includes the Texas panhandle, the natural recharge of the aquifer is very slow and the problem of declining groundwater levels is acute. In fact, in years of average or below-average precipitation, recharge is negligible because all or nearly all of the meager rainfall is returned to the atmosphere by evaporation and transpiration. Thus, it is only during especially wet years that significant recharge of the aquifer occurs, and this averages only about 5 millimeters per year.

Therefore, where intense irrigation has been practiced for an extended period, depletion of groundwater can be severe. Declines in groundwater levels up to 1 meter annually have led to an overall drop in the water table of 15 to 60 meters (50 to 200 feet) in some areas. Under these circumstances, it can be said that the groundwater is literally being "mined." Even if pumping were to cease immediately, it could take hundreds or thousands of years for the groundwater to be fully replenished.

Land Subsidence Caused by Groundwater Withdrawal

As you shall see later in this chapter, surface subsidence can result from natural processes related to groundwater. However, the ground may also sink when water is pumped from wells faster than natural recharge processes can replace it. This effect is particularly pronounced in areas underlain by thick layers of loose sediments. As water is withdrawn, the weight of the overburden packs the sediment grains more tightly together and the ground subsides.

Many areas can be used to illustrate such land subsidence. A classic example in the United States occurred in the San Joaquin Valley of California (Figure 10.9). This important agricultural region relies heavily on irrigation. Land subsidence due to groundwater withdrawal began in the valley in the mid-1920s and locally exceeded 8 meters (28 feet) by 1970. Then, because of the importation of surface water and a decrease in groundwater pumping,

FIGURE 10.8 In some agricultural regions water is pumped from the ground faster than it is replenished. In such instances, groundwater is being treated as a nonrenewable resource. (Photo by Steve Welsh/Gamma–Liaison)

FIGURE 10.9 The shaded area on the map shows California's San Joaquin Valley. The marks on the utility pole in the photo indicate the level of the surrounding land in preceding years. Between 1925 and 1975 this part of the San Joaquin Valley subsided almost 9 meters because of the withdrawal of groundwater and the resulting compaction of sediments. (Photo courtesy of U.S. Geological Survey)

water levels in the aquifer recovered and subsidence ceased. However, during a drought in 1976–1977, heavy groundwater pumping led to renewed subsidence. This time, water levels dropped at a much faster rate than during the previous period because of the reduced storage capacity caused by earlier compaction of material in the aquifer. In all, more than 13,400 square kilometers (5200 square miles) of irrigable land, one-half the entire valley, were affected by subsidence. Damage to structures, including highways, bridges, water lines, and wells, was extensive. Because subsidence changed the gradients of some streams, flooding also became a costly problem. Many other examples of land subsidence due to groundwater pumping occur in the United States and elsewhere in the world.

Groundwater Contamination

The pollution of groundwater is a serious matter, particularly in areas where aquifers provide a large part of the water supply. One common source of groundwater pollution is sewage, which emanates from an ever-increasing number of septic tanks. Other sources are inadequate or broken sewer systems, and farm wastes.

If sewage water, which is contaminated with bacteria, enters the groundwater system, it may become purified through natural processes. The harmful bacteria may be mechanically filtered by the sediment through which the water percolates, destroyed by chemical oxidation, and/or assimilated by other organisms. For purification to occur, however, the aquifer must be of the correct composition. For example, extremely permeable aquifers (such as highly fractured crystalline rock, coarse gravel, or cavernous limestone) have such large openings that contaminated groundwater might travel long distances without being cleansed. In this case, the water flows too rapidly and is not in contact with the surrounding material long enough for purification to occur. This is the problem at Well 1 in Figure 10.10.

On the other hand, when the aquifer is composed of sand or permeable sandstone, the water can sometimes be purified after traveling only a few dozen meters through it. The openings between sand grains are large enough to permit water movement, yet the movement of the water is slow enough to allow ample time for its purification (Well 2, Figure 10.10).

Other sources and types of contamination also threaten groundwater supplies (Figure 10.11). These include widely used substances such as highway salt, fertilizers that are spread across the land surface, and pesticides. In addition, a wide array of chemicals and industrial materials may leak from pipelines, storage tanks, landfills, and holding ponds. Some of these pollutants are classified as *hazardous*, meaning that they are either flammable, corrosive, explosive, or toxic. As rainwater percolates through the soil, it can carry pollutants to the water table. Here they mix with the groundwater and contaminate the supply.

Because groundwater movement is usually slow, polluted water may go undetected for a long time. In fact, most contamination is discovered only after drinking water has been affected and people become ill. By this time, the volume of polluted water may be very large, and even if the source of contamination is removed immediately, the problem is not solved. Although the sources of groundwater contamination are numerous, the solutions are relatively few.

Once the source of the problem has been identified and eliminated, the most common practice is simply to abandon the water supply and allow the pollutants to be flushed away gradually. This is the least costly and easiest solution, but the aquifer must remain unused for many years. To accelerate this process, polluted water is sometimes pumped out and treated. Following removal of the tainted water,

FIGURE 10.10 A. Although the contaminated water has traveled more than 100 meters before reaching Well 1, the water moves too rapidly through the cavernous limestone to be purified. **B.** As the discharge from the septic tank percolates through the permeable sandstone, it is purified in a relatively short distance.

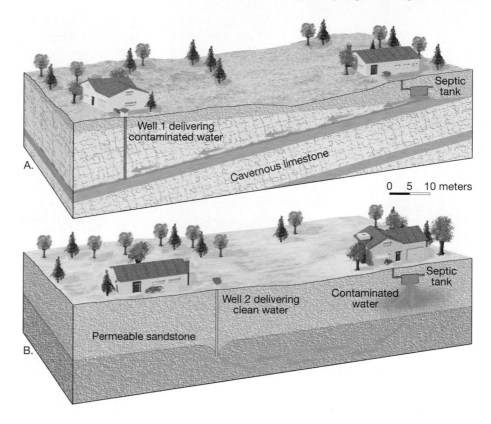

FIGURE 10.11 Sometimes materials leached from landfills find their way into the groundwater. This is just one of several potential sources of groundwater contamination. (Photo by F. Rossotto/The Stock Market)

the aquifer is allowed to recharge naturally or, in some cases, the treated water or other freshwater is pumped back in. This process is costly, time-consuming, and it may be risky because there is no way to be certain that all of the contamination has been removed. Clearly, the most effective solution to groundwater contamination is prevention.

Hot Springs and Geysers

By definition, the water in **hot springs** is 6–9°C (11–16°F) warmer than the mean annual air temperature for the localities where they occur. In the United States alone, there are well over 1000 such springs.

Temperatures in deep mines and oil wells usually rise with increasing depth, an average of about 2°C per 100 meters (1°F per 100 feet). Therefore, when groundwater circulates at great depths, it becomes heated. If it rises to the surface, the water may emerge as a hot spring. The water of some hot springs in the eastern United States is heated in this manner. The great majority (over 95 percent) of the hot springs (and geysers) in the United States are found in the West. The reason for such a distribution is that the source of heat for most hot springs is cooling igneous rock, and it is in the West that igneous activity has been most recent.

FIGURE 10.12 Old Faithful, one of the world's most famous geysers, emits as much as 45,000 liters (almost 12,000 gallons) of hot water and steam about once each hour. (Photo by Dallas and John Heaton/Westlight)

Geysers are intermittent hot springs or fountains where columns of water are ejected with great force at various intervals, often rising 30 to 60 meters (100 to 200 feet) into the air. After the jet of water ceases, a column of steam rushes out, usually with a thunderous roar. Perhaps the most famous geyser in the world is Old Faithful in Yellowstone National Park, which erupts about once each hour (Figure 10.12). The great abundance, diversity, and spectacular nature of Yellowstone's geysers and other thermal features undoubtedly were the primary reason for its becoming the first national park in the United States. Geysers are also found in other parts of the world, notably New Zealand and Iceland. In fact, the Icelandic word *geysa*, to gush, gives us the name *geyser*.

Geysers occur where extensive underground chambers exist within hot igneous rocks. How they operate is shown in Figure 10.13. As relatively cool groundwater enters the chambers, it is heated by the surrounding rock. At the bottom of the chambers, the water is under great pressure because of the weight of the overlying water. Consequently, this water must reach temperatures above 100ºC (212ºF) before it will boil. For example, water at the bottom of a 300-meter (1000-feet) water-filled chamber must reach nearly 230ºC (450ºF) to boil. As the temperature rises, water expands and some flows out at the surface. This loss of water reduces the pressure on the remaining water in the chamber, which lowers the boiling point. A small portion of the water deep within the chamber quickly

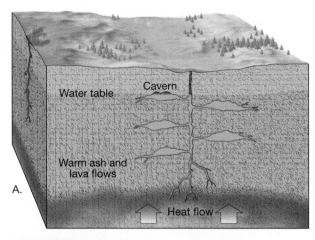

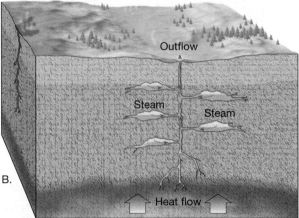

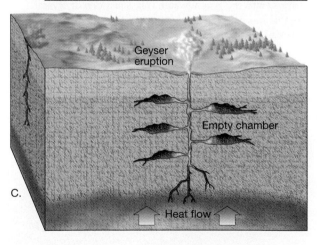

FIGURE 10.13 Idealized diagrams of a geyser. A geyser can form if the heat is not distributed by convection. **A.** In this figure, the water near the bottom is heated to near its boiling point. The boiling point is higher there than at the surface because the weight of the water above increases the pressure. **B.** The water higher in the geyser system is also heated; therefore, it expands and flows out at the top, reducing the pressure on the water at the bottom. **C.** At the reduced pressure on the bottom, boiling occurs. Some of the bottom water flashes into steam, and the expanding steam causes an eruption.

turns to steam and the geyser erupts (Figure 10.13). Following the eruption, cool groundwater again seeps into the chamber and the cycle begins anew.

Groundwater from hot springs and geysers usually contains more material in solution than groundwater from other sources because hot water is a more effective dissolver than cold. When the water contains much dissolved silica, *geyserite* is deposited around the spring. In limestone regions, *travertine*, a form of calcite, is deposited. Some hot springs also contain sulphur. Sulfur-rich spring water tastes bad and emits an unpleasant odor. Undoubtedly Rotten Egg Spring, Nevada, is such a situation.

Geothermal Energy

We harness **geothermal energy** by tapping natural underground reservoirs of steam and hot water. These occur where subsurface temperatures are high due to relatively recent volcanic activity. We put geothermal energy to use in two ways: the steam and hot water are used for heating and to generate electricity.

Iceland is a large volcanic island with current volcanic activity. In Iceland's capital, Reykjavik, steam and hot water are pumped into buildings throughout the city for space heating. They also warm greenhouses, where fruits and vegetables are grown all year. In the United States, localities in several western states use hot water from geothermal sources for space heating.

As for generating electricity geothermally, the Italians were first to do so in 1904, so the idea is not new. By the mid-1990s, the U.S. Geological Survey reported that nearly 200 separate power units in seventeen countries were operating, with a combined capacity of almost 4800 megawatts (million watts). The leaders are the United States, the Philippines, Indonesia, Mexico, Italy, and New Zealand.

The first commercial geothermal power plant in the United States was built in 1960 at The Geysers, north of San Francisco (Figure 10.14). By 1986 development at this location had grown to almost 1800 megawatts, enough to satisfy the needs of San Francisco and Oakland. However, this peak soon passed and the production of electricity began to decline. In addition to The Geysers, geothermal development is occurring elsewhere in the western United States, including Nevada, Utah, and the Imperial Valley in southern California.

What geologic factors favor a geothermal reservoir of commercial value?

1. *A potent source of heat*, such as a large magma chamber deep enough to ensure adequate pressure and slow cooling, yet not so deep that the natural water circulation is inhibited. Such magma chambers are most likely in regions of recent volcanic activity.

2. *Large and porous reservoirs with channels connected to the heat source*, near which water can circulate and then be stored in the reservoir.

3. *A cap of low permeability rocks* that inhibits the flow of water and heat to the surface. A deep, well-insulated reservoir contains much more stored energy than a similar but uninsulated reservoir.

We must recognize that geothermal power is not inexhaustible. When hot fluids are pumped from

FIGURE 10.14 The Geysers, a field of steaming vents 115 kilometers (70 miles) north of San Francisco, California. The natural steam beneath these hills was first tapped to power electrical-generating plants in 1960. (Photo courtesy of Pacific Gas and Electric)

volcanically heated reservoirs, water cannot be replaced and then heated sufficiently to recharge the reservoir. Experience shows that steam and hot water from individual wells usually lasts no more than 10 to 15 years, so more wells must be drilled to maintain power production. Eventually the field is depleted.

As with other alternative methods of power production, geothermal sources are not expected to provide a high percentage of the world's growing energy needs. Nevertheless, in regions where its potential can be developed, its use will no doubt continue to grow.

The Geologic Work of Groundwater

Running water on the surface dramatically erodes soil, sediment, and rock through mechanical abrasion and transport of debris. But the primary erosional work carried out by groundwater is that of *dissolving* rock. Soluble rocks, especially limestone, underlie millions of square kilometers of Earth's surface. Thus, it is in these limestone regions that groundwater carries on its important role as an erosional agent. Limestone is nearly insoluble in pure water, but it is quite easily dissolved by water containing small quantities of carbonic acid. Most natural water contains this weak acid because rainwater readily dissolves carbon dioxide from the air and from decaying plants. Therefore, when groundwater comes in contact with limestone, the carbonic acid reacts with calcite (calcium carbonate) in the rocks to form calcium bicarbonate, a soluble material that is then carried away in solution.

Caverns

Among the most spectacular results of groundwater's erosional handiwork is the creation of limestone **caverns**. In the United States alone, about 17,000 caves have been discovered, and new ones are being found every year. Although most are modest, some have spectacular dimensions. Mammoth Cave in Kentucky and Carlsbad Caverns in southeastern New Mexico are famous examples. The Mammoth Cave system is the most extensive in the world, with more than 540 kilometers (340 miles) of interconnected passages. The dimensions at Carlsbad Caverns are impressive in a different way. Here we find the largest and perhaps most spectacular single chamber. The Big Room at Carlsbad Caverns has an area equivalent to fourteen football fields and enough height to accommodate the U.S. Capitol Building.

Most caverns are created at or just below the water table in the zone of saturation. Here the groundwater follows lines of weakness in the rock, such as joints and bedding planes. As time passes, the dissolving process slowly creates cavities and gradually enlarges them into caverns (Figure 10.15). The material dissolved by groundwater is carried away and discharged into streams.

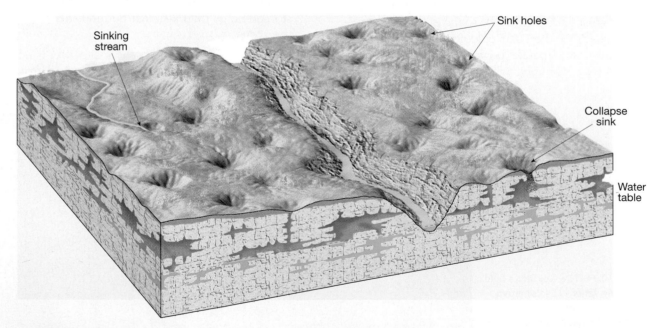

FIGURE 10.15 Development of a karst landscape. During early stages, groundwater percolates through limestone along joints and bedding planes. Solution activity creates and enlarges caverns at and below the water table. Sinkholes develop and surface streams are funneled below ground. With the passage of time, caverns grow larger and the number and size of sinkholes increase. Collapse of caverns and coalescence of sinkholes form larger, flat-floored depressions. Eventually solution activity may remove most of the limestone from the area, leaving only isolated remnants.

In many caves, development has occurred at several levels, with the current cavern-forming activity occurring at the lowest elevation. This situation reflects the close relationship between the formation of major subterranean passages and the river valleys into which they drain. As streams cut their valleys deeper, the water table drops as the elevation of the river drops.

Certainly the features that arouse the greatest curiosity for most cavern visitors are the stone formations that give some caverns a wonderland appearance. These are not *erosional* features like the cavern itself, but are depositional features created by the seemingly endless dripping of water over great spans of time. The calcite that is left behind produces the limestone we call travertine. These cave deposits, however, are also commonly called *dripstone*, an obvious reference to their mode of origin. Although the formation of caverns takes place in the zone of saturation, the deposition of dripstone is not possible until the caverns are above the water table in the zone of aeration. As soon as the chamber is filled with air, the stage is set for the decoration phase of cavern building to begin.

The various dripstone features found in caverns are collectively called **speleothems**, no two of which are exactly alike (Figure 10.16). Perhaps the most familiar speleothems are **stalactites**. These icicle-like pendants hang from the ceiling of the cavern and form where water seeps through cracks above. When the water reaches the air in the cave, some of the carbon dioxide in solution escapes and a residue of calcium carbonate remains. Deposition occurs as a ring around the edge of the water drop. As drop after drop follows, each leaves an infinitesimal trace of calcite behind, and a hollow limestone tube is created. Water then moves through the tube, remains suspended momentarily at the end, contributes a tiny ring of calcite, and falls to the cavern floor.

The stalactite just described is appropriately called a *soda straw* (Figure 10.17). Often the hollow tube of the soda straw becomes plugged or its supply of water increases. In either case, the water is forced to flow, and hence deposit, along the outside of the tube. As deposition continues, the stalactite takes on the more common conical shape.

Speleothems that form on the floor of a cavern and reach upward toward the ceiling are called **stalagmites** (Figure 10.16). The water supplying the calcite for stalagmite growth falls from the ceiling and splatters over the surface. As a result, stalagmites do not have a central tube, and they are usually more massive in appearance and more rounded on their upper ends than stalactites. Given enough time, a downward-growing stalactite and an upward-growing stalagmite may join to form a *column*.

Karst Topography

Many areas of the world have landscapes that to a large extent have been shaped by the dissolving power of groundwater. Such areas are said to exhibit **karst topography**, named for the Krs Plateau located along the northeastern shore of the Adriatic Sea in the border area between Slovenia (formerly a part of Yugoslavia) and Italy where such topography is strikingly developed (Figure 10.15). The most common geologic setting for karst is where limestone is

FIGURE 10.16 Speleothems are of many types, including stalactites, stalagmites, and columns. Chinese Theater, Carlsbad Caverns National Park. (Photo by David Muench Photography)

FIGURE 10.17 A "live" solitary soda straw stalactite. (Photo by Clifford Stroud, National Park Service)

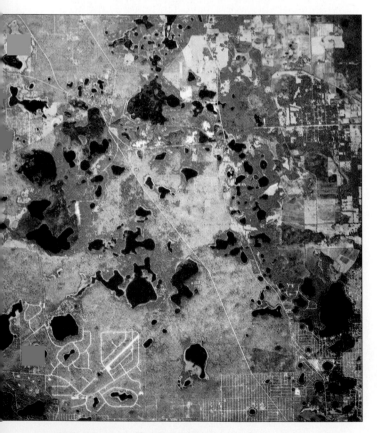

FIGURE 10.18 This high-altitude infrared image shows an area of karst topography in central Florida. The numerous lakes occupy sinkholes. (Courtesy of USDA-ASCS)

present at the surface beneath a mantle of soil. In the United States, karst landscapes occur in Kentucky, Tennessee, southern Indiana, and central and northern Florida. Generally, arid and semiarid areas are too dry to develop karst topography. When solution features exist in such regions, they are likely to be remnants of a time when rainier conditions prevailed.

Karst areas characteristically exhibit an irregular terrain punctuated with many depressions, called **sinkholes** or **sinks** (Figure 10.18). In the limestone areas of Florida, Kentucky, and southern Indiana, there are literally tens of thousands of these depressions varying in depth from just a meter or two to a maximum of more than 50 meters.

Sinkholes commonly form in two ways. Some develop gradually over many years without any physical disturbance to the rock. In these situations, the limestone immediately below the soil is dissolved by downward-seeping rainwater that is freshly charged with carbon dioxide. With time, the bedrock surface is lowered and the fractures into which the water seeps are enlarged. As the fractures grow in size, soil subsides into the widening voids, from which it is removed by groundwater flowing in the passages below. These depressions are usually shallow and have gentle slopes.

By contrast, sinkholes can also form abruptly and without warning when the roof of a cavern collapses under its own weight. Typically, the depressions created in this manner are steep-sided and deep. When they form in populous areas, they may represent a serious geologic hazard.

In addition to a surface pockmarked by sinkholes, karst regions characteristically show a striking lack of surface drainage (streams). Following a rainfall, the runoff is quickly funneled below ground through the sinks. It then flows through caverns until it finally reaches the water table. Where streams do exist at the surface, their paths are usually short. The names of such streams often give a clue to their fate. In the Mammoth Cave area of Kentucky, for example, there is Sinking Creek, Little Sinking Creek, and Sinking Branch. Other sinkholes become plugged with clay and debris to create small lakes or ponds.

The Chapter in Review

The following statements are intended to help you review the primary objectives presented in this chapter.

- As a resource, *groundwater* represents the largest reservoir of freshwater that is readily available to humans. Geologically, the dissolving action of groundwater produces *caves* and *sinkholes*. Groundwater is also an equalizer of stream flow.

- Groundwater is that water which completely fills the pore spaces in sediment and rock in the subsurface *zone of saturation*. The upper limit of this zone is the *water table*. The *zone of aeration* is above the water table where the soil, sediment, and rock are not saturated. Groundwater generally moves within the zone of saturation. The quantity of water that can be stored in a material depends upon its *porosity* (the volume of open spaces). However, the *permeability* (the ability to transmit a fluid through interconnected pore spaces) of a material is the primary factor controlling the movement of groundwater.

- *Springs* occur whenever the water table intersects the land surface and a natural flow of groundwater results. *Wells,* openings bored into the zone of saturation, withdraw groundwater and create roughly conical depressions in the water table known as *cones of depression*. *Artesian wells* occur when water rises above the level at which it was initially encountered.

- Some of the current environmental problems involving groundwater include (1) *overuse* by intense irrigation, (2) *land subsidence* caused by groundwater withdrawal, and (3) *contamination*.

- When groundwater circulates at great depths, it becomes heated. If it rises, the water may emerge as a *hot spring*. Geysers occur when groundwater is heated in underground chambers, expands, and some water quickly changes to steam, causing the geyser to erupt. *Geothermal energy* is harnessed by tapping natural underground reservoirs of steam and hot water.

- Most *caverns* form in limestone at or below the water table when acidic groundwater dissolves rock along lines of weakness, such as joints and bedding planes. The various *dripstone* features found in caverns are collectively called *speleothems*. Landscapes that to a large extent have been shaped by the dissolving power of groundwater exhibit *karst topography,* an irregular terrain punctuated with many depressions, called *sinkholes* or *sinks*.

Key Terms

aquifer (p. 196)
aquitard (p.196)
artesian (p. 198)
cavern (p. 204)
cone of depression (p. 197)
drawdown (p. 197)
effluent stream (p. 194)

geothermal energy (p. 203)
geyser (p. 202)
groundwater (p. 194)
hot spring (p. 201)
influent stream (p. 195)
karst topography (p. 205)
perched water table (p. 197)

permeability (p. 196)
porosity (p. 196)
sinkhole (sink) (p. 206)
speleothem (p. 205)
spring (p. 197)
stalactite (p. 205)
stalagmite (p. 205)
water table (p. 194)

well (p. 197)
zone of aeration (p. 194)
zone of saturation (p. 194)

Questions for Review

1. What percentage of freshwater is groundwater? If glacial ice is excluded and only liquid freshwater is considered, about what percentage is groundwater?

2. Geologically, groundwater is important as an erosional agent. Name another significant geological role of groundwater.

3. Define groundwater and relate it to the water table.

4. What is an effluent stream? How does an influent stream differ?

5. Distinguish between porosity and permeability.

6. What is the difference between an aquitard and an aquifer?

7. Under what circumstances can a material have a high porosity but not be a good aquifer?

8. When an aquitard is situated above the main water table, a localized saturated zone may be created. What term is applied to such a situation?

9. What is the single greatest use of groundwater in the United States?

10. What is meant by the term *artesian*? List two conditions that must be present for artesian wells to exist.

11. What problem is associated with the pumping of groundwater for irrigation in the southern part of the High Plains?

12. Briefly explain what happened in the San Joaquin Valley of California as the result of excessive groundwater withdrawal.

13. Which would be most effective in purifying polluted groundwater: an aquifer composed mainly of coarse gravel, sand, or cavernous limestone?

14. What is meant when a groundwater pollutant is classified as hazardous?

15. What is the source of heat for most hot springs and geysers? How is this reflected in the distribution of these features?

16. Is geothermal power considered an inexhaustible energy source?

17. Name two common speleothems and distinguish between them.

18. Speleothems form in the zone of saturation. True or false? Briefly explain your answer.

19. Areas whose landscapes are largely a reflection of the erosional work of groundwater are said to exhibit what kind of topography?

20. Describe two ways in which sinkholes are created.

Testing What You Have Learned

To test your knowledge of the material presented in this chapter, answer the following questions:

Multiple-Choice Questions

1. The upper limit of the zone of saturation is called the _____.
   a. aquitard   c. aquifer   e. zone of aeration
   b. water table   d. drawdown

2. Landscapes that have been shaped largely by the dissolving power of groundwater are said to exhibit _____ topography.
   a. alluvial   c. eolian   e. glacial
   b. fluvial   d. karst

3. The term _____ is applied to any situation in which groundwater rises in a well above the level where it was initially encountered.
   a. artesian   c. karst   e. saturation
   b. aquifer   d. speleothem

4. Most caverns are created just below the water table in the zone of _____.
   a. saturation   c. aeration   e. depression
   b. sublimation   d. subduction

5. The percentage of the total volume of rock or sediment that consists of open spaces is referred to as the _____ of the material.
   a. permeability   c. density   e. specific gravity
   b. saturation   d. porosity

6. Most natural water contains small quantities of _____ acid.
   a. hydrochloric   c. sulfuric   e. nitric
   b. carbonic   d. acetic

7. Whenever the water table intersects Earth's surface, a natural outflow of groundwater called a(n) _____ results.
   a. sinkhole   c. cavern   e. well
   b. spring   d. aquitard

8. Geothermal energy is used to produce steam and hot water for heating and to _____.
   a. irrigate crops   **c.** build roads   **e.** generate electricity

   b. power ships   **d.** make plastics

9. More than 65 percent of the groundwater used each year is for _____.
   a. industry   **c.** drinking   **e.** swimming

   b. irrigation   **d.** cooling

10. In the zone of _____, the open spaces of materials are filled mainly with air.
    a. aeration   **c.** mobility   **e.** saturation

    b. aquitard   **d.** groundwater

Fill-In Questions

11. During dry periods, the movement of _____ into the channel maintains the flow of a stream.
12. Intermittent hot springs or fountains where columns of hot water and steam are ejected with great force at various intervals are called _____.
13. Without question, _____ represents the largest reservoir of freshwater that is readily available to humans.
14. Icicle-like pendants that hang from the ceiling of a cavern are called _____.
15. Impermeable layers that hinder or prevent groundwater movement are termed _____.

True/False Questions

16. Within the zone of saturation, groundwater does not move. ___
17. A steady rain that falls on a gentle slope composed of permeable materials results in a high percentage of infiltration. ___
18. The primary geologic work carried out by groundwater is that of dissolving rock. ___
19. The water table is usually a subdued replica of the surface topography. ___
20. In the middle latitudes, the level of the water table remains fairly constant during the course of a year. ___

Answers

1.b; 2.d; 3.a; 4.a; 5.d; 6.b; 7.b; 8.e; 9.b; 10.a; 11. groundwater; 12. geysers; 13. groundwater; 14. stalactites; 15. aquitards; 16.F; 17.T; 18.T; 19.T; 20.F

Chapter 11

Glaciers and Glaciation

Focus on Learning

To assist you in learning the important concepts in this chapter, you will find it helpful to focus on the following questions:

- What is a glacier? What are the different types of glaciers? Where are glaciers located?

- How do glaciers move and what are the various processes of glacial erosion?

- What are the features created by glacial erosion and deposition? What materials make up the depositional features?

- What is the evidence for the Ice Age? What are some indirect effects of Ice Age glaciers?

- What are some proposals that attempt to explain the causes of glacial ages?

Valley glaciers continue to modify the alpine landscape of the Alaska Range in Denali National Park, Alaska. (Photo by Michael Collier)

Today, glaciers cover nearly 10 percent of Earth's land surface; however, in the recent geologic past ice sheets were three times more extensive, covering vast areas with ice thousands of meters thick. Many regions still bear the mark of these glaciers. The basic character of such diverse places as the Alps, Cape Cod, and Yosemite Valley was fashioned by now-vanished masses of glacial ice. Moreover, Long Island, the Great Lakes, and the fiords of Norway and Alaska all owe their existence to glaciers. Glaciers, of course, are not just a phenomenon of the geologic past. As we shall see, they are still sculpting and depositing debris in many regions today.

Glaciers: A Part of the Hydrologic Cycle

Earlier we learned that Earth's water is in constant motion. Time and time again the same water is evaporated from the oceans into the atmosphere, precipitated upon the land, and carried by rivers and underground flow back to the sea. However, when precipitation falls at high elevations or high latitudes, the water may not immediately make its way toward the sea. Instead, it may become part of a glacier. Although the ice will eventually melt allowing the water to continue its path to the sea, water can be stored as glacial ice for many tens, hundreds, or even thousands of years.

A **glacier** is a thick ice mass that forms over hundreds or thousands of years. It originates on land from the accumulation, compaction, and recrystallization of snow. A glacier appears to be motionless, but it is not—glaciers move very slowly. Like running water, groundwater, wind, and waves, glaciers are dynamic erosional agents that accumulate, transport, and deposit sediment. Although glaciers are found in many parts of the world today, most are located in remote areas, either near Earth's poles or in high mountains.

Literally thousands of relatively small glaciers exist in lofty mountain areas, where they usually follow valleys originally occupied by streams. Unlike the rivers that previously flowed in these valleys, the glaciers advance slowly, perhaps only a few centimeters per day. Because of their setting, these moving ice masses are termed **valley glaciers** or **alpine glaciers** (Figure 11.1). Each glacier is a stream of ice, bounded by precipitous rock walls, that flows downvalley from a snow accumulation center near its head. Like rivers, valley glaciers can be long or

FIGURE 11.1 Holgate Glacier, in Kenai Fiords National Park near Seward Alaska, is a valley glacier. (Photo by Wolfgang Kaehler)

short, wide or narrow, single or with branching tributaries. Generally the widths of alpine glaciers are small compared to their lengths. In length, some extend for just a fraction of a kilometer, whereas others go on for many tens of kilometers. The west branch of the Hubbard Glacier, for example, runs through 112 kilometers of mountainous terrain in Alaska and the Yukon Territory.

In contrast to valley glaciers, **ice sheets** exist on a much larger scale. These enormous masses flow out in all directions from one or more centers and completely obscure all but the highest areas of underlying terrain. Although many ice sheets have existed in the past, just two achieve this status at present (Figure 11.2). In the Northern Hemisphere, Greenland is covered by an imposing ice sheet averaging nearly 1500 meters thick. It occupies 1.7 million square kilometers, or about 80 percent of this large island. In the Southern Hemisphere, the huge Antarctic Ice sheet attains a maximum thickness of nearly 4300 meters and covers an area of more than

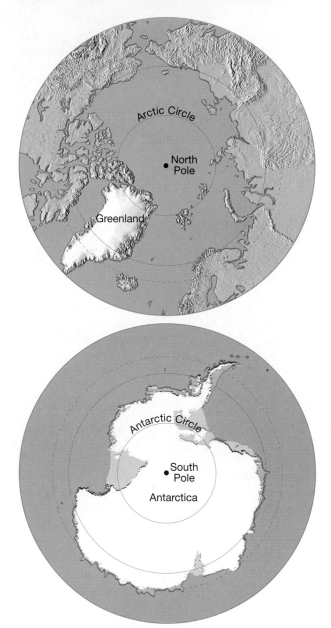

FIGURE 11.2 The only present-day continental ice sheets are those covering Greenland and Antarctica. Their combined areas represent almost 10 percent of Earth's land area. Greenland's ice sheet occupies 1.7 million square kilometers, or about 80 percent of the island. The area of the Antarctic Ice Sheet is almost 14 million square kilometers.

13.9 million square kilometers. Because of the proportions of these huge features, they are often called *continental ice sheets*. Indeed, the combined areas of present-day continental ice sheets represent almost 10 percent of Earth's land area.

In addition to valley glaciers and ice sheets, other types of glaciers are also identified. Covering some uplands and plateaus are masses of glacial ice called **ice caps**. They resemble ice sheets but are much smaller than the continental-scale features. Ice

caps occur in many places, including Iceland and several of the large islands in the Arctic Ocean. Another type, known as **piedmont glaciers**, occupy broad lowlands at the bases of steep mountains and form when one or more valley glaciers emerge from the confining walls of mountain valleys. Here the advancing ice spreads out to form a broad sheet. The size of individual piedmont glaciers varies greatly. Among the largest is the broad Malaspina Glacier along the coast of southern Alaska. It covers more than 5000 square kilometers of the flat coastal plain at the foot of the lofty St. Elias range.

How Glaciers Move

The movement of glacial ice is generally referred to as *flow*. The fact that glacial movement is described in this way seems paradoxical—how can a solid flow? Glacial ice flows in two ways. One mechanism involves plastic movement within the ice. Ice behaves as a brittle solid until the pressure upon it is equivalent to the weight of about 50 meters (165 feet) of ice. Once that load is surpassed, ice behaves as a plastic material and flow begins. A second and often equally important mechanism of glacial movement consists of the whole ice mass slipping along the ground. The lowest portions of most glaciers are thought to move by this sliding process.

The uppermost 50 meters of a glacier are appropriately referred to as the *zone of fracture*. Because there is not enough overlying ice to cause plastic flow, this upper part of the glacier consists of brittle ice. Consequently, the ice in this zone is carried along piggy-back style by the ice below. When the glacier moves over irregular terrain, the zone of fracture is subjected to tension, with cracks called **crevasses** resulting (Figure 11.3). These gaping cracks, which often make travel across glaciers dangerous, may extend to depths of 50 meters (165 feet). Beyond this depth, plastic flow seals them off.

Rates of Glacial Movement

Unlike streamflow, the movement of glaciers is not readily apparent to the casual observer. If we could watch an alpine glacier move, we would see that, like the water in a river, all of the ice in the valley does not move downvalley at an equal rate. Just as friction with the bedrock floor slows the movement of the ice at the bottom of the glacier, the drag created by the valley walls leads to the flow being greatest in the center of the glacier.

This was first demonstrated by experiments in the nineteenth century, in which markers were carefully placed in a straight line across the top of

FIGURE 11.3 Crevasses form in the brittle ice of the zone of fracture. They do not continue down into the zone of flow. Fox Glacier, Westland National Park, New Zealand. (Photo by Tom Till Photography)

FIGURE 11.4 Ice movement and changes in the terminus at Rhone Glacier, Switzerland. In this classic study of a valley glacier, the movement of stakes clearly showed that ice along the sides of the glacier moves slowest. Also notice that even though the ice front was retreating, the ice within the glacier was advancing.

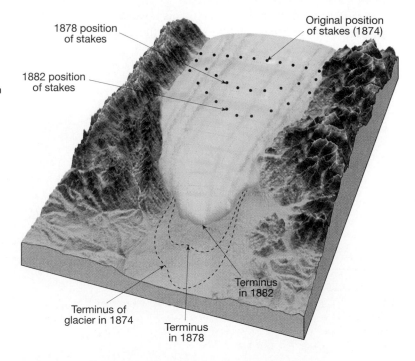

1878 position of stakes

Original position of stakes (1874)

1882 position of stakes

Terminus in 1882

Terminus of glacier in 1874

Terminus in 1878

a valley glacier. Periodically, the positions of the stakes were recorded, revealing the type of movement just described (Figure 11.4).

How rapidly does glacial ice move? Average velocities vary considerably from one glacier to another. Some move so slowly that trees and other vegetation may become well established in the debris that has accumulated on the glacier's surface, whereas others may move at rates of up to several meters per day. For example, Byrd Glacier, an outlet glacier in Antarctica that was the subject of a 10-year-study using satellite images, moved at an average rate of 750 to 800 meters per year (about 2 meters per day). Other glaciers in the study advanced at one-fourth that rate.

The advance of some glaciers is characterized by periods of extremely rapid movements called **surges**. Glaciers that exhibit such movement may flow along in an apparently normal manner, then speed up for a relatively short time before returning to the normal rate again. The flow rates during surges are as much as 100 times the normal rate.

Budget of a Glacier

Snow is the raw material from which glacial ice originates; therefore, glaciers form in areas where more snow falls in winter than melts during the summer. Glaciers are constantly gaining and losing ice. Snow accumulation and ice formation occur in the **zone of accumulation.** Its outer limits are defined by the *snowline.* The elevation of the snowline varies greatly. In polar regions, it may be sea level, whereas in tropical areas, the snowline exists only high in mountain areas, often at altitudes exceeding 4500 meters. Above the snowline, in the zone of accumulation, the addition of snow thickens the glacier and promotes movement. Below the snowline is the **zone of wastage**. Here there is a net loss to the glacier as all of the snow from the previous winter melts, as does some of the glacial ice (Figure 11.5).

In addition to melting, glaciers also waste as large pieces of ice break off the front of the glacier

in a process called **calving**. Calving creates *icebergs* in places where the glacier has reached the sea or a lake. Because icebergs are just slightly less dense than seawater, they float very low in the water, with more than 80 percent of their mass submerged. Along the margins of Antarctica's ice shelves, calving is the primary means by which these masses lose ice. The relatively flat icebergs produced here can be several kilometers across and 600 meters thick. By comparison, thousands of irregularly shaped icebergs are produced by outlet glaciers flowing from the margins of the Greenland Ice Sheet. Many drift southward and find their way into the North Atlantic, where they can pose a hazard to navigation.

Whether the margin of a glacier is advancing, retreating, or remaining stationary depends on the budget of the glacier. The **glacial budget** is the balance, or lack of balance, between accumulation at the upper end of the glacier, and loss at the lower end. This loss is termed **ablation**. If ice accumulation exceeds ablation, the glacial front advances until the two factors balance. When this happens, the terminus of the glacier is stationary. Should a warming trend occur that causes ablation to exceed accumulation,

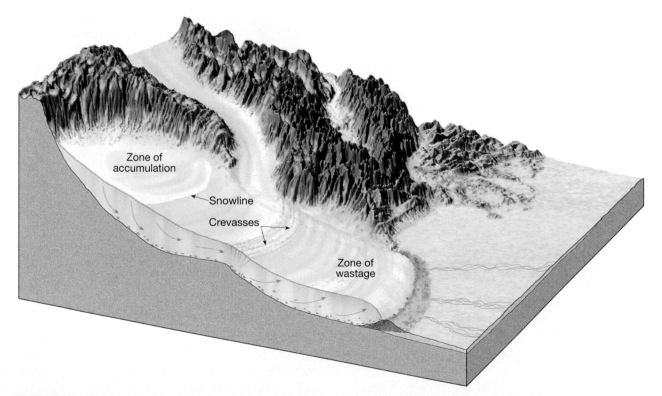

FIGURE 11.5 The snowline separates the zone of accumulation and the zone of wastage. Above the snowline, more snow falls each winter than melts each summer. Below the snowline, the snow from the previous winter completely melts, as does some of the underlying ice. Whether the margin of a glacier advances, retreats, or remains stationary depends on the balance or lack of balance between accumulation and wastage. When a glacier moves across irregular terrain, *crevasses* form in the brittle portion.

the ice front will retreat. As the terminus of the glacier retreats, the extent of the zone of ablation diminishes. Therefore, in time a new balance will be reached between accumulation and ablation, and the ice front will again become stationary.

Whether the margin of a glacier is advancing, retreating, or stationary, the ice within the glacier continues to flow forward. In the case of a receding glacier, the ice still flows forward, but not rapidly enough to offset ablation. This point is illustrated well in Figure 11.4. As the line of stakes within Rhone Glacier continued to move downvalley, the terminus of the glacier slowly retreated upvalley.

Glacial Erosion

Glaciers are capable of great erosion. For anyone who has observed the terminus of an alpine glacier, the evidence of its erosive force is clear. The observer can witness firsthand the release of rock material of various sizes from the ice as it melts. All signs lead to the conclusion that the ice has scraped, scoured, and torn rock from the floor and walls of the valley and carried it downvalley. Indeed, as a medium of sediment transport, ice has no equal. Once rock debris is acquired by the glacier, the enormous competency of ice will not allow the debris to settle out like the load carried by a stream or by the wind. Consequently, glaciers can carry huge blocks that no other erosional agent could possibly budge (Figure 11.6). Although today's glaciers are of limited importance as erosional agents, many landscapes that were modified by the widespread glaciers of the most recent Ice Age still reflect to a high degree the work of ice

Glaciers erode the land primarily in two ways—plucking and abrasion. First, as a glacier flows over a fractured bedrock surface, it loosens and lifts blocks of rock and incorporates them into the ice. This process, known as **plucking**, occurs when meltwater penetrates the cracks and joints of bedrock beneath a glacier and freezes. As the water expands, it exerts tremendous leverage that pries the rock loose. In this manner sediment of all sizes, ranging from particles as fine as flour to blocks as big as houses, becomes part of the glacier's load.

The second major erosional process is **abrasion** (Figure 11.7). As the ice and its load of rock fragments slide over bedrock, they function like sandpaper to smooth and polish the surface below. The pulverized rock produced by the glacial "grist mill" is appropriately called **rock flour**. So much rock flour may be produced that meltwater streams flowing out of a glacier often have the grayish appearance of

FIGURE 11.6 A large, glacially transported boulder in Acadia National Park, Maine. (Photo by Carr Clifton)

FIGURE 11.7 Glacier abrasion produced scratches and grooves in this bedrock exposed at Glacier Bay National Park, Alaska. (Photo by Carr Clifton)

skim milk and offer visible evidence of the grinding power of ice. When the ice at the bottom of a glacier contains large fragments of rock, long scratches and grooves called **glacial striations** may be cut into the bedrock (Figure 11.7). These linear grooves provide clues to the direction of ice flow. By mapping the striations over large areas, patterns of glacial flow can often be reconstructed.

On the other hand, not all abrasive action produces striations. The rock surfaces over which the glacier moves may also become highly polished by the ice and its load of finer particles. The broad expanses of smoothly polished granite in Yosemite National Park provide an excellent example.

As is the case with other agents of erosion, the rate of glacial erosion is highly variable. This differential erosion by ice is largely controlled by four factors: (1) rate of glacial movement; (2) thickness of the ice; (3) shape, abundance, and hardness of the rock fragments contained in the ice at the base of the glacier; and (4) the erodibility of the surface beneath the glacier. Variations in any or all of these factors from time to time and/or from place to place mean that the features, effects, and degree of landscape modification in glaciated regions can vary greatly.

Landforms Created by Glacial Erosion

The erosional effects of valley glaciers and ice sheets are quite different. A visitor to a glaciated mountain region is likely to see a sharp and angular topography. The reason is that as alpine glaciers move downvalley, they tend to accentuate the irregularities of the mountain landscape by creating steeper canyon walls and making bold peaks even more jagged. By contrast, continental ice sheets generally override the terrain and hence subdue rather than accentuate the irregularities they encounter. Although the erosional potential of ice sheets is enormous, landforms carved by these huge ice masses usually do not inspire the same wonderment and awe as do the erosional features created by valley glaciers. Much of the rugged mountain scenery so celebrated for its majestic beauty is the product of erosion by alpine glaciers. Figure 11.8 shows a mountain area before, during, and after glaciation.

Glaciated Valleys

A hike up a glaciated valley reveals a number of striking ice-created features. The valley itself is often a dramatic sight. Unlike streams, which create their own valleys, glaciers take the path of least resistance by following the course of existing stream valleys. Prior to

glaciation, mountain valleys are characteristically narrow and V-shaped because streams are well above base level and are therefore downcutting. However, during glaciation these narrow valleys undergo a transformation as the glacier widens and deepens them, creating a U-shaped **glacial trough** (Figure 11.8C and Figure 11.9). In addition to producing a broader and deeper valley, the glacier also straightens the valley. As ice flows around sharp curves, its great erosional force removes the spurs of land that extend into the valley. The results of this activity are triangular-shaped cliffs called **truncated spurs** (Figure 11.8B).

The intensity of glacial erosion depends in part upon the thickness of the ice. Consequently, main glaciers, also called trunk glaciers, cut their valleys deeper than do their smaller tributary glaciers. Thus, when the glaciers eventually recede, the valleys of tributary glaciers are left standing above the main glacial trough, and are termed **hanging valleys** (Figure 11.8C). Rivers flowing through hanging valleys may produce spectacular waterfalls, such as those in Yosemite National Park (Figure 11.10).

As hikers walk up a glacial trough, they may pass a series of bedrock depressions on the valley floor that were probably formed by plucking and scoured by the abrasive force of the ice. If these depressions are filled with water, they are called **pater noster lakes** (Figure 11.9). The Latin name means "our Father," and is a reference to a string of rosary beads.

At the head of a glacial valley is a very characteristic and often imposing feature associated with an alpine glacier—a **cirque**. These bowl-shaped depressions have precipitous walls on three sides but are open on the downvalley side. The cirque is the focal point of the glacier's growth, because it is the area of snow accumulation and ice formation. Although the origin of cirques is not completely clear, they are believed to begin as irregularities in the mountainside that are subsequently enlarged by the frost wedging and plucking that occur along the sides and bottom of the glacier. After the glacier has melted away, the cirque basin is often occupied by a small lake (Figure 11.11).

Before leaving the topic of glacial troughs and their associated features, one more rather well-known feature should be discussed—fiords. **Fiords** are deep, often spectacular, steep-sided inlets of the sea that are present at high latitudes where mountains are adjacent to the ocean (Figure 11.12). They are drowned glacial troughs that became submerged as the ice left the valley and sea level rose following the Ice Age. The depths of fiords may exceed 1000 meters. However, the great depths of these flooded troughs is only partly explained by the post–Ice Age

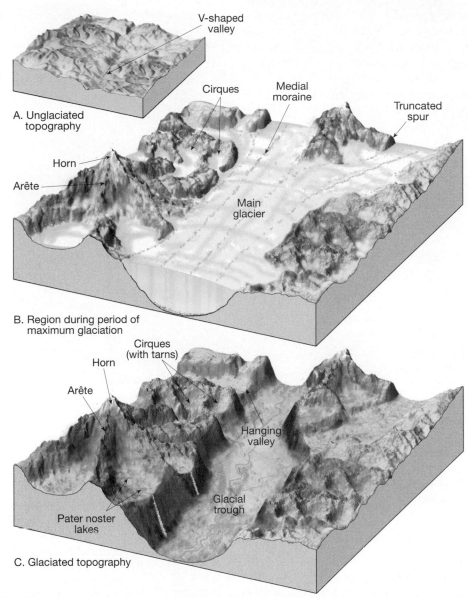

FIGURE 11.8 Erosional landforms created by alpine glaciers. The unglaciated landscape in part **A** is modified by valley glaciers in part **B**. After the ice recedes, in part **C**, the terrain looks very different from before glaciation.

A. Unglaciated topography

B. Region during period of maximum glaciation

C. Glaciated topography

rise in sea level. Unlike the situation governing the downward erosional work of rivers, sea level does not act as base level for glaciers. As a consequence, glaciers are capable of eroding their beds far below the surface of the sea. For example, a 300-meter thick alpine glacier can carve its valley floor more than 250 meters below sea level before downward erosion ceases and the ice begins to float. Norway, British Columbia, Greenland, New Zealand, Chile, and Alaska all have coastlines characterized by fiords.

Arêtes and Horns

A visit to the Alps, the Northern Rockies, or many other scenic mountain landscapes carved by valley glaciers would reveal not only glacial troughs, cirques, pater noster lakes, and the other related

features just discussed. You also would likely see sinuous, sharp-edged ridges called **arêtes** (French for *knife-edge*) and sharp, pyramid-like peaks called **horns** projecting above the surroundings. Both features can originate from the same basic process, the enlargement of cirques produced by plucking and frost action (Figure 11.8 B and C). In the case of the spires of rock called horns, a group of cirques around a single high mountain are responsible. As the cirques enlarge and converge, an isolated horn is produced. The most famous example is the Matterhorn in the Swiss Alps (Figure 11.13).

Arêtes can be formed in a similar manner, except that the cirques are not clustered around a point but rather exist on opposite sides of a divide. As the cirques grow, the divide separating them is reduced

FIGURE 11.9 Prior to glaciation, a mountain valley is typically narrow and V-shaped. During glaciation, an alpine glacier widens, deepens, and straightens the valley, creating the U-shaped glacial trough seen here. The string of lakes is called pater noster lakes. This valley is in Glacier National Park, Montana. (Photo by John Montagne)

to a very narrow knife-like partition. An arête, however, may also be created in another way. When two glaciers occupy parallel valleys, an arête can form when the divide separating the moving tongues of ice is progressively narrowed as the glaciers scour and widen their adjacent valleys.

Roches Moutonnées

In many glaciated landscapes, but most frequently where continental ice sheets have modified the terrain, the ice carves small streamlined hills from protruding bedrock knobs. Such an asymmetrical knob of bedrock is called a **roche moutonnée** (French for *sheep rock*). They are formed when glacial abrasion smoothes the gentle slope facing the oncoming ice sheet and plucking steepens the opposite side as the ice rides over the knob (Figure 11.14). Roches moutonnées indicate the direction of glacial flow, because the gentler slope is always on the side from which the ice advanced.

Glacial Deposits

Glaciers pick up and transport a huge load of debris as they slowly advance across the land. Ultimately these materials are deposited when the ice melts. In regions where glacial sediment is deposited, it can play a truly significant role in forming the physical landscape. For example, in many areas once covered by the ice sheets of the recent Ice Age, the bedrock is rarely exposed because glacial deposits that are tens or even hundreds of meters thick completely mantle the terrain. The general effect of these deposits is to reduce the local relief and thus level the topography. Indeed, much of the familiar country scenery today—rocky pastures in New England, wheat fields in the Dakotas, rolling farmland in the Midwest—results directly from glacial deposition.

Types of Glacial Drift

Long before the theory of an extensive Ice Age was proposed, much of the soil and rock debris covering portions of Europe was recognized as coming from elsewhere. At the time, these foreign materials were believed to have been "drifted" into their present positions by floating ice during an ancient flood. As a consequence, the term *drift* was applied to this sediment. Although rooted in a concept that was not correct, this term was so well established by the time the true glacial origin of the debris became widely recognized that it remained in the glacial vocabulary. Today, **drift** is an all-embracing term for sediments of glacial origin, no matter how, where, or in what form they were deposited.

Glacial drift is divided into two distinct types: (1) materials deposited directly by the glacier, which are known as *till*, and (2) sediments laid down by glacial meltwater, called *stratified drift*. Here is the difference: **till** is deposited as glacial ice melts and drops its load of rock fragments. Unlike moving water and wind, ice cannot sort the sediment it carries; therefore, deposits of till are characteristically unsorted mixtures of many particle sizes (Figure 11.15). **Stratified drift** is sorted according to the size and weight of the fragments. Because ice is not capable of such sorting activity, these sediments are not deposited directly by the glacier. Rather, they reflect the sorting action of glacial meltwater.

Some deposits of stratified drift are made by streams issuing directly from the glacier. Other stratified deposits involve sediment that was originally laid down as till and later picked up, transported, and redeposited by meltwater beyond the margin of the ice. Accumulations of stratified drift often consist largely of sand and gravel, because the meltwater is not capable of moving larger material and because the finer rock flour remains suspended and is commonly carried far from the glacier. An indication that stratified drift consists primarily of sand and gravel can be seen in many areas where these deposits are actively mined as aggregate for road work and other construction projects.

When boulders are found in the till or lying free on the surface, they are called **glacial erratics** if they are different from the bedrock below. Of course, this means that they must have been derived from a source outside the area where they are found. Although the locality of origin for most erratics is unknown, the origin of some can be determined. Therefore, by studying glacial erratics as well as the mineral composition of the

FIGURE 11.10 Bridalveil Falls in Yosemite National Park cascades from a hanging valley into the glacial trough below. (Photo by E. J. Tarbuck)

FIGURE 11.11 Small lakes occupying cirques in the Tombstone Mountains, Yukon Territory, Canada. Cirques are bowl-shaped depressions at the heads of glacial troughs that represent a focal point of ice formation. (Photo by Stephen J. Krasemann/DRK Photo)

FIGURE 11.12 A fiord is a drowned glacial trough. This is one of hundreds of fiords along the coast of Norway. (Photo by H. P. Merten/The Stock Market)

till, geologists can sometimes trace the path of a lobe of ice. In portions of New England as well as other areas, erratics can be seen dotting pastures and farm fields. In some places, these rocks were cleared from fields and piled to make fences and walls (Figure 11.16).

Moraines, Outwash Plains, and Kettles

Perhaps the most widespread features created by glacial deposition are *moraines*, which are simply layers or ridges of till. Several types of moraines are

FIGURE 11.13 Horns are sharp, pyramid-like peaks that are fashioned by alpine glaciers. This example is the famous Matterhorn in the Swiss Alps. (Photo by E. J. Tarbuck)

identified; some are common only to mountain valleys, and others are associated with areas affected by either ice sheets or valley glaciers. Lateral and medial moraines fall in the first category, whereas end moraines and ground moraines are in the second.

The sides of a valley glacier accumulate large quantities of debris from the valley walls. When the glacier wastes away, these materials are left as ridges, called **lateral moraines**, along the sides of the valley (Figure 11.17). **Medial moraines** are formed when two valley glaciers coalesce to form a single ice stream. The till that was once carried along the edges of each glacier joins to form a single dark strip of debris within the newly enlarged glacier. The creation of these dark stripes within the ice stream is one obvious proof that glacial ice moves, because the medial moraine could not form if the ice did not flow downvalley (Figure 11.17).

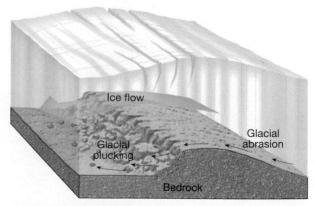

FIGURE 11.14 Roche Moutonnée. The gentle slope was abraded and the steep side was plucked. The ice moved from right to left.

FIGURE 11.15 Glacial till is an unsorted mixture of many different sediment sizes. (Photo by E. J. Tarbuck)

FIGURE 11.16 Land cleared of glacial erratics, which were then piled atop one another to build this stone wall near West Bend, Wisconsin. (Photo by Tom Bean)

End moraines, as the name implies, form at the terminus of a glacier. Here, while the ice front is stationary, the glacier continues to carry in and deposit large quantities of rock debris, creating a ridge of till tens to hundreds of meters high. The end moraine marking the farthest advance of the glacier is called the *terminal moraine*, and those moraines that are formed as the ice front periodically become stationary during retreat are termed *recessional moraines*. As the glacier recedes, a layer of till is laid down, forming a gently undulating surface of **ground moraine**. Ground moraine has a leveling effect, filling in low spots and clogging old stream channels, often leading to a disruption of drainage.

End moraines deposited by the most recent stage of Ice Age glaciation are prominent features in many parts of the Midwest and Northeast. In Wisconsin, the wooded, hilly terrain of the Kettle Moraine near Milwaukee is a particularly picturesque example. A well-known example in the Northeast is Long Island. This linear strip of glacial sediment that extends northeastward from New York City is part of an end moraine complex that stretches from eastern Pennsylvania to Cape Cod, Massachusetts (Figure 11.18).

At the same time that an end moraine is forming, meltwater emerges from the ice in rapidly moving streams. Often they are choked with suspended material and carry a substantial bed load. As the water leaves the glacier, it rapidly loses velocity and much of its bed load is dropped. In this way a broad, ramplike surface of stratified drift is built adjacent to the downstream edge of most end moraines. When the feature is formed in association with an ice sheet, it is termed an **outwash plain**, and when it is confined to a mountain valley, it is usually referred to as a **valley train**. Figure 11.19 shows an outwash plain and other common depositional features.

Often end moraines, outwash plains, and valley trains are pockmarked with basins or depressions known as **kettles** (Figure 11.19). Kettles form when blocks of stagnant ice become buried in drift and

FIGURE 11.17 Lateral moraines form from the accumulation of debris along the sides of a valley glacier. Medial moraines form when the lateral moraines of merging valley glaciers join. Medial moraines could not form if the ice did not advance downvalley. Therefore, these dark stripes are proof that glacial ice moves. (Photo by Austin Post, U.S. Geological Survey)

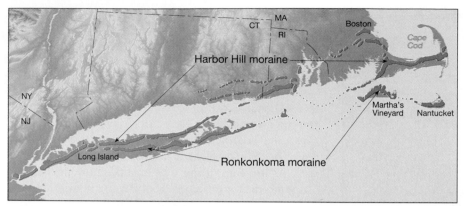

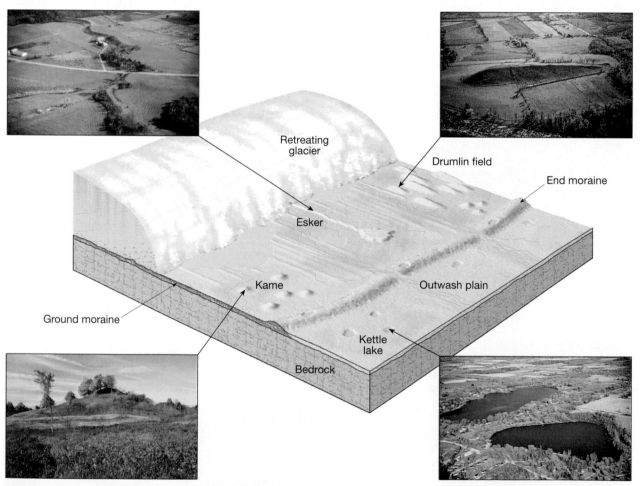

FIGURE 11.19 This hypothetical area illustrates many common depositional landforms. (Photos by Richard Jacobs/JLM Visuals, except upper right, by Ward's Natural Science Establishment, Inc.)

melt, leaving pits in the glacial sediment. Most kettles do not exceed 2 kilometers in diameter and the typical depth of most kettles is less than 10 meters (33 feet). Water often fills the depression and forms a pond or lake. One well-known example is Walden Pond near Concord, Massachusetts. It is here that Henry David Thoreau lived alone for two years in the 1840s and about which he wrote *Walden*, his classic of American literature.

Drumlins, Eskers, and Kames

Drumlins are streamlined asymmetrical hills composed of till (Figure 11.19). They range in height from 15 to 60 meters (50 to 200 feet) and average 0.4

to 0.8 kilometer (.025 to 0.50 mile) in length. The steep side of the hill faces the direction from which the ice advanced, while the gentler slope points in the direction the ice moved. Drumlins are not found singly, but rather occur in clusters, called *drumlin fields*. One such cluster, east of Rochester, New York, is estimated to contain about 10,000 drumlins. Their streamlined shape indicates that they were molded in the zone of flow within an active glacier. It is thought that drumlins originate when glaciers advance over previously deposited drift and reshape the material.

In some areas that were once occupied by glaciers, sinuous ridges composed largely of sand and gravel may be found. These ridges, called **eskers**, are deposits made by streams flowing in tunnels beneath the ice, near the terminus of a glacier (Figure 11.19). They may be several meters high and extend for many kilometers. In some areas they are mined for sand and gravel, and for this reason, eskers are disappearing in some localities.

Kames are steep-sided hills that, like eskers, are composed of sand and gravel (Figure 11.19). Kames originate when glacial meltwater washes sediment into openings and depressions in the stagnant wasting terminus of a glacier. When the ice eventually melts away, the stratified drift is left behind as mounds or hills.

Glaciers of the Past

At various points in the preceding pages, we mentioned the Ice Age, a time when ice sheets and alpine glaciers were far more extensive than they are today. There was a time when the most popular explanation for drift was that the material had been drifted in by means of icebergs or perhaps simply swept across the landscape by a catastrophic flood. However, during the nineteenth century, field investigations by many scientists provided convincing proof that an extensive Ice Age was responsible for these deposits, and for many other features.

By the beginning of the twentieth century, geologists had largely determined the extent of Ice Age glaciation. Further, they discovered that many glaciated regions had not one layer of drift, but several. Close examination of these older deposits showed well-developed zones of chemical weathering and soil formation as well as the remains of plants that require warm temperatures. The evidence was clear: there had not been just one glacial advance but several, each separated by extended periods when climates were as warm or warmer than at present. The Ice Age had not simply been a time when the ice advanced over the land, lingered for a while, and then receded. Rather, the period was a very complex event characterized by a number of advances and withdrawals of glacial ice.

The glacial record on land is punctuated by many erosional gaps. This makes it difficult to reconstruct the episodes of the Ice Age clearly. But sediment on the ocean floor provides an uninterrupted record of climate cycles for this period. Studies of cores drilled from these sea-floor sediments show that glacial/interglacial cycles have occurred about every 100,000 years. About twenty such cycles of cooling and warming were identified for the span we call the Ice Age.

During the glacial age, ice left its imprint on almost 30 percent of Earth's land area, including about 10 million square kilometers of North America, 5 million square kilometers of Europe, and 4 million square kilometers of Siberia (Figure 11.20). The amount of glacial ice in the Northern Hemisphere was roughly twice that of the Southern Hemisphere. The primary reason is that the Southern Hemisphere has little land in the middle latitudes and therefore the southern polar ice could not spread far beyond the margins of Antarctica. By contrast, North America and Eurasia provided great expanses of land for the spread of ice sheets.

Today we know that the Ice Age began between two million and three million years ago. This means that most of the major glacial episodes occurred during a division of the geologic time scale called the **Pleistocene epoch**. Although the Pleistocene is commonly used as a synonym for the Ice Age, this epoch does not encompass it all. The Antarctic Ice Sheet, for example, formed at least 14 million years ago, and, in fact, might be much older.

Glaciers have not been ever-present features throughout Earth's long history. In fact, for most of geologic time, glaciers have been absent. Evidence does indicate that, in addition to the Pleistocene epoch, there were at least three earlier periods of glacial activity: 2 billion, 600 million, and 250 million years ago. However, the most recent period of glaciation is of greatest interest, because the features of many present-day landscapes are a reflection of the work of Pleistocene glaciers.

Some Indirect Effects of Ice Age Glaciers

In addition to the massive erosional and depositional work carried on by Pleistocene glaciers, the ice sheets had other, sometimes profound, effects on the landscape. For example, as the ice advanced and

FIGURE 11.20 Maximum extent of glaciation in the Northern Hemisphere during the Ice Age.

retreated, animals and plants were forced to migrate. This led to stresses that some organisms could not tolerate. Furthermore, many present-day stream courses bear little resemblance to their preglacial routes. The Missouri River once flowed northward toward Hudson Bay in Canada. The Mississippi River followed a path through central Illinois, and the head of the Ohio River reached only as far as Indiana. Some rivers that today carry only a trickle of water but occupy broad channels are a testament to the fact that they once carried torrents of glacial meltwater.

In areas that were centers of ice accumulation, such as Scandinavia and northern Canada, the land has been slowly rising for the past several thousand years. The land had downwarped under the tremendous weight of 3-kilometer-thick masses of ice. Following the removal of this immense load, the crust has been adjusting by gradually rebounding upward ever since.

A far-reaching effect of the Ice Age was the worldwide change in sea level that accompanied each advance and retreat of the ice sheets (see Box 11.1). The snow that nourishes glaciers ultimately comes from moisture evaporated from the oceans.

Therefore, when the ice sheets increased in size, sea level fell and the shoreline shifted seaward (see Figure 11.A in Box 11.1). Estimates suggest that sea level was as much as 100 meters (330 feet) lower than today. Consequently, the Atlantic Coast of the United States was located more than 100 kilometers (60 miles) to the east of New York City. Moreover, France and Britain were joined where the English Channel is today. Alaska and Siberia were connected across the Bering Strait, and Southeast Asia was tied by dry land to the islands of Indonesia.

The formation and growth of ice sheets was an obvious response to significant changes in climate. But the existence of the glaciers themselves triggered climatic changes in the regions beyond their margins. In arid and semiarid areas on all continents, temperatures were lowered, which meant evaporation rates were also lowered. At the same time, precipitation was moderate. This cooler, wetter climate resulted in the formation of many lakes called **pluvial lakes** (from the Latin term *pluvia* meaning "rain"). In North America, pluvial lakes were concentrated in the vast Basin and Range region of Nevada and Utah (Figure

Box 11.1   What If the Ice Melted?

How much water is stored as glacial ice? Estimates by the U.S. Geological Survey indicate that only slightly more than 2 percent of the world's water is accounted for by glaciers. But this small figure may be misleading when the actual amounts of water are considered. The total volume of all valley glaciers is about 210,000 cubic kilometers, comparable to the combined volume of the world's largest saline and freshwater lakes. Furthermore, 80 percent of the world's ice and nearly two-thirds of Earth's freshwater are represented by Antarctica's ice sheet, which covers an area almost one and one-half times that of the United States. If this ice melted, sea level would rise an estimated 60 to 70 meters, and the ocean would inundate many densely populated coastal areas (Figure 11.A). The hydrologic importance of the continent and its ice can be illustrated in another way. If Antarctica's ice sheet were melted at a uniform rate, it could feed (1) the

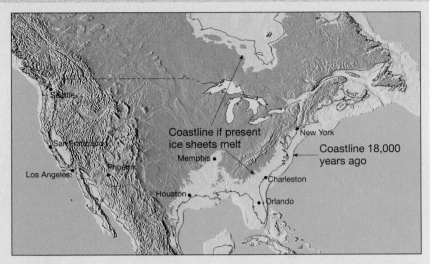

FIGURE 11.A This map of a portion of North America shows the present-day coastline compared to the coastline that existed during the last Ice Age maximum (18,000 years ago) and the coastline that would exist if present ice sheets in Greenland and Antarctica melted. (After R.H. Dott, Jr., and R.L. Battan, *Evolution of the Earth*, New York: McGraw-Hill, 1971. Reprinted by permission of the publisher.)

Mississippi River for more than 50,000 years, (2) all the rivers in the United States for about 17,000 years, (3) The Amazon River for approximately 5,000 years, or (4) all the rivers of the world for about 750 years.

As the foregoing discussion illustrates, the quantity of ice on Earth today is truly immense. However, present glaciers occupy only about one-third the area they did in the very recent geologic past.

11.21). Although most are now gone, a few remain, the largest being Utah's Great Salt Lake.

Causes of Glaciation

A great deal is known about glaciers and glaciation. Much has been learned about glacier formation and movement, the extent of glaciers past and present, and the features created by glaciers, both erosional and depositional. However, scientists have not yet developed a completely satisfactory explanation for the causes of glacial ages.

Any theory that attempts to explain the causes of glacial ages must successfully answer two basic questions. (1) *What causes the onset of glacial conditions?* For continental ice sheets to have formed, average temperatures must have been somewhat lower than at present and perhaps substantially lower than throughout much of geologic time. Thus,

a successful theory would have to account for the cooling that finally leads to glacial conditions. (2) *What caused the alternation of glacial and interglacial stages that have been documented for the Pleistocene epoch?* The first question deals with long-term trends in temperature on a scale of millions of years but this second question relates to much shorter-term changes.

Although the literature of science contains a vast array of hypotheses relating to the possible causes of glacial periods, we will discuss only a few major ideas to summarize current thought.

Plate Tectonics

Probably the most attractive proposal for explaining the fact that extensive glaciations have occurred only a few times in the geologic past comes from the theory of plate tectonics. Not only does this theory provide geologists with explanations about

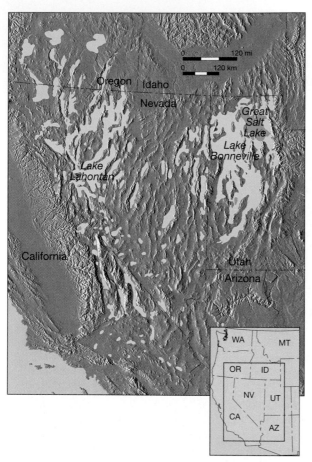

FIGURE 11.21 Pluvial lakes of the Western United States. (After R. F. Flint, *Glacial and Quaternary Geology*, New York: John Wiley & Sons)

many previously misunderstood processes and features, but it also provides a possible explanation for some hitherto unexplainable climatic changes, including the onset of glacial conditions. Because glaciers can form only on land, we know that landmasses must exist somewhere in the higher latitudes before an ice age can commence. Many believe that ice ages have occurred only when Earth's shifting crustal plates have carried the continents from tropical latitudes to more poleward positions.

Glacial features in present-day Africa, Australia, South America, and India indicate that these regions, which are now tropical or subtropical, experienced an Ice Age near the end of the Paleozoic era, about 250 million years ago. However, there is no evidence that ice sheets existed during this same period in what are today the higher latitudes of North America and Eurasia. For many years this puzzled scientists. Was the climate in these relatively tropical latitudes once like it is today in Greenland and Antarctica? Why did glaciers not form in North America and

Eurasia? Until the plate tectonics theory was formulated, there had been no reasonable explanation.

Today, scientists realize that the areas containing these ancient glacial features were joined together as a single supercontinent located at latitudes far to the south of their present positions. Later, this landmass broke apart, and its pieces, each moving on a different plate, drifted toward their present locations (Figure 11.22). It is now understood that during the geologic past, plate movements accounted for many dramatic climatic changes as landmasses shifted in relation to one another and moved to different latitudinal positions.

Changes in oceanic circulation also must have occurred, altering the transport of heat and moisture and consequently the climate as well. Because the rate of plate movement is very slow—a few centimeters per year—appreciable changes in the positions of the continents occur only over great spans of geologic time. Thus, climatic changes brought about by shifting plates are extremely gradual and happen on a scale of millions of years.

Variations in Earth's Orbit

Because climatic changes brought about by moving plates are extremely gradual, the plate tectonics theory cannot be used to explain the alternation between glacial and interglacial climates that occurred during the Pleistocene epoch. Therefore, we must look to some other triggering mechanism that may cause climatic change on a scale of thousands rather than millions of years. Today many scientists strongly suspect that the climatic oscillations that characterized the Pleistocene may be linked to variations in Earth's orbit. This hypothesis was first developed and strongly advocated by the Yugoslavian scientist Milutin Milankovitch and is based on the premise that variations in incoming solar radiation are a principal factor controlling Earth's climate.

Milankovitch formulated a comprehensive mathematical model based on the following elements (Figure 11.23):

1. Variations in the shape (*eccentricity*) of Earth's orbit about the Sun;
2. Changes in *obliquity*; that is, changes in the angle that the axis makes with the plane of Earth's orbit; and
3. The wobbling of Earth's axis, called *precession*.

Using these factors, Milankovitch calculated variations in the receipt of solar energy and the corresponding surface temperature of Earth back into time

FIGURE 11.22 A. The supercontinent Pangaea showing the area covered by glacial ice 300 million years ago. **B.** The continents as they are today. The white areas indicate where evidence of the old ice sheets exists.

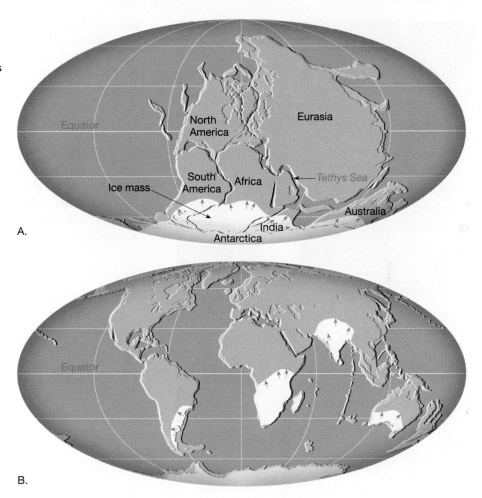

A.

B.

in an attempt to correlate these changes with the climatic fluctuations of the Pleistocene. In explaining climatic changes that result from these three variables, note that they cause little or no variation in the total solar energy reaching the ground. Instead, their impact is felt because they change the degree of contrast between the seasons. Somewhat milder winters in the middle to high latitudes means greater snowfall totals, whereas cooler summers would bring a reduction in snowmelt.

Among the studies that have added credibility to this astronomical hypothesis is one in which deep-sea sediments containing certain climatically sensitive microorganisms were analyzed to establish a chronology of temperature changes going back nearly one-half million years.[*] This time scale of climatic change was then compared to astronomical calculations of eccentricity, obliquity, and precession to determine whether a correlation did indeed exist. Although the study was very involved and mathematically complex, the conclusions were straightforward. The authors found that major variations in climate over the past several hundred thousand years were closely associated with changes in the geometry of Earth's orbit; that is, cycles of climatic change were shown to correspond closely with the periods of obliquity, precession, and orbital eccentricity. More specifically, the authors stated: "It is concluded that changes in the earth's orbital geometry are the fundamental cause of the succession of Quaternary ice ages."[*]

Let us briefly summarize the ideas that were just described. The theory of plate tectonics provides us

[*]J. D. Hays, John Imbrie, and N.J. Shackelton, "Variations in the Earth's Orbit: Pacemaker of the Ice Ages," *Science* 194 (1976): 1121–32.

[*]J. D. Hays et al., p. 1131. The term *Quaternary* refers to the period on the geologic time scale that encompasses the last 1.6 million years.

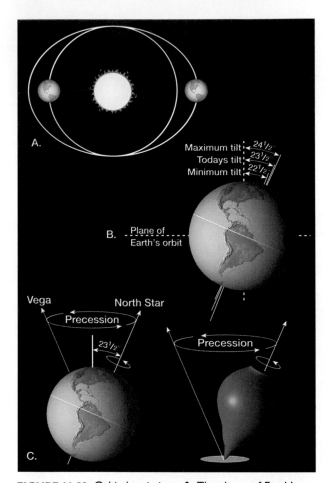

FIGURE 11.23 Orbital variations. **A.** The shape of Earth's orbit changes during a cycle that spans about 100,000 years. It gradually changes from nearly circular to one that is more elliptical and then back again. This diagram greatly exaggerates the amount of change. **B.** Today the axis of rotation is tilted about 23.5° to the plane of Earth's orbit. During a cycle of 41,000 years, this angle varies from 21.5° to 24.5°. **C.** Precession. Earth's axis wobbles like that of a spinning top. Consequently, the axis points to different spots in the sky during a cycle of about 26,000 years.

with an explanation for the widely spaced and non-periodic onset of glacial conditions at various times in the geologic past, while the astronomical model proposed by Milankovitch and supported by the work of J. D. Hays and his colleagues furnishes an explanation for the alternating glacial and interglacial episodes of the Pleistocene.

In conclusion, we emphasize that the ideas just discussed do not represent the only possible explanations for glacial ages. Although interesting and attractive, these proposals are certainly not without critics; nor are they the only possibilities currently under study. Other factors may be, and probably are, involved.

The Chapter in Review

The following statements are intended to help you review the primary objectives presented in this chapter.

- A *glacier* is a thick mass of ice originating on the land as a result of the compaction and recrystallization of snow, and it shows evidence of past or present flow. Today, *valley* or *alpine glaciers* are found in mountain areas where they usually follow valleys that were originally occupied by streams. *Ice sheets* exist on a much larger scale, covering most of Greenland and Antarctica.

- Near the surface of a glacier, in the *zone of fracture,* ice is brittle. However, below about 50 meters, pressure is great, causing ice to *flow* like a *plastic material.* A second important mechanism of glacial movement consists of the entire ice mass *slipping* along the ground.

- The average velocity of glacial movement is generally quite slow, but varies considerably from one glacier to another. The advance of some glaciers is characterized by periods of extremely rapid movements called *surges*.

- Glaciers form in areas where more snow falls in winter than melts during summer. Snow accumulation and ice formation occur in the *zone of accumulation*. Its outer limits are defined by the *snowline*. Below the snowline is the *zone of wastage*, where there is a net loss to the glacier. The *glacial budget* is the balance, or lack of balance, between accumulation at the upper end of the glacier, and loss, called *ablation,* at the lower end.

- Glaciers erode land by *plucking* (lifting pieces of bedrock out of place) and *abrasion* (grinding and scraping of a rock surface). Erosional features produced by valley glaciers include *glacial troughs, hanging valleys, pater noster lakes, fiords, cirques, arêtes, horns,* and *roches moutonnées.*

- Any sediment of glacial origin is called *drift.* The two distinct types of glacial drift are (1) *till,* which is material deposited directly by the ice; and (2) *stratified drift,* which is sediment laid down by meltwater from a glacier.

- The most widespread features created by glacial deposition are layers or ridges of till, called *moraines.* Associated with valley glaciers are *lateral moraines,* formed along the sides of the valley, and *medial moraines,* formed between two valley glaciers that have joined. *End moraines,* which mark the former position of the front of a glacier, and *ground moraine,* an undulating layer of till deposited as the ice front retreats, are common to both valley glaciers and ice sheets. An *outwash plain* is often associated with the end moraine of an ice sheet. A *valley train* may form when the glacier is confined to a valley. Other depositional features include *drumlins* (streamlined asymmetrical hills composed of till), *eskers* (sinuous ridges composed largely of sand and gravel deposited by streams flowing in tunnels beneath the ice, near the terminus of a glacier), and *kames* (steep-sided hills composed of sand and gravel).

- The *Ice Age,* which began two million to three million years ago, was a very complex period characterized by a number of advances and withdrawals of glacial ice. Most of the major glacial episodes occurred during a division of the geologic time scale called the *Pleistocene epoch.* Perhaps the most convincing evidence for the occurrence of several glacial advances during the Ice Age is the widespread existence of *multiple layers of drift* and an uninterrupted record of climate cycles preserved in *sea-floor sediments.* In addition to massive erosional and depositional work, other effects of Ice Age glaciers included the *forced migration* of animals, *changes in stream courses, adjustment of the crust* by rebounding after the removal of the immense load of ice, and *climate changes* caused by the existence of the glaciers themselves. In the sea, the most far-reaching effect of the Ice Age was the *worldwide change* in *sea level* that accompanied each advance and retreat of the ice sheets.

- Any theory that attempts to explain the causes of glacial ages must answer two basic questions: (1) What causes the onset of glacial conditions? and (2) What caused the alternating glacial and interglacial stages that have been documented for the Pleistocene epoch? Two of the many hypotheses for the cause of glacial ages involve (1) plate tectonics and (2) variations in Earth's orbit.

Key Terms

ablation (p. 214)

abrasion (p. 215)

alpine glacier (p. 211)

arête (p. 217)

calving (p. 214)

cirque (p. 216)

crevasse (p. 212)

drift (p. 218)

drumlin (p. 222)

end moraine (p. 221)

esker (p. 223)

fiord (p. 216)

glacial budget (p. 214)

glacial erratic (p. 219)

glacial striations (p. 216)

glacial trough (p. 216)

glacier (p. 211)

ground moraine (p. 221)

hanging valley (p. 216)

horn (p. 217)

ice cap (p. 212)

ice sheet (p. 211)

kame (p. 223)

kettle (p. 221)

lateral moraine (p. 220)

medial moraine (p. 220)

outwash plain (p. 221)

pater noster lakes (p. 216)

piedmont glacier (p. 212)

Pleistocene epoch (p. 223)

plucking (p. 215)

pluvial lake (p. 224)

roche moutonnée (p. 218)

rock flour (p. 215)

stratified drift (p. 219)

surge (p. 213)

till (p. 219)

truncated spur (p. 216)

valley glacier (p. 211)

valley train (p. 221)

zone of accumulation (p. 214)

zone of wastage (p. 214)

Questions for Review

1. What is a glacier? What percentage of Earth's land area do glaciers cover?

2. Each of the following statements refers to a particular type of glacier. Name the type of glacier.

   (a) The term *continental* is often used to describe this type of glacier.
   (b) This type of glacier is also called an *alpine glacier.*
   (c) This is a glacier formed when one or more valley glaciers spreads out at the base of a steep mountain front.
   (d) Greenland is the only example in the Northern Hemisphere.

3. Describe how glaciers fit into the hydrologic cycle. What role do they play in the rock cycle?

4. Describe the two components of glacial flow. At what rates do glaciers move? In a valley glacier does all of the ice move at the same rate? Explain.

5. Why do crevasses form in the upper portion of a glacier but not below a depth of about 50 meters?

6. Under what circumstances will the front of a glacier advance? Retreat? Remain stationary?

7. Describe two basic processes of glacial erosion.

8. How does a glaciated mountain valley differ from a mountain valley that was not glaciated?

9. List and describe the erosional features you might expect to see in an area where alpine glaciers exist or have recently existed.

10. What is glacial drift? What is the difference between till and stratified drift? What general effect do glacial deposits have on the landscape?

11. List the four basic moraine types. What do all moraines have in common? Distinguish between terminal and recessional moraines.

12. List and briefly describe four depositional features other than moraines.

13. Examine the photo of the drumlin in Figure 11.19. From what direction (right or left) did the ice sheet advance in this area?

14. How does a kettle form?

15. About what percentage of Earth's land surface was covered at some time by Pleistocene glaciers? How does this compare to the area presently covered by ice sheets and glaciers? (Check your answer with Question 1).

16. List three indirect effects of Ice Age glaciers.

17. How might plate tectonics help us understand the cause of ice ages? Can plate tectonics explain the alternation between glacial and interglacial climates during the Pleistocene?

Testing What You Have Learned

To test your knowledge of the material presented in this chapter, answer the following questions:

Multiple-Choice Questions

1. The combined areas of present-day continental ice sheets represents almost _____ percent of Earth's land surface.
   a. 5   **c.** 15   **e.** 25
   b. 10   **d.** 20

2. Layers or ridges of till are called _____.
   a. kettles   **c.** moraines   **e.** valley trains
   b. kames   **d.** drumlins

3. Fractures in the uppermost 50 meters of a glacier are called _____.
   a. eskers   **c.** crevasses   **e.** pater nosters
   b. fissures   **d.** ice faults

4. When ice accumulation equals ablation, the front of a glacier will _____.
   a. reverse   **c.** flow downhill   **e.** remain stationary
   b. disappear   **d.** flow uphill

5. Which one of the following is NOT a feature of valley glaciation?
    a. glacial trough   **c.** horn   **e.** fiord
    b. cirque   **d.** levee

6. Which one of the following is deposited directly by a glacier?
    a. till   **c.** loess   **e.** stratified drift
    b. sandstone   **d.** granite

7. Most of the major glacial episodes of the Ice Age occurred during a division of geologic time called the _____.
    a. Mesozoic era   **c.** Eocene epoch   **e.** Hadean eon
    b. Pliocene epoch   **d.** Pleistocene epoch

8. Presently, the two existing ice sheets are those that cover _____ and Antarctica.
    a. Greenland   **c.** northern Canada   **e.** southern Australia
    b. Iceland   **d.** New Zealand

9. The broad, ramplike surface of stratified drift built adjacent to the downstream edge of most end moraines associated with ice sheets is called a(n) _____.
    a. outwash plain   **c.** valley train   **e.** kame
    b. drumlin   **d.** esker

10. The hypothesis that the climatic oscillations that characterized the Pleistocene epoch may be linked to Earth's orbit was first developed by the scientist Milutin _____.
    a. Smith   **c.** Martinich   **e.** McCool
    b. Milankovitch   **d.** Playfair

Fill-in Questions

11. Glacial sediments, no matter how, where, or in what form they were deposited are called _____.
12. The two basic types of glaciers are _____ glaciers and _____ _____.
13. _____ moraines form at the terminus of a glacier.
14. The movement of glacial ice is generally referred to as _____.
15. Two theories that attempt to explain the causes of glaciation involve variations in Earth's orbit and _____ _____.

True/False Questions

16. During the Ice Age, sea level was as much as 100 meters higher than today. ___
17. Stratified drift is sorted according to the size and weight of the fragments. ___
18. Erosion by valley glaciers in mountainous areas tends to produce spectacular features and rugged scenery. ___
19. If ice accumulation exceeds ablation, the glacial front retreats until the two factors balance. ___
20. Under a pressure equivalent to the weight of about 50 meters of ice, glacial ice behaves as a plastic material. ___

Answers

1.b; 2.c; 3.c; 4.e; 5.d; 6.a; 7.d; 8.a; 9.a; 10.b; 11. drift; 12. valley (alpine), ice sheets; 13. End; 14. flow; 15. plate tectonics; 16.F; 17.T; 18.T; 19.F; 20.T.

Deserts and Wind

Focus on Learning

To assist you in learning the important concepts in this chapter, you will find it helpful to focus on the following questions:

- What are some common misconceptions concerning the world's dry regions?

- What are the causes of deserts in both the lower and middle latitudes?

- What are the roles of weathering, water, and wind in arid and semiarid climates?

- How does wind erode?

- What are some depositional features produced by wind?

- How have many of the landscapes in the dry Basin and Range region of the United States evolved?

Mesquite Flat sand dunes in California's Death Valley. (Photo by Carr Clifton)

Desert landscapes frequently appear stark. Their profiles are not softened by a carpet of soil and abundant plant life. Instead, barren rocky outcrops with steep, angular slopes are common. At some places the rocks are tinted orange and red. At others they are gray and brown and streaked with black. For many visitors desert scenery exhibits a striking beauty; to others, the terrain seems bleak. No matter which feeling is elicited, it is clear that deserts are very different from the more humid places where most people live.

As you shall see, arid regions are not dominated by a single geologic process. Rather, the effects of tectonic forces, running water, and wind are all apparent. Because these processes combine in different ways from place to place, the appearance of desert landscapes varies a great deal as well (Figure 12.1).

Common Misconceptions

The word *desert* literally means *deserted* or *unoccupied*. For many dry regions this is a very appropriate description. Yet where water is available, deserts *are occupied* by many people. Nevertheless, the world's dry regions are among the least familiar land areas on Earth outside of the polar realm.

One popular image of deserts is that they consist of mile after mile of drifting sand dunes. However, this is true only for a small percentage of the world's desert area. In the Sahara, the world's largest desert, sand accumulations cover only 10 percent of the surface. In the sandiest of all deserts, the Arabian, about 30 percent is sand covered. A more typical desert surface consists of barren rock or expanses of stony ground.

A second common but incorrect perception of dry lands is that they are practically lifeless. Although reduced in numbers and different in character, plant and animal life is often present. Desert lifeforms differ widely from place to place, yet all have one characteristic in common—they have developed adaptations that make them highly tolerant of drought.

A third misconception is that the world's dry lands are always hot. It is true that many of Earth's highest temperatures have been recorded in desert regions, yet deserts can be quite cold. Dry climates exist from the tropics to the high middle latitudes. Consequently, the temperature regime of a desert depends partly on its latitude. Tropical deserts lack a cold season, but dry regions in the middle latitudes do experience seasonal temperature changes, with extended cold periods common in winter.

FIGURE 12.1 The appearance of desert landscapes varies a great deal from place to place. This rocky desert scene is near the Red Sea in Israel's Timna National Park. (Photo by Tom Till)

Distribution and Causes of Dry Lands

We all recognize that deserts are dry places, but just what is meant by the term *dry*? That is, how much rain defines the boundary between humid and dry regions? Sometimes it is arbitrarily defined by a single rainfall figure—for example, 25 centimeters per year of precipitation. However, the concept of *dryness* is very relative; it refers to *any situation in which water deficiency exists*. Hence climatologists define **dry climate** as one in which yearly precipitation is less than the potential loss of water by evaporation.

Dryness then is related not only to annual rainfall totals but is also a function of evaporation, which, in turn, closely depends upon temperature. As temperatures climb, potential evaporation also increases. Twenty-five centimeters of rain may support only a sparse vegetative cover in Nevada, whereas the same amount of precipitation falling in northern Scandinavia is sufficient to support forests.

The dry regions of the world encompass about 42 million square kilometers, a surprising 30 percent of Earth's land surface. No other climatic group covers so large a land area (see Box 12.1). Within these water-deficient regions, two climatic types are commonly recognized: **desert**, which is *arid*, and **steppe**, which is *semiarid*. The two share many features; their differences are primarily a matter of degree. The steppe is a marginal and more humid variant of the desert and is a transition zone that surrounds the desert and separates it from bordering humid climates. The world map showing the distribution of desert and steppe regions reveals that dry lands are concentrated in the subtropics and in the middle latitudes (Figure 12.2).

Low-Latitude Deserts

The heart of the low-latitude dry climates lies in the vicinities of the Tropics of Cancer and Capricorn. A glance at Figure 12.2 shows a virtually unbroken desert environment stretching for more than 9300 kilometers from the Atlantic coast of North Africa to the dry lands of northwestern India. In addition to this single great expanse, the Northern Hemisphere contains another much smaller area of tropical desert and steppe in northern Mexico and the southwestern United States.

In the Southern Hemisphere, dry climates dominate Australia. Almost 40 percent of the continent is desert, and much of the remainder is steppe. In addition, arid and semiarid areas occur in southern Africa and make a limited appearance in coastal Chile and Peru.

What causes these bands of low-latitude desert? The answer is the global distribution of air pressure and winds. Zones of high air pressure called *subtropical highs* coincide with dry regions in lower latitudes. These pressure systems are characterized by subsiding air currents. When air sinks, it is compressed and warmed. Such conditions are just the opposite of what is needed to produce clouds and precipitation. Consequently, these regions are known for their clear skies, sunshine, and ongoing drought.

Middle-Latitude Deserts

Unlike their low-latitude counterparts, middle-latitude deserts and steppes are not controlled by the subsiding air masses associated with high pressure. Instead, these dry lands exist principally because of their positions in the deep interiors of large landmasses far removed from the ocean, which is the ultimate source of moisture for cloud formation and precipitation.

The presence of high mountains across the paths of prevailing winds further separates these areas from water-bearing maritime air masses. As prevailing winds meet mountain barriers, the air is forced to ascend. When air rises it expands and cools, a process that can produce clouds and precipitation. The windward side of mountains, therefore, often have high precipitation. By contrast, the leeward sides of mountains are usually much drier (Figure 12.3). This situation exists because air reaching the leeward side has lost much of its moisture and, if the air descends, it is compressed and warmed, making cloud formation even less likely. The dry region that results is often referred to as a **rainshadow desert**.

Because most middle-latitude deserts occupy sites on the leeward sides of mountains, they can also be classified as rainshadow deserts. In North America, the Coast Ranges, Sierra Nevada, and Cascades are the foremost mountain barriers to moisture from the Pacific. In Asia, the great Himalayan chain prevents the summertime monsoon flow of moist Indian Ocean air from reaching the interior.

Because the Southern Hemisphere lacks extensive land areas in the middle latitudes, only a small area of desert and steppe occurs in this latitude range, existing primarily near the southern tip of South America in the rainshadow of the towering Andes. Middle-latitude deserts provide an example of how tectonic processes affect climate. Without such mountain-building episodes, wetter climates would prevail where many dry regions exist today.

Geologic Processes in Arid Climates

The angular hills, the sheer canyon walls, and the desert surface of pebbles or sand contrast sharply with the rounded hills and curving slopes of more

Box 12.1   Desertification

On nearly any list of major environmental issues facing the world, one is likely to find reference to the problem of desertification. The term *desertification* means the expansion of desertlike conditions into nondesert areas. Such a transformation can result from natural processes that act gradually over decades, centuries, and millennia. However, in recent years, desertification has come to mean the rapid alteration of land to desertlike conditions as the result of human activities.

Desertification commonly takes place on the margins of deserts. The advancement of desertlike conditions into areas that were previously productive is not a process in which the borders of a desert gradually expand in a uniform manner. Rather, degeneration into desert usually occurs as a patchy transformation of dry but habitable land into dry, uninhabitable land. It results primarily from inappropriate land use and is aided and accelerated by drought. Desertification may be halted during wet years, only to advance rapidly during succeeding dry years.

On marginal land used for crops, natural vegetation is cleared. During periods of drought, crops fail and the unprotected soil is exposed to the forces of erosion. Gullying of slopes and accumulations of sediment in stream channels are visible signs on the landscape, as are the clouds of dust created as topsoil is removed by the wind.

Where crops are not grown, the raising of livestock leads to degradation of the land. Although the modest vegetation associated with marginal lands may be adequate to maintain local wildlife, it cannot support the intensive grazing of large domesticated herds. Overgrazing reduces or eliminates plant cover. When the vegetative cover is destroyed beyond the minimum required for protection of the soil against erosion, the destruction becomes irreversible.

Desertification first received worldwide attention when drought struck a region in Africa called the *Sahel* in the late 1960s (Figure 12.A). During that period and others since, the people in this vast expanse south of the Sahara Desert have suffered malnutrition and death by starvation. Livestock herds have been decimated, and the loss of productive lands has been great. Hundreds of thousands of people have been forced to migrate. As agricultural lands shrink, people must rely on smaller areas for food production. This, in turn, places greater stress on the environment and accelerates the desertificaiton process.

Although human suffering from desertification is most serious in the Sahel, the problem is by no means confined to that region. Desertification exists in other parts of Africa and on every other continent except Antarctica. Each year millions of acres are lost beyond practical hope for reclamation. In every locality, the chief cause of desertification is the stress placed by people on a tenuous environment with fragile soil.

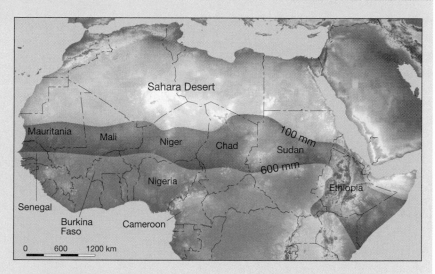

FIGURE 12.A Desertification is most serious along the southern margin of the Sahara in a region known as the Sahel. The lines defining the approximate boundaries of the Sahel represent average rainfall in millimeters.

humid places. Indeed, to a visitor from a humid region, a desert landscape may seem to have been shaped by forces altogether different from those operating in well-watered areas. However, although the contrasts may be striking, they do not reflect different processes. They merely disclose the differing effects of the same processes that operate under contrasting climatic conditions.

Weathering

In humid regions, relatively well-developed soils support an almost continuous cover of vegetation. Here the slopes and rock edges are rounded. Such a landscape reflects the strong influence of chemical weathering in a humid climate. By contrast, much of the weathered debris in deserts consists of unaltered rock and mineral fragments—the results of mechanical

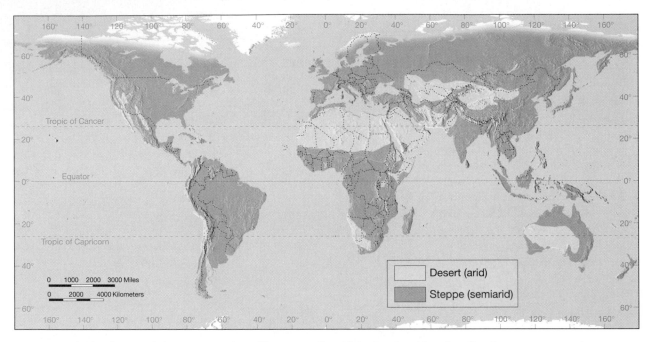

FIGURE 12.2 Arid and semiarid climates cover about 30 percent of Earth's land surface. No other climatic group covers so large an area.

FIGURE 12.3 Many deserts in the middle latitudes are rainshadow deserts. As moving air meets a mountain barrier, it is forced to rise. Clouds and precipitation on the windward side often result. Air descending the leeward side is much drier. The mountains effectively cut the leeward side off from the source of moisture, producing a rainshadow desert.

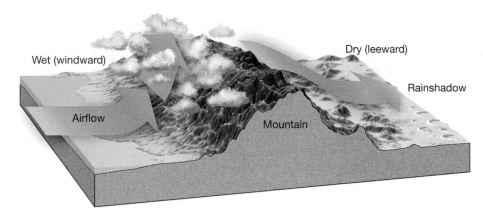

weathering processes. In dry lands, rock weathering of any type is greatly reduced because of the lack of moisture and the scarcity of organic acids from decaying plants. Chemical weathering, however, is not completely lacking in deserts. Over long spans of time, clays and thin soils do form and many iron-bearing silicate minerals oxidize, producing the rust-colored stain found tinting some desert landscapes.

The Role of Water

Permanent streams are normal in humid regions but practically all desert streams are dry most of the time (Figure 12.4A). Desert streams are said to be **ephemeral**, which means that they carry water only in response to specific episodes of rainfall. A typical ephemeral stream may flow only a few days

or perhaps just a few hours during the year. In some years the channel may carry no water at all.

This fact is obvious even to the casual observer who, while traveling in a dry region, notices the number of bridges with no streams beneath them or the number of dips in the road where dry channels cross. However, when the rare heavy showers do come, so much rain falls in such a short time that all of it cannot soak in. Because desert vegetative cover is sparse, runoff is largely unhindered and consequently rapid, often creating flash floods along valley floors (Figure 12.4B). Such floods, however, are quite unlike floods in humid regions. A flood on a river like the Mississippi may take several days to reach its crest and then subside. But desert floods arrive suddenly and subside quickly. Because much

A.

B.

FIGURE 12.4 A. Most of the time, desert stream channels are dry. **B.** An ephemeral stream shortly after a heavy shower. Although such floods are short-lived, large amounts of erosion occur. (Photos by E. J. Tarbuck)

of the surface material is not anchored by vegetation, the amount of erosional work that occurs during a single short-lived rain event is impressive.

In the dry western United States, different names are used for ephemeral streams. Two of the most common are *wash* and *arroyo*. In other parts of the world, a dry desert stream might be called a *wadi* (Arabia and North Africa), a *donga* (South America), or a *nullah* (India).

Unlike the drainage in humid regions, stream courses in arid regions are seldom well integrated. That is, desert streams lack an extensive system of tributaries. In fact, a basic characteristic of deserts is that most of the streams that originate in them are small and die out before reaching the sea. Because the water table is usually far below the surface, few desert streams can draw upon it. Without a steady supply of water, the combination of evaporation and infiltration soon depletes the stream.

The few permanent streams that do cross arid regions, such as the Colorado and Nile rivers, originate *outside* the desert, often in well-watered mountains. Here the water supply must be great to compensate for the losses occurring as the stream crosses the desert. For example, after the Nile leaves the lakes and mountains of central Africa that are its source, it traverses almost 3000 kilometers of the Sahara *without a single tributary*.

It should be emphasized that running water, although an infrequent occurrence, nevertheless does most of the erosional work in deserts. This is contrary to a common belief that wind is the most important erosional agent sculpturing desert landscapes. Although wind erosion is indeed more significant in dry areas than elsewhere, most desert landforms are

nevertheless carved by running water. As you will see shortly, the main role of wind is in the transportation and deposition of sediment, which creates and shapes the ridges and mounds we call dunes.

Transportation of Sediment by Wind

Moving air, like moving water, is turbulent and able to pick up loose debris and transport it to other locations. Just as in a stream, the velocity of wind increases with height above the surface. Also like a stream, wind transports fine particles in suspension while heavier ones are carried as bed load. However, the transport of sediment by wind differs from that by running water in two significant ways. First, wind has a low density compared to water; thus, it is not capable of picking up and transporting coarse materials. Second, because wind is not confined to channels, it can spread sediment over large areas, as well as high into the atmosphere.

Bed Load

The **bed load** carried by wind consists of sand grains. Observations in the field and experiments using wind tunnels indicate that windblown sand moves by skipping and bouncing along the surface—a process termed **saltation**. The term is not a reference to salt, but instead derives from the Latin word meaning "to jump."

The movement of sand grains begins when wind reaches a velocity sufficient to overcome the inertia of the resting particles. At first the sand rolls along the surface. When a moving sand grain strikes

FIGURE 12.5 A cloud of saltating sand grains moving up the gentle slope of a dune. (Photo by Stephen Trimble)

another grain, one or both of them may jump into the air. Once in the air, the grains are carried forward by the wind until gravity pulls them back toward the surface. When the sand hits the surface, it either bounces back into the air or dislodges other grains, which then jump upward. In this manner, a chain reaction is established, filling the air near the ground with saltating sand grains in a short period of time (Figure 12.5).

Bouncing sand grains never travel far from the surface. Even when winds are very strong, the height of the saltating sand seldom exceeds one meter and usually is no greater than one-half meter.

Suspended Load

Unlike sand, finer dust particles can be swept high into the atmosphere by the wind. Because dust is often composed of rather flat particles that have large surface areas compared to their weight, it is relatively easy for turbulent air to counterbalance the pull of gravity and keep these fine particles airborne for hours or even days. Although both silt and clay can be carried in suspension, silt commonly makes up the bulk of the **suspended load** because the reduced level of chemical weathering in deserts produces only small amounts of clay.

Fine particles are easily carried by the wind, but they are not so easily picked up to begin with. The reason is that the wind velocity is practically zero within a very thin layer close to the ground. Thus, the wind cannot lift the sediment by itself. Instead, the dust must be ejected or spattered into the moving currents of air by bouncing sand grains or other disturbances. This idea is illustrated nicely by a dry unpaved country road on a windy day. Left undisturbed, little dust is raised by the wind. However, as a car or truck moves over the road, the layer of silt is kicked up, creating a thick cloud of dust.

Although the suspended load is usually deposited relatively near its source, high winds are capable of carrying large quantities of dust great distances (see Box 12.2). In the 1930s, silt picked up in Kansas was transported to New England and beyond into the North Atlantic. Similarly, dust blown from the Sahara has been traced as far as the West Indies.

Wind Erosion

Compared to running water and glaciers, wind is a relatively insignificant erosional agent. Recall that even in deserts, most erosion is performed by intermittent running water, not by the wind. Wind erosion is more effective in arid lands than in humid areas because in humid regions moisture binds particles together and vegetation anchors the soil. For wind to be effective, dryness and scanty vegetation are important prerequisites.

FIGURE 12.6 This photo was taken north of Granville, North Dakota, in July 1936, during a prolonged drought. Strong winds removed the soil that was not anchored by vegetation. The mounds are 1.2 meters (4 feet) high and show the level of the land prior to deflation. (Photo courtesy of the State Historical Society of North Dakota)

| Box 12.2 | Dust Bowl: Soil Erosion in the Great Plains |

During a span of dry years in the 1930s, large dust storms plagued the Great Plains. Because of the size and severity of these storms, the region came to be called the "Dust Bowl," and the time period, the "dirty thirties." The heart of the Dust Bowl consisted of nearly 100 million acres in the panhandles of Texas and Oklahoma, as well as adjacent parts of Colorado, New Mexico, and Kansas (Figure 12.B). To a lesser extent, dust storms were also a problem over much of the Great Plains, from North Dakota to west central Texas.

At times dust storms were so severe that they were called "black blizzards" and "black rollers" because visibility was reduced to only a few feet. Examples of storms that lasted for hours and stripped huge volumes of topsoil from the land are numerous. In the spring of 1934, a wind storm that lasted for a day and a half created a dust cloud that extended for 2000 kilometers (1200 miles). As the sediment moved east, "muddy rains" were experienced in New York, and "black snows" in Vermont. Less than a year later, another storm carried dust more than 3 kilometers (2 miles) into the atmosphere and transported it 3000 kilometers from its source in Colorado to create twilight conditions in the middle of the day in parts of New England and New York.

FIGURE 12.B Dust blackens the sky on May 21, 1937, near Elkhart, Kansas. It was because of storms like this that portions of the Great Plains were called the "Dust Bowl" in the 1930s. (Photo reproduced from the collection of the Library of Congress)

What caused the Dust Bowl? Clearly, the fact that portions of the Great Plains experience some of North America's strongest winds is important. However, it was the expansion of agriculture that set the stage for the disastrous period of soil erosion. Mechanization allowed the rapid transformation of the grass covered prairies of this semiarid region into farms. Between the 1870s and 1930, the area of cultivation in the region expanded nearly tenfold, from about 10 million acres to more than 100 million acres.

As long as precipitation was adequate, the soil remained in place.

However, when a prolonged drought struck in the 1930s, the unprotected fields were vulnerable to the wind. The results were severe soil loss, crop failures, and economic hardship.

Beginning in 1939, a return to rainier conditions brought relief. Moreover, farming practices that were designed to reduce soil loss by wind had also been instituted. Although dust storms are less numerous and not as severe as in the "dirty thirties," soil erosion by strong winds still occurs periodically whenever the combination of drought and unprotected soil exists.

When such circumstances exist, wind may pick up, transport, and deposit great quantities of fine sediment. During the 1930s, parts of the Great Plains experienced great dust storms. The plowing under of the natural vegetative cover for farming, followed by severe drought, made the land ripe for wind erosion and led to the area being labeled the Dust Bowl (see Box 12.2).

Deflation, Blowouts, and Desert Pavement

One way that wind erodes is by **deflation**, the lifting and removal of loose material. Although the effects of deflation are sometimes difficult to notice because

the entire surface is being lowered at the same time, they can be significant. In portions of the 1930s Dust Bowl, vast areas of land were lowered by as much as one meter in only a few years (Figure 12.6).

The most noticeable results of deflation in some places are shallow depressions called **blowouts**. In the Great Plains region, from Texas north to Montana, thousands of blowouts are visible on the landscape. They range from small dimples less than a meter deep and 3 meters wide to depressions that approach 50 meters in depth and several kilometers across. The factor that controls the depths of these basins (that is, acts

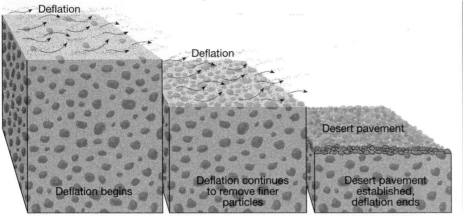

FIGURE 12.7 Formation of desert pavement. Coarse particles gradually become concentrated into a tightly packed layer as deflation lowers the surface by removing sand and silt. If left undisturbed, desert pavement will protect the surface from further deflation. (Photo by Scott T. Smith)

as base level) is the local water table. When blowouts are lowered to the water table, damp ground and vegetation prevent further deflation.

In portions of many deserts the surface is a layer of coarse pebbles and gravels too large to be moved by the wind. Such a layer, callcd **desert pavement**, is created as the wind lowers the surface by removing fine material until eventually only a continuous cover of coarse sediment remains (Figure 12.7). Once desert pavement becomes established, a process that may take hundreds of years, the surface is protected from further deflation if left undisturbed. However, as the layer is only one or two stones thick, disruption by vehicles or animals can dislodge the pavement and expose the fine-grained material below. If this happens, the surface is once again subject to deflation.

Wind Abrasion

Like glaciers and streams, wind also erodes by **abrasion**. In dry regions and along some beaches, windblown sand cuts and polishes exposed rock surfaces. It also creates interestingly shaped stones called **ventifacts** (Figure 12.8). The side of the stone exposed to the prevailing wind is abraded, leaving it polished, pitted, and with sharp edges. If the wind is not consistently from one direction, or if the pebble becomes reoriented it may have several faceted surfaces.

Unfortunately, abrasion is often credited for accomplishments beyond its actual capabilities. Such features as balanced rocks that stand high atop narrow pedestals and intricate detailing on tall pinnacles are not the results of abrasion. Because sand seldom travels more than a meter above the surface, the wind's sandblasting effect is obviously quite limited in vertical extent.

Wind Deposits

Although wind is relatively unimportant in producing *erosional* landforms, there are significant *depositional* landforms created by the wind in some regions. Accumulations of windblown sediment are particularly conspicuous in the world's dry lands and along many sandy coasts. Wind deposits are of two distinctive types: (1) mounds and ridges of sand from the wind's bed load, which we call dunes, and (2) extensive blankets of silt, called loess, that once were carried in suspension.

Sand Deposits

As is the case with running water, wind drops its load of sediment when its velocity falls and the

FIGURE 12.8 Ventifacts are rocks that are polished and shaped by sandblasting. (Photo by Stephen Trimble)

energy available for transport diminishes. Thus, sand begins to accumulate wherever an obstruction across the path of the wind slows its movement. Unlike many deposits of silt, which form blanketlike layers over large areas, winds commonly deposit sand in mounds or ridges called **dunes** (Figure 12.9).

As moving air encounters an object, such as a clump of vegetation or a rock, the wind sweeps around and over it, leaving a shadow of slower moving air behind the obstacle as well as a smaller zone of quieter air just in front of the obstacle. Some of the saltating sand grains moving with the wind come to rest in these wind shadows. As the accumulation of sand continues, it becomes a more imposing barrier to the wind and thus a more efficient trap for even more sand. If there is a sufficient supply of sand and the wind blows steadily for a long enough time, the mound of sand grows into a dune.

The profile of a dune shows an asymmetrical shape with the leeward (sheltered) slope being steep and the windward slope more gently inclined. Sand moves up the gentle slope on the windward side by saltation. Just beyond the crest of the dune, where the wind velocity is reduced, the sand accumulates. As more sand collects, the slope steepens and eventually some of it slides or slumps under the pull of gravity. In this way the leeward slope of the dune, called the **slip face**, maintains an angle of about 34 degrees, the angle of repose for loose dry sand.

(Recall from Chapter 8 that the angle of repose is the steepest angle at which loose material remains stable.) Continued sand accumulation, coupled with periodic slides down the slip face, results in the slow migration of the dune in the direction of air movement (see Figure 12.9).

As sand is deposited on the slip face, layers form that are inclined in the direction the wind is blowing. These sloping layers are called **cross beds**. When the dunes are eventually buried under other layers of sediment and become part of the sedimentary rock record, their asymmetrical shape is destroyed, but the cross beds remain as testimony to their origin. Nowhere is cross-bedding more prominent than in the sandstone walls of Zion Canyon in southern Utah (Figure 12.10).

Types of Sand Dunes

Dunes are not just random heaps of wind-blown sediment. Rather, they are accumulations that usually assume surprisingly consistent patterns (Figure 12.11). A broad assortment of dune forms exists; so to simplify and provide some order, several major types are recognized. Of course, gradations exist among different forms as well as irregularly shaped dunes that do not fit easily into any category. Several factors influence the form and size that dunes ultimately assume. These include wind direction and velocity, availability of sand, and the amount of vegetation present.

FIGURE 12.9 Sand sliding down the steep slip face of a dune, in White Sands National Monument, New Mexico. (Photo by Michael Collier)

FIGURE 12.10 Dunes commonly have an asymmetrical shape. The steeper leeward side is called the *slip face*. Sand grains deposited on the slip face create the *cross-bedding* of the dunes. A complex pattern develops in response to changes in prevailing winds. The sandstone walls in Zion Canyon, Utah, exhibit excellent cross-bedding. (Photo by E. J. Tarbuck)

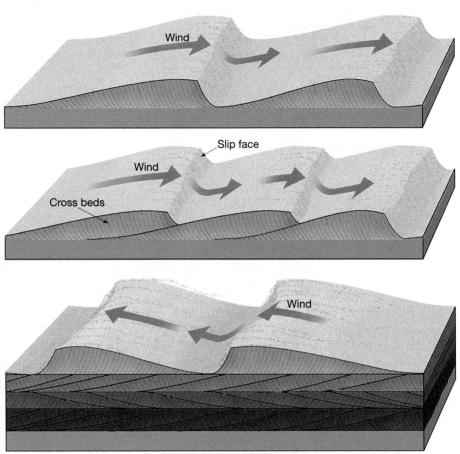

Barchan Dunes Solitary sand dunes shaped like crescents and with their tips pointing downwind are called **barchan dunes** (Figure 12.11A). These dunes form where supplies of sand are limited and the surface is relatively flat, hard, and lacking vegetation. They migrate slowly with the wind at a rate of up to 15 meters per year. Their size is usually modest, with the largest barchans reaching heights of about 30 meters while the maximum spread between their horns approaches 300 meters. When the wind direction is nearly constant, the crescent form of these dunes is nearly symmetrical. However, when the wind direction is not perfectly fixed, one tip becomes larger than the other.

Transverse Dunes In regions where the prevailing winds are steady, sand is plentiful, and vegetation is sparse or absent, the dunes form a series of long ridges that are separated by troughs and oriented at right angles to the prevailing wind. Because of this orientation, they are termed **transverse dunes** (Figure 12.11B). Typically, many coastal dunes are of this type. In addition, transverse dunes are common in many arid regions where the extensive surface of wavy sand is sometimes called a *sand sea*. In some parts of the Sahara and Arabian deserts, transverse dunes reach heights of 200 meters, are 1 to 3 kilometers across, and can extend for distances of 100 kilometers or more.

There is a relatively common dune form that is intermediate between isolated barchans and extensive waves of transverse dunes. Such dunes, called **barchanoid dunes**, form scalloped rows of sand oriented at right angles to the wind (Figure 12.11C). The rows resemble a series of barchans that have been positioned side by side. Visitors exploring the gypsum dunes at White Sands National Monument, New Mexico, will recognize this form.

Longitudinal Dunes **Longitudinal dunes** are long ridges of sand that form more or less parallel to the prevailing wind and where sand supplies are limited (Figure 12.11D). Apparently the prevailing wind direction must vary somewhat, but still remain in the same quadrant of the compass. Although the smaller types are only 3 or 4 meters high and several dozens of meters long, in some large deserts longitudinal dunes can reach great size. For example, in portions of North Africa, Arabia, and central Australia, these dunes can approach a height of 100 meters and extend for distances of more than 100 kilometers (62 miles).

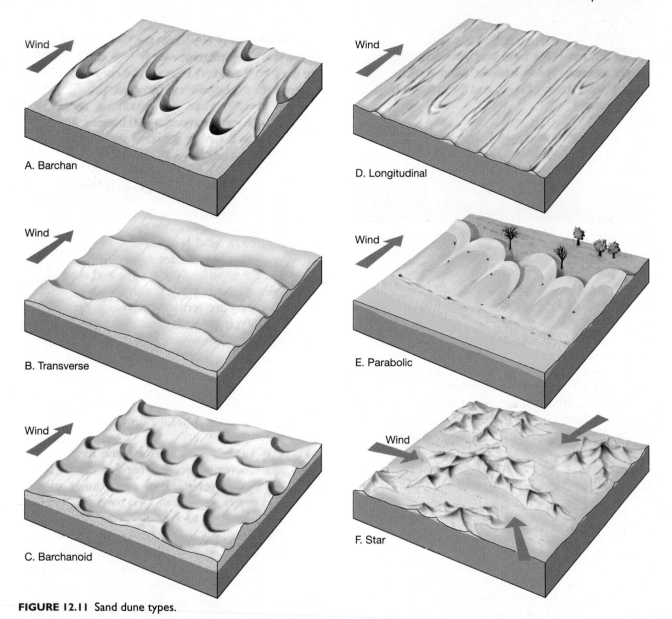

Wind

A. Barchan

Wind

B. Transverse

Wind

C. Barchanoid

Wind

D. Longitudinal

Wind

E. Parabolic

Wind

F. Star

FIGURE 12.11 Sand dune types.

Parabolic Dunes Unlike the other dunes that have been described thus far, **parabolic dunes** form where vegetation partially covers the sand. The shape of these dunes resembles the shape of barchans except that their tips point into the wind rather than downwind (Figure 12.11E). Parabolic dunes often form along coasts where there are strong onshore winds and abundant sand. If the sand's sparse vegetative cover is disturbed at some spot, deflation creates a blowout. Sand is then transported out of the depression and deposited as a curved rim that grows higher as deflation enlarges the blowout.

Star Dunes Confined largely to parts of the Sahara and Arabian deserts, **star dunes** are isolated hills of sand that exhibit a complex form (Figure 12.11F).

Their name is derived from the fact that the bases of these dunes resemble multipointed stars. Usually three or four sharp-crested ridges diverge from a central high point that in some cases may approach a height of 90 meters. As their form suggests, star dunes develop where wind directions are variable.

Loess Deposits

In some parts of the world the surface topography is mantled with deposits of windblown silt, called **loess**. Over periods of perhaps thousands of years dust storms deposited this material. When loess is breached by streams or road cuts, it tends to maintain vertical cliffs and lacks any visible layers, as you can see in Figure 12.12.

FIGURE 12.12 This vertical loess bluff near the Mississippi River in southern Illinois is about 3 meters high. (Photo by James E. Patterson)

The distribution of loess worldwide indicates that there are two principal sources for this sediment: deserts and glacial deposits. The thickest and most extensive deposits of loess on Earth occur in western and northern China. They were blown here from the extensive desert basins of Central Asia. Accumulations of 30 meters are common and thickness of more than 100 meters have been measured. It is this fine, buff-colored sediment that gives the Yellow River (Hwang Ho) and the adjacent Yellow Sea their names.

In the United States, deposits of loess are significant in many areas, including South Dakota, Nebraska, Iowa, Missouri, and Illinois, as well as portions of the Columbia Plateau in the Pacific Northwest. The correlation between the distribution of loess and important farming regions in the Midwest and eastern Washington State is not just a coincidence, because soils derived from this wind-deposited sediment are among the most fertile in the world.

Unlike the deposits in China, which originated in deserts, the loess in the United States and Europe is an indirect product of glaciation, for its source is deposits of stratified drift. During the retreat of the glacial ice, many river valleys were choked with sediment deposited by meltwater. Strong westerly winds sweeping across the barren floodplains picked up the finer sediment and dropped it as a blanket on the eastern sides of the valleys.

Basin and Range: The Evolution of a Desert Landscape

Because arid regions typically lack permanent streams, they are characterized as having **interior drainage**. This means that they have a discontinuous pattern of intermittent streams that do not flow out of the desert to the ocean. In the United States, the dry Basin and Range region provides an excellent example. The region includes southern Oregon, all of Nevada, western Utah, southeastern California, southern Arizona, and southern New Mexico. The name Basin and Range is an apt description for this almost 800,000-square-kilometer region, since it is characterized by more than 200 relatively small mountain ranges that rise 900 to 1500 meters above the basins that separate them.

In this region, as in others like it around the world, most erosion occurs without reference to the ocean (ultimate base level), because the interior drainage never reaches the sea. Even where permanent streams flow to the ocean, few tributaries exist, and thus only a narrow strip of land adjacent to the stream has sea level as its ultimate level of land reduction.

The block models in Figure 12.13 depict how the landscape has evolved in the Basin and Range region. During and following uplift of the mountains, running water begins carving the elevated mass and depositing large quantities of debris in the basin. In this early stage, relief is greatest, and as erosion lowers the mountains and sediment fills the basins, elevation differences diminish (Figure 12.13A).

When the occasional torrents of water produced by sporadic rains move down the mountain canyons, they are heavily loaded with sediment. Emerging from the confines of the canyon, the runoff spreads over the gentler slopes at the base of the mountains and quickly loses velocity. Consequently, most of its load is dumped within a short distance. The result is a cone of debris known as an **alluvial fan** at the mouth of a canyon (Figure 12.14). Over the years, a fan enlarges, eventually coalescing with fans from adjacent canyons to produce an apron of sediment (*bajada*) along the mountain front (Figure 12.13B).

On the rare occasions of abundant rainfall, streams may flow across the alluvial fans to the center of the basin, converting the basin floor into a shallow **playa lake**. Playa lakes last only a few days or weeks, before evaporation and infiltration remove the water. The dry, flat lake bed that remains is termed a *playa*.

Playas occasionally become encrusted with salts left behind by evaporation. These precipitated salts may be uncommon. A case in point is the sodium borate (better known as borax) mined from ancient playa lake deposits in Death Valley, California.

With the ongoing erosion of the mountain mass and the accompanying sedimentation, the local relief continues to diminish. Eventually nearly the entire mountain mass is gone. Thus, by the late stages of erosion, the mountain areas are reduced to a few large bedrock knobs (called

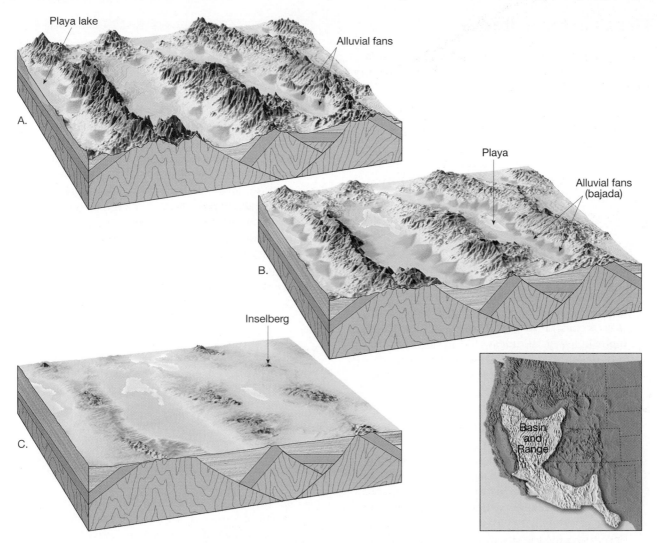

Playa lake

Alluvial fans

A.

Playa

Alluvial fans
(bajada)

B.

Inselberg

C.

Basin
and
Range

FIGURE 12.13 Stages of landscape evolution in a mountainous desert such as the Basin and Range region of the West. As erosion of the mountains and deposition in the basins continue, relief diminishes. **A.** Early stage. **B.** Middle stage. **C.** Late stage.

inselbergs) projecting above the sediment-filled basin (Figure 12.13C).

Each of the stages of landscape evolution in an arid climate depicted in Figure 12.13 can be observed in the Basin and Range region. Recently uplifted mountains in an early stage of erosion are found in southern Oregon and northern Nevada. Death Valley, California, and southern Nevada fit into the more advanced middle stage, while the late stage, with its inselbergs can be seen in southern Arizona.

FIGURE 12.14 Alluvial fans develop where the gradient of a stream changes abruptly from steep to flat. Such a situation exists in Death Valley, California, where streams emerge from the mountains into a flat basin. As a result, Death Valley, has many large alluvial fans. (Photo by Michael Collier)

The Chapter in Review

The following statements are intended to help you review the primary objectives presented in this chapter.

- The word *desert* literally means deserted or unoccupied. Three common misconceptions about deserts are (1) that they consist of mile after mile of drifting sand dunes, (2) they are practically lifeless, and (3) the world's dry lands are always hot.

- The *concept of dryness is relative;* it refers to any situation in which a water deficiency exists. Dry regions encompass about 30 percent of Earth's land surface. Two climatic types are commonly recognized: *desert,* which is arid, and *steppe* (a marginal and more humid variant of desert), which is semiarid. *Low-latitude deserts* coincide with the zones of subtropical highs in lower latitudes. On the other hand, *middle-latitude deserts* exist principally because of their positions in the deep interiors of large landmasses far removed from the ocean.

- The same geologic processes that operate in humid regions also operate in deserts, but under contrasting climatic conditions. In dry lands *rock weathering of any type is greatly reduced* because of the lack of moisture and the scarcity of organic acids from decaying plants. Much of the weathered debris in deserts is the result of *mechanical weathering*. Practically all desert streams are dry most of the time and are said to be *ephemeral*. Stream courses in deserts are seldom well integrated and lack an extensive system of tributaries. Nevertheless, *running water is responsible for most of the erosional work in a desert*. Although wind erosion is more significant in dry areas than elsewhere, the main role of wind in a desert is in the transportation and deposition of sediment.

- The transport of sediment by wind differs from that by running water in two ways. First, wind has a low density compared to water; thus, it is not capable of picking up and transporting coarse materials. Second, because wind is not confined to channels, it can spread sediment over large areas. The *bed load* of wind consists of sand grains skipping and bouncing along the surface in a process termed *saltation*. Fine dust particles are capable of being carried by the wind great distances as *suspended load*.

- Compared to running water and glaciers, wind is a relatively insignificant erosional agent. *Deflation,* the lifting and removal of loose material, often produces shallow depressions called *blowouts*. In portions of many deserts the surface is a layer of coarse pebbles and gravels, called *desert pavement,* too large to be moved by the wind. Wind also erodes by *abrasion,* often creating interestingly shaped stones called *ventifacts*. Because sand seldom travels more than a meter above the surface, the effect of abrasion is obviously limited in vertical extent.

- Wind deposits are of two distinct types: (1) *mounds and ridges of sand,* called *dunes,* which are formed from sediment that is carried as part of the wind's bed load; and (2) extensive *blankets of silt,* called *loess,* that once were carried by wind in *suspension*. The profile of a dune shows an asymmetrical shape with the leeward (sheltered) slope being steep and the windward slope more gently inclined. The *types of sand dunes* include (1) *barchan dunes;* (2) *transverse dunes;* (3) *barchanoid dunes;* (4) *longitudinal dunes;* (5) *parabolic dunes;* and (6) *star dunes*. The thickest and most extensive deposits of loess occur in western and northern China. Unlike the deposits in China, which originated in deserts, the loess in the United States and Europe is an indirect product of glaciation.

- Because arid regions typically lack permanent streams, they are characterized as having *interior drainage*. Many of the landscapes of the Basin and Range region of the western and southwestern United States are the result of streams eroding uplifted mountain blocks and depositing the sediment in interior basins. *Alluvial fans, playas,* and *playa lakes* are features often associated with these landscapes. In the late stages of erosion, the mountain areas are reduced to a few large bedrock knobs, called *inselbergs,* projecting above the sediment-filled basin.

Key Terms

abrasion (p. 240)
alluvial fan (p. 244)
barchan dune (p. 242)
barchanoid dune (p. 242)
bed load (p. 237)
blowout (p. 239)
cross beds (p. 241)

deflation (p. 239)
desert (p. 234)
desert pavement (p. 240)
dry climate (p. 234)
dune (p. 241)
ephemeral stream
 (p. 236)
interior drainage (p. 244)

loess (p. 242)
longitudinal dune
 (p. 242)
parabolic dune (p. 243)
playa lake (p. 244)
rainshadow desert
 (p. 234)
saltation (p. 237)

slip face (p. 241)
star dune (p. 243)
steppe (p. 234)
suspended load (p. 238)
transverse dune (p. 242)
ventifact (p. 240)

Questions for Review

1. Most deserts consist of mile after mile of drifting sand dunes. True or false? Provide some examples to support your answer.

2. How extensive are the desert and steppe regions of Earth?

3. What is the primary cause of subtropical deserts? Of middle-latitude deserts?

4. In which hemisphere (Northern or Southern) are middle-latitude deserts most common?

5. Why is rock weathering less in deserts than in humid climates?

6. What is the most important erosional agent in deserts?

7. What term refers to the process by which desert-like conditions expand into areas that were previously productive? Is this strictly a natural process?

8. Describe the way in which wind transports sand. During very strong winds, how high above the surface can sand be carried?

9. Why is wind erosion relatively more important in arid regions than in humid areas?

10. What factor limits the depths of blowouts?

11. How do sand dunes migrate?

12. Indicate which type of dune is associated with each of the statements below.

   a. Dunes whose tips point into the wind.
   b. Long sand ridges oriented at right angles to the wind.
   c. Often form along coasts where strong winds create a blowout.
   d. Solitary dunes whose tips point downward.
   e. Long sand ridges that are oriented more or less parallel to the prevailing wind.
   f. An isolated dune consisting of three or four sharp-crested ridges diverging from a central high point.
   g. Scalloped rows of sand oriented at right angles to the wind.

13. What is loess? Where are deposits of loess found? What are the origins of this sediment?

14. Why is sea level (ultimate base level) not a significant factor influencing erosion in desert regions?

15. Describe the features and characteristics associated with each of the stages in the evolution of a mountainous desert. Where in the United States can these stages be observed?

Testing What You Have Learned

To test your knowledge of the material presented in this chapter, answer the following questions:

Multiple-Choice Questions

1. About what percentage of Earth's land surface do dry regions encompass?
   a. 10%   c. 30%   e. 50%
   b. 20%   d. 40%

2. In the Basin and Range region, which stage of erosion is characterized by a few inselbergs projecting above the sediment-filled basin?
   a. early   c. rejuvenated   e. intermediate
   b. middle   d. late

3. Most of the erosional work in deserts is accomplished by _____.
   a. running water   c. wind   e. ice

   b. heat   d. groundwater

4. The lifting and removal of loose material by wind is referred to as _____.
   a. abrasion   c. saltation   e. cavitation

   b. deflation   d. blowout

5. Desert streams that carry water only in response to specific episodes of rainfall are said to be _____.
   a. rejuvenated   c. semiarid   e. ephemeral

   b. cyclic   d. transverse

6. Sand dunes shaped like crescents and with their tips pointing downwind are called _____ dunes.
   a. parabolic   c. barchan   e. longitudinal

   b. star   d. transverse

7. Many arid areas, including the Basin and Range region of the United States, are characterized by this type of drainage.
   a. slow   c. swampy   e. exterior

   b. circular   d. interior

8. Desert _____ is a surface consisting of a layer of coarse pebbles and gravels too large to be moved by wind.
   a. pavement   c. inselberg   e. ventifact

   b. playa   d. dune

9. Many middle latitude deserts may also be characterized as _____ deserts.
   a. orographic   c. highland   e. parabolic

   b. rainshadow   d. ephemeral

10. Wind deposits can be either dunes or extensive blankets of silt, called _____.
    a. drift   c. alluvium   e. laterite

    b. loess   d. till

Fill-in Questions

11. Windblown sand moves by skipping and bouncing along the surface in a process called _____.
12. In regions where the prevailing winds are steady, sand is plentiful, and vegetation is sparse or absent, _____ dunes form a series of long ridges that are separated by troughs and oriented at right angles to the prevailing wind.
13. The two climatic types commonly recognized within the dry regions of the world are _____, or arid, and _____, or semiarid.
14. A(n) _____ fan is a cone of debris deposited by a stream at the mouth of a canyon.
15. The thickest and most extensive deposits of loess on Earth occur in western and northern _____.

True/False Questions

16. Middle-latitude deserts and steppes exist principally because of their positions beneath zones of high pressure known as subtropical highs. ___
17. When compared to running water and glaciers, wind is a relatively insignificant erosional agent. ___
18. All deserts are practically lifeless. ___
19. The steepest slope of a dune is located on the side facing the wind. ___
20. Streams in arid regions have few tributaries. ___

Answers

1.c; 2.d; 3.a; 4.b; 5.e; 6.c; 7.d; 8.a; 9.b; 10.b; 11. saltation; 12. transverse; 13. desert, steppe; 14. alluvial; 15. China; 16.F; 17.T; 18.F; 19.F; 20.T

Shorelines

Focus on Learning

To assist you in learning the important concepts in this chapter, you will find it helpful to focus on the following questions:

- What factors influence the height, length, and period of a wave?

- What are the two types of wind-generated waves and the motion of water particles within each?

- How do waves erode?

- What are some typical features produced by wave erosion and from sediment deposited by beach drift and longshore currents?

- What are the local factors that influence shoreline erosion and some basic responses to shoreline erosion problems?

- How do emergent and submergent coasts differ in their formation and characteristic features?

- How are tides produced?

Waves are a powerful erosional force along shorelines. Montana del Oro State Park, Claifornia. (Photo by Carr Clifton)

The restless waters of the ocean are constantly in motion. Winds generate surface currents, the gravity of the Moon and Sun produce tides, and density differences create deep-ocean circulation. Further, waves carry the energy from storms to distant shores, where their impact erodes the land.

Nowhere is the restless nature of the ocean's water more noticeable than along the shore—the dynamic interface among air, land, and sea. Here we can observe the rhythmic rise and fall of tides and see waves constantly rolling in and breaking. Sometimes the waves are low and gentle. At other times, they pound the shore with awesome fury.

Although it may not be readily apparent, the shoreline is constantly being shaped and modified by the moving ocean waters. For example, along Cape Cod, Massachusetts, wave activity is eroding cliffs of poorly consolidated glacial sediments so aggressively that the cliffs are retreating inland at up to 1 meter per year. By contrast, at Point Reyes, California, the far more durable bedrock cliffs are less susceptible to wave attack and are therefore retreating much more slowly. Along both coasts, wave activity is moving sediment near the shore and building narrow sand bars that protrude into and across some bays.

However, the nature of present-day shorelines is not just the result of the relentless attack on the land

by the sea. Indeed, the shore has a complex character that results from multiple geologic processes. For example, practically all coastal areas were affected by the worldwide rise in sea level that accompanied the melting of ice sheets at the close of the Pleistocene epoch. As the sea encroached landward, the shoreline retreated, becoming superimposed upon existing landscapes that had resulted from such diverse processes as stream erosion, glaciation, volcanic activity, and the forces of mountain building.

Today, the coastal zone is experiencing intensive human activity. Unfortunately, people often treat the shoreline as if it were a stable platform on which structures can safely be built. This attitude inevitably leads to conflicts between people and nature. As we shall see, many coastal landforms, especially beaches and barrier islands, are relatively fragile, short-lived geological features that are inappropriate sites for development.

Waves

Wind-generated waves provide most of the energy that shapes and modifies shorelines. Where the land and sea meet, waves that may have traveled unimpeded for hundreds or thousands of kilometers suddenly encounter a barrier that will not allow them to advance

FIGURE 13.1 When waves break against the shore, the force of the water can be powerful and the erosional work that is accomplished can be great. (Photo by H. Richard Johnston/Tony Stone Images)

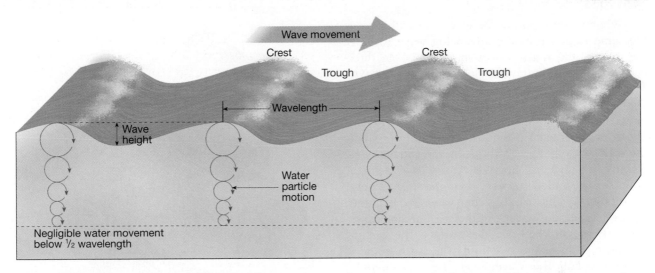

FIGURE 13.2 This diagram illustrates the basic parts of a wave as well as the movement of water particles with the passage of a wave. Negligible water movement occurs below a depth equal to one-half the wavelength (the level of the dashed line).

further. Stated another way, the shore is where a practically irrisistible force confronts an almost immovable object. The conflict that results is never-ending and sometimes dramatic (Figure 13.1).

How Waves Form

It all begins when the wind blows over the ocean. Friction exists between the moving air and the water surface, so the wind drags the water, transferring energy to it. Thus, the undulations of the water surface, called *waves*, derive their energy and motion entirely from the wind. If a breeze of less than 3 kilometers (2 miles) per hour starts to blow across still water, small wavelets appear almost instantly. When the breeze dies, the ripples disappear as suddenly as they formed. However, if the wind exceeds 3 kilometers per hour, more stable waves gradually form and progress with the wind.

All waves are described in terms of the characteristics illustrated in Figure 13.2. The tops of the waves are *crests*, which are separated by *troughs*. The vertical distance between trough and crest is the **wave height**. The horizontal distance separating successive crests is the **wave length**. The **wave period** is the time interval between the passage of two successive crests at a stationary point.

The height, length, and period that are eventually achieved by a wave depend upon three factors: (1) wind speed; (2) length of time the wind has blown; and (3) *fetch*, the distance that the wind has traveled across the open water. As the quantity of energy transferred from the wind to the water increases, the height and steepness of the waves increase as well. Eventually a critical point is reached where waves grow so tall that they topple over, forming ocean breakers called *whitecaps*.

For a particular wind speed there is a maximum fetch and duration of wind beyond which waves will not longer increase in size. When the maximum fetch and duration are reached for a given wind velocity, the waves are said to be "fully developed." The reason that waves can grow no further is that they are losing as much energy through the breaking of whitecaps as they are receiving from the wind.

When the wind stops or changes direction, or the waves leave the stormy area where they were created, they continue on without relation to local winds. The waves also undergo a gradual change to *swells* that are lower in height and longer in length and may carry the storm's energy to distant shores. Because many independent wave systems exist at the same time, the sea surface acquires a complex and irregular pattern. Hence, the sea waves we watch from the shore are usually a mixture of swells from faraway storms and waves created by local winds.

How Waves Move

When observing waves, always remember that you are watching *energy* travel through a medium (water). If you make waves by tossing a pebble into a pond, or by splashing in a pool, or by blowing across the surface of an aquarium, you are imparting *energy* to the water, and the waves you see are just the visible evidence of the energy passing through. In the open sea, it is the wave energy that moves forward, not the water itself. Each water particle moves in a nearly circular path during the passage of a wave (see Figure 13.2). As a wave passes, a water particle returns almost to its original position. The circular orbits followed by the water particles at the surface have a diameter equal to the wave height.

When water is part of the wave crest, it moves in the same direction as the advancing wave form. In the trough, the water moves in the opposite direction. This is demonstrated by observing the behavior of a floating object as a wave passes. The toy boat in Figure 13.3 merely seems to bob up and down and sway slightly to and fro without advancing appreciably from its original position. (The wind does drag the water slightly forward, causing the surface circulation of the ocean.) For this reason, waves in the open sea are called *waves of oscillation*. The energy contributed by the wind to the water is transmitted not only along the surface of the sea but also downward. However, beneath the surface the circular motion rapidly diminishes until at a depth equal to about one-half the wavelength the movement of water particles becomes negligible. This is shown by the rapidly diminishing diameters of water-particle orbits in Figure 13.2.

As long as a wave is in deep water it is unaffected by water depth. However, when a wave approaches the shore, the water becomes shallower and influences wave behavior. The wave begins to "feel bottom" at a water depth equal to about one-half its wavelength. Such depths interfere with water movement at the base of the wave and slow its advance. As the wave continues to advance toward the shore, the slightly faster waves farther out to sea catch up, decreasing the wavelength. As the speed and length of the wave diminish, the wave steadily grows higher. Finally a critical point is reached when the steep wave front is unable to support the wave, and it collapses, or *breaks* (Figure 13.4).

What had been a wave of oscillation now becomes a *wave of translation* in which the water advances up the shore. The turbulent water created by breaking waves is called **surf**. On the landward margin of the surf zone the turbulent sheet of water from collapsing breakers, called *swash*, moves up the slope of the beach. When the energy of the swash has been expended, the water flows back down the beach toward the surf zone as *backwash*.

Wave Erosion

During calm weather, wave action is at a minimum. However, just as streams do most of their work during floods, so waves do most of their work during storms. The impact of high, storm-induced waves against the shore can be awesome (Figure 13.1). Each breaking wave may hurl thousands of tons of water against the land, sometimes making the ground tremble. The pressures exerted by Atlantic

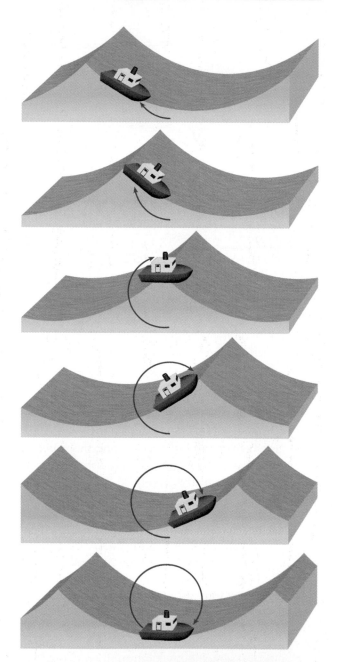

FIGURE 13.3 The movements of the toy boat show that the wave energy advances, but the water itself advances only slightly from its original position. In this sequence, the wave moves from left to right as the boat (and the water in which it is floating) rotates in a circular motion. The boat moves slightly to the left up the front of the approaching wave, then after reaching the crest, slides to the right down the back of the wave.

waves in wintertime, for example, average nearly 10,000 kilograms per square meter (more than 2000 pounds per square foot). The force during storms is even greater. During one such storm in Scotland, for instance, a 1350-ton portion of a steel and concrete breakwater was ripped from the rest of the structure

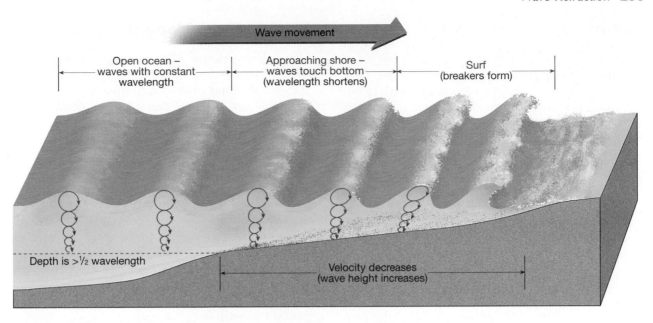

Wave movement

| Open ocean – waves with constant wavelength | Approaching shore – waves touch bottom (wavelength shortens) | Surf (breakers form) |

Depth is >½ wavelength

Velocity decreases (wave height increases)

FIGURE 13.4 Changes that occur when a wave moves onto shore.

and moved toward shore. Five years later, the 2600-ton unit that replaced the first met a similar fate. There are many such stories that demonstrate the great force of breaking waves.

It is no wonder, then, that cracks and crevices are quickly opened in cliffs, seawalls, breakwaters, and anything else that is subjected to these enormous shocks. Water is forced into every opening, greatly compressing air trapped in the cracks. When the wave subsides, the air expands rapidly, dislodging rock fragments and enlarging and extending fractures.

In addition to the erosion caused by wave impact and pressure, **abrasion**, the sawing and grinding action of water armed with rock fragments, is also important. In fact, abrasion is probably more intense in the surf zone than in any other environment. Smooth, rounded stones and pebbles along the shore are obvious reminders of the grinding action of rock against rock in the surf zone. Further, such fragments are used as "tools" by the waves as they cut horizontally into the land (Figure 13.5).

Along shorelines composed of unconsolidated material rather than hard rock, the rate of erosion by breaking waves can be extraordinary. In parts of Britain, where waves have the easy task of eroding glacial deposits of sand, gravel, and clay, the coast has been worn back 3 to 5 kilometers (2 to 3 miles) since Roman times (2000 years ago), sweeping away many villages and ancient landmarks.

Wave Refraction

The bending of waves, called **wave refraction**, plays an important part in shoreline processes. It affects the distribution of energy along the shore, and thus strongly influences where and to what degree erosion, sediment transport, and deposition will take place.

FIGURE 13.5 Cliff undercut by wave erosion along the Oregon coast. (Photo by E.J. Tarbuck)

Waves seldom approach the shore straight on. Rather, most waves move toward the shore at an angle. When they reach the shallow water of a smoothly sloping bottom, however, they are bent and tend to become parallel to the shore. Such bending occurs because the part of the wave nearest the shore touches bottom and slows first, while the end that is still in deep water continues forward at its full speed. The net result is a wave front that may approach nearly parallel to the shore regardless of the original direction of the wave.

Because of refraction, wave impact is concentrated against the sides and ends of headlands that project into the water, whereas wave attack is weakened in bays. This differential wave attack along irregular coastlines is illustrated in Figure 13.6. Because the waves reach the shallow water in front of the headland sooner than they do in adjacent bays, they are bent more nearly parallel to the protruding land and strike it from all three sides. By contrast, refraction in the bays causes waves to diverge and expend less energy. In these zones of weakened wave activity, sediments can accumulate and form sheltered sandy beaches. Over a long period, erosion of the headlands and deposition in the bays will straighten an irregular shoreline.

Moving Sand Along the Beach

Although waves are refracted, most still reach the shore at an angle, however slight. Consequently, the uprush of water from each breaking wave (the swash) is not head-on, but oblique. However, the backwash is straight down the slope of the beach. The effect of this pattern of water movement is to transport particles of sediment in a zigzag pattern along the beach (Figures 13.7). This movement, called **beach drift**, can transport sand and pebbles hundreds or even thousands of meters each day.

Oblique waves also produce currents within the surf zone that flow parallel to the shore. Because the water here is turbulent, these **longshore currents** easily move the fine suspended sand, and roll larger sand and gravel along the bottom. When the sediment transported by longshore currents is added to the quantity moved by beach drift, the total can be very large. At Sandy Hook, New Jersey, for example, the quantity of sand transported along the shore over a 48-year period averaged almost 750,000 tons annually. For a 10-year period at Oxnard, California, more than 1.5 million tons of sediment moved along the shore each year.

There should be little wonder that beaches have been characterized as "rivers of sand." At any point along a beach there is likely to be more sediment that was derived elsewhere than material eroded from the

FIGURE 13.6 Because of wave refraction, the greatest erosional power is concentrated on the headlands. In the bays, the force of the waves is much weaker.

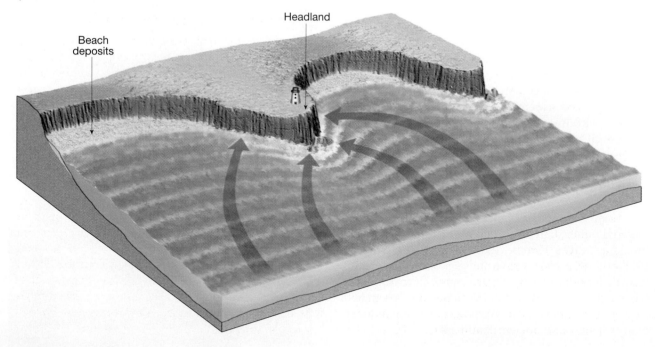

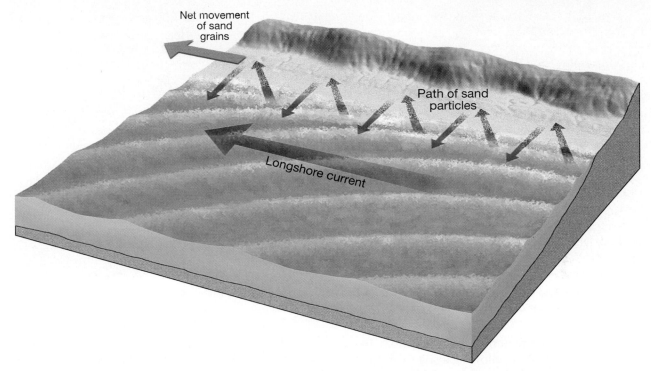

Net movement of sand grains

Path of sand particles

Longshore current

FIGURE 13.7 Beach drift and longshore currents are created by obliquely breaking waves. Beach drift occurs as incoming waves carry sand obliquely up the beach, while the water from spent waves carries it directly down the slope of the beach. Similar movements occur offshore in the surf zone to create the longshore current. These processes transport large quantities of material along the beach and in the surf zone.

land immediately behind it. It is also worth noting that much of the sediment composing beaches is not wave-eroded debris. In many areas the major source of material is sediment-laden rivers that discharge into the ocean. For that reason, if it were not for beach drift and longshore currents to distribute the sand, many beaches would be nearly sandless.

Shoreline Features

Shoreline features vary, depending on the rocks of the shore, currents, wave intensity, and whether the coast is stable, sinking, or rising. This section summarizes several shoreline features, some of which are shown in Figure 13.12.

Wave-Cut Cliffs and Platforms

Whether along the rugged and irregular New England coast or along the steep shorelines of the West Coast, the effects of wave erosion are often easily seen. **Wave-cut cliffs**, as their name implies, originate by the cutting action of the surf against the base of coastal land. As erosion progresses, rocks overhanging the notch at the base of the cliff crumble into the surf and the cliff retreats. A relatively flat, benchlike surface, the **wave-cut platform**, is left

behind by the receding cliff (Figure 13.8). The platform broadens as wave attack continues. Some debris produced by the breaking waves remains along the water's edge as part of the beach, while the remainder is transported farther seaward.

Arches, Stacks, Spits, and Bars

As explained, headlands that extend into the sea are vigorously attacked by waves because of refraction. The surf erodes the rock selectively, wearing away the softer or more highly fractured rock at the fastest rate. At first, sea caves may form. When two caves on opposite sides of a headland unite, a **sea arch** results (Figure 13.9). Finally, the arch falls in, leaving an isolated remnant, or **sea stack**, on the wave-cut platform (Figure 13.9). Eventually it too will be consumed by the action of the waves.

Where beach drift and longshore currents are active, several features related to the movement of sediment along the shore may develop. **Spits** are elongated ridges of sand that project from the land into the mouth of an adjacent bay. Often the end in the water hooks landward in response to wave-generated currents (Figure 13.10). The term **baymouth bar** is applied to a sand bar that completely crosses a bay, sealing it off from the open ocean (Figure

FIGURE 13.8 Elevated wave-cut platform along the California coast near San Francisco (right). A new platform is being created at the base of the cliff (left foreground). (Photo by John S. Shelton)

13.10). Such a feature tends to form across bays where currents are weak, allowing a spit to extend to the other side. A **tombolo**, a ridge of sand that connects an island to the mainland or to another island, forms in much the same manner as does a spit.

Barrier Islands

The Atlantic and Gulf Coastal Plains are relatively flat and slope gently seaward. The shore zone is characterized by **barrier islands**. These low ridges of sand parallel the coast at distances from 3 to 30 kilometers offshore. From Cape Cod, Massachusetts, to Padre Island, Texas, nearly 300 barrier islands rim the coast (Figure 13.11).

Most barrier islands are from 1 to 5 kilometers wide and between 15 and 30 kilometers long. The highest features are sand dunes, which usually reach heights of 5 to 10 meters. The lagoons that separate these narrow islands from the shore are zones of relatively quiet water that allow small craft traveling between New York and northern Florida to avoid the rough waters of the North Atlantic.

Barrier islands probably originate in several ways. Some originate as spits that were subsequently severed from the mainland by wave erosion or by the general rise in sea level following the last episode of glaciation. Others are created when turbulent waters in the line of breakers heap up sand that has been

FIGURE 13.9 Sea stack (left) and sea arch (right) at Cabo San Lucas, Mexico. (Photo by Natalie Forbes/Tony Stone Images)

FIGURE 13.10 High-altitude image of a well-developed spit and baymouth bar along the coast of Martha's Vineyard, Massachusetts. Also notice the tidal delta in the lagoon adjacent to the inlet through the baymouth bar. (Photo courtesy of USDA-ASCS)

scoured from the bottom. Also some barrier islands may be former sand dune ridges that originated along the shore during the last glacial period, when sea level was lower. As the ice sheets melted, sea level rose and flooded the area behind the beach-dune complex.

The Evolving Shore

A shoreline continually undergoes modification regardless of its initial configuration. At first most coastlines are irregular, although the degree of and reason for the irregularity may differ considerably from place to place. Along a coastline of varied geology, the pounding surf may at first increase its irregularity because the waves will erode the weaker rocks more easily than the stronger ones. However, if a shoreline remains stable, marine erosion and deposition will eventually produce a straighter, more regular coast.

Figure 13.12 illustrates the evolution of an initially irregular coast. As waves erode the headlands, creating cliffs and a wave-cut platform, sediment is carried along the shore. Some material is deposited in the bays, while other debris is formed into spits and baymouth bars. At the same time rivers fill the bays with sediment. Ultimately, a generally straight, smooth coast results.

Shoreline Erosion Problems

Compared with natural hazards such as earthquakes, volcanic eruptions, and landslides, shoreline erosion appears to be a more continuous and predictable

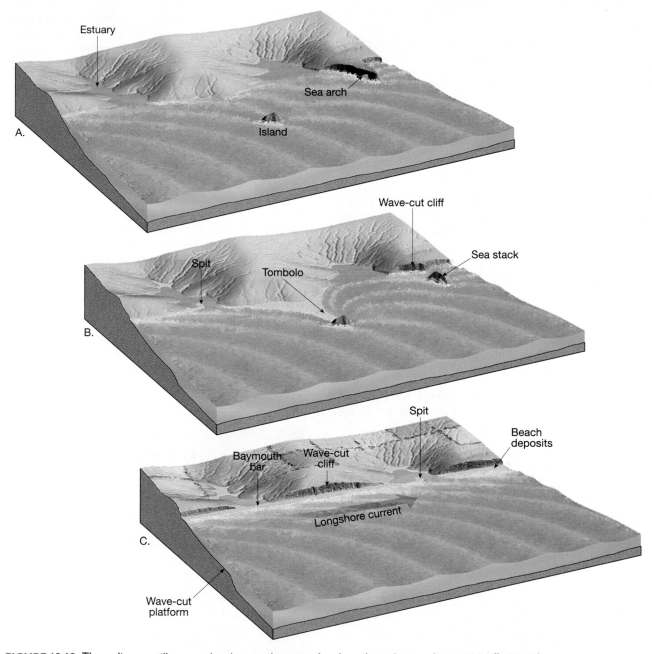

FIGURE 13.12 These diagrams illustrate the changes that can take place through time along an initially irregular coastline that remains relatively stable. The coastline shown in part **A** gradually evolves to **B** and then **C**. The diagrams also serve to illustrate many of the features described in the section on the shoreline features.

During this century, growing affluence and increasing demands for recreation have brought unprecedented development to many coastal areas. As the number and value of buildings has increased, so too have efforts to protect property from storm waves. Also, controlling the natural migration of sand is an ongoing struggle in many coastal areas. Such interference can result in unwanted changes that are difficult and expensive to correct.

Groins

To maintain or widen beaches that are losing sand, **groins** are sometimes constructed. A groin is a barrier built at a right angle to the beach to trap sand that is moving parallel to the shore (Figure 13.13). These structures often do their job so effectively that the longshore current beyond the groin becomes sand-starved. As a result, the current erodes sand from the beach on the leeward side of the groin.

FIGURE 13.13 A series of groins at Ship Bottom, New Jersey. Because groins trap sand on the upcurrent side, the movement of sand along this coast, caused by the longshore current, must be toward the bottom of the photograph. (Photo by John S. Shelton)

To offset this effect, property owners downcurrent from the structure may erect a groin on their property. In this manner, the number of groins multiplies. An example of such proliferation is the shoreline of New Jersey, where hundreds of these structures have been built. Because it has been shown that groins often do not provide a satisfactory solution, they are no longer the preferred method of keeping beach erosion in check.

Breakwaters and Seawalls

In some coastal areas a **breakwater** may be constructed parallel to the shoreline. The purpose of such a structure is to protect boats from the force of large breaking waves by creating a quiet water zone near the shore. However, when this is done, the reduced wave activity along the shore behind the structure may allow sand to accumulate. If this happens, the marina will eventually fill with sand while the downstream beach erodes and retreats. At Santa Monica, California, where the building of a breakwater created such a problem, the city had to install a dredge to remove sand from the protected quiet water zone and deposit it down the beach where longshore currents and beach drift could recirculate the sand (Figure 13.14).

As development has moved ever closer to the beach, seawalls are sometimes built to defend property from the force of breaking waves. **Seawalls** are simply massive barriers intended to prevent waves from reaching the areas behind the wall. Waves

expend much of their energy as they move across an open beach. Seawalls cut this process short by reflecting the force of unspent waves seaward. As a consequence, the beach to the seaward side of the seawall experiences significant erosion and may, in some instances, be eliminated entirely. Once the width of the beach is reduced, the seawall is subjected to even greater pounding by the waves. Eventually this battering will cause the wall to fail, and a larger, more expensive wall must be built to take its place.

The wisdom of building temporary protective structures along shorelines is increasingly questioned. The feelings of many coastal scientists are expressed in the following excerpt from a position paper that grew out of a conference on America's Eroding Shoreline:

It is now clear that halting the receding shoreline with protective structures benefits only a few and seriously degrades or destroys the natural beach and the value it holds for the majority. Protective structures divert the ocean's energy temporarily from private properties, but usually refocus the energy on the adjacent natural beaches. Many interrupt the natural sand flow in coastal currents, robbing many beaches of vital sand replacement.[*]

---

[*]"Strategy for Beach Preservation Proposed," *Geotimes* 30 (No. 12, December 1985): 15.

FIGURE 13.14 Aerial view of a breakwater at Santa Monica, California. The breakwater appears as a faint line in the water behind which many boats are anchored. The construction of the breakwater disrupted longshore transport and caused the seaward growth of the beach. (Photo by John S. Shelton)

Beach Nourishment

Beach nourishment represents another approach to stabilizing shoreline sands. As the term implies, this practice simply involves the addition of large quantities of sand to the beach system. By building the beaches seaward, beach quality and storm protection are both improved. Beach nourishment, however, is not a permanent solution to the problem of shrinking beaches. The same processes that removed the sand in the first place will eventually remove the replacement sand as well. In addition, beach nourishment is very expensive. When beach nourishment was used to renew 24 kilometers of Miami Beach, the cost was $64 million. Here the restoration must be redone every 10 to 12 years.

In some instances, beach nourishment can lead to unwanted environmental effects. For example, beach replenishment at Waikiki Beach, Hawaii, involved replacing coarse calcareous sand with softer, muddier calcareous sand. Destruction of the soft beach sand by breaking waves increased the water's turbidity and killed offshore coral reefs. At Miami Beach, increased turbidity also damaged local coral communities.

So far, two basic responses to shoreline erosion problems have been considered: (1) the building of structures, such as groins and seawalls, to hold the shoreline in place, and (2) the addition of sand to replenish eroding beaches. However, a third option is also available, and that is to relocate buildings away from the beach. For many areas, the wisdom of adding new sand to the beaches and/or building stronger and stronger seawalls is indeed questionable. When this is the case, abandonment and relocation are the alternatives.

Contrasting the Atlantic and Pacific Coasts

The shoreline along the Pacific Coast of the United States is strikingly different from that characterizing the Atlantic and Gulf Coast regions. Some of the differences are related to plate tectonics. The West Coast represents the leading edge of the North American plate, and because of this, it experiences active uplift and deformation. By contrast, the East Coast is a tectonically quiet region that is far from any active plate margin. Because of this basic geological difference, the nature of shoreline erosion problems along America's opposite coasts is different.

Atlantic and Gulf Coasts During this century growing affluence and increasing demand for recreation have brought unprecedented development to many coastal areas. Much of this development has occurred on barrier islands. Typically, barrier islands consist of a wide beach that is backed by dunes and separated from the mainland by marshy lagoons. The broad expanses of sand and exposure to the ocean have made barrier islands exceedingly attractive sites for development. Unfortunately, development has taken place more rapidly than our understanding of barrier island dynamics.

Because barrier islands face the open ocean, they receive the full force of major storms that strike the coast. When a storm occurs, the barriers absorb the energy of the waves primarily through the movement of sand. Frank Lowenstein describes this process and the dilemma that results:

Waves may move sand from the beach to offshore areas or, conversely, into the dunes; they may erode the dunes, depositing sand onto the beach or carrying it out to sea; or they may carry sand from the beach and the dunes into the marshes behind the barrier, a process known of overwash. The common factor is movement. Just as a flexible reed may survive a wind that destroys an oak tree, so the barriers survive hurricanes and nor'easters not through unyielding strength but by giving before the storm.

This picture changes when a barrier is developed for homes or a resort. Storm waves that previously rushed harmlessly through gaps between the dunes now encounter buildings and roadways. Moreover, because the dynamic nature of the barrier is readily perceived only during storms, homeowners tend to attribute damage to a particular storm, rather than to the basic mobility of coastal barriers. With their homes or investments at stake, local residents are more likely to seek to hold the sand in place and the waves at bay than to admit that development was improperly placed to begin with. [*]

Pacific Coast In contrast to the broad, gently sloping coastal plains of the East, much of the Pacific Coast is characterized by relatively narrow beaches that are backed by steep cliffs and mountain ranges. Recall that America's western margin is a more rugged and tectonically active region than the East. Because uplift continues, the apparent rise in sea level in the West is not so readily apparent. Nevertheless, like the shoreline erosion problems facing the East's barrier islands, West Coast difficulties also stem largely from the alteration of a natural system by people.

A major problem facing the Pacific shoreline, and especially portions of southern California, is a significant narrowing of many beaches. The bulk of the

---

[*] "Beaches or Bedrooms—The Choice as Sea Level Rises," *Oceanus* 28 (3) (Fall 1985): 22.

sand on many of these beaches is supplied by rivers that transport it from the mountainous regions to the coast. Over the years this natural flow of material to the coast has been interrupted by dams built for irrigation and flood control. The reservoirs effectively trap the sand that would otherwise nourish the beach environment. When the beaches were wider, they served to protect the cliffs behind them from the force of storm waves. Now, however, the waves move across the narrowed beaches without losing much energy and cause more rapid erosion of the sea cliffs.

Although the retreat of the cliffs provides material to replace some of the sand impounded behind dams, it also endangers homes and roads built on the bluffs. In addition, development atop the cliffs aggravates the problem. Urbanization increases runoff, which, if not carefully controlled, can result in serious bluff erosion. Watering lawns and gardens adds significant quantities of water to the slope. This water percolates downward toward the base of the cliff where it may emerge in small seeps. This action reduces the slope's stability and facilitates mass wasting.

Shoreline erosion along the Pacific Coast varies considerably from one year to the next, largely because of the sporadic occurrence of storms. As a consequence, when the infrequent but serious episodes of erosion occur, the damage is often blamed on the unusual storms and not on coastal development or the sediment-trapping dams that may be great distances away. If, as predicted, sea level rises at an increasing rate in the years to come, increased shoreline erosion and sea cliff retreat should be expected along many parts of the Pacific Coast (see Box 13.1).

Emergent and Submergent Coasts

The great variety of shorelines demonstrates their complexity. Indeed, to understand any particular coastal area, you must consider rock types, size and direction of waves, number of storms, range between high and low tides, and submarine profile. Moreover, recent tectonic events and changes in sea level must also be taken into account. These many variables make shoreline classification difficult.

Many geologists classify coasts based on changes that have occurred with respect to sea level. This commonly used classification divides coasts into two very general categories: emergent and submergent. **Emergent coasts** develop either because an area experiences uplift or as a result of a drop in sea level. Conversely, **submergent coasts** are created when sea level rises or the land adjacent to the sea subsides.

In some areas the coast is clearly emergent because rising land or a falling water level expose wave-cut cliffs and platforms above sea level. Excellent examples include portions of coastal California where uplift has occurred in the recent geological past. The elevated wave-cut platform shown in Figure 13.8 illustrates this. In the case of the Palos Verdes Hills, south of Los Angeles, seven different terrace levels exist, indicating seven episodes of uplift. The ever-persistent sea is now cutting a new platform at the base of the cliff. If uplift follows, it too will become an elevated marine terrace.

Other examples of emergent coasts include regions that were once buried beneath great ice sheets. When glaciers were present, their weight depressed the crust, and when the ice melted, the crust began gradually to spring back. As a result, prehistoric shoreline features today are found high above sea level. The Hudson Bay region of Canada is one such area, portions of which are still rising at a rate of more than 1 centimeter per year.

In contrast to the preceding examples, other coastal areas show definite signs of submergence. The shoreline of a coast that has been submerged in the relatively recent past is often highly irregular because the sea typically floods the lower reaches of river valleys flowing into the ocean. The ridges separating the valleys, however, remain above sea level and project into the sea as headlands. These drowned river mouths, which are called **estuaries**, characterize many coasts today. Along the Atlantic coast, the Chesapeake and Delaware bays are examples of estuaries created by submergence (Figure 13.15). The picturesque coast of Maine, particularly in the vicinity of Acadia National Park, is another excellent example of an area that was flooded by the postglacial rise in sea level and transformed into a highly irregular submerged coastline.

Keep in mind that most coasts have a complicated geologic history. With respect to sea level, many have at various times emerged and then submerged again. Each time they retain some of the features created during the previous situation.

Tides

Tides are daily changes in the elevation of the ocean surface. Their rhythmic rise and fall along coastlines have been known since antiquity. Other than waves, they are the easiest ocean movements to observe. An exceptional example of extreme daily tides is shown in Figure 13.16 (see Box 13.2).

Although known for centuries, tides were not explained satisfactorily until Sir Isaac Newton applied

Box 13.1 # Is Global Warming Causing Sea Level to Rise?

The shifting dynamic nature of barrier islands and the ineffectiveness of most shoreline protection measures are now relatively well-established facts. Unfortunately, recent research, which indicates that sea level is rising, has compounded this already distressing situation. Studies indicate that sea level has risen between 10 and 15 centimeters over the past century. Furthermore, some investigators predict an accelerated sea-level rise in the years to come—30 cm or more by the middle of the next century. Although such a vertical change may seem modest, many coastal geologists believe that any given rise in sea level along the gently sloping Atlantic and Gulf coasts will cause form 10 to 1000 times as much horizontal shoreline retreat (Figure 13.A)

The idea that sea level will continue to rise in the coming decades is linked to the results of climatic studies that predict a global warming trend. Such predictions are based on the now well-established fact that the carbon dioxide (CO_2) content of the atmosphere has been rising at an accelerating rate for more than a century. The CO_2 is added primarily as a by-product of the combustion of ever-increasing quantities of fossil fuels.

If we assume that the use of fossil fuels will continue to rise at projected rates, current estimates indicate that the atmosphere's CO_2 content will grow by an additional 40 percent or more by some time in the second half of the next century. The importance of CO_2 lies in the fact that it traps a portion of the radiation emitted by Earth and thereby keeps the air near the surface warmer than it would be

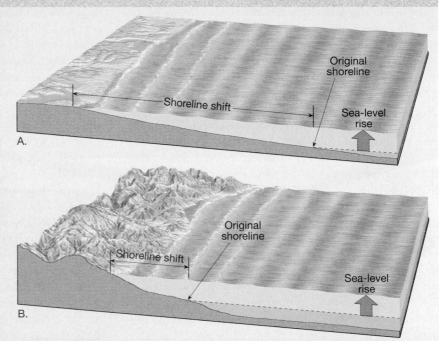

FIGURE 13.A The slope of a shoreline is critical to determining the degree to which sea-level changes will affect it. **A.** When the slope is gentle, small changes in sea level cause a substantial shift. **B.** The same sea-level rise along a steep coast results in only a small shoreline shift.

without CO_2. Because CO_2 is an important heat-absorbing gas, an increase in the air's CO_2 content is believed to contribute to higher atmospheric temperatures. In addition, other trace gases generated by human activities also play a role.

How is a warmer atmosphere related to a global rise in sea level? First, higher temperatures can cause glacial ice to melt. About one-half of the 10- to 15-centimeter rise in sea level over the past century is attributed to the melting of small glaciers and ice caps. Second, a warmer atmosphere causes an increase in ocean volume through thermal expansion. That is, higher air temperatures raise the temperature of the upper layers of the

ocean, which, in turn, causes the water to expand and sea level to rise. It is also believed that a warmer ocean may spur storm development. Of course, an increase in storm activity would compound an already serious problem in many coastal areas.

As rising sea level is a gradual phenomenon, it may be overlooked by coastal residents as a significant contributor to shoreline erosion problems. Rather, the blame is assigned to other forces, especially storm activity. Although a given storm may be the immediate cause, the magnitude of its destruction may result from the relatively small sea-level rise that allowed the storm's power to cross a much greater land area.

the law of gravitation to them. Newton showed that there is a mutual attractive force between two bodies, as between Earth and the Moon. Because the atmosphere and the ocean both are fluids, and free

to move, both are deformed by this force. Hence, ocean tides result from the gravitational attraction exerted upon Earth by the Moon, and to a lesser extent by the Sun.

distance. In this case, the two objects are the Moon and Earth. Because the force of gravity decreases with distance, the Moon's gravitational pull on Earth is slightly greater on the near side of Earth than on the far side. The result of this differential pulling is to stretch (elongate) the solid Earth very slightly. In contrast, the world ocean, which is mobile, is deformed quite dramatically by this effect to produce the two opposing tidal bulges.

Because the position of the Moon changes only moderately in a single day, the tidal bulges remain in place while Earth rotates "through" them. For this reason, if you stand on the seashore for 24 hours, Earth will rotate you through alternating areas of deeper and shallower water. As you are carried into regions of deeper water, the tide rises, and as you are carried away, the tide falls. Therefore, during one day you experience two high tides and two low tides.

Further, the tidal bulges migrate as the Moon revolves around Earth every 29 days. As a result, the

FIGURE 13.15 Estuaries along the East Coast of the United States. The lower portions of many river valleys were submerged by the rise in sea level that followed the end of the Ice Age. Chesapeake Bay and Delaware Bay are especially prominent examples. Also note the prominent string of barrier islands that parallels the Atlantic coast.

Causes of Tides

To illustrate how tides are produced, we will assume Earth is a rotating sphere covered to a uniform depth with water (Figure 13.17). It is easy to see how the Moon's gravitational force can cause the water to bulge on the side of Earth nearest the Moon. In addition, however, an equally large tidal bulge is produced on the side of Earth directly opposite the Moon.

Both tidal bulges are caused, as Newton discovered, by the pull of gravity. Gravity is inversely proportional to the square of the distance between two objects, meaning simply that it quickly weakens with

FIGURE 13.16 High tide and low tide on Nova Scotia's Minas Basin in the Bay of Fundy. The areas exposed during low tide and flooded during high tide are called *tidal flats*. Tidal flats here are extensive. (Courtesy of Nova Scotia Dept. of Tourism)

Box 13.2 Tidal Power

Several methods of generating electrical energy from the oceans have been proposed, but the ocean's energy potential remains largely untapped. The development of tidal power is the principal example of energy production from the ocean.

Tides have been used as a source of power for centuries. Beginning in the twelfth century, water wheels driven by the tides were used to power gristmills and sawmills. During the seventeenth and eighteenth centuries, much of Boston's flour was produced at a tidal mill. Today, far greater energy demands must be satisfied, and more sophisticated ways of utilizing the force created by the perpetual rise and fall of the ocean must be employed.

Tidal power is harnessed by constructing a dam across the mouth of a bay or an estuary in a coastal area having a large tidal range (Figure 13.B). The narrow opening between the bay and the open ocean magnifies the variations in water level that occur as the tides rise and fall. The strong in-and-out flow that results at such a site is then used to drive turbines and electrical generators.

Tidal energy utilization is exemplified by the tidal power plant at the mouth of the Rance River in France. By far the largest yet constructed, this plant went into operation in 1966 and produces enough power to satisfy the needs of Brittany and also contribute to the demands of other regions. Much smaller experimental facilities have been built near Murmansk in Russia and near Taliang in China, and on the Bay of Fundy in the Canadian province of Nova Scotia (Figure 13.C). The United States has not yet tapped its tidal power potential, although a site at Passamaquoddy Bay in Maine, where the tidal range approaches 15 meters (50 feet), has been under review for many years.

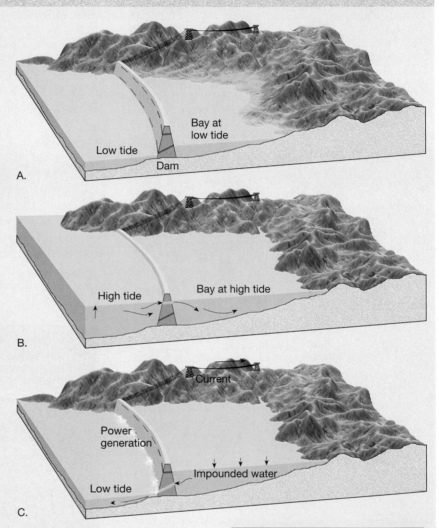

FIGURE 13.B Simplified diagram showing the principle of the tidal dam.

Along most of the world's coasts it is not possible to harness tidal energy. If the tidal range is less than 8 meters (25 feet) or if narrow, enclosed bays are absent, tidal power development is uneconomical. For this reason, the tides will never provide a very high portion of our ever-increasing electrical energy requirements. Nevertheless, the development of tidal power may be worth pursuing at feasible sites because electricity produced by the tides consumes no exhaustible fuels and creates no noxious wastes.

FIGURE 13.C The first tidal power generating station in North America opened in 1984 at Annapolis Royal, Nova Scotia. This pilot plant on Canada's Bay of Fundy is being used to assess the merits and drawbacks of building a larger facility. (Photo by Stephen J. Krasemann/Photo Researchers)

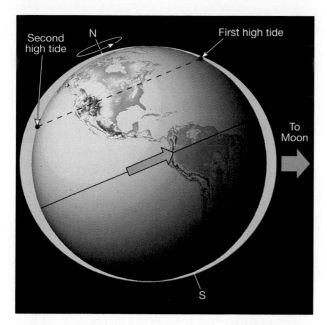

FIGURE 13.17 Tides on an Earth that is covered to a uniform depth with water. Depending on the Moon's position, tidal bulges may be inclined to the equator. In this situation an observer will experience two unequal high tides.

tides, like the time of moonrise, shift about 50 minutes later each day. After 29 days, the cycle is complete and a new one begins.

There may be an inequality between the high tides during a given day. Depending on the Moon's position, the tidal bulges may be inclined to the equator as in Figure 13.17. This figure illustrates that the first high tide experienced by an observer in the Northern Hemisphere is considerably lower than the high tide half a day later. On the other hand, a Southern Hemisphere observer would experience the opposite effect.

Spring and Neap Tides

The Sun also influences the tides. It is far larger than the Moon, but because the Sun is so far away, the effect is considerably less than that of the Moon. In fact, the tide-generating potential of the Sun is slightly less than half that of the Moon.

Near the times of new and full moons, the Sun and Moon are aligned and their forces are added together (Figure 13.18A). Accordingly, the combined gravity of these two tide-producing bodies cause higher tidal bulges (high tides) and lower tidal troughs (low tides). These are called the **spring tides**. Spring tides create the largest daily tidal range, that is, the largest variation between high and low tides. Conversely, at about the time of the first and third quarters of the Moon, the gravitational forces of the Moon and Sun act on Earth at right angles, and

each partially offsets the influence of the other (Figure 13.18B). As a result, the daily tidal range is less. These are called **neap tides.**

So far we have explained the basic causes and patterns of tides. But keep in mind these theoretical considerations cannot be used to predict either the height or the time of actual tides at a particular place. The shape of coastlines and the configuration of ocean basins greatly influence the tides. Consequently, tides at various locations respond differently to the tide-producing forces. This being the case, the nature of the tide at any place can be determined most accurately by actual observation. The predictions in tidal tables and the tidal data on nautical charts are based on such observations.

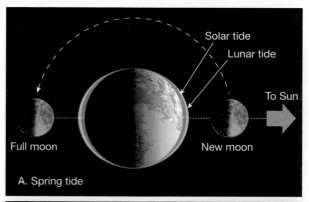

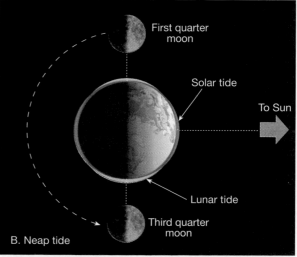

FIGURE 13.18 Relationship of the Moon and Sun to Earth during **A.** spring tides and **B.** neap tides.

The Chapter in Review

The following statements are intended to help you review the primary objectives presented in this chapter.

- The three factors that influence the *height, wavelength,* and *period* of a wave are (1) *wind speed,* (2) *length of time the wind has blown,* and (3) *fetch,* the distance that the wind has traveled across the open water.

- The two types of wind-generated waves are (1) *waves of oscillation,* which are waves in the open sea in which the wave form advances as the water particles move in circular orbits, and (2) *waves of translation,* the turbulent advance of water formed near the shore as waves of oscillation collapse, or *break,* and form *surf.*

- Wave erosion is caused by *wave impact, pressure,* and *abrasion* (the sawing and grinding action of water armed with rock fragments). The bending of waves is called *wave refraction.* Owing to refraction, wave impact is concentrated against the sides and ends of headlands.

- Most waves reach the shore at an angle. The uprush (swash) and backwash of water from each breaking wave moves the sediment in a zigzag pattern along the beach. This movement, called *beach drift,* can transport sand and pebbles hundreds or even thousands of meters each day. Oblique waves also produce *longshore currents* within the surf zone that flow parallel to the shore.

- Features produced by *shoreline erosion* include *wave-cut cliffs* (which originate from the cutting action of the surf against the base of coastal land), *wave-cut platforms* (relatively flat, benchlike surfaces left behind by receding cliffs), *sea arches* (formed when a headland is eroded and two caves from opposite sides unite), and *sea stacks* (formed when the roof of a sea arch collapses).

- Some of the depositional features formed when sediment is moved by beach drift and longshore currents are *spits* (elongated ridges of sand that project from the land into the mouth of an adjacent bay), *baymouth bars* (sand bars that completely cross a bay), and *tombolos* (ridges of sand that connect an island to the mainland or to another island). Along the Atlantic and Gulf Coastal Plains, the shore zone is characterized by *barrier islands,* low ridges of

sand that parallel the coast at distances from 3 to 30 kilometers offshore.

- Local factors that influence shoreline erosion are (1) the proximity of a coast to sediment-laden rivers, (2) the degree of tectonic activity, (3) the topography and composition of the land, (4) prevailing winds and weather patterns, and (5) the configuration of the coastline and nearshore areas.

- Three basic responses to shoreline erosion problems are (1) building *structures* such as *groins* (short walls built at a right angle to the shore to trap moving sand), *breakwaters* (structures built parallel to the shoreline to protect it from the force of large breaking waves), and *seawalls* (barriers constructed to prevent waves from reaching the area behind the wall) to hold the shoreline in place; (2) *beach nourishment,* which involves the addition of sand to replenish eroding beaches; and (3) *relocating* buildings away from the beach.

- Because of basic geological differences, the *nature of shoreline erosion problems along America's Pacific and Atlantic coasts is very different.* Much of the development along the Atlantic and Gulf coasts has occurred on barrier islands, which receive the full force of major storms. A majority of the Pacific Coast is characterized by narrow beaches backed by steep cliffs and mountain ranges. A major problem facing the Pacific shoreline is a narrowing of beaches caused because the natural flow of materials to the coast has been interrupted by dams built for irrigation and flood control.

- One frequently used classification of coasts is based upon changes that have occurred with respect to sea level. *Emergent coasts,* often with wave-cut cliffs and wave-cut platforms above sea level, develop either because an area experiences uplift or as a result of a drop in sea level. Conversely, *submergent coasts,* with their drowned river mouths, called *estuaries,* are created when sea level rises or the land adjacent to the sea subsides.

- *Tides,* the daily rise and fall in the elevation of the ocean surface, are caused by the *gravitational attraction* of the Moon and, to a lesser extent, by the Sun. Near the times of new and full moons, the Sun and Moon are aligned, and their gravitational forces are added together to produce especially high and low tides. These are called the *spring tides*. Conversely, at about the times of the first and third quarters of the Moon, when the gravitational forces of the Moon and Sun are at right angles, the daily tidal range is less. These are called *neap tides*.

Key Terms

abrasion (p. 253)
barrier island (p. 256)
baymouth bar (p. 255)
beach drift (p. 254)
beach nourishment (p. 260)
breakwater (p. 259)
emergent coast (p. 261)

estuary (p. 261)
groin (p. 258)
longshore current (p. 254)
neap tide (p. 265)
sea arch (p. 255)
sea stack (p. 255)
seawall (p. 259)

spit (p. 255)
spring tide (p. 265)
submergent coast (p. 261)
surf (p. 252)
tide (p. 261)
tombolo (p. 256)
wave-cut cliff (p. 255)

wave-cut platform (p. 255)
wave height (p. 251)
wavelength (p. 251)
wave period (p. 251)
wave refraction (p. 253)

Questions for Review

1. List three factors that determine the height, length, and period of a wave.
2. Describe the motion of a water particle as a wave passes.
3. Explain what happens when a wave breaks.
4. Describe two ways in which waves cause erosion.
5. What is wave refraction? What is the effect of this process along irregular coastlines?
6. Why are beaches often called "rivers of sand"?
7. Describe the formation of the following features: wave-cut cliff, wave-cut platform, sea stack, spit, baymouth bar, tombolo.
8. List three possible ways that barrier islands form.
9. For what purpose is a groin built? Why might the building of one groin lead to the building of others?
10. How might a seawall increase beach erosion?
11. It is believed that global temperatures will be increasing in the decades to come. How can a warmer atmosphere lead to rise in sea level? (See Box 13.1)
12. Relate the damming of rivers to the shrinking of beaches at some locations along the West Coast of the United States. Why do narrower beaches lead to accelerated sea-cliff retreat?
13. What observable features would lead you to classify a coastal area as emergent?
14. Are estuaries associated with submergent or emergent coasts? Why?
15. Discuss the origin of ocean tides.
16. Explain why an observer can experience two unequal high tides during one day.
17. How does the Sun influence tides?
18. What advantages does tidal power production offer? Is it likely that tides will ever provide a significant proportion of the world's electrical energy requirements? (See Box 13.2.)

Testing What You Have Learned

To test your knowledge of the material presented in this chapter, answer the following questions:

Multiple-Choice Questions

1. Especially high and low tides that occur near the times of new and full moons are called _____ tides.
   a. spring
   b. fall
   c. neap
   d. ebb
   e. super

2. Waves derive their energy and motion entirely from _____.
   a. salinity   **c.** wind   **e.** tsunamis
   b. tectonics   **d.** gravity

3. A(n) _____ is a barrier built at a right angle to the beach to trap sand that is moving parallel to the shore.
   a. breakwater   **c.** baymouth bar   **e.** seawall
   b. groin   **d.** spit

4. Which one of the following is NOT commonly used to describe a wave?
   a. wave height   **c.** wave period
   b. wavelength   **d.** wave bar

5. The Atlantic and Gulf coasts are characterized by low ridges of sand that parallel the coast at distances of 3 to 30 kilometers called _____.
   a. tombolos   **c.** sea stacks   **e.** coral reefs
   b. dune arcs   **d.** barrier islands

6. The sawing and grinding action of water armed with rock fragments is referred to as _____.
   a. fetch   **c.** translation   **e.** swash
   b. oscillation   **d.** abrasion

7. Which one of the following is NOT a feature produced by wave erosion?
   a. spit   **c.** sea stack   **e.** wave-cut platform
   b. sea arch   **d.** wave-cut cliff

8. A deep-water wave begins to "feel bottom" at a water depth equal to about _____ its wavelength.
   a. one-fourth   **c.** one-half   **e.** twice
   b. one-third   **d.** two-thirds

9. Drowned river mouths associated with a submergent coast are called _____.
   a. spits   **c.** tarns   **e.** estuaries
   b. baymouth bars   **d.** deltas

10. The gravitational pull on Earth by the Moon causes _____ tidal bulge(s) to form in Earth's world ocean.
    a. one   **c.** three   **e.** five
    b. two   **d.** four

Fill-in Questions

11. The turbulent water created by breaking waves is called _____.
12. At about the time of the first and third quarters of the moon, _____ tides produce low daily tidal ranges.
13. Currents within the surf zone that flow parallel to the shore are referred to as _____ currents.
14. The bending of waves, called wave _____, along a coast influences where and to what degree erosion, sediment transport, and deposition will take place.
15. Based on changes that have occurred with respect to sea level, coasts are often classified into two types: _____ coasts and _____ coasts.

True/False Questions

16. Because the Sun is much larger than the Moon, its gravity influences the tides more than the Moon's gravity. ____
17. *Run* is the term used to describe the distance traveled by wind across the open water. ____
18. In the open sea, it is the wave form that moves forward, not the water itself. ____
19. A major problem facing the Pacific shoreline, and especially portions of southern California, is a significant narrowing of many beaches. ____
20. If an irregular shoreline remains stable, marine erosion and deposition will eventually produce a straighter, more regular coast. ____

Answers

1.a; 2.c; 3.b; 4.d; 5.d; 6.d; 7.a; 8.c; 9.e; 10.b; 11. surf; 12. neap; 13. longshore; 14. refraction; 15. emergent, submergent; 16.F; 17.F; 18.T; 19.T; 20.T.

The Ocean Floor

Sea arch along the north coast of the
island of Crete. (Photo by Tom Till
Photography)

Focus on Learning

*To assist you in learning the
important concepts in this
chapter, you will find it helpful to
focus on the following questions:*

- What is the extent of the world
  ocean?

- What are the three major
  topographic units of the ocean
  basins and the major features
  associated with each?

- What is the relation between
  submarine canyons, turbidity
  currents, and turbidites?

- How do coral reefs and atolls form?

- How do the three broad
  categories of seafloor sediments
  originate?

How deep is the ocean? How much of Earth is covered by the global sea? These are basic questions, but they went unanswered for thousands of years. Interest in the oceans undoubtedly dates back to ancient times, but it was not until rather recently that these seemingly simple questions began to be answered. The beginning of this chapter deals with these answers. Then the remainder of the chapter provides a look at Earth beneath the sea. Suppose that all of the water were drained from the ocean. What would we see? Plains? Mountains? Canyons? Plateaus? Indeed, the ocean conceals all of these features, and more. In fact, the topography of the ocean floor is as varied as that of any continent. And what about the carpet of sediment that covers much of the seafloor? Where did it come from and what can we learn by examining it?

The Vast World Ocean

A glance at a globe or a view of Earth from space reveals a planet dominated by the world ocean (Figure 14.1). Indeed, it is for this reason that Earth is often referred to as the *blue planet*. The area of Earth is about 510 million square kilometers (197 million square miles). Of this total, approximately 360 million square kilometers (140 million square miles), or 71 percent, is represented by oceans and marginal seas (meaning seas around the ocean's margin, like the Mediterranean Sea and Caribbean Sea). Continents and islands comprise the remaining 29 percent, or 150 million square kilometers (58 million square miles).

By studying a globe or world map, it is readily apparent that the continents and oceans are not evenly divided between the Northern and Southern hemispheres (Figure 14.1). When we compute the percentages of land and water in the Northern Hemisphere, we find that nearly 61 percent of the surface is water, whereas about 39 percent is land. In the Southern Hemisphere, on the other hand, almost 81 percent of the surface is water, and only 19 percent is land. It is no wonder then that the Northern Hemisphere is called the *land hemisphere*, and the Southern Hemisphere the *water hemisphere*.

Figure 14.2A shows the distribution of land and water in the Northern and Southern hemispheres by way of a graph. Between latitudes 45 degrees north and 70 degrees north, there is actually more land than water, whereas between 40 degrees south and 65 degrees south there is almost no land to interrupt the oceanic and atmospheric circulation.

FIGURE 14.1 These views of Earth show the uneven distribution of land and water between the Northern and Southern hemispheres. Almost 81 percent of the Southern Hemisphere is covered by the oceans—20 percent more than the Northern Hemisphere.

The volume of the ocean basins is many times greater than the volume of the continents above sea level. In fact, the volume of all land above sea level is only 1/18 that of the ocean.

An obvious difference between continents and the ocean basins is their relative levels. The average elevation of the continents above sea level is about 840 meters (2750 feet), whereas the average depth of the oceans is more than 4.5 times this figure—3800 meters (12,500 feet). If Earth's solid mass were perfectly smooth (level) and spherical, the oceans would cover it to a uniform depth of more than 2000 meters (1.2 miles).

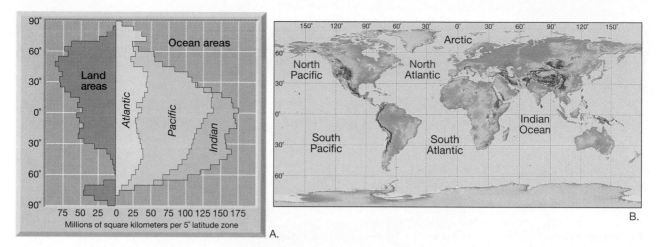

FIGURE 14.2 Distribution of land and water. **A.** The graph shows the amount of land and water in each 5-degree latitude belt. **B.** The world map provides a more familiar view.

A comparison of the three major oceans reveals that the Pacific is by far the largest; it is nearly as large as the Atlantic and Indian oceans combined (Figure 14.2). The Pacific contains slightly more than half of the water in the world ocean, and because it includes few shallow seas along its margins, it has the greatest average depth—3940 meters (12,900 feet).

The Atlantic, bounded by almost parallel continental margins, is a relatively narrow ocean when compared to the Pacific. When the Arctic Ocean is included, the Atlantic has the greatest north–south extent and connects the two polar regions. Because the Atlantic has many shallow adjacent seas, including the Caribbean, Gulf of Mexico, Baltic, and Mediterranean, as well as wide continental shelves along its borders, it is the shallowest of the three oceans, with an average depth of 3310 meters (10,860 feet).

As Figure 14.2 shows, the Indian Ocean covers less area than the Pacific or the Atlantic. Unlike the others, it is largely a Southern Hemisphere water body.

Earth Beneath the Sea

If all water were removed from the ocean basins, what kind of surface would be revealed? It would not be quiet, subdued topography as once was thought, but a surface characterized by mountains, deep canyons, and flat plains. In fact, the scenery would be just as varied as that on the continents.

The ocean's vast expanse first became apparent through voyages of discovery in the fifteenth and sixteenth centuries. An understanding of the ocean floor's varied topography did not unfold until much later with the historic 3-1/2-year voyage of the H.M.S. *Challenger*. From December 1872 to May 1876, the

Challenger expedition made the first, and perhaps still most comprehensive, study of the global ocean ever attempted by one agency. The 110,000-kilometer (68,000-mile) trip took the ship and its crew of scientists to every ocean except the Arctic. Throughout the voyage they sampled the depth of water by laboriously lowering a weighted line overboard. Not many years later, the knowledge gained by the *Challenger* of the ocean's great depths and varied topography was further expanded with the laying of transatlantic cables. However, as long as ocean depths had to be measured with weighted lines, our knowledge of the seafloor remained limited. Then, in the 1920s a technological breakthrough occurred with the invention of electronic depth-sounding equipment—the echo sounder.

The **echo sounder** works by transmitting sound waves toward the ocean bottom (Figure 14.3). A sensitive receiver intercepts the echo reflected from the bottom, and a clock precisely measures how long it took for the sound waves to make the round trip to the bottom and back up to the ship. We know how fast sound moves in water (about 1500 meters or 5000 feet per second). So if sound takes 1 second to make the round trip from the ship to the bottom and back, you know the distance traveled was about 1500 meters for the round trip, so the distance to the bottom is half of that, or 750 meters.

The depths determined from continuous monitoring of these echoes are plotted to produce a profile of the ocean floor. Since the invention of the echo sounder, millions of kilometers of sonic depth determinations have provided a more complete and detailed view of the ocean floor (Figure 14.4).

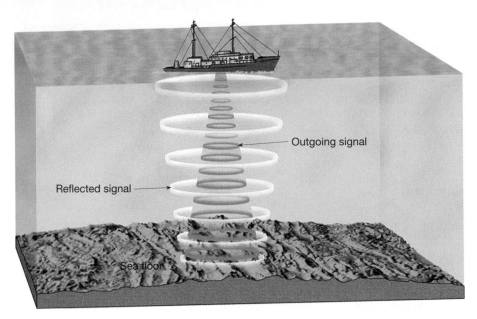

FIGURE 14.3 An echo sounder determines water depth by measuring the time required for a sonic wave to travel from ship to the seafloor and back. The speed of sound in water is 1500 m/sec. Therefore, water depth = 1/2 (1500 m/sec X echo travel time).

Oceanographers studying the topography of the ocean basins have delineated three major units: *continental margins,* the *ocean basin floor,* and *mid-ocean ridges.* The map in Figure 14.5 shows these provinces for the North Atlantic, and the profile at the bottom of the illustration shows the varied topography. Such profiles usually have their vertical dimension exaggerated many times to make topographic features more conspicuous. Because of this, the slopes in the profile of the seafloor in Figure 14.5 appear to be much steeper than they actually are.

Although information about the ocean floor is substantial and growing every day, scientists have nevertheless only scratched the surface. Ironically, in many respects we know more about the details of the surface of Venus than of the floor of the ocean! For example, in 1991 and 1992 the *Magellan* spacecraft used special radar to scan and map more than 90 percent of the surface of Venus. By contrast, in the 500 years since Ferdinand Magellan circumnavigated the world's oceans, oceanographers have scanned *less than 10 percent* of the deep ocean floor at the resolution provided by the *Magellan* spacecraft.

Why have details about the ocean floor come so slowly? To the present day, efforts to map the seafloor accurately are made from ships equipped with echo-sounding equipment because radar waves (which are used to map our neighboring planets) do not travel far through seawater. In addition, ships move across the ocean at a mere 18 kilometers per hour rather than orbiting a planet at 19,500 kilometers per hour.

Continental Margins

We will now look at the three major topographic regions of the seafloor (continental margin, ocean basin floor, and mid-ocean ridge). To keep your perspective during this discussion, refer to Figures 14.4, 14.5 and 14.6.

Continental Shelf

The zones that collectively make up the **continental margin** include the continental shelf, continental slope, and continental rise (Figure 14.6). The first of these parts, the **continental shelf,** is a gently sloping submerged surface extending from the shoreline toward the deep-ocean basin. It is a flooded extension of the continent.

The continental shelf varies greatly in width, as you can see in Figures 14.4 and 14.5. Almost nonexistent along some continents, the shelf may extend seaward as far as 1500 kilometers (900 miles) along others. On the average, the continental shelf is about 80 kilometers (50 miles) wide and 130 meters (423 feet) deep at its seaward edge, the *shelf break.* The average inclination of the continental shelf is less than one-tenth of one degree, a drop of only about 2 meters per kilometer (10 feet per mile). The slope is so slight that it appears horizontal.

The continental shelves represent 7.5 percent of the total ocean area, which is equivalent to about 18 percent of Earth's total land area. These areas have taken on increased economic and political significance because they contain important mineral deposits, including large reservoirs of petroleum and natural gas, as well as huge sand and gravel deposits. The waters of the continental shelf also contain many important fishing grounds that are significant sources of food.

When compared with many parts of the deep-ocean floor, the surface of the continental shelf is relatively featureless. This is not to say that the shelves are completely smooth. In many regions, long valleys run from the coastline into deeper waters. Many of these *shelf valleys* are seaward extensions of river valleys on the adjacent landmass. Such valleys were carved by streams during the Pleistocene epoch (Ice Age).

During this time, great quantities of water were stored in vast ice sheets on the continent. The result was that sea level dropped by 100 meters (330 feet) or more, exposing large portions of the continental shelves (see Figure 11.A, p. 225). Because of the lower sea level, rivers extended their courses, and land-dwelling plants and animals moved onto the newly exposed portions of the continents. Today these areas are again covered by the sea and inhabited by marine organisms. Scientists dredging along the eastern coast of North America have discovered the remains of numerous land dwellers, including mammoths, mastodons, and horses. Bottom sampling has also revealed that freshwater peat bogs existed, adding to the evidence that the continental shelves were once above sea level.

Continental Slope and Rise

Marking the seaward edge of the continental shelf is the **continental slope** (Figure 14.6). It leads into deep water and has a steep gradient compared to the continental shelf. While the gradient varies from place to place, it has an average drop of about 70 meters per kilometer (370 feet per mile). The continental slope is the true edge of the continent.

In regions where trenches do not exist, the steep continental slope merges into a more gradual incline known as the **continental rise** (Figure 14.6). Here the gradient lessens to between 4 and 8 meters per kilometer (20 to 40 feet per mile). Whereas the width of the continental slope averages about 20 kilometers (12 miles), the continental rise may extend for hundreds of kilometers. This feature consists of a thick accumulation of sediment that moved downslope from the continental shelf to the deep-ocean floor. The sediments are delivered to the base of the continental slope by *turbidity currents* that follow submarine canyons (Figure 14.7). (We will discuss these shortly.) When these muddy currents emerge from

the mouth of a canyon onto the relatively flat ocean floor, they deposit sediment that forms a **deep-sea fan.** Deep-sea fans have the same basic shape as alluvial fans, which form at the foot of steep mountain slopes on land. As fans from adjacent submarine canyons grow, they coalesce to produce the continuous apron of sediment at the base of the continental slope.

Along some mountainous coasts, the continental slope does not grade gently into the continental rise, but descends abruptly into a deep-ocean trench that parallels the landmass. In such cases, the shelf is very narrow or does not exist at all. The side of the trench and the continental slope are essentially the same feature and grade into the adjacent mountains that tower thousands of meters above sea level. These narrow continental margins are primarily located around the Pacific Ocean where plates are converging. The stress between colliding plates results in deformation of the continental margin. An example of this activity is found along the west coast of South America (see Figure 14.4). Here the vertical distance from the peaks of the Andes Mountains to the floor of the deep Peru–Chile Trench bordering the continent exceeds 12,000 meters.

Submarine Canyons and Turbidity Currents

About 100 kilometers south of San Francisco, deep beneath the waters of California's Monterey Bay, is Monterey Canyon. In places the canyon walls of this 100-kilometer-long chasm gradually drop 2300 meters, nearly 400 meters more than the highest cliff in the Grand Canyon. Monterey Canyon is one of many impressive examples of deep, steep-sided valleys known as **submarine canyons** that originate on the continental slope and may extend to depths as great as 3 kilometers (Figure 14.7). Although some of these canyons appear to be the seaward extensions of river valleys, many others do not line up in this manner.

Furthermore, because these canyons extend to depths far below the maximum lowering of sea level during the Ice Age, we cannot attribute their formation to stream erosion. These features must be created by some process that operates far below the ocean surface. Most available information seems to favor the view that submarine canyons have been excavated by turbidity currents (see Box 14.1).

Turbidity currents are downslope movements of dense, sediment-laden water. They are created when sand and mud on the continental shelf and slope are dislodged and thrown into suspension.

FIGURE 14.4 The topography of Earth's solid surface is shown on this and the following page. (The World Ocean Floor by Bruce C. Heezen and Marie Tharp/Copyright © 1977 by Marie Tharp)

WORLD OCEAN FLOOR
BY BRUCE C. HEEZEN AND MARIE THARP

Based on Research and Exploration Initiated and Supported by the

UNITED STATES NAVY   OFFICE OF NAVAL RESEARCH

1977

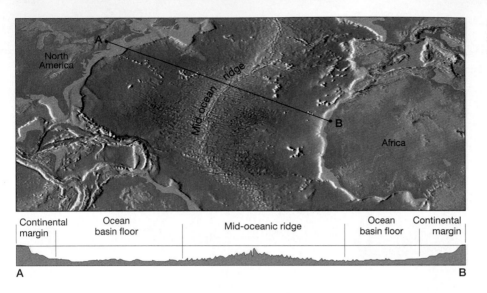

FIGURE 14.5 This map and profile show the major topographic divisions of the North Atlantic.

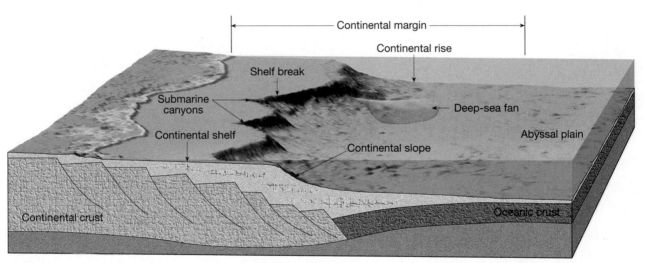

FIGURE 14.6 Schematic view showing the parts of the continental margin.

Because the mud-choked water is denser than normal sea water, it flows downslope, eroding and accumulating more sediment as it goes. The erosional work repeatedly carried on by these muddy torrents is thought to be the major force in the excavation of most submarine canyons.

Turbidity currents usually originate along the continental slope and continue across the continental rise, still cutting channels. Eventually they lose momentum and come to rest along the ocean basin floor (Figure 14.7). As these currents slow, suspended sediments begin to settle out. First, the coarser sand is dropped, followed by successively fine accumulations of silt and then clay. Consequently, these deposits, called **turbidites,** are characterized by a decrease in sediment grain size from bottom to top, a phenomenon known as **graded bedding** (Figure 14.7).

Although there is still more to be learned about the complex workings of turbidity currents, it has been well established that they are a very important mechanism of sediment transport in the ocean. By the action of turbidity currents, submarine canyons are created and sediments are carried to the deep-ocean floor.

The Ocean Basin Floor

Between the continental margin and the mid-ocean ridge system lies the ocean basin floor (see Figure 14.5). The size of this region—almost 30 percent of Earth's surface—is roughly comparable to the percentage of the surface that presently projects above the sea as land. On the ocean basin floor, we find dramatically deep grooves, which are *ocean trenches;* remarkably flat regions, known as *abyssal plains;* and steep-sided volcanic peaks,

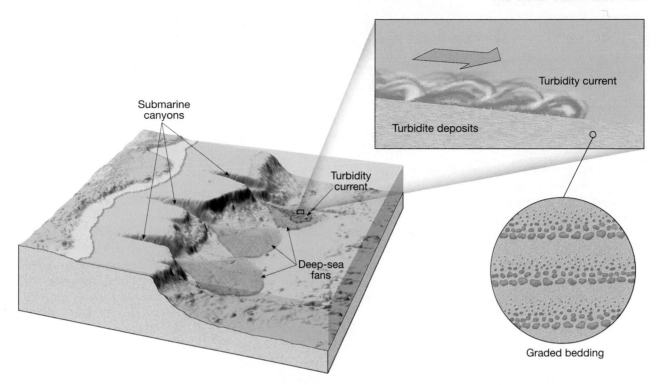

FIGURE 14.7 Turbidity currents move downslope, eroding the continental marin to enlarge submarine canyons. These sediment-laden density currents eventually lose momentum and deposit their loads of sediment as deep-sea fans. Beds deposited by these currents are called *turbidites*. Each event produces a single bed characterized by a decrease in sediment size from bottom to top, a feature known as a *graded bed*.

called *seamounts*. We will now examine each of these features.

Deep-Ocean Trenches

Deep-ocean trenches are long, narrow troughs that are the deepest parts of the ocean. Most trenches are located in the Pacific Ocean where some approach or exceed 10,000 meters (33,000 feet) in depth (Figure.14.8). A portion of one, the Challenger Deep in the Mariana trench, is more than 11,000 meters (36,000 feet) below sea level. Table 14.1 presents dimensions of some of the larger trenches.

Although deep-ocean trenches represent only a very small portion of the ocean floor area, they are very significant geologically. Trenches are the sites where moving crustal plates plunge back into the mantle. In addition to the earthquakes created as one plate descends beneath another, melting of the subducting plate is also associated with trench regions. Thus, trenches in the open ocean are often paralleled by volcanic island arcs. Moreover, volcanic mountains, such as those making up a portion of the Andes, are situated adjacent to trenches that parallel the continental margins.

Abyssal Plains

Abyssal plains are incredibly flat features; in fact, these regions are likely the most level places on Earth. The abyssal plain found off the coast of Argentina, for example, has less than 3 meters (10 feet) of relief over a distance exceeding 1300 kilometers (800 miles). The monotonous topography of abyssal plains is occasionally interrupted by the protruding summit of a buried volcanic structure.

By employing seismic profilers, instruments whose signals penetrate far below the ocean floor, researchers have shown that abyssal plains consist of thick accumulations of sediment that were deposited atop the low, rough portions of the ocean floor. The nature of the sediment indicates that these plains consist primarily of sediments transported far out to sea by turbidity currents. These deposits are interbedded with sediments composed of minute clay-sized particles that continuously settle onto the ocean floor.

Abyssal plains occur in all of the oceans. However, they are more widespread where there are no deep-ocean trenches adjacent to the continents. Because the Atlantic Ocean has fewer trenches to act as traps for the sediments carried down the continental slope, it has more extensive abyssal plains than the Pacific.

Box 14.1 — Evidence for the Existence of Turbidity Currents

For many years the existence of turbidity currents in the ocean was a matter of considerable debate among marine geologists. Not until the 1950s did the speculation begin to subside. Two lines of evidence helped establish turbidity currents as important mechanisms of submarine erosion and sediment transportation.

The first important evidence came from records of a rather severe earthquake that took place off the coast of Newfoundland in 1929 and resulted in the breakage of 13 transatlantic telephone and telegraph cables. At the time, it was presumed that the tremor had caused the multiple breaks. However, when the data were analyzed, it appeared that this was not the case.

After plotting the locations of the breaks on a map, researchers saw that all the breaks had occurred along the steep continental slope and the gentler continental rise. As the time of each break was known from information provided by automatic recorders, a pattern of what had happened could be deduced.

The breaks high up on the continental slope took place first, almost concurrently with the earthquake. The other breaks happened in succession, the last occurring 13 hours later, some 720 kilometers (450 miles) from the source of the quake (Figure 14.A).

The breaks downslope had obviously taken place too long after the tremor to have been caused by the shock of the earthquake. The existence of a turbidity current, triggered by the quake, thus appeared as a plausible alternative. As the avalanche of sediment-choked water raced downslope, it snapped the cables in its path. Invesigators calculated that the current reached speeds approaching 80 kilometers (50 miles) per hour on the continental slope and about 24 kilometers (15 miles) per hour on the more gently sloping continental rise.

A second compelling line of evidence relating turbidity currents to submarine erosion and transportation of sediment came from the examination of deep-sea sediment samples. These cores show that extensive graded beds of sand, silt, and clay exist in the quiet waters of the deep ocean. Some samples also include fragments of plants and animals that live only in the shallower waters of the continental shelves. No mechanism other than turbidity currents has been identified to satisfactorily explain the existence of these deposits.

More recently, turbidity currents have been measured directly. In one study, instruments were deployed in Bute Inlet, a deep fiord along the coast of British Columbia in western Canada. The inlet served as a natural laboratory for the study of turbidity current dynamics. The year-long program found turbidity currents to be common events.

Although the majority were low-velocity flows that covered limited distances, occasional faster-moving, large-scale currents were also measured. These larger events moved significant amounts of sand down the gently sloping floor of the fiord for distances of 40 to 50 kilometers. The dense currents of sediment in this investigation were linked to submarine landslides in a delta at the head of the fiord.

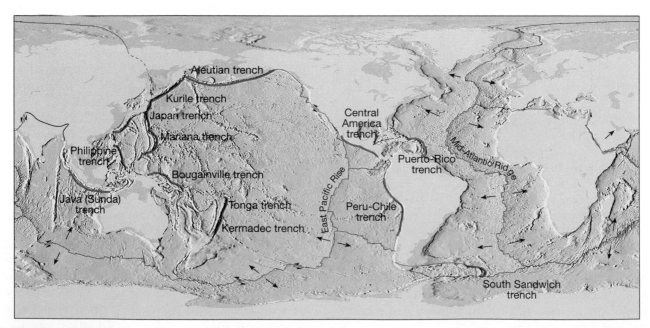

FIGURE 14.8 Distribution of the world's major oceanic trenches.

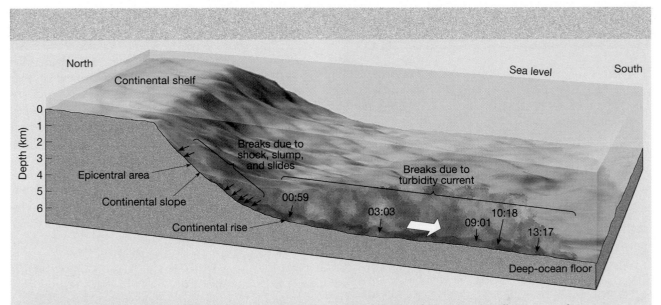

FIGURE 14.A Profile of the seafloor showing the events of the November 18, 1929, earthquake off the coast of Newfoundland. The arrows point to cable breaks, the numbers show times of breaks in hours and minutes after the earthquake. Vertical scale is greatly exaggerated. (After B.C. Heezen and M. Ewing, "Turbidity Currents and Submarine Slump and the 1929 Grand Banks Earthquake," *American Journal of Science* 250:867)

TABLE 14.1 Dimensions of Some Deep-Ocean Trenches

Trench	Depth (kilometers)	Average Width (kilometers)	Length (kilometers)
Aleutian	7.7	50	3700
Japan	8.4	100	800
Java	7.5	80	4500
Kurile–Kamchatka	10.5	120	2200
Mariana	11.0	70	2550
Central America	6.7	40	2800
Peru–Chile	8.1	100	5900
Philippine	10.5	60	1400
Puerto Rico	8.4	120	1550
South Sandwich	8.4	90	1450
Tonga	10.8	55	1400

Seamounts

Dotting the ocean floors are isolated volcanic peaks called **seamounts** that may rise hundreds of meters above the seafloor. Although these steep-sided conical peaks are found on the floors of all the oceans, the greatest number have been identified in the Pacific (see Figure 14.4).

Some, like the Emperor Seamount chain that stretches from the Hawaiian Islands to the Aleutian Trench, form in association with volcanic hot spots. Others are born near oceanic ridges, divergent plate boundaries where the plates of the lithosphere move apart. If the volcano grows large enough before being carried from the zone that nourishes it, the structure emerges as an island. Examples in the Atlantic include the Azores, Ascension, Tristan da Cunha, and St. Helena.

During the time they exist as islands, some of these volcanoes are eroded to near sea level by running water and wave action. Over a span of millions of years the islands gradually sink as the moving plate slowly carries them away from the elevated oceanic ridge or hot spot where they originated. These submerged, flat-topped seamounts are called **guyots.**

Mid-Ocean Ridges

Our knowledge of **mid-ocean ridges,** the sites of seafloor spreading, comes from soundings taken of the ocean floor, core samples obtained from deep-sea drilling, and visual inspection using deep-diving submersibles (Figure 14.9). We also have firsthand inspection of slices of ocean floor that have been shoved up onto dry land by plate movements. Ocean ridge systems are characterized by an elevated position, extensive faulting, and numerous volcanic structures that have developed on the newly formed crust. This is quite evident as you follow the path of mid-ocean ridges in Figure 14.4.

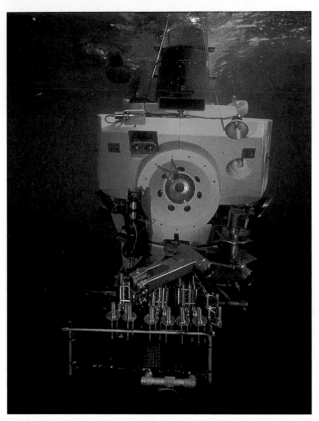

FIGURE 14.9 The deep-diving submersible *Alvin* is 7.6 meters long, weighs 16 tons, has a cruising speed of 1 knot, and can reach depths as great as 4000 meters. A pilot and two scientific observers are along during a normal 6- to 10-hour dive. Vessels such as this are important tools for detecting and studying the fine-scale features of the ocean floor. (Courtesy of Rod Catanach, Woods Hole Oceanographic Institution)

Mid-ocean ridges exist in all major oceans and represent more than 20 percent of Earth's surface. They are certainly the most prominent topographic features in the oceans, for they form an almost continuous mountain range, which extends for about 65,000 kilometers (40,000 miles) in a manner similar to the seam on a baseball. Although ocean ridges stand above the adjacent deep-ocean basins, they are much different from the mountains found on the continents. Rather than containing thick sequences of folded and faulted sedimentary rocks, oceanic ridges consist of layer upon layer of basaltic rocks that have been faulted and uplifted. The term *ridge* may also be misleading, for these features are not narrow, but have widths from 500 to 5000 kilometers. In places they can occupy as much as one-half of the total ocean floor area.

Ridges are broad features, yet much of the geologic activity occurs along a relatively narrow region on the ridge crest. This belt, called the **rift zone,** is the site where magma from the asthenosphere moves upward to create new slivers of oceanic crust. Active rift zones have frequent but generally weak earthquakes and more rapid heat flow than other crustal segments. Here vertical displacement of large slabs of oceanic crust caused by faulting and the growth of volcanic piles contribute to the characteristically rugged topography of the oceanic ridge system. Further, the rocks along the ridge axis appear very fresh and are nearly void of sediment. Away from the ridge axis the topography becomes more subdued, and the thickness of the sediments and the depth of the water both increase. Gradually the ridge system grades into the less rugged sediment-laden regions of the deep-ocean floor.

During seafloor spreading, new material is added about equally to the two diverging plates. Hence, we would expect new ocean floor to grow symmetrically on both sides of a centrally located ridge. Indeed, the ridge systems of the Atlantic and Indian oceans are located near the middles of these water bodies and as a consequence are named mid-ocean ridges (Figure 14.4). However, the East Pacific Rise is situated far from the center of the Pacific Ocean. Despite uniform spreading along the East Pacific Rise, much of the Pacific Basin that once lay east of this spreading center has been overridden by the westward migration of the American plate.

Partly because of its accessibility to both American and European scientists, the Mid-Atlantic Ridge has been studied more thoroughly than other ridge systems. The Mid-Atlantic Ridge is a submerged mountain range standing 2500 to 3000 meters (8200 to 9800 feet) above the adjacent floor of the deep-ocean basins. In a few places, such as Iceland, the ridge actually extends above sea level (Figure 14.4). Throughout most of its length, however, this divergent plate boundary lies 2500 meters below sea level.

Another prominent feature of the Mid-Atlantic Ridge is a deep linear valley extending along the ridge axis. In places, this rift valley is deeper than the Grand Canyon of the Colorado River and two or three times as wide. The name *rift valley* has been applied to this feature because it is strikingly similar to continental rift valleys such as those in East Africa. An examination of Figure 14.4 reveals that this central rift is broken into sections that are offset by transform faults.

Coral Reefs and Atolls

Coral reefs are among the most picturesque features in the ocean. They are constructed primarily from the skeletal remains and secretions of corals and certain algae, built up over thousands of years. The term *coral reef* is somewhat misleading in that it makes no mention of the skeletons of many other small animals and plants found inside the branching framework built by the corals, or the fact that limy secretions of algae help bind the entire structure together.

Coral thrive in the warm water of the Pacific and Indian oceans, so most reefs are in these oceans, although a few occur elsewhere. Reef-building corals grow best in waters with an average annual temperature of about 24°C (75°F). They can survive neither sudden temperature changes nor prolonged exposure to temperatures below 18°C (65°F). In addition, these reef-builders require clear sunlit water. For this reason, the limiting depth of active reef growth is about 45 meters (150 feet).

In 1831, the naturalist Charles Darwin set out aboard the British ship H.M.S. *Beagle* on its famous 5-year expedition that circumnavigated the globe. One outcome of Darwin's studies during the voyage was his hypothesis on the formation of coral islands, called **atolls.** As Figure 14.10 illustrates, atolls consist of a continuous or broken ring of coral reef surrounding a central lagoon.

Atolls present a paradox: how can corals, which require warm, shallow, sunlit water no deeper than 45 meters, create structures that reach thousands of meters to the floor of the ocean? The essence of Darwin's hypothesis was that coral reefs form on the flanks of sinking volcanic islands. As an island slowly sinks, the corals continue to build the reef complex upward (Figure 14.11).

Thus atolls, like guyots, are thought to owe their existence to the gradual sinking of oceanic crust. Drilling on atolls has revealed that volcanic rock does indeed underlie the thick coral reef structure. This finding confirmed Darwin's explanation.

Seafloor Sediments

Except for a few areas, such as near the crests of mid-ocean ridges, the ocean floor is mantled with sediment. Part of this material has been deposited by turbidity currents, and the rest has slowly settled to the bottom from above. The thickness of this carpet of debris varies greatly. In some trenches, which act as traps for sediments originating on the continental margin, accumulations may approach 10 kilometers (6 miles). In general, however, sediment accumulations are considerably less. In the Pacific Ocean, uncompacted sediment measures about 600 meters (2000 feet) or less, whereas on the floor of the Atlantic, the thickness varies from 500 to 1000 meters (1500 to 3000 feet).

Although deposits of sand-sized particles are found on the deep-ocean floor, mud is the most common sediment covering this region. Muds also predominate on the continental shelf and slope, but the sediments in these areas are coarser overall because of greater quantities of sand.

Seafloor sediments can be classified according to their origin into three broad categories: (1) terrigenous ("derived from land") sediment, (2) biogenous ("derived from organisms") sediment, and (3) hydrogenous ("derived from water") sediment. Although we discuss each category separately,

FIGURE 14.10 A. An aerial view of Tetiaroa Atoll in the Pacific. The light blue waters of the relatively shallow lagoon contrast with the dark blue color of the deep ocean surrounding the atoll. (Photo by Douglas Peebles Photography.) **B.** View from space of a group of atolls in the Pacific Ocean. (Courtesy of NASA)

A.

B.

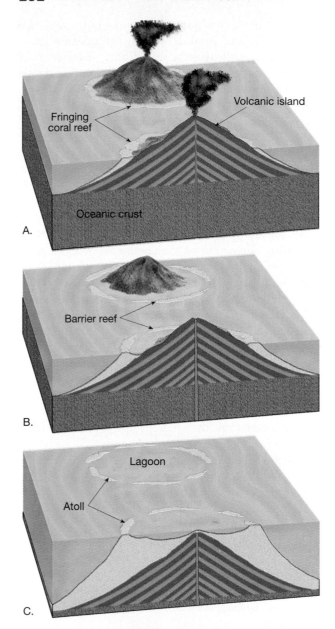

A.

Fringing coral reef

Volcanic island

Oceanic crust

B.

Barrier reef

C.

Lagoon

Atoll

FIGURE 14.11 Formation of a coral atoll caused by the gradual sinking of oceanic crust.

remember that all seafloor sediments are mixtures. No body of sediment comes entirely from a single source.

Terrigenous Sediment

Terrigenous sediment consists primarily of mineral grains that were weathered from continental rocks and transported to the ocean. The sand-sized particles settle near shore. Because the very smallest particles take years to settle to the ocean floor, they may be carried for thousands of kilometers by ocean currents. As a consequence, virtually every area of the ocean receives some terrigenous sediment. The rate at which this sediment accumulates on the deep-ocean floor is indeed very slow. From 5000 to 50,000 years are necessary for a 1-centimeter layer to form. Conversely, on the continental margins near the mouths of large rivers, terrigenous sediment accumulates rapidly. In the Gulf of Mexico, for example, the sediment has reached a depth of many kilometers.

Because fine particles remain suspended in the water for a very long time, there is ample opportunity for chemical reactions to occur. Because of this, the colors of deep-sea sediments are often red or brown. This results when iron in the particle or in the water reacts with dissolved oxygen in the water and produces a coating of iron oxide (rust).

Biogenous Sediment

Biogenous sediment consists of shells and skeletons of marine animals and plants (Figure 14.12). This debris is produced mostly by microscopic organisms living in the sunlit waters near the ocean surface. The remains continually "rain" down upon the seafloor.

The most common biogenous sediments are known as *calcareous* ($CaCO_3$) *oozes,* and as the name implies, they have the consistency of thick mud. These sediments are produced by organisms that inhabit warm surface waters. When calcareous hard parts slowly sink through a cool layer of water,

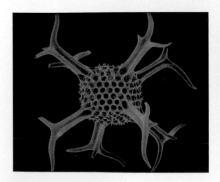

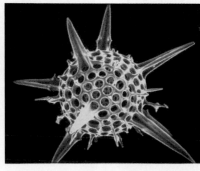

FIGURE 14.12 Microscopic radiolarian hard parts are examples of biogenous sediments. These photomicrographs have been enlarged *hundreds* of times. (Photo by Manfred Kage/Peter Arnold, Inc.)

they begin to dissolve. This results because cold sea-water contains more carbon dioxide and is thus more acidic than warm water. In seawater deeper than about 4500 meters (15,000 feet), calcareous shells will completely dissolve before they reach bottom. Consequently, calcareous ooze does not accumulate where depths are great.

Other biogenous sediments include *siliceous* (SiO_2) *oozes.* They are composed primarily of skeletons of diatoms (single-celled algae) and radiolarians (single-celled animals). Other sediments are derived from the bones, teeth, and scales of fish and other marine organisms.

Hydrogenous Sediment

Hydrogenous sediment consists of minerals that crystallize directly from seawater through various chemical reactions. For example, some limestones are formed when calcium carbonate precipitates directly from the water; however, most limestone is composed of biogenous sediment.

One type of hydrogenous sediment is significant for its economic potential. **Manganese nodules** are rounded blackish lumps composed of a complex mixture of minerals that form very slowly on the floor of the ocean basins (Figure 14.13). In fact, their formation rate is one of the slowest chemical reactions known.

Although manganese nodules may contain more than 20 percent manganese, the interest in them as a potential resource lies in the fact that other more valuable metals may be concentrated in them. In addition to manganese, these nodules may contain significant iron, copper, nickel, and cobalt. All regions containing nodules, however, are not equally

FIGURE 14.13 Manganese nodules photographed at a depth of 2909 fathoms (5323 meters) beneath the *Robert Conrad* south of Tahiti. (Photo courtesy of Lawrence Sullivan/Lamont-Doherty Earth Observatory)

good potential sites for mining. Possible mining locations must have abundant nodules and contain the economically optimum mix of cobalt, copper, and nickel. Sites meeting these criteria are relatively limited. Furthermore, before such areas prove to be valuable commercial sources for these metals, the technical problems of extracting nodules from the floor of the deep-ocean basins must be solved. Finally, the political and legal ramifications of mining the sea bed must be worked out.

The Chapter in Review

The following statements are intended to help you review the primary objectives presented in this chapter.

- *Earth is a planet dominated by oceans.* Seventy-one percent of Earth's surface area consists of oceans and marginal seas. In the Southern Hemisphere, often called the *water hemisphere,* about 81 percent of the surface is water. Of the three major oceans, Pacific, Atlantic, and Indian, the *Pacific Ocean* is the *largest,* contains slightly *more than half of the water* in the world ocean, and has the *greatest average depth*— 3940 meters (12,900 feet).

- Ocean depths are determined using an *echo sounder,* a device carried by a ship that bounces sound off the ocean floor. The time it takes for the sound waves to make the round trip to the bottom and back to the ship is directly related to the depth. Continuous data from the echoes are plotted to produce a profile of the ocean floor.

- Oceanographers studying the topography of the ocean basins have delineated three major units: *continental margins,* the *ocean basin floor,* and *mid-ocean ridges.*

- The zones that collectively make up the continental margin include the *continental shelf* (a gently sloping, submerged surface extending from the shoreline toward the deep-ocean basin), *continental slope* (the true edge of the continent, which has a steep slope that leads from the continental shelf into deep water), and in regions where trenches do not exist, the steep continental slope merges into a more gradual incline known as the *continental rise.* The continental rise consists of sediments that have moved downslope from the continental shelf to the deep-ocean floor.

- *Submarine canyons* are deep, steep-sided valleys that originate on the continental slope and may extend to depths of 3 kilometers. Some of these canyons appear to be the seaward extensions of river valleys. However, most information seems to favor the view that many submarine canyons have been excavated by *turbidity currents* (downslope movements of dense, sediment-laded water). *Turbidites,* sediments deposited by turbidity currents, are characterized by a decrease in sediment grain size from bottom to top, a phenomenon known as *graded bedding.*

- The ocean basin floor lies between the continental margin and the mid-ocean ridge system. The features of the ocean basin floor include *deep-ocean trenches* (long, narrow troughs that are the deepest parts of the ocean, and where moving crustal plates descend back into the mantle), *abyssal plains* (among the most level places on Earth, consisting of thick accumulations of sediments that were deposited atop the low, rough portions of the ocean floor by turbidity currents), and *seamounts* (isolated, steep-sided volcanic peaks on the ocean floor that

originate near oceanic ridges or in association with volcanic hot spots).

- *Mid-ocean ridges,* the sites of seafloor spreading, are found in all major oceans and represent more than 20 percent of Earth's surface. These broad features are certainly the most prominent features in the oceans, for they form an almost continuous mountain range. Ridges are characterized by an *elevated position, extensive faulting,* and *volcanic structures* that have developed on newly formed oceanic crust. Most of the geologic activity associated with ridges occurs along a narrow region on the ridge crest, called the *rift zone,* where magma from the asthenosphere moves upward to create new slivers of oceanic crust.

- *Coral reefs,* which are confined largely to the warm, sunlit waters of the Pacific and Indian oceans, are constructed over thousands of years primarily from the accumulation of skeletal remains and secretions of corals and certain algae. A coral island, called an *atoll,* consists of a continuous or broken ring of coral reef surrounding a central lagoon. Atolls form from corals that grow on the flanks of sinking volcanic islands, where the corals continue to build the reef complex upward as the island slowly sinks.

- *There are three broad categories of seafloor sediments. Terrigenous sediment* consists primarily of mineral grains that were weathered from continental rocks and transported to the ocean. *Biogenous sediment* consists of shells and skeletons of marine animals and plants. *Hydrogenous sediment* includes minerals that crystallize directly from seawater through various chemical reactions.

Key Terms

abyssal plain (p. 277)
atoll (p. 281)
biogenous sediment (p. 282)
continental margin (p. 272)
continental rise (p. 273)

continental shelf (p. 272)
continental slope (p. 273)
coral reef (p. 280)
deep-ocean trench (p. 277)
deep-sea fan (p. 273)
echo sounder (p. 271)

graded bedding (p. 276)
guyot (p. 279)
hydrogenous sediment (p. 283)
manganese nodule (p. 283)
mid-ocean ridge (p. 279)

rift zone (p. 280)
seamount (p. 279)
submarine canyon (p. 273)
terrigenous sediment (p. 282)
turbidite (p. 276)
turbidity current (p. 273)

Questions for Review

1. How does the area covered by the oceans compare with that of the continents? Describe the distribution of land and water on Earth.

2. Answer the following questions about the Atlantic, Pacific, and Indian oceans:

    (a) Which is the largest in area? Which is smallest?
    (b) Which has the greatest north-south extent?
    (c) Which is shallowest? Explain why its average depth is least.

3. How does the average depth of the ocean compare to the average elevation of the continents?

4. Assuming that the average speed of sound waves in water is 1500 meters per second, determine the water depth if the signal sent out by an echo sounder requires 6 seconds to strike bottom and return to the recorder (see Figure 14.3).

5. List the three major subdivisions of the continental margin. Which subdivision is considered a flooded extension of the continent? Which has the steepest slope?

6. How does the continental margin along the west coast of South America differ from the continental margin along the east coast of North America?

7. Defend or rebut the statement "Most submarine canyons found on the continental slope and rise were formed during the Ice Age when rivers extended their valleys seaward."

8. What are turbidites? What is meant by the term *graded bedding?*

9. Discuss the evidence that helped confirm the existence of turbidity currents in the ocean and establish them as significant mechanisms of erosion and sediment transport. (See Box 14.1)

10. Why are abyssal plains more extensive on the floor of the Atlantic than on the floor of the Pacific?

11. How are mid-ocean ridges and deep-ocean trenches related to seafloor spreading?

12. What is an atoll? Describe Darwin's proposal on the origin of atolls. Was it ever confirmed?

13. Distinguish among the three basic types of seafloor sediment.

14. If you were to examine recently deposited biogenous sediment taken from a depth in excess of 4500 meters (15,000 feet), would it more likely be rich in calcareous materials or siliceous materials? Explain.

Testing What You Have Learned

To test your knowledge of the material presented in this chapter, answer the following questions:

Multiple-Choice Questions

1. Downslope movements of dense, sediment-laden ocean water are called _____ currents.
    a. mud   **b.** gravity   **c.** turbidity   **d.** velocity   **e.** submarine

2. Which one of the following types of seafloor sediments includes calcareous and siliceous oozes?
    a. ridge   **b.** terrigenous   **c.** biogenous   **d.** hydrogenous   **e.** lithogenous

3. What percentage of Earth's surface is covered by oceans and marginal seas?
    a. 29%   **b.** 43%   **c.** 67%   **d.** 71%   **e.** 82%

4. The _____ Ocean is by far the largest of the oceans.
    a. Pacific   **b.** Atlantic   **c.** Arctic   **d.** Indian

5. Isolated, flat-topped volcanic peaks found on the ocean floor are called _____.
    a. guyots   **b.** cirques   **c.** ridges   **d.** atolls   **e.** rifts

6. Marking the seaward edge of the continental shelf is the continental _____.
    a. rise   **b.** trough   **c.** ridge   **d.** slope   **e.** canyon

7. Using an echo sounder, scientists can determine the depth of water by measuring the time it takes for _____ to reflect off the ocean floor.
    a. light   **b.** radio waves   **c.** tsunamis   **d.** microwaves   **e.** sound

8. The sites where moving crustal plates plunge back into the mantle are _____.
   a. seamounts   **b.** abyssal plains   **c.** continental slopes   **d.** deep-ocean trenches   **e.** mid-ocean ridges

9. _____ plains are among the most level places on Earth.
   a. Basin   **b.** Abyssal   **c.** Rift   **d.** Mid-ocean   **e.** Shelf

10. Currents that emerge from the mouths of submarine canyons onto the relatively flat ocean floor often deposit sediments that form triangular shaped features called _____.
   a. submarine   **b.** seamounts   **c.** continental rifts   **d.** turbidity currents   **e.** deep-sea fans deltas

Fill-In Questions

11. The zones that collectively make up the continental margin include the continental _____, continental _____, and the continental _____.

12. The _____ Hemisphere is often referred to as the "land hemisphere."

13. Most available information favors the view that submarine canyons have been excavated by _____ currents.

14. The three major topographical units of the ocean basins are the continental _____, the ocean basin _____, and the mid-ocean _____.

15. Dotting the ocean floor are isolated, steep-sided, conical volcanic peaks called _____.

True/False Questions

16. The continental shelf represents the true edge of the continent. _____

17. Mid-ocean ridges consist of layer upon layer of folded and faulted sedimentary rocks. _____

18. Rift zones, located on the crests of mid-ocean ridges, are the active sites of seafloor spreading. _____

19. Mud is the most common sediment found on the deep-ocean floor. _____

20. Turbidites are characterized by a decrease in sediment grain size from bottom to top, a phenomenon known as cross-bedding. _____

Answers

1. c; 2. c; 3. d; 4. a; 5. a; 6. d; 7. e; 8. d; 9. b; 10. e; 11. shelf, slope, rise; 12. Northern; 13. turbidity; 14. margins, floor, ridges; 15. seamounts; 16. F; 17. F; 18. T; 19. T; 20. F

Earthquakes and Earth's Interior

Focus on Learning

To assist you in learning the important concepts in this chapter, you will find it helpful to focus on the following questions:

- What is an earthquake?
- What are the types of earthquake waves?
- How is the epicenter of an earthquake determined?
- Where are the principal earthquake zones on Earth?
- How is earthquake magnitude expressed by the Richter scale?
- What are the four major zones of Earth's interior?
- How do continental crust and oceanic crust differ?

In January 1995, a strong earthquake damaged this and numerous other buildings in Kobe, Japan. (Photo by Haruyoshi Yamaguchi/SYGMA)

What is an earthquake? How does a seismograph record the location of a "quake"? Can earthquakes ever be predicted? If we could view Earth's interior, what would it look like? In this chapter we shall answer these and other questions. The study of earthquakes is important, not only because of the devastating effect that some earthquakes have on us, but also because they furnish clues about the structure of Earth's interior.

On October 17, 1989, at 5:04 P.M. Pacific Daylight Time, millions of television viewers around the world were settling in to watch the third game of the World Series. Instead, they saw their television sets go black as tremors hit San Francisco's Candlestick Park. Although the earthquake was centered in a remote section of the Santa Cruz Mountains, 100 kilometers to the south, major damage occurred in the Marina District of San Francisco.

The most tragic result of the violent shaking was the collapse of some double-decked sections of Interstate 880, also known as the Nimitz Freeway. The ground motions caused the upper deck to sway, shattering the concrete support columns along a mile-long section of the freeway. The upper deck then collapsed onto the lower roadway, flattening cars as if they were aluminum beverage cans. This earthquake, named the Loma Prieta quake for its point of origin, claimed 67 lives.

In mid-January 1994, less than five years after the Loma Prieta earthquake devastated portions of the San Francisco Bay area, a major earthquake struck the Northridge area of Los Angeles. Although it was not the fabled "Big One," the moderate 6.6 to 6.9 magnitude earthquake left 51 dead, over 5000 injured, and tens of thousands of households without water and electricity. In total, damage in excess of $15 billion was attributed to an apparently unknown fault that ruptured at a depth of 14 kilometers (9 miles) beneath Northridge (Figure 15.1).

The Northridge earthquake began at 4:31 A.M. and lasted roughly 40 seconds. During this brief period, the quake terrorized the entire Los Angeles area. In the three-story Northridge Meadows apartment complex, 16 people died when sections of the upper floors collapsed onto the first-floor units. Nearly 300 schools were seriously damaged and a dozen major roadways buckled. Among these were two of California's major arteries—the Golden State Freeway (Interstate 5), where an overpass collapsed completely and blocked the roadway, and the Santa Monica Freeway. Fortunately, these roadways had practically no traffic at this early morning hour.

In nearby Granada Hills, broken gas lines were set ablaze while the streets were flooded from broken water mains. Seventy homes burned in the Sylmar area. A 64-car freight train derailed, including

FIGURE 15.1 This parking deck in Northridge, California, collapsed during an earthquake in January 1994. (Photo by Spencer Grant/Gamma–Liaison)

FIGURE 15.2 San Francisco in flames after the 1906 earthquake. (Reproduced from the collection of the Library of Congress)

some cars carrying hazardous cargo. But it is remarkable that the destruction was not greater. Unquestionably, the upgrading of structures to meet the requirements of building codes developed for this earthquake-prone area helped minimize what could have been a much greater human tragedy.

Across the Pacific from California lies Japan, no stranger to earthquakes, and among the most "quake-proofed" countries in the world. Yet at 5:46 A.M. on January 24, 1995, much of the "quake-proofing" proved futile as more than 5000 people perished in a 7.2-magnitude tremor centered near Kobe, the country's sixth-largest city (see chapter-opening photo).

Over 30,000 earthquakes that are strong enough to be felt occur worldwide annually. Fortunately, most are minor tremors and do very little damage. Generally, only about 75 significant earthquakes take place each year, and many of these occur in remote regions. However, occasionally a large earthquake occurs near a large population center. Under these conditions, an earthquake is among the most destructive natural forces on Earth (see Box 15.1).

The shaking of the ground, coupled with the liquefaction of some soils, wreaks havoc on buildings and other structures. In addition, when a quake occurs in a populated area, power and gas lines are often ruptured, causing numerous fires. In the famous 1906 San Francisco earthquake, much of the damage was caused by fires (Figure 15.2). They quickly became uncontrollable when broken water mains left firefighters with only trickles of water.

What Is an Earthquake?

An **earthquake** is the vibration of Earth produced by the rapid release of energy. Most often earthquakes are caused by slippage along a fault in Earth's crust. The energy released radiates in all directions from its source, the **focus,** in the form of waves. These waves are analogous to those produced when a stone is dropped into a calm pond (Figure 15.3). Just as the impact of the stone sets water waves in motion, an earthquake generates seismic waves that radiate throughout the Earth. Even though the energy dissipates rapidly with increasing distance from the focus, sensitive instruments located throughout the world record the event.

Earthquakes and Faults

The tremendous energy released by atomic explosions or by volcanic eruptions can produce an earthquake, but these events are relatively weak and infrequent. What mechanism produces a destructive earthquake? Ample evidence exists that Earth is not a static planet. We know that Earth's crust has been uplifted at times, because we have found numerous ancient wave-cut benches many meters above the level of the highest tides. Other

BOX 15.1 Damaging Earthquakes East of the Rockies

The majority of earthquakes occur near plate boundaries, as exemplified by California and Japan. However, areas distant from plate boundaries are not necessarily immune. A team of seismologists recently estimated that the probability of a damaging earthquake east of the Rocky Mountains during the next 30 years is roughly two-thirds as likely as an earthquake of comparable damage in California. Like all earthquake risk assessments, this prediction is based in part on the geographic distribution and average rate of earthquake occurrences in these regions.

At least six major earthquakes have occurred in the central and eastern United States since colonial times. Three of them, having estimated Richter magnitudes of 7.5, 7.3, and 7.8, were centered near the Mississippi River Valley in southeastern Missouri. Occurring over a 3-month period in December 1811, January 1812, and February 1812, these earthquakes and numerous smaller tremors destroyed the town of New Madrid, Missouri. They also triggered massive landslides, damaged a six-state area, altered the course of the Mississippi River, and enlarged Tennessee's Reelfoot Lake.

The distance over which these earthquakes were felt is truly remarkable. Chimneys were downed in Cincinnati and Richmond, and even Boston residents, 1770 kilometers (1100 miles) to the northeast, felt the tremor. Although the destruction from the New Madrid earthquakes was slight compared to the Loma Prieta earthquake of 1989, the Midwest in the early 1800s was sparsely populated. Memphis, near the epicenter, had not yet been established, and St. Louis was a small frontier town. Other damaging earthquakes— Aurora, Illinois (1909) and Valentine, Texas (1931)—remind us that the cen-

FIGURE 15.A Damage to Charleston, South Carolina, caused by the August 31, 1886, earthquake. Damage ranged from toppled chimneys and broken plaster to total collapse. (Photo courtesy of U.S. Geological Survey)

tral United States is vulnerable.

The greatest historical earthquake in the eastern states occurred in Charleston, South Carolina, in 1886. This 1-minute event caused 60 deaths, numerous injuries, and great economic loss within 200 kilometers (120 miles) of Charleston. Within 8 minutes, strong vibrations shook the upper floors of buildings in Chicago and St. Louis, causing people to rush outdoors. In Charleston alone over a hundred buildings were destroyed, and 90 percent of the remaining structures were damaged. It was difficult to find a chimney that was still standing (Figure 15.A).

New England and adjacent areas have experienced sizable shocks since colonial times including the 1683 quake in Plymouth and the 1755 quake in Cambridge, Massachusetts. Since records have been kept, New York State has experienced over 300 earthquakes large enough to be felt.

These eastern and central earthquakes occur far less frequently than do those in California. Yet the shocks east of the Rockies have generally pro-

duced structural damage over a larger area than tremors of similar magnitude in California. The reason is that the underlying bedrock in the central and eastern United States is older and more rigid. As a result, seismic waves travel greater distances with less attenuation than in the western United States. For similar earthquakes, the region of maximum ground motion in the East may be up to 10 times larger than in the West. Consequently, the higher rate of earthquakes in the West is partly balanced by more widespread damage in the East.

Despite recent geologic history, Memphis, the largest population center in the area of the New Madrid earthquake, lacks adequate provision for earthquakes in its building code. Worse, Memphis rests on unconsolidated floodplain deposits, so its buildings are more susceptible to damage. If an earthquake the size of the 1811–1812 New Madrid events were to strike the Memphis area today, it would cause casualties in the thousands and damage in the tens of billions of dollars.

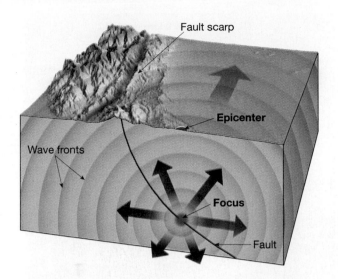

FIGURE 15.3 The focus of all earthquakes is located at depth. The surface location directly above it is called the epicenter.

FIGURE 15.4 Slippage along a fault produced an offset in this orange grove east of Calexico, California. (Photo by John S. Shelton)

regions exhibit evidence of extensive subsidence. In addition to these vertical displacements, offsets in fence lines, roads, and other structures indicate that horizontal movement is common (Figure 15.4). These movements are usually associated with large fractures in Earth's crust called **faults.**

Most of the motion along faults can be satisfactorily explained by the plate tectonics theory. This theory states that large slabs of Earth's crust are in continual slow motion. These mobile plates interact with neighboring plates, straining and deforming the rocks at their edges. In fact, it is along faults associated with plate boundaries that most earthquakes occur. Furthermore, earthquakes are repetitive: as soon as one is over, the continuous motion of the plates resumes, adding strain to the rocks until they fail again.

Elastic Rebound

The actual mechanism of earthquake generation eluded geologists until H. F. Reid of Johns Hopkins University conducted a study following the great 1906 San Francisco earthquake. The earthquake was accompanied by horizontal surface displacements of several meters along the northern portion of the San Andreas fault. This 1300-kilometer (780-mile) fracture runs north-south through southern California. It is a large fault zone that separates two great sections of Earth's crust, the North American plate and the Pacific plate. Field investigations determined that, during this single earthquake, the Pacific plate lurched as much as 4.7 meters (15 feet) northward past the adjacent North American plate.

The mechanism for earthquake formation that Reid deduced from this information is illustrated in

Figure 15.5. In part A of the figure, you see an existing fault, or break in the rock. In part B, tectonic forces ever so slowly deform the crustal rocks on both sides of the fault, as demonstrated by the bent features. Under these conditions, rocks are bending and storing elastic energy, much like a wooden stick does if bent. Eventually, the frictional resistance holding the rocks together is overcome. As slippage occurs at the weakest point (the focus), displacement will exert stress farther along the fault, where additional slippage will occur until most of the built-up strain is released (Figure 15.5C). This slippage allows the deformed rock to "snap back." The vibrations we know as an earthquake occur as the rock elastically returns to its original shape. The "springing back" of the rock was termed **elastic rebound** by Reid, because the rock behaves elastically, much like a stretched rubber band does when it is released.

In summary, most earthquakes are produced by the rapid release of elastic energy stored in rock that has been subjected to great stress. Once the strength of the rock is exceeded, it suddenly ruptures, causing the vibrations of an earthquake. Earthquakes also occur along existing fault surfaces whenever the frictional forces on the fault surfaces are overcome.

The San Andreas is undoubtedly the most studied fault system in the world. Over the years, investigations have shown that displacement occurs along discrete segments that are 100 to 200 kilometers long. Further, each fault segment behaves somewhat differently from the others. Some portions of the San Andreas exhibit a slow, gradual displacement known as *fault creep,* which occurs relatively smoothly, and therefore with little noticeable seismic activity. Other segments regularly slip, producing small earthquakes.

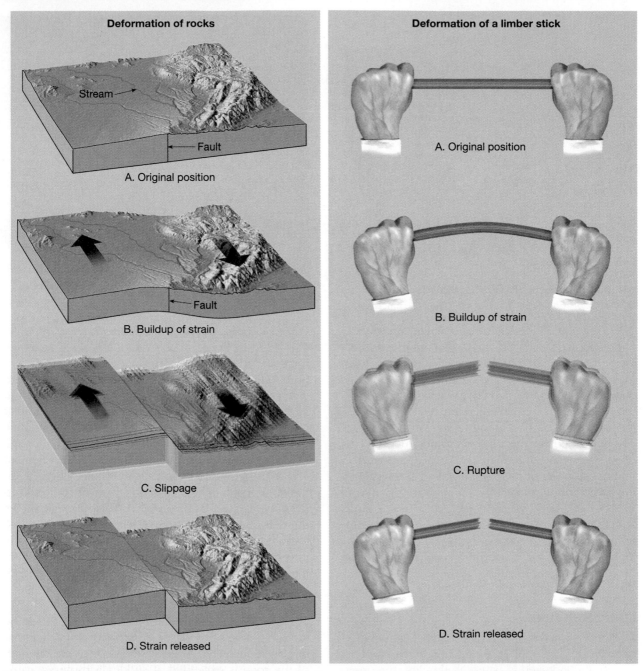

Deformation of rocks

Stream

Fault

A. Original position

Fault

B. Buildup of strain

C. Slippage

D. Strain released

Deformation of a limber stick

A. Original position

B. Buildup of strain

C. Rupture

D. Strain released

FIGURE 15.5 Elastic rebound. As rock is deformed it bends, storing elastic energy. Once the rock is strained beyond its breaking point it ruptures, releasing the stored-up energy in the form of earthquake waves.

Still other segments remain locked and store elastic energy for hundreds of years before rupturing in great earthquakes. The latter process is described as *stick-slip* motion, because the fault exhibits alternating periods of locked behavior followed by sudden slippage. It is estimated that great earthquakes should occur about every 50 to 200 years along those sections of the San Andreas fault that exhibit stick-slip motion. This knowledge is useful when assigning a potential earthquake risk to a given segment of the fault zone.

Not all movement along faults is horizontal.

Vertical displacement, in which one side is lifted higher in relation to the other, is also common. Figure 15.6 shows a *fault scarp* (cliff) produced by such vertical movement. Further, many earthquakes occur at such great depths that no displacement is evident at the surface.

Foreshocks and Aftershocks

The intense vibrations of the 1906 San Francisco earthquake lasted about 40 seconds. Although most of the displacement along the fault occurred in this

FIGURE 15.6 Fault scarp produced from vertical movement during the 1964 Alaskan earthquake. (Courtesy of U.S. Geological Survey)

uplift

rather short period, additional movements along this and other nearby faults lasted for several days following the main quake. The adjustments that follow a major earthquake often generate smaller earthquakes called **aftershocks.** Although these aftershocks are usually much weaker than the main earthquake, they can sometimes destroy already badly weakened structures. This occurred, for example, during a 1988 earthquake in Armenia. A large aftershock of magnitude 5.8 collapsed many structures that had been weakened by the main tremor. That disaster killed about 25,000 people.

In addition, small earthquakes called **fore-shocks** often precede a major earthquake by days or, in some cases, by as much as several years. Monitoring of these foreshocks has been used as a means of predicting forthcoming major earthquakes, with mixed success. We will consider the topic of earthquake prediction in a later section of this chapter.

Tectonic Forces and Earthquakes

It is important to understand that the tectonic forces creating the strain that was eventually released during the 1906 San Francisco earthquake are still active. Currently, laser beams are used to measure the relative motion between the opposite sides of this fault. These measurements reveal a displacement of 2 to 5 centimeters (1 to 2 inches) per year. Although this seems slow, it produces substantial movement over millions of years. To illustrate, in 30 million years, this rate of displacement would slide the western portion of California northward so that Los Angeles, on the Pacific plate, would be adjacent to San Francisco on the North American plate! More important in the short term, a displacement of just 2 centimeters per year produces 2 meters of offset every 100 years. Consequently, the 4 meters of displacement produced during the 1906 San Francisco earthquake should occur

at least every 200 years along this segment of the fault zone. This fact lies behind California's concern for making buildings earthquake-resistant, in anticipation of the inevitable "big one."

Earthquake Waves

The study of earthquake waves, **seismology**, dates back to attempts by the Chinese almost 2000 years ago to determine the direction to the source of each earthquake. Modern **seismographs** are instruments that record earthquake waves. Their principle is simple. A weight is freely suspended from a support that is attached to bedrock (Figure 15.7). When waves from an earthquake reach the instrument, the inertia of the weight keeps it stationary, while Earth and the support vibrate. The movement of Earth in relation to the stationary weight is recorded on a rotating drum. (*Inertia* is the tendency of a stationary object to hold still, or a moving object to stay in motion.)

Modern seismographs amplify and record ground motion, producing a trace as shown in Figure 15.8. These records, called **seismograms** reveal that seismic waves are elastic energy. This energy radiates outward in all directions from the focus, as you saw in Figure 15.3. The transmission of this energy can be compared to the shaking of gelatin in a bowl that is jarred. Seismograms reveal that two main types of seismic waves are generated by the slippage of a rock mass. Some travel along Earth's outer layer, and are called **surface waves.** Others travel through Earth's interior and are called **body waves**. Body waves are further divided into **primary waves (P waves)** and **secondary waves (S waves).**

Body waves are divided into P and S waves by their mode of travel through intervening materials. P waves are "push-pull" waves—they push (compress) and pull (expand) rocks in the direction the wave is

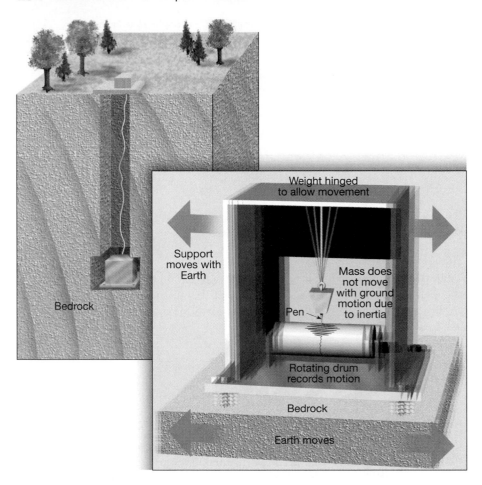

FIGURE 15.7 Principle of the seismograph. The inertia of the suspended mass tends to keep it motionless, while the recording drum, which is anchored to bedrock, vibrates in response to seismic waves. Thus, the stationary mass provides a reference point from which to measure the amount of displacement occurring as the seismic wave passes through the ground below.

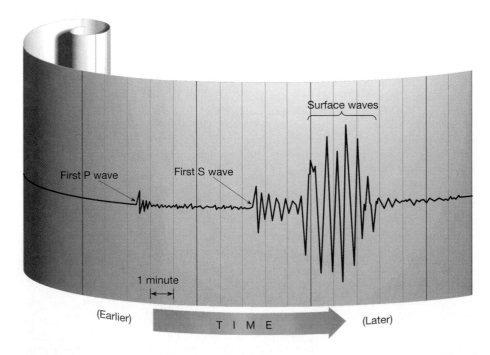

FIGURE 15.8 Typical seismic record. Note the time interval (about 5 minutes) between the arrival of the first P waves and the arrival of the first S waves.

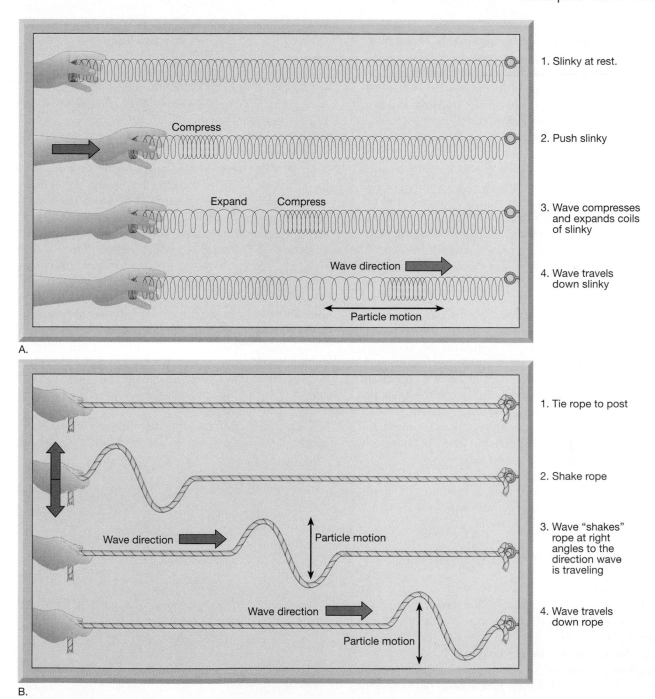

1. Slinky at rest.

2. Push slinky

Compress

3. Wave compresses and expands coils of slinky

Expand   Compress

4. Wave travels down slinky

Wave direction

Particle motion

A.

1. Tie rope to post

2. Shake rope

Wave direction

3. Wave "shakes" rope at right angles to the direction wave is traveling

Particle motion

Wave direction

4. Wave travels down rope

Particle motion

B.

FIGURE 15.9 Types of seismic waves and their characteristic motion. **A.** P waves compress, causing the particles in the material to vibrate back and forth in the same direction as the waves move. **B.** S waves cause particles to oscillate at right angles to the direction of wave motion.

traveling (Figure 15.9A). Imagine holding someone by the shoulders and shaking that person. This push-pull movement is how P waves move through Earth. This wave motion is analogous to that generated by human vocal cords as they move air to create sound. Solids, liquids, and gases resist a change in volume when compressed and will elastically spring back once the force is removed. Therefore, P waves, which are compressional waves, can travel through all these materials.

S waves, on the other hand, "shake" the particles at right angles to their direction of travel. This can be illustrated by fastening one end of a rope and shaking the other end, as shown in Figure 15.9B. Unlike

P waves, which temporarily change the *volume* of the intervening material by alternately compressing and expanding it, S waves temporarily change the *shape* of the material that transmits them. Because fluids (gases and liquids) do not respond elastically to changes in shape, they will not transmit S waves.

The motion of surface waves is somewhat more complex. As surface waves travel along the ground, they cause the ground and anything resting upon it to move, much like ocean swells toss a ship. In addition to their up-and-down motion, surface waves have a side-to-side motion similar to an S wave oriented in a horizontal plane. This latter motion is particularly damaging to the foundations of structures.

By observing a "typical" seismic record, as shown in Figure 15.8, you can see a major difference among these seismic waves: P waves arrive at the recording station first; then S waves; and then surface waves. This is a consequence of their speeds. To illustrate, the velocity of P waves through granite within the crust is about 6 kilometers per second. S waves under the same conditions travel at 3.5 kilometers per second. Differences in density and elastic properties of the rock greatly influence the velocities of these waves. Generally, in any solid material, P waves travel about 1.7 times faster than S waves, and surface waves can be expected to travel at 90 percent of the velocity of the S waves.

As you shall see, seismic waves allow us to determine the location and magnitude of earthquakes. In addition, seismic waves provide us with a tool for probing Earth's interior.

Finding Earthquake Epicenters

Recall that the *focus* is the place within Earth where earthquake waves originate. The **epicenter** is the location on the surface directly above the focus (see Figure 15.3).

The difference in velocities of P and S waves provides a method for locating the epicenter. The principle used is analogous to a race between two autos, one faster than the other. The P wave always wins the race, arriving ahead of the S wave. But, the greater the length of the race, the greater will be the difference in the arrival times at the finish line (the seismic station). Therefore, the greater the interval measured on a seismogram between the arrival of the first P wave and the first S wave, the greater the distance to the earthquake source.

A system for locating earthquake epicenters was developed by using seismograms from earthquakes whose epicenters could be easily pinpointed from physical evidence. From these

seismograms, travel-time graphs were constructed (Figure 15.10). The first travel-time graphs were greatly improved when seismograms became available from nuclear explosions, because the precise location and time of detonation were known.

Using the sample seismogram in Figure 15.8 and the travel-time curves in Figure 15.10, we can determine the distance separating the recording station from the earthquake in two steps: (1) determine the time interval between the arrival of the first P wave and the first S wave, and (2) find on the travel-time graph the equivalent time spread between the P and S wave curves. From this information, we can determine that this earthquake occurred 3800 kilometers (2350 miles) from the recording instrument.

Now we know the *distance,* but what *direction?* The epicenter could be in any direction from the seismic station. As shown in Figure 15.11, the precise location can be found when the distance is known from three or more different seismic stations. On a globe, we draw a circle around each seismic station. Each circle represents the epicenter distance for each station. The point where the three circles intersect is the epicenter of the quake. This method is called triangulation.

About 95 percent of the energy released by earthquakes originates in a few relatively narrow

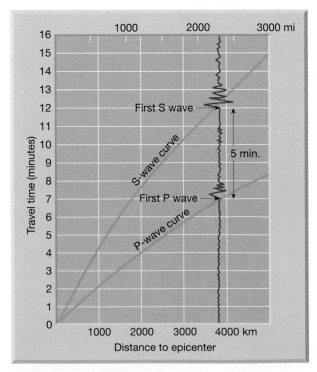

FIGURE 15.10 A travel-time graph is used to determine the distance to the epicenter. The difference in arrival time of the first P wave and the first S wave in the example is 5 minutes. Thus, the epicenter is roughly 3800 kilometers (2350 miles) away.

FIGURE 15.11 Earthquake epicenter is located using the distance obtained from three-seismic stations.

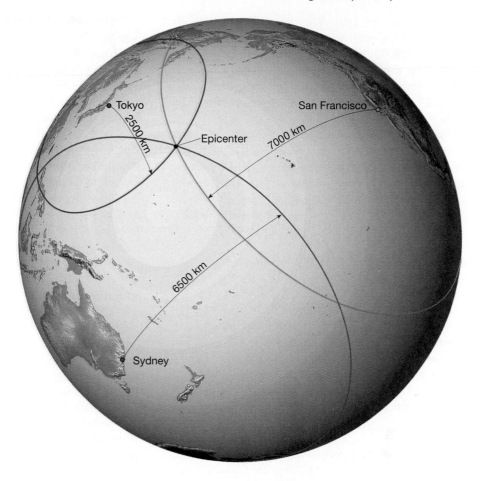

FIGURE 15.12 Distribution of the 14,229 earthquakes with magnitudes equal to or greater than 5 for the period 1980–1990. (Data from National Geophysical Data Center/NOAA)

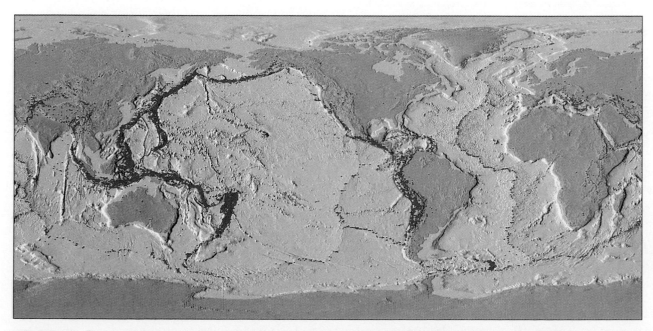

zones (Figure 15.12). The greatest energy is released along a path around the outer edge of the Pacific Ocean known as the *circum-Pacific belt*. Included in this zone are regions of great seismic activity, such as Japan, the Philippines, Chile, and numerous volcanic island chains, as exemplified by Alaska's Aleutian Islands.

Figure 15.12 reveals another continuous belt that extends for thousands of kilometers through the world's oceans. This zone coincides with the oceanic ridge system, an area of frequent but low-intensity seismic activity. By comparing this figure with Figure 16.8, you can see a close correlation between the location of earthquake epicenters and plate boundaries.

Earthquake Intensity and Magnitude

Until a century ago, earthquake size and strength were described subjectively, making accurate classification of earthquake intensity difficult. Then, in 1902, Giuseppe Mercalli developed a fairly reliable intensity scale based on damage to various types of structures. The U.S. Coast and Geodetic Survey uses a modification of this scale today (Table 15.1).

TABLE 15.1 Modified Mercalli intensity scale.

I	Not felt except by a very few under especially favorable circumstances.
II	Felt only by a few persons at rest, especially on upper floors of buildings.
III	Felt quite noticeably indoors, especially on upper floors of buildings, but many people do not recognize it as an earthquake.
IV	During the day felt indoors by many, outdoors by few. Sensation like heavy truck striking building.
V	Felt by nearly everyone, many awakened. Disturbances of trees, poles, and other tall objects sometimes noticed.
VI	Felt by all; many frightened and run outdoors. Some heavy furniture moved; few instances of fallen plaster or damaged chimneys. Damage slight.
VII	Everybody runs outdoors. Damage negligible in buildings of good design and construction; slight to moderate in well-built ordinary structures; considerable in poorly built or badly designed structures.
VIII	Damage slight in specially designed structures; considerable in ordinary substantial buildings with partial collapse; great in poorly built structures. (Fall of chimneys, factory stacks, columns, monuments, walls.)
IX	Damage considerable in specially designed structures. Buildings shifted off foundations. Ground cracked conspicuously.
X	Some well-built wooden structures destroyed. Most masonry and frame structures destroyed. Ground badly cracked.
XI	Few, if any (masonry) structures remain standing. Bridges destroyed. Broad fissures in ground.
XII	Damage total. Waves seen on ground surfaces. Objects thrown upward into air.

The **Mercalli intensity scale** assesses the damage from a quake at a specific location. Please note that earthquake intensity depends not only on the strength of the earthquake but also on other factors, such as distance from the epicenter, the nature of surface materials, and building design. A modest 6.9-Richter-magnitude earthquake in Armenia in 1988 was very destructive, mainly because of inferior building construction. A 1985 Mexico City quake was deadly because of the soft sediment upon which the city rests. Thus, the destruction wrought by earthquakes is very meaningful to people living there, but it is not a true measure of the earthquake's actual strength. Further, many earthquakes occur beneath the sea or at great depths in the crust and are not felt.

In 1935, Charles Richter of the California Institute of Technology introduced the concept of earthquake **magnitude**. Today, a refined **Richter scale** is used worldwide to describe earthquake magnitude. Richter magnitude is determined by measuring the amplitude of the largest wave recorded on the seismogram (see Figure 15.8). For seismic stations worldwide to obtain the same magnitude for a given earthquake, adjustments are made for the weakening of seismic waves with distance as they move away from the focus and for the sensitivity of the recording instrument.

The largest earthquakes ever recorded had Richter magnitudes near 8.6. These great shocks released energy roughly equivalent to the detonation of one billion tons of TNT. Conversely, earthquakes with a Richter magnitude of less than 2.0 are usually not felt by humans. Table 15.2 shows how earthquake magnitudes and their effects are related.

Earthquakes vary enormously in strength, and great earthquakes produce traces having wave amplitudes that are thousands of times larger than those generated by weak tremors. To accommodate this wide variation, Richter could not use a linear scale, but instead used a *logarithmic scale* to express magnitude. On this scale, an increase of one magnitude means a *tenfold* increase in wave amplitude. Thus, the amplitude of the largest surface wave for a 5.3-magnitude earthquake is 10 times greater than the wave amplitude produced by an earthquake having a magnitude of 4.3.

More important, each unit of Richter magnitude equates to roughly a *30-fold energy increase*. Thus, an earthquake with a magnitude of 6.5 releases 30 times more energy than one with a magnitude of 5.5, and roughly 900 times (30 × 30) more energy than a 4.5-magnitude quake. A great earthquake with a magnitude or 8.5 releases millions of times more energy than the smallest earthquake felt by humans.

TABLE 15.2 Earthquake magnitudes and expected world incidence.

Richter Magnitudes	Effects Near Epicenter	Estimated Number per Year
<2.0	Generally not felt, but recorded.	600,000
2.0–2.9	Potentially perceptible.	300,000
3.0–3.9	Felt by some.	49,000
4.0–4.9	Felt by most.	6200
5.0–5.9	Damaging shocks.	800
6.0–6.9	Destructive in populous regions.	266
7.0–7.9	Major earthquakes. Inflict serious damage.	18
≥8.0	Great earthquakes. Destroy communities near epicenter.	1.4

SOURCE: *Earthquake Information Bulletin* and others.

TABLE 15.3 Some notable earthquakes.

Year	Location	Deaths (est.)	Magnitude	Comments
1290	Chihli (Hopei), China	100,000		
1556	Shensi, China	830,000		Possibly the greatest natural disaster.
1737	Calcutta, India	300,000		
1755	Lisbon, Portugal	70,000		Tsunami damage extensive.
*1811–1812	New Madrid, Missouri	Few		Three major earthquakes.
*1886	Charleston, South Carolina	60		Greatest historical earthquake in the eastern United States.
*1906	San Francisco, California	1500	8.1–8.2	Fires caused extensive damage.
1908	Messina, Italy	120,000		
1920	Kansu, China	180,000		
1923	Tokyo, Japan	143,000	7.9	Fire caused extensive destruction.
1960	Southern Chile	5700	8.5–8.6	Possibly the largest-magnitude earthquake ever recorded.
*1964	Alaska	131	8.3–8.4	
1970	Peru	66,000	7.8	Great rockslide.
*1971	San Fernando, California	65	6.5	Damage exceeded $1 billion.
1975	Liaoning Province, China	Few	7.5	First major earthquake to be predicted.
1976	Tangshan, China	240,000	7.6	Not predicted.
1985	Mexico City	9500	8.1	Major damage occurred 400 km from epicenter.
1988	Armenia	25,000	6.9	Poor construction practices contributed to destruction.
*1989	San Francisco Bay area	62	7.1	Damages exceeded $6 billion.
1990	Northwestern Iran	50,000	7.3	Landslides and poor construction practices caused great damage.
*1994	Northridge, California	61	6.7	Damages in excess of $15 billion.
1995	Kobe, Japan	5472	6.9	Damage estimated to exceed $100 billion.

*U.S. earthquakes.
SOURCE: U.S. National Oceanic and Atmospheric Administration.

Knowing this should dispel the notion that a moderate earthquake acts as a "pressure relief valve," decreasing the chances for a major quake in the same region. A moderate quake certainly relieves some strain, but thousands of moderate tremors would be needed to release the equivalent energy of one "great" earthquake.

Some of the world's major earthquakes and their corresponding Richter magnitudes are listed in Table 15.3.

Destruction from Earthquakes

The most violent earthquake to jar North America this century—the Good Friday Alaskan Earthquake—occurred in 1964. Felt throughout that state, the earthquake had a Richter magnitude of 8.3–8.4 and reportedly lasted 3 to 4 minutes. This event left 131 people dead, thousands homeless, and the economy of the state badly disrupted because it occurred near major towns and seaports (Figure 15.13). Had the

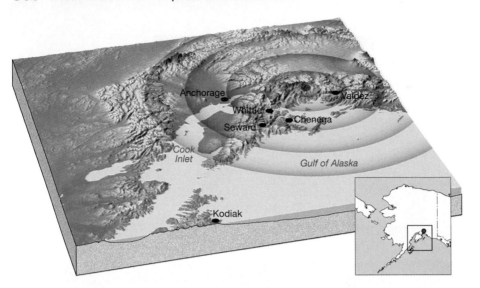

FIGURE 15.13 Region most affected by the Good Friday earthquake of 1964. Note the epicenter (red dot). (After U.S. Geological Survey)

schools and business districts been open on this holiday, the toll surely would have been higher. Within 24 hours of the initial shock, 28 aftershocks were recorded, 10 of which exceeded a Richter magnitude of 6.

Destruction from Seismic Vibrations

The 1964 Alaskan earthquake provided geologists with new insights into the role of ground shaking as a destructive force. Recall that the structural damage from earthquake waves depends on several factors, including (1) the amplitude, (2) duration of the vibrations, (3) the nature of the material upon which the structure rests, and (4) the design of the structure.

All multistory structures in Anchorage were damaged by the vibrations, but the more flexible wood-frame residential buildings fared best. Figure 15.14

FIGURE 15.14 Damage to the five-story J C Penney Co. building, Anchorage, Alaska. Very little structural damage was incurred by the adjacent building. (Courtesy of NOAA)

offers a striking example of how construction variations affect earthquake damage. You can see that the steel-frame building on the left withstood the vibrations, whereas the relatively rigid concrete structure was badly damaged.

Most large structures in Anchorage were damaged even though they were built to conform to the earthquake provisions of the Uniform Building Code of California. Perhaps some of that destruction can be attributed to the unusually long duration of this earthquake (estimated at 3 to 4 minutes). (The San Francisco earthquake of 1906 was felt for about 40 seconds; the strong vibrations of the 1989 Loma Prieta earthquake lasted less than 15 seconds.)

The buildings in Anchorage are situated on soft, unconsolidated sediments, which amplify vibrations more than does solid bedrock. Thus, damage in Anchorage was more severe than in Whittier, which rests on a firm foundation of granite. However, Whittier was damaged by a seismic sea wave (described in the next section).

The 1985 Mexican earthquake gave seismologists and engineers a vivid reminder of what had been learned following the 1964 Alaskan earthquake. In central Mexico City, nearly 400 kilometers (250 miles) from the epicenter, the vibrations intensified to five times that experienced in outlying districts, owing to the city's construction on soft sediments, remnants of an ancient lake bed (Figure 15.15).

Where unconsolidated materials are saturated with water, earthquakes can generate a phenomenon known as **liquefaction.** Under these conditions, what had been a stable soil turns into a fluid that is no longer capable of supporting buildings or other structures (Figure 15.16). As a result, underground objects such as storage tanks and sewer lines may float

FIGURE 15.15 During the 1985 Mexican earthquake, multistory buildings swayed back and forth as much as 1 meter. Many, including the hotel shown here, collapsed or were seriously damaged. (Photo by James L. Beck)

toward the surface of their newly liquefied environment. Buildings and other structures may settle and collapse. During the Loma Prieta earthquake, in San Francisco's Marina District, foundations failed and geysers of sand and water shot from the ground, indicating that liquefaction had occurred (Figure 15.17).

Tsunami

Most deaths associated with the 1964 Alaskan quake were caused by **seismic sea waves**, or **tsunami.** These destructive waves often are called "tidal waves" by the media. However, this name is wrong, for these waves are generated by earthquakes, not by the tidal effect of the Moon or Sun. The name *tsunami* is Japanese for "harbor wave," for Japanese harbors have suffered from many of them.

Most tsunami result from vertical displacement of the ocean floor during an earthquake (Figure 15.18.) Once formed, a tsunami resembles the ripples created when a pebble is dropped into a pond. In contrast to ripples, tsunami advance across the ocean at speeds between 500 and 950 kilometers (300 and 600 miles) per hour. Despite this, a tsunami in the open ocean can pass undetected because its height is usually less than one meter and the distance between wave crests is great, ranging from 100 to 700 kilometers. However, upon entering shallower coastal water,

these destructive waves are slowed and the water begins to pile up to heights that occasionally exceed 30 meters (100 feet), as shown in Figure 15.18. As the crest of a tsunami approaches shore, it appears as a rapid rise in sea level with a turbulent and chaotic surface. Tsunami can be very destructive (Figure 15.19).

Usually the first warning of an approaching tsunami is a rather rapid withdrawal of water from

FIGURE 15.16 Effects of liquefaction. This tilted building rests on unconsolidated sediment that imitated quicksand during the 1985 Mexican earthquake. (Photo by James L. Beck)

FIGURE 15.17 These "mud volcanoes" were produced by the Loma Prieta earthquake of 1989. They formed when geysers of sand and water shot from the ground, an indication that liquefaction occurred. (Photo by Richard Hilton, courtesy of Dennis Fox)

beaches. Coastal residents have learned to heed this warning and move to higher ground, for about 5 to 30 minutes later the retreat of water is followed by a surge capable of extending hundreds of meters inland. In a successive fashion, each surge is followed by a rapid oceanward retreat of the water. These waves are separated by intervals of between 10 and 60 minutes.

They are able to traverse thousands of kilometers of the ocean before their energy is dissipated.

The tsunami generated in the 1964 Alaskan earthquake heavily damaged communities along the Gulf of Alaska and killed 107. By contrast, only 9 people died in Anchorage as a direct result of the vibrations.

Tsunami damage following the Alaskan earthquake extended along much of the west coast of North America. Despite a one-hour warning, 12 persons perished in Crescent City, California, from the fifth wave. The first wave crested about 4 meters (13 feet) above low tide and was followed by three progressively smaller waves. Believing that the tsunami had ceased, people returned to the shore, only to be met by the fifth and most devastating wave. Superimposed upon high tide, it crested about 6 meters higher than the level of low tide.

Although most tsunami are generated by earthquakes, a volcanic eruption in the ocean can generate this destructive phenomenon as well. For example, the 1883 volcanic explosion of Krakatoa, an island in Indonesia, generated a tsunami that drowned some 36,000 coastal residents of Java and Sumatra.

Landslides and Ground Subsidence

In the 1964 Alaskan earthquake, the greatest damage to structures was from landslides and ground subsidence triggered by the vibrations. At Valdez and Seward, the violent shaking caused river-delta materials to experience liquefaction; the subsequent slumping carried both waterfronts away. Because the disaster could happen again, the entire town of Valdez was relocated about 7 kilometers away on more stable ground. In Valdez, 31 persons on a dock died when it slid into the sea.

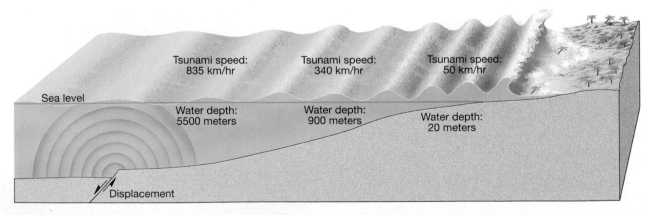

FIGURE 15.18 Schematic drawing of a tsunami generated by displacement of the ocean floor. The speed of a wave moving across the ocean correlates with ocean depth. As shown, waves moving in deep water advance at speeds in excess of 800 kilometers per hour. Speed gradually slows to 50 kilometers per hour at depths of 20 meters. Decreasing depth slows the movement of the wave column. As waves slow in shallow water, they grow in height until they topple and rush onto shore with tremendous force. The size and spacing of these swells are not to scale.

FIGURE 15.19 A man stands before a wall of water about to engulf him at Hilo, Hawaii, on April 1, 1946. This tsunami, which originated in the Aleutian Islands near Alaska, was still powerful enough when it hit Hawaii to rise 9 to 16 meters (30 to 55 feet). The *S.S. Brigham Victory*, from which this photograph was taken, managed to survive the onslaught, but 159 people in Hawaii, including the man seen here, were killed. (Photo courtesy of Water Resources Center Archives, University of California, Berkeley.)

Most of the damage in Anchorage was attributed to landslides. Many homes were destroyed in Turnagain Heights when a layer of clay lost its strength and over 200 acres of land slid toward the ocean (Figure 15.20). A portion of this landslide has been left in its natural condition as a reminder of this destructive event. The site was named "Earthquake Park." Downtown Anchorage was also disrupted as sections of the main business district dropped by as much as 3 meters (10 feet).

FIGURE 15.20 Photo of a small portion of the Turnagain Heights slide. (Photo courtesy of U.S. Geological Survey)

Fire

The 1906 San Francisco earthquake reminds us of the formidable threat of fire. The central city contained mostly large, older wooden structures and brick buildings. The greatest destruction was caused by fires that started when gas and electrical lines were severed. The fires raged uncontrolled for three days and devastated over 500 city blocks (see Figure 15.2). The problem was compounded by the initial ground shaking, which broke the city's water lines into hundreds of unconnected pieces.

The fire was finally contained when buildings were dynamited along a wide boulevard to create a *fire break,* the same strategy used in fighting a forest fire. Although only a few deaths were attributed to the fires, such is not always the case. A 1923 earthquake in Japan (their worst quake prior to the 1995 Kobe tremor) triggered an estimated 250 fires, which devastated the city of Yokohama and destroyed more than half the homes in Tokyo. Over 100,000 deaths were attributed to the fires, which were driven by unusually high winds.

Can Earthquakes Be Predicted?

The vibrations that shook Northridge, California, in 1994 inflicted 51 deaths and about $15 billion in damage (Figure 15.21). This was from a brief earthquake (about 40 seconds) of moderate rating (6.6 to 6.9 on the Richter scale). Seismologists warn that earthquakes of comparable or greater strength will occur along the

FIGURE 15.21 Damage to Interstate 5 during the January 17, 1994, Northridge, California, earthquake. (Photo by Bill Nation/SYGMA)

San Andreas fault, which cuts a 1300-kilometer path through the state (Figure 15.22). Following the 1995 earthquake in Kobe, Japan, a U.S. Geological Survey physicist cautioned: "Kobe is almost a dress rehearsal for an earthquake on the Hayward fault (a fault parallel to the San Andreas, near San Francisco)." The obvious question is, can earthquakes be predicted?

Substantial research to predict earthquakes is under way in Japan, the United States, China, and Russia—countries where earthquake risk is high. This research is striving to identify phenomena that precede major earthquakes. In California, for example, a pattern has been observed to precede some earthquakes: uplift or subsidence of the land, changes in the movements along a fault zone, and a period of seismic quiescence, often followed by renewed activity. Japanese scientists are studying peculiar animal behavior that may precede a quake. Others are examining changes in groundwater levels and radio waves.

Short-Range Predictions

Although no consistent method of short-range prediction has yet been devised, one notable success was the foretelling in 1975 of a 7.3-magnitude earthquake in Liaoning Province of China. Here, for the first and only time, seismologists forecast a large earthquake that was about to destroy a major city. By evacuating some three million residents from their unreinforced masonry structures, tens of thousands of lives were spared. But tragically, one year later, a devastating earthquake in China was *not* predicted, and more than 200,000 people were killed.

The Chinese have also issued false alarms. In a province near Hong Kong, people left their dwellings for over a month, but no earthquake followed. Can you imagine the debate that would precede an order to evacuate a large city in the United States, such as Los Angeles or San Francisco? The cost of evacuating millions of people, arranging for living accommodations, and providing for their lost work time and wages, would have to be weighed against the earthquake's probability.

Long-Range Predictions

Long-range forecasts are based on the premise that earthquakes are repetitive or cyclical, like the weather. In other words, as soon as one earthquake is over, the continuing motions of Earth's plates begin to build strain in the rocks again, until they fail once more. This has led seismologists to study the history of earthquakes, for patterns, so their occurrences might be predicted.

One U.S. Geological Survey study gives the probability of a rupture occurring along various segments of the San Andreas fault for the 30 years between 1988 and 2018 (Figure 15.22). From this study, the

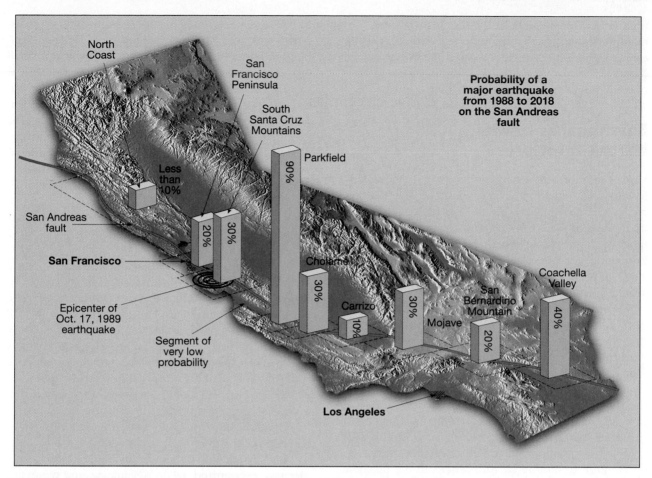

FIGURE 15.22 Probability of a major earthquake from 1988 to 2018 on the San Andreas fault.

Santa Cruz Mountains area was given a 30 percent probability of producing a 6.5-magnitude earthquake during this time period. In fact, it produced the Loma Prieta quake in 1989, of 7.1 magnitude.

The region given the highest probability (90 percent) of generating a quake is the Parkfield section. This area has been called the "Old Faithful" of earthquake zones because activity here has been very regular since record keeping began in 1857. The U.S. Geological Survey has established an elaborate monitoring network of creep meters, tiltmeters, and bore hole strain meters that measure the accumulation and release of strain in the rocks. Moreover, 70 seismographs have been installed to record foreshocks as well as main events. A network of laser distance-measuring devices records movement across the fault. The object is to identify ground displacements that may precede a sizable rupture.

One section of the San Andreas between Parkfield and the Santa Cruz Mountains is given a very low probability of generating an earthquake. This area has experienced very little seismic activity in historical times; rather, it exhibits a slow, continual

movement known as *fault creep*. Such movement is beneficial because it prevents strain from building to high levels in the rocks.

Californians understandably worry about the next "Big One." Unfortunately, seismologists cannot make a specific prediction, and can only point to southern California as the most likely location. The reason is that the southern portion of the San Andreas fault has not had a major earthquake since 1857, and strain is slowly building.

In addition to the attention given to California, new concerns are being brought to the remainder of the nation as well (see Box 15.1). For example, Charleston, South Carolina, was hit by a major earthquake in 1886 and could be a site for renewed activity. In addition, three major earthquakes struck the New Madrid, Missouri, region in 1811–1812. Although this area, near the confluence of the Mississippi and Ohio rivers, was sparsely populated in the early nineteenth century, this is no longer the case. This portion of the Mississippi Valley now has significant agricultural and urban development with Memphis, Tennessee, being the major city in the region. A 1985 federal study concluded that a 7.6-magnitude earthquake in this area could cause an estimated 2500 deaths, collapse 3000 structures, cause $25 billion in damages, and displace a quarter of a million people in Memphis alone.

Earthquakes Reveal Earth's Interior

Earth's interior lies just below us, yet its accessibility to direct observation is very limited. The deepest well yet drilled has penetrated Earth's crust only 13 kilometers (8 miles), less than 0.2 percent of the distance to the planet's center. Consequently, most knowledge of our planet's interior comes from the study of P and S waves that travel through Earth and vibrate the surface at some distant point.

From seismograms of earthquakes, seismologists accurately *measure the time* required for seismic waves to travel from the focus of an earthquake to a seismographic station. We know that P waves travel faster, and we use this fact to determine distances to earthquake epicenters. But the time required for P and S waves to travel through Earth also depends on the properties of the rock materials encountered, so seismologists search for variations in travel times that cannot be accounted for simply by differences in the distance traveled. These variations correspond to changes in rock properties.

Changes in rock properties indicate that Earth has four major layers: (1) the **crust,** a very thin outer layer; (2) the **mantle**, a rocky layer beneath the crust (thickness: 2885 kilometers or 1789 miles); (3) the **outer core**, a layer that exhibits the characteristics of a mobile liquid (thickness: about 2270 kilometers or 1407 miles); and (4) the **inner core,** a solid metallic sphere (radius: about 1216 kilometers or 754 miles). Figure 15.23 shows these four layers.

Discovering Earth's Structure

The story of how seismologists discovered Earth's core-and-layers structure is interesting. In 1909, a pioneering Yugoslavian seismologist, Andrija Mohorovičić, presented the first convincing evidence for layering within Earth. By studying seismic records, he found that the velocity of seismic waves increases abruptly below about 50 kilometers depth. This boundary separates the crust from the underlying mantle and is known as the **Mohorovičić discontinuity** in his honor. For reasons that are obvious, the name for this boundary was quickly shortened to **Moho.**

A few years later, another boundary was discovered by the German seismologist Beno Gutenberg. Generally, seismic waves from even small earthquakes are strong enough to travel around the world. This is why a seismograph in Antarctica can record earthquakes in California or Italy. But Gutenberg observed that P waves diminish and eventually die out about 105 degrees around the globe from an earthquake. Then, about 140 degrees away, the P waves reappear, but about 2 minutes later than would be expected, based on the distance traveled. This belt, where direct P waves are absent, is about 35 degrees wide and has been named the **shadow zone.** Figure 15.24 illustrates how this works.

Gutenberg realized that the shadow zone could be explained if Earth contained a core composed of material unlike the overlying mantle. The core must somehow hinder the transmission of P waves in a manner similar to the light rays being blocked by an opaque object which casts a shadow. However, rather than actually stopping the P waves, the shadow zone bends them, as shown in Figure 15.24. It was further learned that S waves could not travel through the core. Therefore, geologists concluded that at least a portion of this region is liquid.

In 1936, the last major subdivision of Earth's interior was recognized when seismic waves were discovered to reflect from a boundary within the core. Hence, Earth has a core-within-a-core. The size of the inner core was not accurately calculated until the early 1960s, when underground nuclear tests were conducted in Nevada. Because the precise locations and times of the explosions were known, echoes from seismic waves that bounced off the inner core provided an accurate means of determining its size.

So, Earth's basic structure is that of a layered ball: inner core, outer core, mantle, and crust. However, we call your attention to a most important zone within the upper mantle. This region, called the **asthenosphere**, begins at a depth of about 100 kilometers (60 miles) and ends at a depth of around 350 kilometers (over 200 miles). It may extend down to a depth of 700 kilometers (over 400 miles). Two views of the asthenosphere are presented in Figure 15.23.

In the asthenosphere, the velocity of S waves decreases, indicating to seismologists that this zone consists of hot, weak rock that is easily deformed. In addition, as much as 10 percent of the material within the asthenosphere may be molten. Overall, we say that the asthenosphere is *plastic,* and this is the key to the ability of Earth's crust to slowly migrate in pieces.

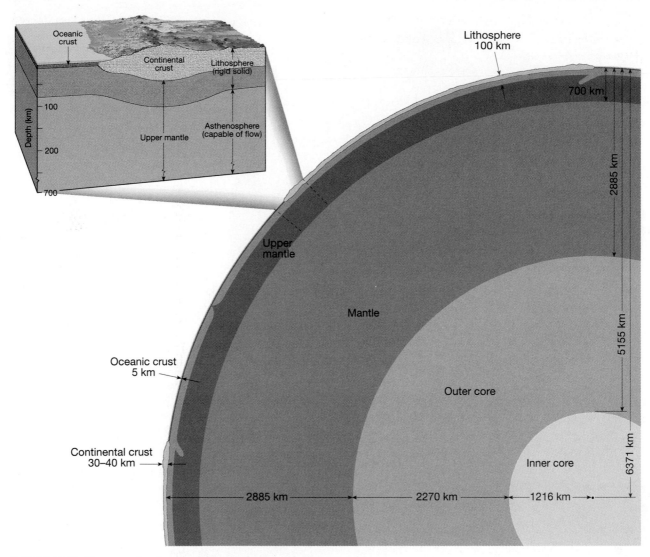

FIGURE 15.23 Cross-sectional view of Earth showing internal structure.

Situated above the asthenosphere is a cool, rigid layer called the **lithosphere** (Figure 15.25), about 100 kilometers thick, which includes the entire crust as well as the uppermost mantle. The lithosphere is defined as that layer of Earth cool enough to behave like a rigid solid. It is thought that the plastic asthenosphere facilitates motion of Earth's rigid lithosphere. This concept, known as the theory of *plate tectonics,* is the topic of the next chapter.

Discovering Earth's Composition

We have examined Earth's structure, so let us now look at the composition of each layer. Composition tells us much about how our planet has developed over its estimated age of 4.6 billion years.

Earth's crust varies in thickness, exceeding 70 kilometers in some mountainous regions and being thinner than 5 kilometers in some oceanic areas (see

Figure 15.23). Early seismic data indicated that the continental crust, which is mostly made of lighter, granitic rocks, is quite different in composition from denser oceanic crust. Until recently, however, scientists had only seismic evidence from which to determine the composition of oceanic crust, because it lies beneath an average of 3 kilometers of water as well as hundreds of meters of sediment. The deep-sea drilling ship *Glomar Challenger* made possible the recovery of ocean-floor samples. The samples were of basaltic composition—very different from the rocks that make up the continents.

Our knowledge of the rocks of the mantle and core is much more speculative. However, we do have some clues. Recall that some of the lava that reaches Earth's surface originates in the partially melted asthenosphere, within the mantle. In the laboratory, experiments have shown that partial melting

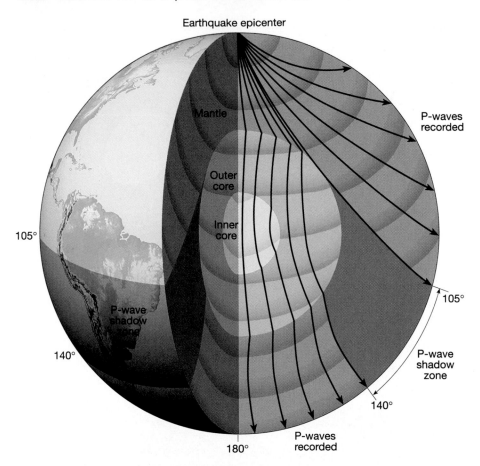

Earthquake epicenter

FIGURE 15.24 The abrupt change in physical properties at the mantle–core boundary causes the wave paths to bend sharply. This abrupt change in wave direction results in a shadow zone for P waves between about 105 and 140 degrees.

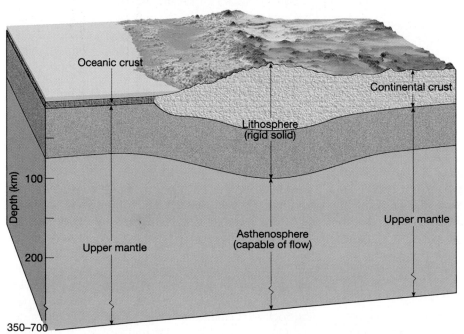

FIGURE 15.25 Relative positions of the asthenosphere and lithosphere.

of a rock called *peridotite* results in a melt that has a basaltic composition similar to lava that emerges during volcanic activity of oceanic islands. Denser rocks like peridotite are thought to make up the mantle and provide the lava for oceanic eruptions.

Surprisingly, meteorites or "shooting stars" that collide with Earth provide evidence of Earth's inner composition. Because meteorites are part of the solar system they are assumed to be representative samples. Their composition ranges from metallic meteorites made of iron and nickel to stony meteorites composed of dense rock similar to peridotite.

Because Earth's crust contains a much smaller percentage of iron than do meteorites, geologists believe that the dense iron, and other dense metals, literally sank toward Earth's center during the planet's early history. By the same token, lighter substances may have floated to the surface, creating the less-dense crust. Thus, Earth's core is thought to be mainly dense iron and nickel, similar to metallic meteorites, whereas the surrounding mantle is believed to be composed of rocks similar to stony meteorites.

The concept of a molten iron outer core is further supported by Earth's magnetic field. Our planet acts as a large magnet. The most widely accepted mechanism explaining why Earth has a magnetic field requires that the core be made of a material that conducts electricity, such as iron, and that is mobile enough to allow circulation. Both of these conditions are met by the model of Earth's core that was established on the basis of seismic data.

Not only does an iron core explain Earth's magnetic field but it also explains the high density of inner Earth, about 13.5 times that of water. Even under the extreme pressure at those depths, average crustal rocks with densities 2.8 times that of water would not have the density calculated for the core. But iron, which is 3 times more dense than crustal rocks, has the required density.

In summary, although earthquakes can be very destructive, much of our knowledge about Earth's interior comes from the study of these phenomena. As our knowledge of earthquakes and their causes improves, we learn more of our planet's internal workings and how to reduce the consequences of tremors. In the next chapter, you will see that most earthquakes originate at the boundaries of Earth's great lithospheric plates.

The Chapter in Review

The following statements are intended to help you review the primary objectives presented in this chapter.

- *Earthquakes* are vibrations of Earth produced by the rapid release of energy from rocks that rupture because they have been subjected to stresses beyond their limit. This energy, which takes the form of waves, radiates in all directions from the earthquake's source, called the *focus*. The movements that produce most earthquakes occur along large fractures, called *faults*, that are usually associated with plate boundaries.

- Along a fault, rocks store energy as they are bent. As slippage occurs at the weakest point (the focus), displacement will exert stress farther along a fault, where additional slippage will occur until most of the built-up strain is released. An earthquake occurs as the rock elastically returns to its original shape. The "springing back" of the rock is termed *elastic rebound*. Small earthquakes, called *foreshocks*, often precede a major earthquake. The adjustments that follow a major earthquake often generate smaller earthquakes called *aftershocks*.

- Two main types of *seismic waves* are generated during an earthquake: (1) *surface waves,* which travel along the outer layer of Earth; and (2) *body waves,* which travel through Earth's interior. Body waves are further divided into *primary,* or *P,*

waves, which push (compress) and pull (dilate) rocks in the direction the wave is traveling, and *secondary,* or *S, waves,* which "shake" the particles in rock at right angles to their direction of travel. P waves can travel through solids, liquids, and gases. Fluids (gases and, liquids) will not transmit S waves. In any solid material, P waves travel about 1.7 times faster than do S waves.

- The location on Earth's surface directly above the focus of an earthquake is the *epicenter*. An epicenter is determined using the difference in velocities of P and S waves. Using the difference in arrival times between P and S waves, the distance separating a recording station from the earthquake can be determined. When the distances are known from three or more seismic stations, the epicenter can be located using a method called *triangulation*.

- *A close correlation exists between earthquake epicenters and plate boundaries.* The principal earthquake epicenter zones are along the outer margin of the Pacific Ocean, known as the *circum-Pacific belt,* and through the world's oceans along the *oceanic ridge system*.

- Earthquake *intensity* depends not only on the strength of the earthquake but also on other factors, such as distance from the epicenter, the nature of surface materials, and building design. The *Mercalli intensity scale* assesses the damage from a quake at a specific location. Using the *Richter scale,* the *magnitude* (a measure of the total amount of energy released) of an earthquake is determined by measuring the *amplitude* (maximum displacement) of the largest seismic wave recorded. A logarithmic scale is used to express magnitude, in which a tenfold increase in recorded wave amplitude corresponds to an increase of 1 on the magnitude scale. Each unit of Richter magnitude equates to roughly a 30-fold energy increase.

- The most obvious factors determining the amount of destruction accompanying an earthquake are the magnitude of the earthquake and the proximity of the quake to a populated area. Structural damage attributable to earthquake vibrations depends on several factors, including (1) wave amplitudes, (2) the duration of the vibrations, (3) the nature of the material upon which the structure rests, and (4) the design of the structure. Secondary effects of earthquakes include *tsunamis,* landslides, ground subsidence, and fire.

- Substantial research to predict earthquakes is underway in Japan, the United States, China, and Russia—countries where earthquake risk is high. No reliable method of short-range prediction has yet been devised. Long-range forecasts are based on the premise that earthquakes are repetitive or cyclical. Seismologists study the history of earthquakes for patterns, so their occurrences might be predicted.

- As indicated by the behavior of P and S waves as they travel through Earth, the four major zones of Earth's interior are the (1) *crust* (the very thin outer layer), (2) *mantle* (a rocky layer located below the crust with a thickness of 2885 kilometers), (3) *outer core* (a layer about 2270 kilometers thick, which exhibits the characteristics of a mobile liquid), and (4) *inner core* (a solid metallic sphere with a radius of about 1216 kilometers). A very important zone in the upper mantle is the *asthenosphere.* Overall, the asthenosphere is plastic, which is a key to the ability of Earth's crust to slowly migrate in pieces. Situated above the asthenosphere is a cool, rigid layer called the *lithosphere,* which includes the entire crust as well as the uppermost mantle.

- The *continental crust* is primarily made of *granitic* rocks, whereas the *oceanic crust* is of *basaltic* composition. Rocks similar to *peridotite* are thought to make up the *mantle.* The *core* is made up mainly of *iron* and *nickel.* An iron core explains the high density of Earth's interior as well as Earth's magnetic field.

Key Terms

aftershock (p. 293)
asthenosphere (p. 306)
body wave (p. 293)
crust (p. 306)
earthquake (p. 289)
elastic rebound (p. 291)
epicenter (p. 296)
fault (p. 291)

focus (p. 289)
foreshock (p. 293)
inner core (p. 306)
liquefaction (p. 300)
lithosphere (p. 307)
magnitude (p. 298)
mantle (p. 306)
Mercalli intensity scale
 (p. 298)

Mohorovičić discontinuity (Moho) (p. 306)
outer core (p. 306)
primary (P) wave (p. 293)
Richter scale (p. 298)
secondary (S) wave
 (p. 293)
seismic sea wave
 (tsunami) (p. 301)

seismogram (p. 293)
seismograph (p. 293)
seismology (p. 293)
shadow zone (p. 306)
surface wave (p. 293)

Questions for Review

1. What is an earthquake? Under what circumstances do earthquakes occur?
2. How are faults, foci, and epicenters related?
3. Who was first to explain the actual mechanism by which earthquakes are generated?
4. Explain what is meant by elastic rebound.

5. Faults that are experiencing no active creep may be considered "safe." Rebut or defend this statement.
6. Describe the principle of a seismograph.
7. Using Figure 15.10, determine the distance between an earthquake and a seismic station if

the first S wave arrives 3 minutes after the first P wave.

8. List the major differences between P and S waves.

9. Which type of seismic wave causes the greatest destruction to buildings?

10. Most strong earthquakes occur in a zone on the globe known as the _____.

11. What factor contributed most to the extensive damage that occurred in the central portion of Mexico City during the 1985 earthquake?

12. The 1988 Armenian earthquake had a Richter magnitude of 6.9, far less than the great quakes in Alaska in 1964 and San Francisco in 1906. Nevertheless, the loss of life was far greater in the Armenian event. Why?

13. An earthquake measuring 7 on the Richter scale releases about _____ times more energy than an earthquake with a magnitude of 6.

14. List four factors that affect the amount of destruction caused by seismic vibrations.

15. In addition to the destruction created directly by seismic vibrations, list three other types of destruction associated with earthquakes.

16. Distinguish between the Mercalli scale and the Richter scale.

17. What is a tsunami? How is one generated?

18. Cite some reasons why an earthquake with a moderate magnitude might cause more extensive damage than a quake with a high magnitude.

19. What evidence do we have that Earth's outer core is molten?

20. Contrast the physical makeup of the asthenosphere and the lithosphere.

21. Why are meteorites considered important clues to the composition of Earth's interior?

22. Describe the composition (mineral makeup) of the following:

   (a) continental crust   (c) mantle
   (b) oceanic crust   (d) core

Testing What You Have Learned

To test your knowledge of the material presented in this chapter, answer the following questions:

Multiple Choice Questions

1. Each unit of magnitude increase on the Richter scale equates to roughly a _____ increase in the energy released.
   a. 2-fold   c. 15-fold   e. 30-fold
   b. 10-fold   d. 20-fold

2. Which one of the following lists the order in which earthquake waves will be recorded by a seismograph located several hundred miles from an earthquake epicenter?
   a. P, S, surface   c. surface, S, P   e. P, surface, S
   b. S, P, surface   d. surface, P, S

3. The movements that cause earthquakes are usually associated with large fractures called _____.
   a. domes   c. faults   e. joints
   b. anticlines   d. tsunamis

4. Earthquakes with magnitudes less than _____ are usually not felt by humans.
   a. two   c. six   e. ten
   b. four   d. eight

5. Which one of the following is true of P waves?
   a. stopped by liquid   c. compression wave   e. b. and c.
   b. greatest velocity   d. a. and c.

6. Earth has a _____ field because the core is made of a material that conducts electricity, such as iron, and which is mobile enough to allow circulation.
   a. force   **c.** radio   **e.** magnetic
   b. interior   **d.** nuclear

7. The zone in the upper mantle consisting of hot, weak rock that is easily deformed is the _____.
   a. lithosphere   **c.** hydrosphere   **e.** crust
   b. asthenosphere   **d.** shadow zone

8. The difference in _____ of P and S waves provides a method for locating the epicenter of an earthquake.
   a. densities   **c.** velocities   **e.** reflections
   b. radiation   **d.** directions

9. The _____ is the rigid layer of Earth that includes the crust as well as the upper mantle.
   a. asthenosphere   **c.** shadow zone   **e.** Moho
   b. lithosphere   **d.** outer core

10. Seismic sea waves are also called _____.
    a. tsunami   **c.** breakers   **e.** currents
    b. swells   **d.** surf

Fill-In Questions

11. The actual place of origin of an earthquake within Earth is called the _____, while the location on Earth's surface directly above is referred to as the _____.

12. The greatest energy released by earthquakes originates around the outer edge of the _____ Ocean known as the _____ belt.

13. The continental crust consists mainly of _____ rocks, whereas the oceanic crust is of _____ composition.

14. Small earthquakes that precede a major earthquake are called _____, while adjustments that follow a major earthquake often generate smaller quakes referred to as _____.

15. The four major zones of Earth's interior, from the surface to center, are the _____, _____, _____, and _____.

True/False Questions

16. Most earthquakes are produced by the rapid release of plastic energy stored in rock that has been subjected to great stress. _____

17. Most of the motions that produce earthquakes can be satisfactorily explained by the plate tectonics theory. _____

18. The slow, gradual displacement, with little noticeable seismic activity, along a fault is known as fault creep. _____

19. Where unconsolidated materials are saturated with water, earthquakes may produce a phenomenon known as liquefaction. _____

20. No consistent method of short-range prediction of earthquakes has yet been devised. _____

Answers

1.e; 2.a; 3.c; 4.a; 5.e; 6.e; 7.b; 8.c; 9.b; 10.a; 11. focus, epicenter; 12. Pacific, circum-Pacific; 13. granitic, basaltic; 14. foreshocks, aftershocks; 15. crust, mantle, outer core, inner core; 16.F; 17.T; 18.T; 19.T; 20.T.

Plate Tectonics

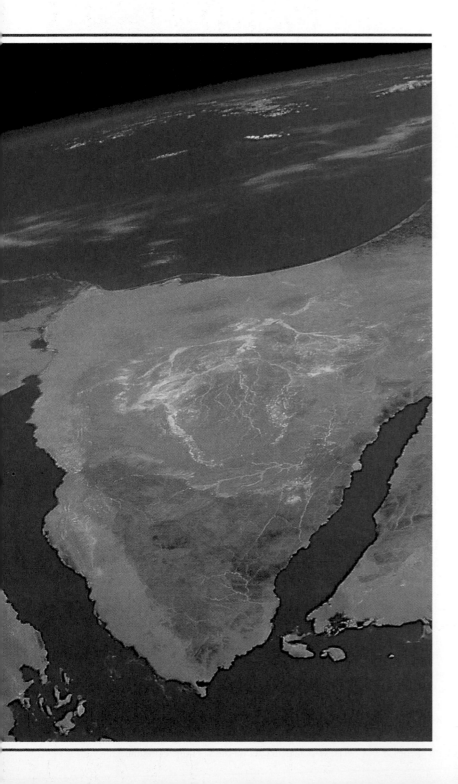

Focus on Learning

To assist you in learning the important concepts in this chapter, you will find it helpful to focus on the following questions:

- What evidence was used to support the continental drift hypothesis?

- What was one of the main objections to the continental drift hypothesis?

- What is the theory of plate tectonics?

- In what major way does the plate tectonics theory depart from the continental drift hypothesis?

- What are the three types of plate boundaries?

- What is the evidence used to support the plate tectonics theory?

- What models have been proposed to explain the driving mechanism for plate motion?

Chapter 16

The Gulf of Suez (left) and the Gulf of Aqaba (right) are rift zones caused by plate movements. (Courtesy of NASA/Tom Stack and Associates)

FIGURE 16.1 Mount Williamson in California's Sierra Nevada, John Muir Wilderness. (Photo by Carr Clifton)

Will California eventually "slide" into the ocean, as some predict? Have continents really "drifted" apart over the centuries? Answers to these questions and many others that have intrigued geologists for decades are now being provided by an exciting theory on large-scale movements taking place within Earth. This theory, called plate tectonics, represents the real frontier of geology, and its implications are so far-reaching that it can be considered the framework from which most other geological processes should be viewed.

Early in this century, most geologists thought that the geographic positions of the ocean basins and continents were fixed. During the last few decades, however, vast amounts of new data have dramatically changed our understanding of the nature and workings of our planet. Earth scientists now realize that the continents gradually migrate across the globe. Where landmasses split apart, new ocean basins are created between the diverging blocks. Meanwhile, older portions of the seafloor are carried back into the mantle in regions where trenches occur in the deep ocean floor. Because of these movements, segments of continental material eventually collide and form Earth's great mountain ranges (Figure 16.1). In short, a revolutionary new model of Earth's tectonic* processes has emerged.

This profound reversal of scientific understanding has been appropriately described as a scientific revolution. Like other scientific revolutions, considerable time elapsed between the idea's inception and its general acceptance. The revolution began early in the twentieth century as a relatively straightforward proposal that the continents drift about the face of Earth. After many years of heated debate, the idea of drifting continents was rejected by the vast majority of geologists as improbable. However, during the 1950s and 1960s, new evidence rekindled interest in this proposal. By 1968, these new developments led to the unfolding of a far more encompassing theory than continental drift—a theory known as *plate tectonics*.

Continental Drift: An Idea Before Its Time

The idea that continents, particularly South America and Africa, fit together like pieces of a jigsaw puzzle originated with improved world maps. However, little significance was given this idea until 1915, when Alfred Wegener, a German climatologist and geophysicist, published *The Origin of Continents and Oceans*. In this book, Wegener set forth his radical hypothesis of **continental drift**.**

---

*Tectonics refers to the deformation of Earth's crust and results in the formation of structural features such as mountains.

**Wegener's ideas were actually preceded by those of an American geologist, F. B. Taylor, who in 1910 published a paper on continental drift. Taylor's paper provided little supporting evidence for continental drift, which may have been the reason that it had a relatively small impact on the scientific community.

He suggested that a supercontinent he called **Pangaea** (meaning "all land") once existed (Figure 16.2). He further hypothesized that, about 200 million years ago, this supercontinent began breaking into smaller continents, which then "drifted" to their present positions (see Figure 16.25).

Wegener and others collected substantial evidence to support these claims. The fit of South America and Africa, fossils, rock structures, and ancient climates all seemed to support the idea that these now-separate landmasses were once joined. Let us examine their evidence.

Evidence: The Continental Jigsaw Puzzle

Like a few others before him, Wegener first suspected that the continents might have been joined when he noticed the remarkable similarity between the coastlines on opposite sides of the South Atlantic. However, his use of present-day shorelines to make a fit of the continents was challenged immediately by other Earth scientists. These opponents correctly argued that shorelines are continually modified by erosional processes, and even if continental displacement had taken place, a good fit today would be unlikely. Wegener appeared to be aware of this problem, and, in fact, his original jigsaw of the continents was only very crude.

A much better approximation of the true outer boundary of the continents is the continental shelf.

Today, the seaward edge of the continental shelf lies submerged, several hundred meters below sea level. In the early 1960s, scientists produced a map that attempted to fit the edges of the continental shelves at a depth of 900 meters. The remarkable fit that was obtained is shown in Figure 16.3. Although the continents overlap in a few places, these are regions where streams have deposited large quantities of sediment, thus enlarging the continental shelves. The overall fit was even better than the supporters of continental drift suspected it would be.

Evidence: Fossils Match Across the Seas

Although Wegener was intrigued by the remarkable jigsaw fit on opposite sides of the Atlantic, he at first thought the idea of a mobile Earth improbable. Not until he came across an article citing fossil evidence for the existence of a land bridge connecting South America and Africa did he begin to take his own idea seriously. Through a search of the literature, Wegener learned that most paleontologists were in agreement that some type of land connection was needed to explain the existence of identical fossils on the widely separated landmasses.

To add credibility to his argument for the existence of the supercontinent of Pangaea, Wegener cited documented cases of several fossil organisms that had been found on different landmasses but which could not have crossed the vast oceans

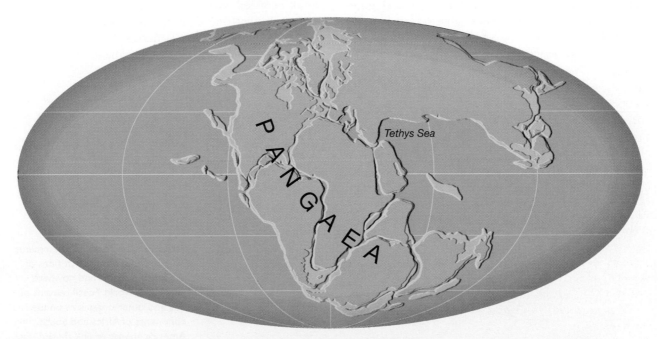

FIGURE 16.2 Reconstruction of Pangaea as it is thought to have appeared 200 million years ago. (After R. S. Deitz and J.C. Holden. *Journal of Geophysical Research 75*: 4943. Copyright by American Geophysical Union)

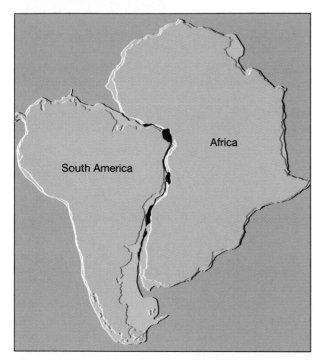

FIGURE 16.3 This shows the best fit of South America and Africa along the continental slope at a depth of 500 fathoms (about 900 meters). The areas where continental blocks overlap appear in brown. (After A. G. Smith. "Continental Drift." In *Understanding the Earth*, edited by I.G. Gass. Courtesy of Artemis Press.)

not the case, Wegener argued that South America and Africa must have been joined—somehow.

Wegener also cited the distribution of the fossil fern *Glossopteris* as evidence for the existence of Pangaea. This plant, identified by its large seeds that could not be blown very far, was known to be widely dispersed among Africa, Australia, India, and South America during the late Paleozoic era. Later, fossil remains of *Glossopteris* were discovered in Antarctica as well. Wegener knew that these seed ferns and associated flora grew only in a subpolar climate; therefore, he concluded that these landmasses must have been joined as they presently include climatic regions that are too diverse to support such flora. For Wegener, fossils proved without question that a supercontinent had existed.

In his book, Wegener also cited the distribution of present-day organisms as evidence to support the concept of drifting continents. For example, modern organisms with similar ancestries clearly had to evolve in isolation during the last few tens of millions of years. Most obvious of these are the Australian marsupials, which have a direct fossil link to the marsupial opossums found in the Americas.

How did scientists explain the discovery of identical fossil organisms separated by thousands of kilometers of open ocean? The idea of land bridges was the most widely accepted solution to the problem of migration (Figure 16.5). We know, for example, that during the most recent glacial period, the lowering of sea level allowed animals to cross the narrow Bering Strait between Asia and North America. Was it possible, then, that one or more land bridges once connected Africa and South America? We are now quite certain that land bridges of this magnitude did not

presently separating the continents. The classic example is *Mesosaurus,* a presumably aquatic, snaggle-toothed reptile whose fossil remains are limited to eastern South America and southern Africa (Figure 16.4). If *Mesosaurus* had been able to swim well enough to cross the vast South Atlantic Ocean, its remains should be more widely distributed. As this is

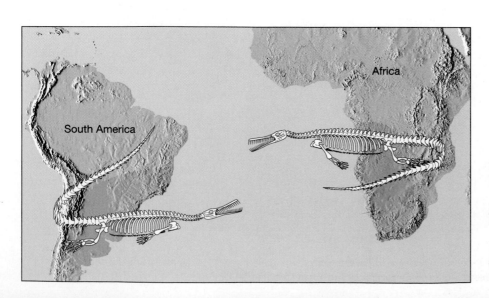

FIGURE 16.4 Fossils of *Mesosaurus* have been found on both sides of the South Atlantic and nowhere else in the world. Fossil remains of this and other organisms on the continents of Africa and South America appear to link these landmasses during the late Paleozoic and early Mesozoic eras.

FIGURE 16.5 These sketches by John Holden illustrate various explanations for the occurrence of similar species on landmasses that are presently separated by vast oceans. (Reprinted with permission of John Holden)

RAFTING

ISTHMIAN LINKS

ISLAND STEPPING STONES

CONTINENTAL DRIFT

exist, for their remnants should still lie below sea level. But they are nowhere to be found.

Evidence: Rock Types and Structures Match

Anyone who has worked a picture puzzle knows that, in addition to the pieces fitting together, the picture must be continuous as well. The "picture" that must match in the "Continental Drift Puzzle" is one of rock types and mountain belts on the continents. If the continents were once together, the rocks found in a particular region on one continent should closely match in age and type those in corresponding positions on the matching continent.

Such evidence exists in the form of several mountain belts that terminate at one coastline, only to reappear on a landmass across the ocean. For instance, the mountain belt that includes the Appalachians trends northeastward through the eastern United States and disappears off the coast of Newfoundland (Figure 16.6A). Mountains of comparable age and structure are found in the British Isles and Scandinavia. When these landmasses are reassembled as in Figure 16.6B, the mountain chains form a nearly continuous belt. Numerous other rock structures exist that appear to have formed at the same time and were subsequently split apart.

Wegener was very satisfied that the similarities in rock structure on both sides of the Atlantic linked these landmasses. In his own words, "It is just as if we were to refit the torn pieces of a newspaper by matching their edges and then check whether the lines of print run smoothly across. If they do, there is nothing left but to conclude that the pieces were in fact joined in this way."[*]

Evidence: Ancient Climates

Because Alfred Wegener was a climatologist by training, he was keenly interested in obtaining paleoclimatic (ancient climatic) data in support of continental drift. His efforts were rewarded when he found evidence for dramatic climatic changes. For instance, glacial deposits indicate that, near the end of the Paleozoic era (between 220 million and 300 million years ago), ice sheets covered extensive areas of the Southern Hemisphere. Layers of glacial till were found in southern Africa and South America, as well as in India and Australia. Below these beds of glacial debris lay striated and grooved bedrock. In some locations, the striations and grooves indicated that the ice had moved from what is now the sea onto land. Much of the land area containing evidence of this late Paleozoic glaciation presently lies within 30 degrees of the equator in a subtropical or tropical climate.

Could Earth have gone through a period sufficiently cold to have generated extensive continental glaciers in what is presently a tropical region?

---

[*]Alfred Wegener, *The Origin of Continents and Oceans.* Translated from the 4th revised German edition of 1929 by J. Birman (London: Methuen, 1966).

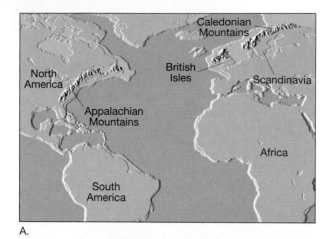

A.

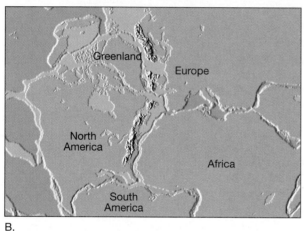

B.

FIGURE 16.6 Matching mountain range across the North Atlantic. **A.** The Appalachian Mountains trend along the eastern flank of North America and disappear off the coast of Newfoundland. Mountains of comparable age and structure are found in the British Isles and Scandinavia. **B.** When these landmasses are placed in their predrift locations, these ancient mountain chains form a nearly continuous belt. These folded mountain belts formed roughly 300 million years ago as the landmasses collided during the formation of the supercontinent of Pangaea.

Wegener rejected this explanation because, during the late Paleozoic, large swamps existed in the Northern Hemisphere. The lush vegetation of these swamps eventually became the major coal fields of the eastern United States, Europe, and Siberia.

Fossils from these coal fields indicate that the tree ferns which produced the coal deposits had large fronds. This indicates a tropical setting. Furthermore, unlike trees in colder climates, the tree trunks lacked growth rings. Growth rings do not form in tropical plants because there are minimal seasonal fluctuations in temperature.

Wegener believed that a better explanation for the paleoclimatic regimes he observed is provided

by fitting together the landmasses as a supercontinent, with South Africa centered over the South Pole (Figure 16.7). This would account for the conditions necessary to generate extensive expanses of glacial ice over much of the Southern Hemisphere. At the same time, this geography would place the northern landmasses nearer the tropics and account for their vast coal deposits.

Wegener was so convinced that his explanation was correct that he wrote, "This evidence is so compelling that by comparison all other criteria must take a back seat."

How does a glacier develop in hot, arid Australia? How do land animals migrate across wide expanses of open water? As compelling as this evidence may have been, 50 years passed before most of the scientific community would accept it and the logical conclusions to which it led.

The Great Debate

Wegener's proposal did not attract much open criticism until 1924 when his book was translated into English. From this time on, until his death in 1930, his drift hypothesis encountered a great deal of hostile criticism. To quote the respected American geologist T. C. Chamberlin, "Wegener's hypothesis…takes considerable liberty with our globe, and is less bound by restrictions or tied down by awkward, ugly facts than most of its rival theories. Its appeal seems to lie in the fact that it plays a game in which there are few restrictive rules and no sharply drawn code of conduct."

One of the main objections to Wegener's hypothesis stemmed from his inability to provide a mechanism that was capable of moving the continents across the globe. Wegener proposed two possible energy sources. One of these, the tidal influence of the Moon, was presumed by Wegener to be strong enough to give the continents a westward motion. However, the prominent physicist Harold Jeffreys quickly countered with the argument that tidal friction of the magnitude needed to displace the continents would bring Earth's rotation to a halt in a matter of a few years. Further, Wegener proposed that the larger and sturdier continents broke through the oceanic crust, much like ice breakers cut through ice. However, no evidence existed to suggest that the ocean floor was weak enough to permit passage of the continents without themselves being appreciably deformed in the process.

Although most of Wegener's contemporaries opposed his views, even to the point of open ridicule, a few considered his ideas plausible. For

FIGURE 16.7 A. The supercontinent Pangaea showing the area covered by glacial ice 300 million years ago. **B.** The continents as they are today. The shading outlines areas where evidence of the old ice sheets exists.

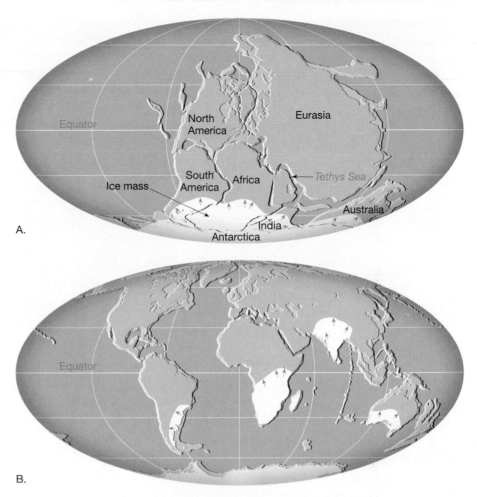

Plate Tectonics: A Modern Version of an Old Idea

these few geologists who continued the search for additional evidence, the exciting concept of continents in motion held their interest. Others viewed continental drift as a solution to previously unexplainable observations.

During the years that followed Wegener's proposal, major strides in technology permitted mapping of the ocean floor. Moreover, extensive data on seismic activity and Earth's magnetic field became available. By 1968, these developments led to the unfolding of a far more encompassing theory than continental drift, known as **plate tectonics.** The implications of plate tectonics are so far-reaching that this theory is today the framework within which to view most geologic processes.

The theory of plate tectonics holds that Earth's outer shell consists of about twenty rigid slabs called **plates.** They are in continuous slow motion relative to each other (Figure 16.8). The largest is the Pacific plate, which is located mostly beneath the ocean. An exception is a small sliver of North America that includes southwestern California and Mexico's Baja Peninsula.

Notice in Figure 16.8 that all of the other large plates include both continental and oceanic crust—a major departure from Wegener's continental drift hypothesis, which proposed that the continents moved through the ocean floor, not with it. Many smaller plates, on the other hand, consist exclusively of oceanic material; an example is the Nazca plate, located off the west coast of South America.

Recall that Earth's rigid outer shell is called the *lithosphere* and consists of both crustal rocks and a portion of the upper mantle. Also recall that the lithosphere varies in thickness. In general, lithospheric plates are thinnest in the ocean basins, where their thicknesses vary from as little as 10 kilometers at the ocean ridges to as much as 100 kilometers in the deep-ocean basins. By contrast, continental lithosphere is generally 100 to 150 kilometers thick and may extend to 250 kilometers in some regions. Beneath the lithosphere is the hotter and weaker zone known as the *asthenosphere*. The

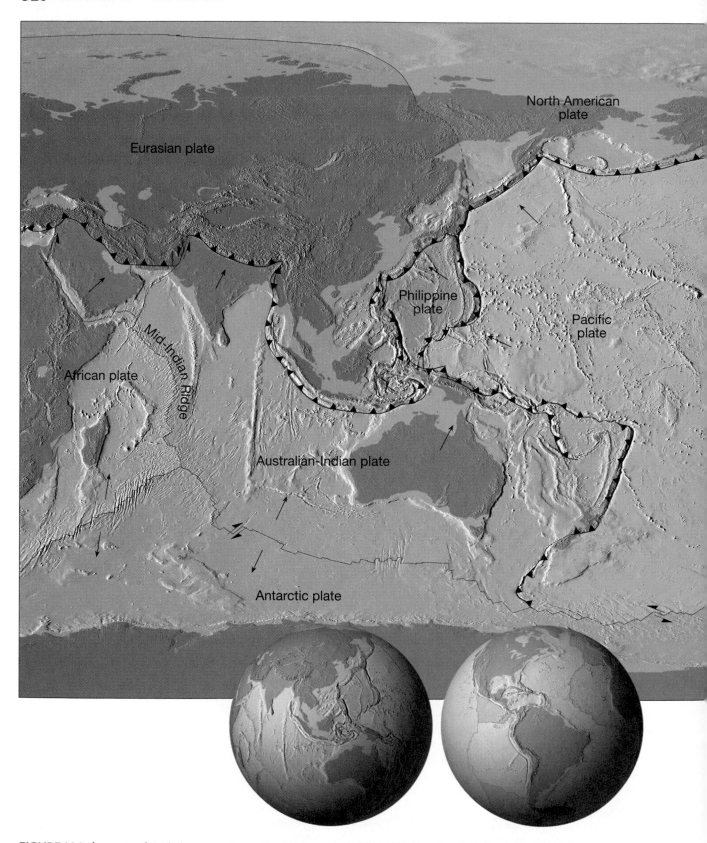

FIGURE 16.8 A mosaic of rigid plates constitutes Earth's outer shell. (After W.B. Hamilton, U.S. Geological Survey)

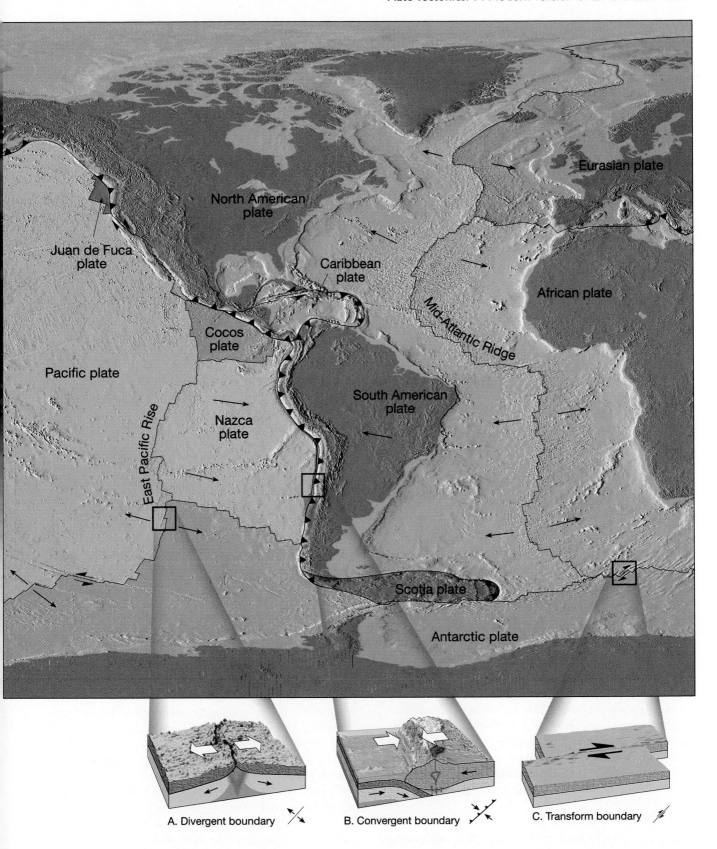

Juan de Fuca plate

North American plate

Caribbean plate

Eurasian plate

African plate

Mid-Atlantic Ridge

Cocos plate

Pacific plate

Nazca plate

East Pacific Rise

South American plate

Scotia plate

Antarctic plate

A. Divergent boundary

B. Convergent boundary

C. Transform boundary

weak nature of the rock within the asthenosphere allows for motion in Earth's rigid outer shell.

A main assumption of plate tectonics theory is that lithospheric plates are rigid. Therefore, the distance between two places on the same plate does not change. For example, as the plates move, the distance between New York and Denver, located on the same plate, remains unchanged. However, the distance between New York and London, which are located on different plates, is continually changing, by centimeters per year. Because each plate moves as a distinct unit, all major interactions between plates occur along *plate boundaries*. Not surprisingly, most of Earth's seismic activity, volcanism, and mountain building occur along these dynamic margins, which we examine next.

Plate Boundaries

Plate boundaries are clearly shown in Figure 16.8. But all of them are concealed under the ocean or beneath rocks and soil on land. How were the boundaries located?

For some time now, earthquake and volcanic activity have been known to be concentrated along narrow zones around the globe. Best known is the so-called *Ring of Fire* that encircles the Pacific. Thus, the first approximations of plate margins relied on the distribution of earthquake and volcanic activity. Later work revealed three distinct types of plate boundaries, differentiated by the movement each exhibits (Figure 16.9). These are:

1. **Divergent boundaries**—where plates move apart, resulting in upwelling of material from the mantle to create new seafloor.

2. **Convergent boundaries**—where plates move together, causing one of the slabs of lithosphere to be consumed into the mantle as it descends beneath an overriding plate.

3. **Transform boundaries**—where plates grind past each other without creating or destroying lithosphere.

Each plate is bounded by a combination of these zones, as you can see in Figure 16.8. For example, the Nazca plate has a divergent boundary on the west, a convergent boundary on the east, and numerous small transform faults that offset segments of divergent boundaries on the north and south.

Divergent Boundaries

Most divergent boundaries, where plate spreading occurs, are situated along the crests of oceanic ridges (Figure 16.10). Here, as the plates move away from the

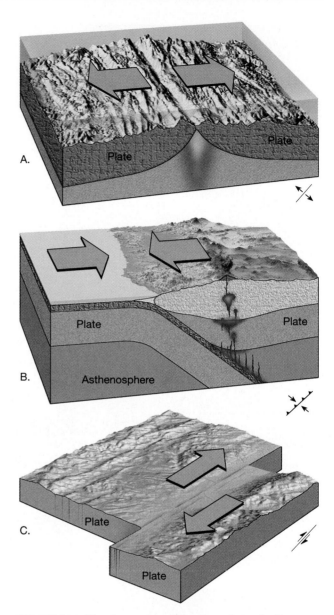

FIGURE 16.9 The three types of plate boundaries. **A.** Divergent boundary. **B.** Convergent boundary. **C.** Transform boundary.

ridge axis, the fractures created are immediately filled with molten rock that oozes up from the hot asthenosphere. This material cools slowly to produce new slivers of seafloor. In a continuous manner, successive plate spreading and upwelling of magma add new oceanic crust (lithosphere) between the diverging plates.

This mechanism is called **seafloor spreading**. It has produced the floor of the Atlantic Ocean during the past 165 million years. The typical rate of spreading at these ridges ranges between 2 and 10 centimeters per year, and averages about 6 centimeters (2 inches) per year. Because new rock is added equally to the trailing edges of both diverging plates, the overall rate of ocean floor growth is twice the spreading

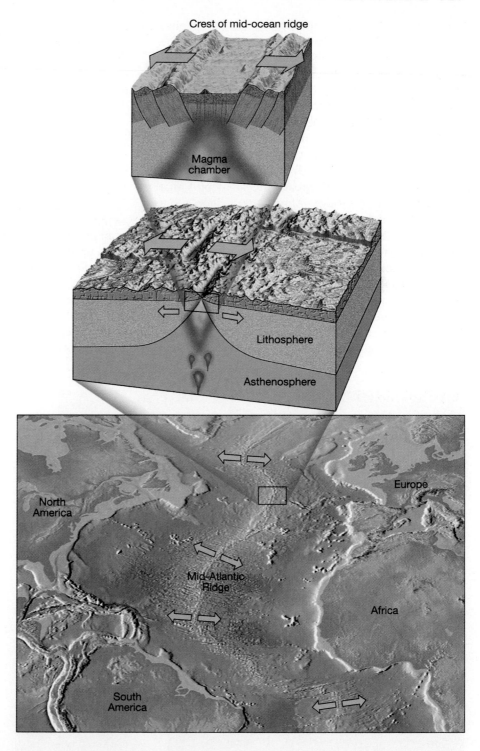

FIGURE 16.10 Most divergent plate boundaries are situated along the crests of oceanic ridges.

rate. Despite these slow rates, the Atlantic Ocean basin could have opened and closed *more than ten times* during the nearly 5-billion-year history of our planet!

Our knowledge of oceanic ridge systems, where seafloor spreading occurs, comes from depth soundings taken of the ocean floor, core samples obtained from deep-sea drilling, visual inspection using submersible seacraft, and even firsthand inspection of slices of ocean floor which have been shoved up onto dry land. Because of its accessibility, the Mid-Atlantic Ridge has been studied more thoroughly than other ridge systems. The Mid-Atlantic Ridge is a gigantic submerged mountain range standing 2500 to 3000 meters (8200 to 10,000 feet) above the adjacent deep-ocean basins. It extends southward from the Arctic Ocean to beyond the southern tip of

Africa. In a few places, the Mid-Atlantic Ridge has actually grown above sea level to form islands, the largest of which is Iceland. Throughout most of its length, however, this divergent boundary lies 2500 meters below sea level.

The buoyant nature of the upwelling magma is the primary reason for the elevated position of the Mid-Atlantic Ridge. As the newly formed lithosphere travels away from the spreading center, it gradually cools and contracts. This thermal contraction accounts in part for the greater ocean depths that exist away from the ridge. Almost 100 million years must pass before cooling and contraction cease completely. By this time, rock that was once a part of the majestic ocean mountain system becomes part of the deep-ocean basin.

Not all spreading centers have existed as long as the Mid-Atlantic Ridge and not all are found in the middle of large oceans. The Red Sea is believed to be the site of a recently formed divergent boundary. Here, the Arabian Peninsula separated from Africa and began to move toward the northeast (Figure 16.11). Consequently, the Red Sea is providing oceanographers with a view of how the Atlantic Ocean may have looked in its infancy. Another narrow, linear sea produced by seafloor spreading in the recent geologic past is the Gulf of California.

Most spreading centers are somewhere on the sea floor, but a few exist on continents. When a spreading center develops within a continent, the landmass may split into smaller segments, just as Wegener had proposed for the breakup of Pangaea. The fragmentation of a continent is thought to be associated with the upward movement of hot rock from below. The effect of this activity is to force the crust upward directly above the hot rising plume. This stretches the crust and causes numerous tensional cracks, as shown in Figure 16.12A.

As the plates move from the area of upwelling, the broken slabs are displaced downward, creating downfaulted valleys called **rifts** or **rift valleys** (Figure 16.12B). As the spreading continues, the rift valley will

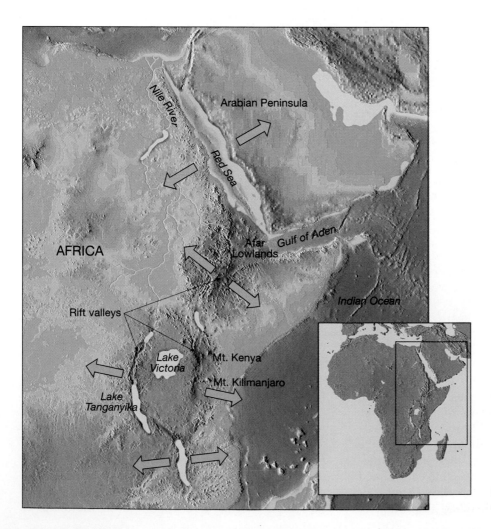

FIGURE 16.11 East African rift valleys and associated features.

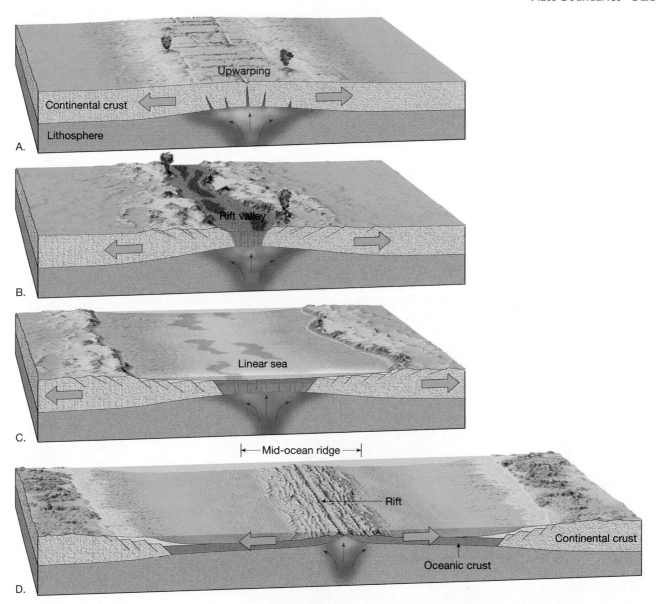

FIGURE 16.12 A. Rising magma forces the crust upward, causing numerous cracks in the rigid lithosphere. **B.** As the crust is pulled apart, large slabs of rock sink, generating a rift zone. **C.** Further spreading generates a narrow sea. **D.** Eventually, an expansive ocean basin and ridge system are created.

lengthen and deepen, eventually extending out into the ocean. At this point the valley will become a narrow linear sea with an outlet to the ocean, similar to the Red Sea today (Figure 16.12C). The zone of rifting will remain the site of igneous activity, continually generating new seafloor in an ever-expanding ocean basin (Figure 16.12D).

The East African rift valleys represent the initial stage in the breakup of a continent as just described (Figure 16.11). The extensive volcanic activity believed to accompany continental rifting is exemplified by large volcanic mountains such as

Kilimanjaro and Mount Kenya. If the rift valleys in Africa remain active, East Africa will eventually part from the mainland in much the same way the Arabian Peninsula did just a few million years ago. However, not all rift valleys develop into full-fledged spreading centers. Running through the central United States is an aborted rift zone extending from Lake Superior to Kansas. This once-active rift valley is filled with rock that was extruded onto the crust more than a billion years ago. Why one rift valley continues to develop while others are abandoned is not yet known.

Convergent Boundaries

At spreading centers, new lithosphere is continually being generated. However, because the total surface area of Earth remains essentially constant, lithosphere must also be consumed. The zone of plate convergence is the site where lithosphere is reabsorbed, or subducted, into the mantle. When two plates collide, the leading edge of one is bent downward, allowing it to descend beneath the other. The typical angle of descent ranges from 35 to nearly 90 degrees from the surface.

Although all convergent zones are basically similar, the nature of plate collisions is influenced greatly by the type of crustal material involved. Convergence can occur between one oceanic and one continental plate, between two oceanic plates, or between two continental plates. All these situations are shown in Figure 16.13.

Oceanic-Continental Convergence Whenever the leading edge of a plate capped with continental crust converges with oceanic crust, the less dense continental material remains "floating," while the more dense oceanic slab sinks into the asthenosphere. The region where an oceanic plate descends into the asthenosphere because of convergence is called a **subduction zone.** As the oceanic plate slides beneath the overriding plate, the oceanic plate bends, thereby producing a **deep-ocean trench** adjacent to the zone

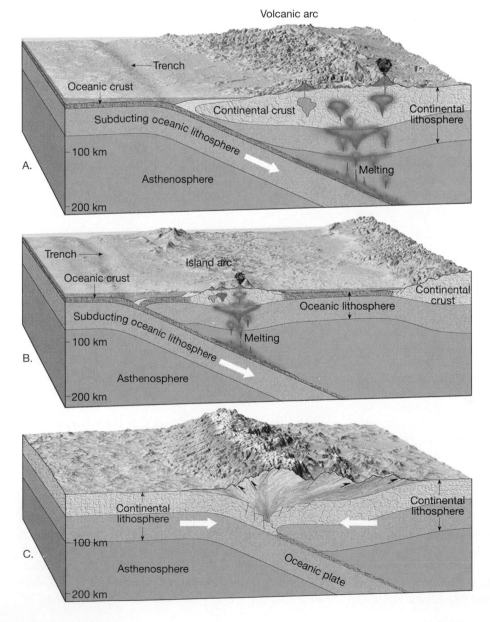

FIGURE 16.13 Three types of convergent plate boundaries.
A. Oceanic-continental.
B. Oceanic-oceanic.
C. Continental-continental.

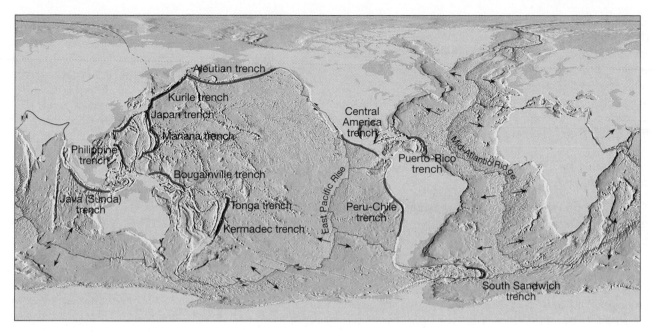

FIGURE 16.14 Distribution of the world's oceanic trenches, ridge system, and transform faults. Where transform faults offset ridge segments, they permit the ridge to change direction (curve), as can be seen in the Atlantic Ocean.

of subduction (Figure 16.13A). Trenches formed in this manner may be thousands of kilometers long and 8 to 11 kilometers deep.

When the descending oceanic plate reaches a depth of about 100 kilometers, partial melting of the water-rich oceanic crust and some of the overlying mantle takes place. The newly formed magma created in this manner is less dense than the surrounding mantle rocks and, when sufficient quantities have gathered the molten rock will slowly rise (Figure 16.13B). Most of the rising magma will intrude into the overlying continental crust where it will cool and crystallize at depth. However, some of the magma may migrate to the surface where it can give rise to numerous and occasionally explosive volcanic eruptions. The volcanic Andes Mountains are believed to have been produced by such activity when the Nazca plate melted as it plunged beneath the continent of South America (see Figure 16.8). The frequent earthquakes that occur within the Andes testify to the activity beneath our view.

Mountains such as the Andes that are believed to be produced in part by volcanic activity associated with the subduction of oceanic lithosphere are called **volcanic arcs** (Figure 16.13A). Two of these volcanic arcs are in the western United States. One, the Cascade Range of Washington and Oregon, consists of several well-known volcanic mountains, including Mounts Rainier, Shasta, and St. Helens. As continuing eruptions of Mount St. Helens testify, the Cascade Range is still active. The magma here arises from the melting of a small remaining segment of the Juan de

Fuca plate (see Figure 16.8). The second volcanic arc is the Sierra Nevada, in which Yosemite National Park is located. The Sierra Nevada system is the older of the two and has been inactive for several million years, as evidenced by the absence of volcanic cones. Here erosion has stripped away most of the obvious traces of volcanic activity and left exposed the large, crystallized magma chambers that once fed lofty volcanoes.

Oceanic-Oceanic Convergence When two oceanic slabs converge, one descends beneath the other, initiating volcanic activity in a manner similar to that which occurs at an oceanic-continental convergent boundary. However, in this case, the volcanoes form on the ocean floor rather than on the continents (Figure 16.13B). If this volcanic activity is sustained, it will eventually build crust upward until it emerges from the ocean depths as dry land. In the early stages, this newly formed land consists of a chain of small volcanic islands called an **island arc.** The Aleutian, Mariana, and Tonga islands exemplify such features. Island arcs such as these are generally located a few hundred kilometers from an ocean trench where active subduction of the lithosphere is occurring. Adjacent to the island arcs just mentioned are the Aleutian trench, Mariana trench, and the Tonga trench (Figure 16.14).

Over an extended period, numerous episodes of volcanic activity build large piles of lava on the ocean floor. This gradually increases the size and elevation of the developing arc. This growth, in turn, increases the amount of eroded sediments added to the seafloor. Some of these sediments reach the trench. Here, they are deformed and metamorphosed by the compressional forces exerted by the two converging plates. The result of these diverse activities is the development of a mature island arc composed of a complex system of volcanic rocks, folded and metamorphosed sedimentary rocks, and intrusive igneous rocks. Examples of mature island arc systems are the Alaskan Peninsula, the Philippines, and Japan (Figure 16.14).

Continental-Continental Convergence When two plates carrying continental crust converge, neither plate will subduct beneath the other because of the low density, and thus the buoyant nature, of continental rocks. The result is a collision between the two continental blocks (Figure 16.13C). Such a collision occurred when the once-separated continent of India "rammed" into Asia and produced the

Himalayas, perhaps the most spectacular mountain range on Earth (Figure 16.15). During this collision, the continental crust buckled, fractured, and was generally shortened. In addition to the Himalayas, several other complex mountain systems, including the Alps, Appalachians, and Urals, are thought to have formed in this manner.

Prior to a continental collision, the landmasses involved are separated by an ocean basin (Figure 16.15, top). As the continental blocks converge, the intervening seafloor is subducted beneath one of the plates. The partial melting of the descending oceanic slab and mantle rocks generates a volcanic arc. Erosion of the newly formed volcanic arc adds large quantities of sediment to the already sediment-laden continental margin. Eventually, as the intervening seafloor is consumed, these continental masses collide. This squeezes, folds, and generally deforms the sediments as if they were placed in a gigantic vise.

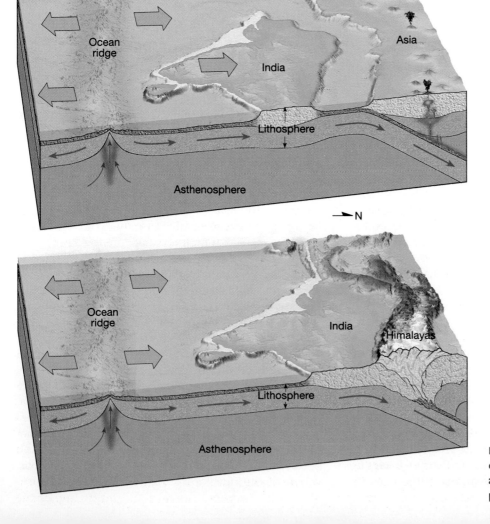

FIGURE 16.15 The ongoing collision of India and Asia, starting about 45 million years ago, produced the majestic Himalayas.

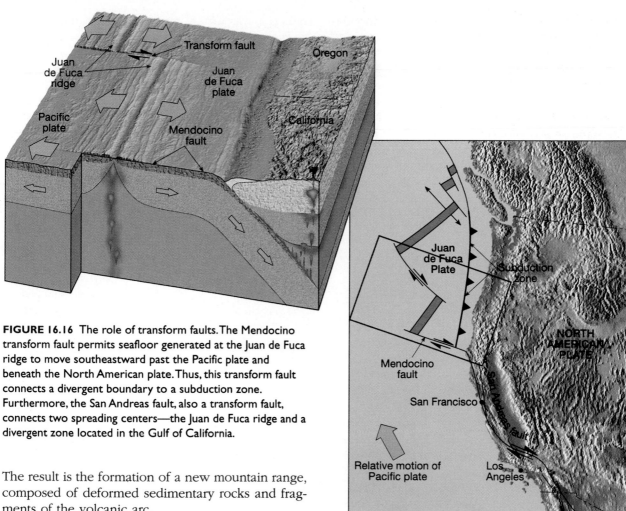

FIGURE 16.16 The role of transform faults. The Mendocino transform fault permits seafloor generated at the Juan de Fuca ridge to move southeastward past the Pacific plate and beneath the North American plate. Thus, this transform fault connects a divergent boundary to a subduction zone. Furthermore, the San Andreas fault, also a transform fault, connects two spreading centers—the Juan de Fuca ridge and a divergent zone located in the Gulf of California.

The result is the formation of a new mountain range, composed of deformed sedimentary rocks and fragments of the volcanic arc.

Transform Boundaries

The third type of plate boundary is the transform fault, where plates grind past one another without the production of new crust, as occurs along oceanic ridges, or without the destruction of crust, as occurs at subduction zones. Transform faults roughly parallel the direction of plate movement. They were first identified where they join offset segments of the oceanic ridge system (Figure 16.14).

Transform faults provide the means by which the oceanic crust created at the ridge crests can be transported to its site of destruction: the deep-ocean trenches. Figure 16.16 illustrates this activity. Notice that the Juan de Fuca plate moves in a southeasterly direction, eventually being subducted under the west coast of the United States. The southern end of this relatively small plate is bounded by the Mendocino transform fault. This transform boundary connects the active spreading center to the subduction zone. Therefore, the fault facilitates the movement of the crustal material created at the ridge crest to its destination beneath the North American continent.

These special faults are called transform faults because the relative motion of the plates can be changed, or transformed, along them. As we saw in the preceding example, divergence occurring at a spreading center can be transformed into convergence at a subduction zone. Because transform faults connect convergent and divergent boundaries in various combinations, other changes in relative plate motion are possible along transform faults.

Most transform faults are located in oceanic crust. However, a few, including California's famous San Andreas fault, are situated within continents (Figure 16.16). Along the San Andreas fault, the Pacific plate is moving toward the northwest, past the North American plate. If this movement continues for millions of years, that part of California west of the fault zone, including the Baja Peninsula, will become an island off the west coast of the United States and Canada. It could eventually reach Alaska. However, a more immediate concern is the earthquakes triggered by movements along this fault system.

Testing the Plate Tectonics Model

With the birth of the plate tectonics model, researchers from all of the Earth sciences began testing it. Some of the evidence supporting continental drift and seafloor spreading has already been presented. Some of the evidence that was instrumental in solidifying the support for this new concept follows. Note that some of the evidence was not new;

rather, it was a new interpretation of old data that swayed the tide of opinion.

Evidence: Paleomagnetism

Probably the most persuasive evidence to the geologic community for the acceptance of the plate tectonics theory comes from the study of Earth's magnetic field. Anyone who has used a compass to find direction knows that the magnetic field has a north pole and a south pole. These magnetic poles align closely, but not exactly, with the geographic poles. (The geographic poles are simply the top and

Robert Dietz and John Holden, who reconstructed Pangaea, also have extrapolated present-day plate movements 50 million years into the future. The map shows where they envision Earth's landmasses will be if present plate movements persist (Figure 16.A).

In North America you can see that the Baja Peninsula and the portion of southern California west of the San Andreas fault will have slid northward to Canada. If this takes place, Los Angeles

and San Francisco will pass each other in about 10 million years, and in about 60 million years, Los Angeles will begin to descend into the Aleutian trench.

In Africa, a new sea is emerging as East Africa parts company with the mainland. Africa also is moving slowly into Europe, perhaps initiating the next major mountain-building stage on our dynamic planet. Meanwhile, the Arabian Peninsula continues to diverge from Africa, allowing the Red Sea to widen and closing the Persian Gulf.

In 50 million years, Australia will be

astride the equator and, with New Guinea, will be on a collision course with Asia. Meanwhile, North America and South America are separating, and the Atlantic and Indian oceans continue to grow at the expense of the Pacific Ocean.

These projections must be viewed with caution because many assumptions are required for these events to unfold as described. Nevertheless, changes in the shapes and positions of continents will undoubtedly occur for millions of years to come.

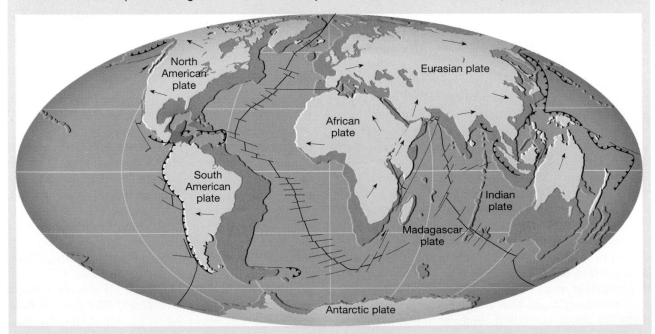

FIGURE 16.A The world as it may look 50 million years from now. (From "The Breakup of Pangaea," Robert S. Dietz and John C. Holden. Copyright 1970 by Scientific American, Inc. All rights reserved)

bottom of the spinning sphere we live on, the points through which passes the imaginary axis of rotation.)

In many respects the magnetic field is very much like that produced by a simple bar magnet. Invisible lines of force pass through Earth and extend from one pole to the other. A compass needle, itself a small magnet free to move about, becomes aligned with these lines of force and thus points toward the magnetic poles.

The technique used to study ancient magnetic fields relies on the fact that certain rocks contain minerals that serve as fossil compasses. These iron-rich minerals, such as magnetite, are abundant in lava flows of basaltic composition. When heated above a certain temperature called the *Curie point*, these magnetic minerals lose their magnetism.

However, when these iron-rich grains cool below their Curie point (about 580ºC) they become magnetized in the direction parallel to the existing magnetic field. Once the minerals solidify, the magnetism they possess will remain "frozen" in this position. In this regard, they behave much like a compass needle inasmuch as they "point" toward the existing magnetic poles. Then, if the rock is moved, or if the magnetic pole changes position, the rock magnetism will, in most instances, retain its original alignment. Rocks formed thousands or millions of years ago thus "remember" the location of the magnetic poles at the time of their formation and are said to possess fossil magnetism, or **paleomagnetism.**

Polar Wandering A study of lava flows conducted in Europe in the 1950s led to an amazing discovery. The magnetic alignment in the iron-rich minerals in lava flows of different ages was found to vary widely. A plot of the apparent positions of the magnetic north pole revealed that, during the past 500 million years, the location of the pole had gradually wandered from a spot near Hawaii northward through eastern Siberia and finally to its present site (Figure 16.17A). This was clear evidence that either the magnetic poles had migrated through time, an idea known as **polar wandering** or that the lava flows had moved—in other words, the continents had drifted.

Although the magnetic poles are known to move, studies of the magnetic field indicated that the average positions of the magnetic poles correspond closely to the positions of the geographic poles. This is consistent with our knowledge of Earth's magnetic field, which is generated in part by the rotation of Earth about its axis. If the geographic poles do not wander appreciably, which we believe is true, neither can the magnetic poles. Therefore, a more acceptable explanation for the apparent polar wandering is provided by the plate tectonics theory. *If the magnetic*

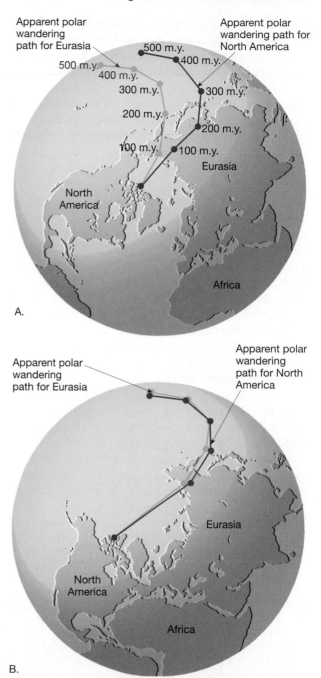

FIGURE 16.17 Simplified apparent polar wandering paths as established from North American and Eurasian paleomagnetic data. **A.** The more westerly path determined from North American data is thought to have been caused by the westward drift of North America by about 24 degrees from Eurasia. **B.** The positions of the wandering paths when the landmasses are reassembled in their predrift locations.

poles remain stationary, their apparent movement was produced by the drifting of the continents.

Further evidence for plate tectonics came a few years later when polar wandering curves were constructed for North America and Europe (Figure

16.17A). To nearly everyone's surprise, the curves for North America and Europe had similar paths, except that they were separated by about 24 degrees of longitude. When these rocks solidified, could there have been two magnetic north poles which migrated parallel to each other? This is very unlikely. The differences in these migration paths, however, can be reconciled if the two presently separated continents are placed next to one another, as we now believe they were prior to the opening of the Atlantic Ocean (Figure 16.17B).

Magnetic Reversals and Seafloor Spreading

Another discovery came when geophysicists learned that Earth's magnetic field periodically reverses polarity; that is, the north magnetic pole becomes the south magnetic pole, and vice versa. A rock solidifying during one of the periods of reverse polarity will be magnetized with the polarity opposite that of rocks being formed today.

When rocks exhibit the same magnetism as the present magnetic field, they are said to possess **normal polarity,** while those rocks exhibiting the opposite magnetism are said to have **reverse polarity**. Evidence for magnetic reversals was obtained from lavas and sediments from around the world. Once the concept of magnetic reversals was confirmed, researchers set out to establish a time scale for polarity reversals. There are many areas where volcanic activity has occurred sporadically

for periods of millions of years (Figure 16.18). The task was to measure the directions of paleomagnetism in numerous lava flows of various ages. These data were collected from several places and were used to determine the dates when the polarity of Earth's magnetic field changed. Figure 16.19 shows the time scale of the polarity reversals established for the last few million years.

A significant relationship was uncovered between the magnetic reversals and the seafloor spreading hypothesis. Very sensitive instruments called *magnetometers* were towed by research vessels across a segment of the ocean floor located off the west coast of the United states. Here workers from the Scripps Institute of Oceanography discovered alternating strips of high- and low-intensity magnetism that trended in roughly a north-south direction. This relatively simple pattern of magnetic variation defied explanation until 1963, when it was tied to the concept of seafloor spreading. The strips of high-intensity magnetism are regions where the paleomagnetism of the ocean crust is of the normal type. Consequently, these positively magnetized rocks *enhance* the existing magnetic field. Conversely, the low-intensity strips represent regions where the ocean crust is polarized in the reverse direction and, therefore, *weaken* the existing magnetic field. But how do parallel strips of normally and reversely magnetized rock become distributed across the ocean floor?

As new basalt is added to the ocean floor at the oceanic ridges, it becomes magnetized according to

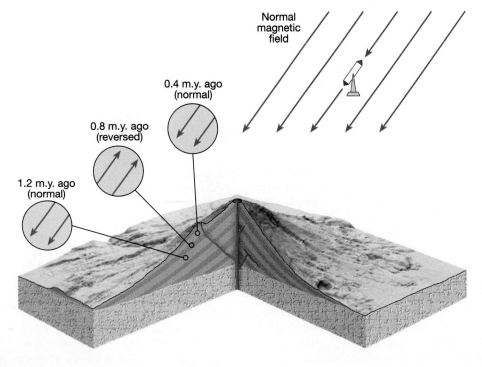

Normal magnetic field

0.4 m.y. ago (normal)

0.8 m.y. ago (reversed)

1.2 m.y. ago (normal)

FIGURE 16.18 Schematic illustration of paleomagnetism preserved in lava flows of various ages. Data such as these from various locales were used to establish the time scale of polarity reversals shown in Figure 16.19.

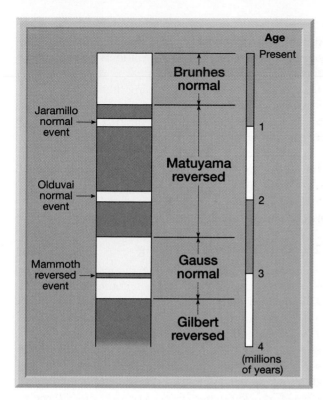

FIGURE 16.19 Time scale of Earth's magnetic field in the recent past. This time scale was developed by establishing the magnetic polarity for lava flows of known age. (Data from Allen Cox and G.B. Dalrymple)

Jaramillo normal event

Olduvai normal event

Mammoth reversed event

Brunhes normal

Matuyama reversed

Gauss normal

Gilbert reversed

Age
Present

1

2

3

4
(millions of years)

the existing magnetic field (Figure 16.20). As new rock is added in approximately equal amounts to the trailing edges of both plates, we should expect strips of equal size and polarity to parallel both sides of the ocean ridges, as shown in Figure 16.20C. This explanation of the alternating strips of normal and reverse polarity, which lay as mirror images across the ocean ridges, was the strongest evidence so far presented in support of the concept of seafloor spreading.

Now that the dates of the most recent magnetic reversals have been established, the rate at which spreading occurs at the various ridges can be determined accurately. In the Pacific Ocean, for example, the magnetic strips are much wider for corresponding time intervals than those of the Atlantic Ocean. Hence, we conclude that a faster spreading rate exists for the spreading center of the Pacific as compared to the Atlantic. When we apply absolute dates to these magnetic events, we find that the spreading rate for the North Atlantic Ridge is only 1 or 2 centimeters per year (note that each side spreads at this rate). The rate is somewhat faster for the South Atlantic. The spreading rates for the East Pacific Rise generally range between 3 and 8 centimeters per year, with a maximum rate of about 10 centimeters per year in one segment. Thus, we have a magnetic tape recorder that records changes in Earth's magnetic field. This

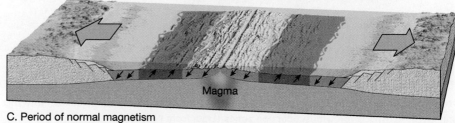
A. Period of normal magnetism

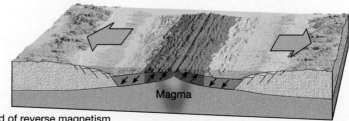
B. Period of reverse magnetism

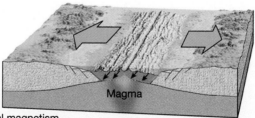
C. Period of normal magnetism

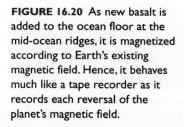
FIGURE 16.20 As new basalt is added to the ocean floor at the mid-ocean ridges, it is magnetized according to Earth's existing magnetic field. Hence, it behaves much like a tape recorder as it records each reversal of the planet's magnetic field.

recorder also permits us to determine the rate of seafloor spreading.

Evidence: Earthquake Patterns

By 1968, the basic outline of global tectonics was firmly established. In this same year, three seismologists at Lamont-Doherty Earth Observatory published papers demonstrating how successfully the new plate tectonics model accounted for the global distribution of earthquakes (Figure 16.21). In particular, these scientists were able to account for the close association between deep-focus earthquakes and ocean trenches. Furthermore, the absence of deep-focus earthquakes along the oceanic ridge system was shown to be consistent with the new theory.

The close association between plate boundaries and earthquakes can be seen by comparing the distribution of earthquakes shown in Figure 16.21 with the map of plate boundaries in Figure 16.8 and trenches in Figure 16.14. In trench regions where dense slabs of lithosphere plunge into the mantle, this association is especially striking. When the depths of earthquake foci and their locations within the trench systems are plotted, an interesting pattern emerges. Figure 16.22, which shows the distribution of earthquakes in the vicinity of the Japan trench, is an example. Here most shallow-focus earthquakes occur within, or adjacent to, the trench, whereas intermediate- and deep-focus earthquakes occur toward the mainland.

In the plate tectonics model, deep-ocean trenches are produced where cold, dense slabs of oceanic lithosphere plunge into the mantle. Shallow-focus earthquakes are produced as the descending plate interacts with the overriding lithosphere. As the slab descends farther into the asthenosphere, deeper-focus earthquakes are generated (Figure 16.22). Because the earthquakes occur within the rigid subducting plate rather than in the "plastic" mantle, they provide a method for tracking the plate's descent. Very few earthquakes have been recorded below 700 kilometers (435 miles), possibly because the slab has been heated sufficiently to lose its rigidity.

Evidence: Ocean Drilling

Some of the most convincing evidence confirming the plate tectonics theory has come from drilling directly into ocean-floor sediment. From 1968 until 1983, the source of these important data was the Deep Sea Drilling Project, an international program sponsored by several major oceanographic institutions and the National Science Foundation. A new drilling ship was built. The *Glomar Challenger* represented a significant technological breakthrough, because this ship could lower drill pipe thousands of meters to the ocean floor and then drill hundreds of meters into the sediments and underlying basaltic crust. In the South Atlantic, at several sites, holes were drilled through the entire thickness of sediments to the basaltic rock below. An important objective

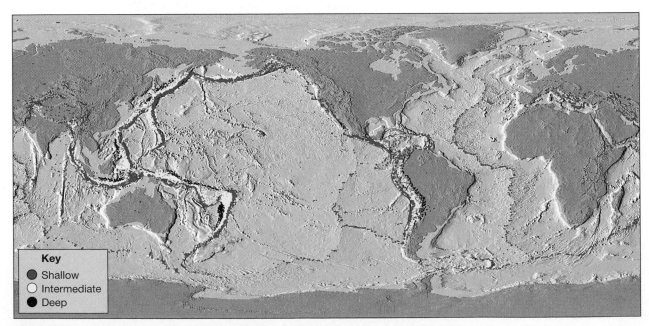

FIGURE 16.21 Distribution of shallow-, intermediate-, and deep-focus earthquakes. Note that deep-focus earthquakes only occur in association with subduction zones. (Data from NOAA)

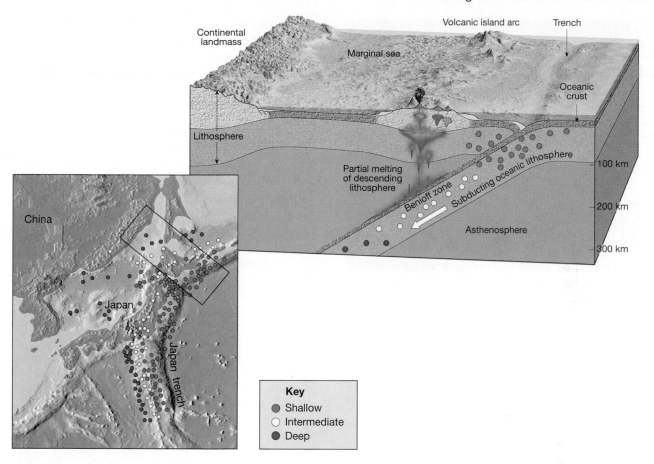

FIGURE 16.22 Distribution of earthquake foci in the vicinity of the Japan trench. Note that intermediate- and deep-focus earthquakes occur only within the sinking slab of oceanic lithosphere. (Data from NOAA)

was to gather samples of sediment from just above the igneous crust as a means of dating the seafloor at each site. (Radiometric dates of the ocean crust itself are unreliable because seawater alters basalt.)

When the oldest sediment from each drill site was plotted against its distance from the ridge crest, it was revealed that the age of the sediment increased with increasing distance from the ridge. This finding agreed with the seafloor spreading hypothesis, which predicted that the youngest oceanic crust would be found at the ridge crest, and that the oldest oceanic crust would be at the continental margins.

The data from the Deep Sea Drilling Project also reinforced the idea that the ocean basins are geologically youthful, because no sediment with an age in excess of 160 million years was found. By comparison, some continental crust has been dated at 3.9 billion years.

During its 15 years of operation, the *Glomar Challenger* drilled 1092 holes and obtained more than 96 kilometers (60 miles) of invaluable core samples. The Ocean Drilling Program has succeeded the Deep Sea Drilling Project and, like its predecessor, it

is a major international program. A more technologically advanced drilling ship, the *JOIDES Resolution,* now continues the work of the *Glomar Challenger* (Figure 16.23).

Evidence: Hot Spots

Mapping of seafloor volcanoes called *seamounts* in the Pacific revealed a chain of volcanic structures extending from the Hawaiian Islands to Midway Island and then continuing northward toward the Aleutian trench (Figure 16.24). Radiometric dates of volcanoes in this chain revealed that the volcanoes increase in age with increasing distance from Hawaii. Suiko Seamount, which is located near the Aleutian trench, is 65 million years old, Midway Island is 27 million years old, and the island of Hawaii built up from the seafloor less than a million years ago (Figure 16.24).

Researchers have proposed that a rising plume of mantle material is located below the island of Hawaii. Melting of this hot rock as it enters the low-pressure environment near the surface generates a volcanic area or **hot spot**. Presumably, as the Pacific plate moved over the hot spot, successive volcanic

FIGURE 16.23 The *JOIDES Resolution*, the drilling ship of the Ocean Drilling Program. This modern drilling ship has replaced the *Glomar Challenger* in the important work of sampling the floors of the world's oceans. JOIDES is an acronym for Joint Oceanographic Institutions for Deep Earth Sampling. (Photo courtesy of Ocean Drilling Program)

mountains have been built. The age of each volcano indicates the time when it was situated over the relatively stationary mantle plume. This pattern is shown in Figure 16.24. Kauai is the oldest of the large islands in the Hawaiian chain. Five million years ago, when it was positioned over the hot spot, Kauai was the only Hawaiian Island in existence (Figure 16.24). Visible evidence of the age of Kauai can be seen by examining its extinct volcanoes, which have been eroded into jagged peaks and vast canyons. By contrast, the south slopes of the relatively youthful island of Hawaii consist of fresh lava flows, and two of Hawaii's volcanoes, Mauna Loa and Kilauea, remain active.

This evidence supports the fact that the plates do indeed move relative to Earth's interior. The hot spot "tracks" also trace the direction of plate motion. Notice, for example, in Figure 16.24 that the Hawaiian Island–Emperor Seamount chain bends. This particular bend in the trace occurred about 40 million years ago when the motion of the Pacific plate changed from nearly due north to a northwesterly path.

Pangaea: Before and After

Robert Dietz and John Holden have projected the gross details of the migrations of individual continents over the past 500 million years. By extrapolating plate motion back in time using such evidence as the orientation of volcanic structures left behind on moving plates, the distribution and movements of transform faults, and paleomagnetism, Dietz and Holden were able to reconstruct Pangaea (Figure 16.25A). The use of radiometric dating helped them establish the time frame for the formation and eventual breakup of Pangaea, and the relatively stationary positions of hot spots through time helped to fix the locations of the continents.

Breakup of Pangaea

The fragmentation of Pangaea began about 200 million years ago. Figure 16.25 illustrates the breakup and subsequent paths taken by the landmasses involved. As we can readily see in Figure 16.25B, two major rifts initiated the breakup. The rift zone between North America and Africa generated numerous outpourings of Jurassic-age basalts, which are presently visible along the eastern seaboard of the United States. Radiometric dating of these basalts indicates that rifting occurred between 200 million and 165 million years ago. This date can be used as the birth date of this section of the North Atlantic. The rift that formed in the southern landmass of Gondwanaland developed a Y-shaped fracture that sent India on a northward journey and simultaneously separated South America-Africa from Australia-Antarctica.

Figure 16.25C illustrates the position of the continents 135 million years ago, about the time Africa and South America began splitting apart to form the South Atlantic. India can be seen halfway into its journey to Asia, while the southern portion of the North Atlantic has widened greatly. By the beginning of the Cenozoic, about 65 million years ago, Madagascar had separated from Africa, and the South Atlantic had emerged as a full-fledged ocean (Figure 16.25D). At this juncture, India had drifted over a hot spot that generated numerous fluid basalt flows across a region in western India now called the Deccan Plateau. These lava flows are very similar to those that make up the Columbia Plateau in Pacific Northwest.

The current map (Figure 16.25E) shows India in contact with Asia, an event that began about 45 million years ago and created the highest mountains on Earth, the Himalayas, along with the Tibetan Highlands. It is interesting to note that the *average* height of Tibet is 5000 meters, higher than any spot in the contiguous United States. India's continued northward migration is believed to cause the numerous and often destructive earthquakes plaguing that part of the world.

By comparing Figures 16.25D and 16.25E, we can see that the separation of Greenland from Eurasia was a recent event in geologic history. Also notice the recent formation of the Baja Peninsula along with

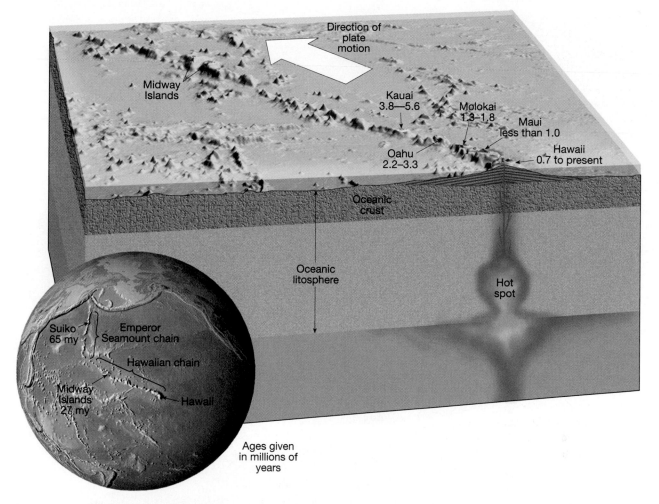

FIGURE 16.24 The chain of islands and seamounts that extends from Hawaii to the Aleutian trench results from the movement of the Pacific plate over an apparently stationary hot spot. Radiometric dating of the Hawaiian Islands shows that the volcanic activity decreases in age toward the island of Hawaii.

the Gulf of California. This event is thought to have occurred less than 10 million years ago.

Before Pangaea

Prior to the formation of Pangaea, the landmasses had probably gone through several episodes of fragmentation similar to what we see happening today. Also like today, these ancient continents moved away from each other only to collide again at some other location. During the period between 500 million and 225 million years ago, the fragments of an earlier dispersal began collecting to form the continent of Pangaea. Evidence of these earlier continental collisions include the Ural Mountains of Russia and the Appalachian Mountains, which flank the east coast of North America.

Available evidence indicates that about 500 million years ago the northern continent of Laurasia was fragmented into three major sections—North America, northern Europe (southern Europe was part of Africa), and Siberia—with each section separated by a sizable ocean. The southern continent of Gondwanaland probably was intact and lay near the South Pole. The first collision is believed to have occurred as North America and Europe closed the pre-North Atlantic. This activity resulted in the formation of the northern Appalachians. Parts of the floor of the former ocean can be seen today high above sea level in Nova Scotia.

The sliver of eastern Canada and the United States that lies seaward of the zone of this collision is truly a gift from Europe. It is also thought that before North America and Europe collided, part of Scotland, Ireland, and Norway were attached to the North American plate. While North America and Europe were joining, Siberia was closing the gap between itself and Europe, which lay farther to the west. This closing culminated about 300 million to

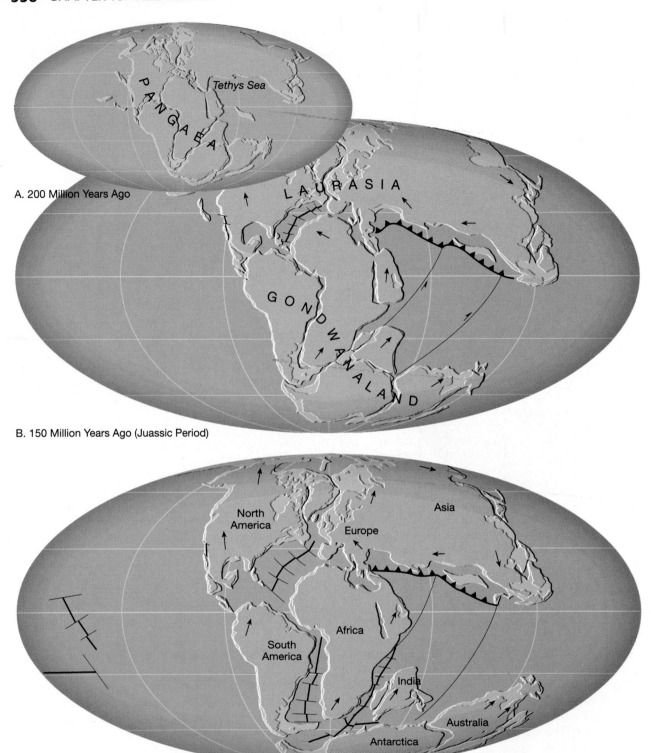

A. 200 Million Years Ago

B. 150 Million Years Ago (Juassic Period)

C. 100 Million Years Ago (Cretaceous Period)

FIGURE 16.25 Several views of the breakup of Pangaea over a period of 200 million years. (After D. Walsh and C. Scotese)

350 million years ago in the formation of the Ural Mountains. The consolidation of these landmasses completed the northern continent of Laurasia.

During the next 50 million years the northern and southern landmasses converged, producing the supercontinent of Pangaea. At this time (about 250

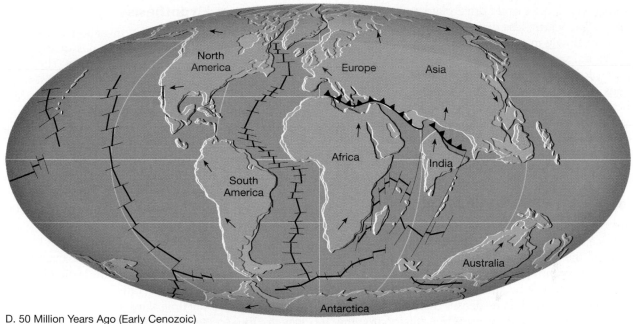

D. 50 Million Years Ago (Early Cenozoic)

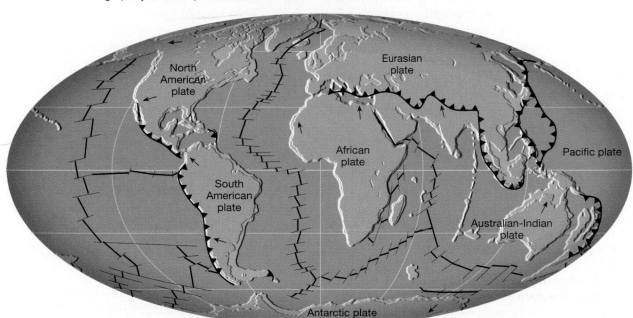

E. Present

million to 300 million years ago), Africa and North America collided to produce the southern Appalachians (see Box 16.1).

The Driving Mechanism

The plate tectonics theory *describes* plate motion and the *effects* of this motion. Because we accept the theory does not mean that we *understand the forces* that move the plates. In fact, none of the driving

mechanisms yet proposed can account for all of the major facets of plate motion. Nevertheless, it is clear that the unequal distribution of heat within Earth is the underlying driving force for plate movement.

Convection Current Hypothesis

One of the first models to explain the movements of plates was originally proposed as a possible driving mechanism for continental drift. Adapted to plate tectonics, this hypothesis suggests that large convection currents within the mantle drive plate motion (Figure 16.26A). The warm, less dense material of the mantle rises very slowly in the regions of oceanic ridges. As the material spreads laterally, it might drag the lithosphere along. Eventually, the material cools and begins to sink back into the mantle, where it is reheated. Partly because of its simplicity, this proposal had wide appeal. However, modern research techniques reveal that the flow of material in the mantle is far more complex than such simple convection cells.

Slab-Push and Slab-Pull Hypotheses

Many other mechanisms that may contribute to or influence plate motion have been suggested. One relies on the fact that, as a newly formed slab of oceanic crust moves away from the ridge crest, it gradually cools and becomes denser. Eventually, the cold oceanic slab becomes denser than the asthenosphere and begins to descend. When this occurs, the dense sinking slab pulls the trailing lithosphere along. This so-called slab-pull hypothesis is similar to another model which suggests that the elevated position of an oceanic ridge could cause the lithosphere to slide under the influence of gravity. However, some ridge systems are subdued, which would reduce the effectiveness of the slab-push model. Further, some ocean basins, notably the Atlantic, lack subduction zones; thus, the slab-pull mechanism cannot adequately explain the spreading occurring at all ridges. Nevertheless, the slab-push and slab-pull phenomena appear to be active in some ridge-trench systems (Figure 16.26B).

Hot Plumes Hypothesis

One version of the thermal convection model suggests that relatively narrow hot plumes of rock contribute to plate motion (Figure 16.26C). These hot plumes are presumed to extend upward from the vicinity of the mantle–core boundary. Upon reaching the lithosphere they spread laterally and facilitate plate motion away from the zone of upwelling. These mantle plumes reveal themselves as long-lived volcanic areas (hot spots) in such places as Iceland. A dozen or so hot spots have been identified along ridge systems where they may contribute to plate divergence. Recall, however, that many hot spots, including the one that generated the Hawaiian Islands, are not located in ridge areas.

The downward limbs of these convection cells are the cold, dense subducting lithospheric plates (Figure 16.26C). Moreover, advocates of this view suggest that subducting slabs may descend all the way to the core–mantle boundary. However, convincing seismic evidence for the existence of these sheetlike structures below 700 kilometers is lacking.

Although there is still much to be learned about the mechanisms that cause plates to move, some facts are clear. The unequal distribution of heat in Earth generates some type of thermal convection in the mantle that ultimately drives plate motion. Whether the upwelling is mainly in the form of rising limbs of convection currents or cylindrical plumes of various sizes and shapes is yet to be determined. Furthermore, the descending lithospheric plates are active components of downwelling, and they serve to transport cold material into the mantle.

The Chapter in Review

The following statements are intended to help you review the primary objectives presented in this chapter.

- In the early 1900s *Alfred Wegener* set forth his *continental drift* hypothesis. One of its major tenets was that a supercontinent called *Pangaea* began breaking apart into smaller continents about 200 million years ago. The smaller continental fragments then "drifted" to their present positions. To support the claim that the now-separate continents were once joined, Wegener and others used the *fit of South America and Africa, fossil evidence, rock types and structures,* and *ancient climates.* One of the main objections to the continental drift hypothesis was its inability to provide an acceptable mechanism for the movement of continents.

- The theory of *plate tectonics,* a far more encompassing theory than continental drift, holds that Earth's outer shell, called the *lithosphere,* consists of about twenty rigid segments called *plates* that are in motion relative to each other. Most of Earth's *seismic activity, volcanism,* and *mountain building* occur along the dynamic margins of these plates.

- A major departure of the plate tectonics theory from the continental drift hypothesis is that large plates contain both continental and ocean crust and the entire plate moves. By contrast, in continental drift, Wegener proposed that the sturdier continents "drifted" by breaking through the oceanic crust, much like ice breakers cut through ice.

FIGURE 16.26 Proposed models of the driving force for plate tectonics. **A.** Large convection cells in the mantle may carry the lithosphere in a conveyor-belt fashion. **B.** Slab-pull results because a subducting slab is more dense than the underlying material. Slab-push is a form of gravity sliding caused by the elevated position of lithosphere at a ridge crest. **C.** The hot plume model suggests that all upward convection is confined to a few narrow plumes, while the downward limbs of these convection cells are the cold, dense subducting oceanic plates.

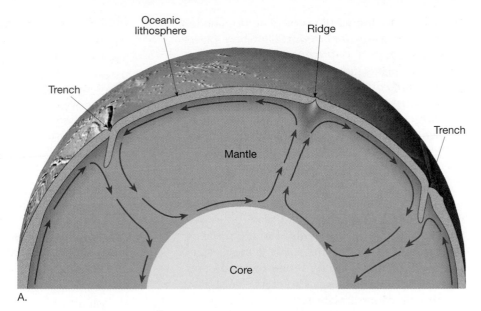

A.

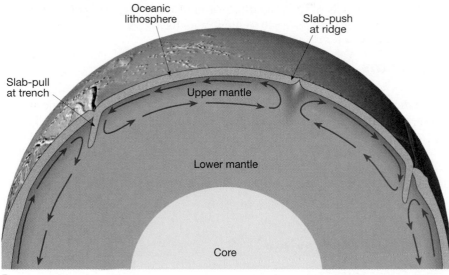

B.

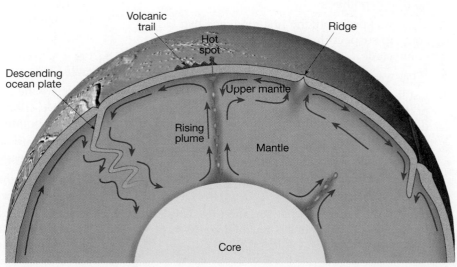

C.

- The three distinct types of plate boundaries are (1) *divergent boundaries*—where plates move apart; (2) *convergent boundaries*—where plates move together as in *oceanic-continental convergence, oceanic-oceanic convergence,* or *continental-continental convergence*; and (3) *transform boundaries*—where plates slide past each other.

- The theory of plate tectonics is supported by (1) *paleomagnetism,* the direction and intensity of Earth's magnetism in the geologic past; (2) the global distribution of *earthquakes* and their close association with plate boundaries; (3) the ages of *sediments* from the floors of the deep-ocean basins; and (4) the existence of island groups that formed over *hot spots* and provide a frame of reference for tracing the direction of plate motion.

- The gross details of the migrations of individual continents over the past 500 million years have been reconstructed. *Pangaea began breaking apart about 200 million years ago.* North America separated from Africa between 200 million and 165 million years ago. Evidence indicates that prior to the formation of Pangaea the landmasses had probably gone through several episodes of fragmentation similar to what we see happening today.

- Several models for the driving mechanism of plates have been proposed. One model, the *convection current hypothesis,* involves large *convection cells* within the mantle carrying the overlying plates. Another, the *slab-push and slab-pull hypothesis,* proposes that dense *oceanic material descends and pulls* the lithosphere along. A third model, the *hot plumes hypothesis,* suggests that relatively narrow hot plumes of rock within the mantle contribute to plate motion. No single driving mechanism can account for all major facets of plate motion.

Key Terms

continental drift (p. 314)
convergent boundary (p. 322)
deep-ocean trench (p. 326)
divergent boundary (p. 322)

hot spot (p. 335)
island arc (p. 327)
normal polarity (p. 332)
paleomagnetism (p. 331)
Pangaea (p. 315)
plate (p. 319)

plate tectonics (p. 319)
polar wandering (p. 331)
reverse polarity (p. 332)
rift (rift valley) (p. 324)
seafloor spreading (p. 322)

subduction zone (p. 326)
transform boundary (p. 322)
volcanic arc (p. 327)

Questions for Review

1. Who is credited with developing the continental drift hypothesis?

2. What was probably the first evidence that led some to suspect the continents were once connected?

3. What was Pangaea?

4. List the evidence that Wegener and his supporters gathered to substantiate the continental drift hypothesis.

5. Explain why the discovery of the fossil remains of *Mesosaurus* in both South America and Africa, but nowhere else, supports the continental drift hypothesis.

6. Early in this century, what was the prevailing view of how land animals migrated across vast expanses of ocean?

7. How did Wegener account for the existence of glaciers in the southern landmasses, while at the same time areas in North America, Europe, and Siberia supported lush tropical swamps?

8. On what basis were plate boundaries first established?

9. What are the three major types of plate boundaries? Describe the relative plate motion at each of these boundaries.

10. What is seafloor spreading? Where is active seafloor spreading occurring today?

11. What is a subduction zone? With what type of plate boundary is it associated?

12. Where is lithosphere being consumed? Why must the production and destruction of lithosphere be going on at approximately the same rate?

13. Briefly describe how the Himalaya Mountains formed.

14. Differentiate between transform faults and the other two types of plate boundaries.

15. Some predict that California will sink into the ocean. Is this idea consistent with the theory of plate tectonics?

16. Define the term *paleomagnetism*.

17. How does the continental drift hypothesis account for the apparent wandering of Earth's magnetic poles?

18. Describe the distribution of earthquake epicenters and foci depths as they relate to oceanic trench systems.

19. What is the age of the oldest sediments recovered by deep-ocean drilling? How do the ages of these sediments compare to the ages of the oldest continental rocks?

20. How do hot spots and the plate tectonics theory account for the fact that the Hawaiian Islands vary in age?

21. With what type of plate boundary are the following places or features associated (be as specific as possible): Himalayas, Aleutian Islands, Red Sea, Andes Mountains, San Andreas fault, Iceland, Japan, Mount St. Helens?

Testing What You Have Learned

To test your knowledge of the material presented in this chapter, answer the following questions:

Multiple Choice Questions

1. Earth's outer shell consists of about twenty rigid segments called _____.
   - **a.** continents
   - **b.** basins
   - **c.** plates
   - **d.** layers
   - **e.** crustal pieces

2. Which of the following is NOT a distinct type of plate boundary?
   - **a.** transform
   - **b.** convergent
   - **c.** zonal
   - **d.** divergent
   - **e.** both a. and c.

3. Which one of the following was NOT used to support the continental drift hypothesis?
   - **a.** fossils
   - **b.** paleomagnetism
   - **c.** ancient climates
   - **d.** fit of continents
   - **e.** rock types and structures

4. One of the major tenets of the continental drift hypothesis suggested the existence of a supercontinent called _____.
   - **a.** Pangaea
   - **b.** Antaustria
   - **c.** Euroam
   - **d.** Soafric
   - **e.** Paneur

5. Researchers have proposed that the Hawaiian Islands are forming over a plume of rising mantle material called a(n) _____ spot.
   - **a.** hot
   - **b.** core
   - **c.** convergent
   - **d.** mantle
   - **c.** island

6. Which one of the following provides a method for tracking a plate's descent into the "plastic" mantle along a deep-ocean trench?
   - **a.** echo soundings
   - **b.** earthquakes
   - **c.** surveying
   - **d.** drilling
   - **e.** magnetism

7. The theory of plate tectonics does NOT help explain the origin and location of which one of the following?
   - **a.** earthquakes
   - **b.** mountains
   - **c.** volcanoes
   - **d.** ocean tides
   - **e.** major seafloor features

8. As plates move apart along oceanic ridges, broken slabs are displaced downward, creating downfaulted valleys called _____ valleys.
   a. trend   **c.** hanging   **e.** rift
   b. displaced   **d.** matching

9. The basic outline of the theory of plate tectonics was firmly established by the late _____.
   a. 1800s   **c.** 1940s   **e.** 1980s
   b. 1920s   **d.** 1960s

10. Earth's rigid outer shell, consisting of crustal rocks and a portion of the upper mantle, is referred to as the _____.
    a. asthenosphere   **c.** rift zone   **e.** lithosphere
    b. core   **d.** transform zone

Fill-In Questions

11. The basic idea of _____ is that Earth's rigid outer shell is made of several large segments that are slowly moving.
12. The region where an oceanic plate descends into the asthenosphere because of convergence is referred to as a(n) _____ zone.
13. When rocks exhibit the same magnetism as the present magnetic field, they are said to possess _____ polarity.
14. _____ _____ is the mechanism responsible for producing ocean floor material at the crest of oceanic ridges.
15. Earth's rigid outer shell lies over a hotter, weaker zone known as the _____.

True/False Questions

16. One result of plate tectonics is that ocean basins are all older than continents. _____
17. According to the plate tectonics theory, continents are moving through the ocean basins. _____
18. The continental drift hypothesis was used by Alfred Wegener to explain the existence of identical fossils on widely separated landmasses. _____
19. The average rate of spreading along ocean ridges is about 60 centimeters per year. _____
20. Mountain ranges such as the Himalayas, Alps, and Appalachians formed in response to continental-continental convergence. _____

Answers

1.c; 2.c; 3.b; 4.a; 5.a; 6.b; 7.d; 8.e; 9.d; 10.e; 11. plate tectonics; 12. subduction; 13. normal; 14. Seafloor spreading; 15. asthenosphere; 16.F; 17.F; 18.T; 19.F; 20.T

Mountain Building

Mount Sheffels, San Juan Mountains, Colorado. (Photo by Carr Clifton)

Focus on Learning

To assist you in learning the important concepts in this chapter, you will find it helpful to focus on the following questions:

- How is the concept of isostasy related to mountain building?

- What are the two basic types of rock deformation?

- What are the most common types of folds and faults? How does each type form?

- Based on structural characteristics, what are the four main types of mountains?

- What are the mountain types associated with the various kinds of convergent plate boundaries?

Mountains provide some of the most spectacular scenery on our planet. This splendor has been captured by poets, painters, and songwriters alike. Geologists believe that at some time all continental regions were mountainous masses and have concluded that the continents grow by the addition of mountains to their flanks. Consequently, by unraveling the secrets of mountain formation, geologists will have taken a major step in understanding the evolution of Earth. If continents do indeed grow by adding mountains to their flanks, how do geologists explain the existence of mountains (the Urals, for example) that are located in the interior of a landmass? To answer this and other related questions, this chapter attempts to piece together the sequence of events that is believed to generate these lofty structures.

Mountains often are spectacular features that rise several hundred meters or more above the surrounding terrain (Figure 17.1). Some occur as isolated masses; the volcanic cone Kilimanjaro, for example, stands almost 6000 meters (20,000 feet) above sea level, overlooking the grasslands of East Africa. Other peaks are parts of extensive mountain belts, such as the American Cordillera, which runs continuously from the tip of South America through Alaska, including the Andes and the Rocky Mountains. Chains such as the Himalayas consist of youthful, towering peaks that are still rising, whereas others, including the Appalachian Mountains in the eastern United States, are much older and have been eroded much lower than their original lofty height.

The name for the processes that collectively produce a mountain system is **orogenesis**, from the Greek *oros* ("mountain") and *genesis* ("to come into being"). Mountain systems show evidence of enormous forces that have bent (folded), broken (faulted), and generally deformed large sections of Earth's crust.

The first encompassing explanation of orogenesis came a little over three decades ago as part of the plate tectonics theory. The idea of crustal plates in slow motion, sometimes converging with great force, has opened many new and exciting avenues to geologists. Before examining mountain building according to the plate tectonics model, however, it will be advantageous to view some basic principles: the processes of crustal uplifting and rock deformation.

Crustal Uplift

The fossilized shells of marine invertebrates are often found at high elevations in mountains, an indication that the sedimentary rock in the mountain was once below sea level. This is convincing evidence that some dramatic changes have occurred between the time these animals died and the discovery of their fossilized remains!

FIGURE 17.1 Mount Kerkeslin in Alberta's Jasper National Park is part of the Canadian Rockies. (Photo by Carr Clifton)

Evidence for crustal uplift such as this is common in rocks, which form the "geologic record." Many examples exist along the coast of the western United States. When a coastal area remains undisturbed for an extended period, wave action cuts a gently sloping platform. In parts of California, ancient wave-cut platforms can now be found as terraces, hundreds of meters above sea level (Figure 17.2). Each terrace represents a period when that area was at sea level. Such evidence of crustal uplift is easy to find, but unfortunately the *reasons* for uplift are not so easily determined.

Isostasy

We know that the force of gravity must play an important role in determining the elevation of the land. In particular, the less dense crust is believed to float on top of the denser and more easily deformed rocks of the mantle. This concept—a floating crust in gravitational balance—is called **isostasy**.

Perhaps the easiest way to grasp the concept of isostasy is to envision a series of wooden blocks of different heights floating in water, as shown in Figure 17.3. Note that the taller wooden blocks float higher in water than do the shorter blocks. In a similar manner, mountain belts stand higher above the surface and have "roots" that extend deeper into the supporting material below (Figure 17.4A). The average thickness of the continental crust is about 35 kilometers, but crustal thicknesses for some mountain chains exceeds 70 kilometers. Thus, mountains exist where unusually thick blocks of light crustal material are supported from below by the dense rocks of the mantle.

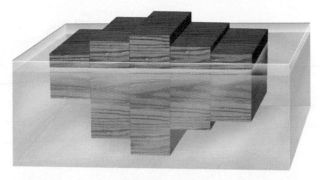

FIGURE 17.3 This drawing illustrates how wooden blocks of different thicknesses float in water. In a similar manner, thick sections of crustal material float higher than thinner crustal slabs.

Carrying this idea one step further, the crust beneath the oceans must be thinner than that of the continents, because its elevation is lower. Not only is this true, but oceanic rocks also have a greater density than do continental rocks, another factor contributing to their lower position.

If the concept of isostasy is correctly applied to mountains, we should expect that adding weight to the crust makes it respond by subsiding, and when that weight is removed, crustal uplifting occurs. (Visualize what happens to a ship as cargo is being loaded—the ship sinks deeper—and unloaded—the ship rises.)

Isostatic Adjustment

Evidence for this type of movement exists, strongly supporting the concept of **isostatic adjustment**. A classic example is provided by Ice Age glaciers. When continental ice sheets occupied portions of North America during the Pleistocene epoch, the added weight of the 3-kilometer-thick masses of ice caused downwarping of Earth's crust by hundreds of meters. In the 8000 to 10,000 years since the last ice sheets vanished, uplifting of as much as 330 meters has occurred in the Hudson Bay region of Canada, where the thickest ice had accumulated.

As the foregoing examples illustrate, isostatic adjustment can account for considerable *vertical* crustal movement. Thus, we can understand why, as erosion slowly lowers the summits of mountains, the crust gradually rises in response to the reduced load. The processes of uplifting and erosion will continue together until the mountain block reaches "normal" crustal thickness (Figure 17.4). When this occurs, the mountains will have been eroded to near sea level and the deeply buried portions of the mountains will become exposed at the surface. In addition, as the mountains succumb to erosion, sediment derived from the shrinking masses is deposited on the adjacent continental margin, causing the margin to subside (Figure 17.4 B, C).

FIGURE 17.2 Former wave-cut platforms now exist as a series of elevated terraces on the western side of San Clemente Island off the southern California coast. Once at sea level, the highest terraces are now about 400 meters (1320 feet) above it. (Photo by John S. Shelton)

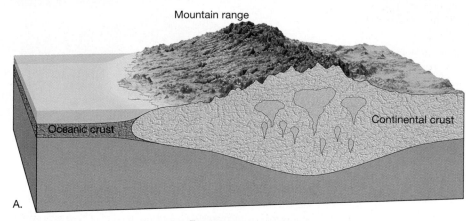

Mountain range

Oceanic crust

Continental crust

A.

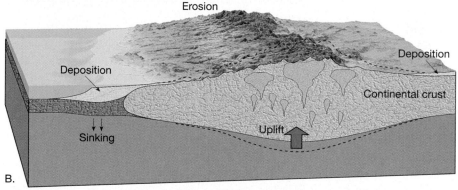

Erosion

Deposition

Deposition

Continental crust

Uplift

Sinking

B.

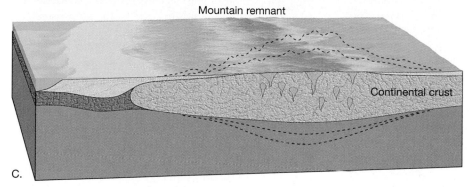

Mountain remnant

Continental crust

C.

FIGURE 17.4 This sequence illustrates how the combined effect of erosion and isostatic adjustment results in a thinning of the crust in mountainous regions. **A.** When mountains are young, the continental crust is thickest. **B.** As erosion lowers the mountains, the crust rises in response to the reduced load. **C.** Erosion and uplift continue until the mountains reach "normal" crustal thickness.

To summarize, we have learned that Earth's major mountain chains consist of unusually thick sections of deformed crustal material that have been elevated above the surrounding terrain. The support for these massive features comes from the buoyancy of light crustal material supported by denser mantle rocks. As erosion lowers the peaks, isostatic adjustment gradually raises the mountains in response. Eventually, the deepest portions of the mountains rise to the shallower depths of the surrounding crust. The question still to be answered is: How do these thickened sections of Earth's crust come into existence in the first place? We attempt to answer this question later in the chapter.

Rock Deformation

When rocks are subjected to stresses greater than their own strength, they begin to deform, usually by folding or fracturing (Figure 17.5). It is easy to visualize how individual rocks break. But how are large rock units, many meters thick and many kilometers long, bent into intricate folds without being appreciably broken during the process? In an attempt to answer this, geologists used the laboratory. Here rocks were subjected to stresses under conditions that simulated those existing at various depths within the crust.

Although each rock type deforms somewhat differently, the general characteristics of rock deformation were determined from these experiments.

FIGURE 17.5 Deformed Paleozoic strata west of Teakettle Junction in Death Valley, California. (Photo by Michael Collier)

Geologists discovered that when stress is applied, rocks first respond by deforming elastically. Changes resulting from *elastic deformation* are reversible; that is, like a rubber band, the rock will return to nearly its original size and shape when the stress is removed.

However, once the *elastic limit* is surpassed, rocks either deform plastically or fracture. *Plastic deformation* results in permanent changes; that is, the size and shape of a rock unit are altered through folding and flowing. Laboratory experiments confirmed that, at the high temperatures and pressures that are typical deep in Earth, most rocks deform plastically once their elastic limit is surpassed.

Rocks tested under surface conditions also deform elastically, but once they exceed their elastic limit, most behave like a brittle solid and fracture. Recall that the energy for most earthquakes comes from stored elastic energy that is released as rock fractures and snaps back to its original shape.

One factor that researchers cannot duplicate in the laboratory is *geologic time*. We know that if stress is applied quickly, as with a hammer, rocks tend to fracture. On the other hand, identical rocks may deform plastically if stress is applied over an extended period. For example, marble benches have been known to sag under their own weight over a period of a hundred years or so. In nature, even modest forces applied over long periods surely play an important role in the deformation of rock.

Folds

During mountain building, flat-lying sedimentary and volcanic rocks are often bent into a series of wavelike undulations called **folds** (Figure 17.6) Folds in sedimentary strata are much like those that would form if you pushed on one edge of a carpet, rumpling it. In nature, folds vary widely in size and shape. Some folds are broad features in which rock units hundreds of meters thick are slightly warped. Others are very tight, microscopic structures found in metamorphic rocks. Size differences notwithstanding, most folds are the result of *compressional stresses* that result in the shortening and thickening of the crust. (Occasionally, folds are found singly, but most often they occur as a series of undulations.)

The two most common types of folds are anticlines and synclines (Figure 17.6). An **anticline** is most commonly formed by the upfolding or arching, of rock layers. Anticlines are sometimes spectacularly displayed where highways have been cut through deformed strata. Often found in association with anticlines are downfolds, or troughs, called **synclines**. Anticlines and synclines vary in symmetry. They may be *symmetrical*, *asymmetrical*, or, if one limb has been tilted beyond vertical, *overturned* (Figure 17.6).

Folds do not continue on forever; rather, their ends die out, much like wrinkles in cloth. These folds are said to be *plunging*, since the axis of the fold plunges to the ground (Figure 17.7). Figure 17.8A shows an example of a plunging anticline and the pattern produced when erosion removes the upper layers of this structure and exposes its interior. Note that the outcrop pattern of an anticline points in the direction it is plunging, whereas the opposite is true for synclines.

A good example of all this—the topography that results when erosional forces wear down folded sedimentary strata for millions of years—is the Valley and Ridge Province of the Appalachians. Here, resistant sandstone beds remain as imposing ridges, separated by valleys that have been cut into more easily eroded shale or limestone beds.

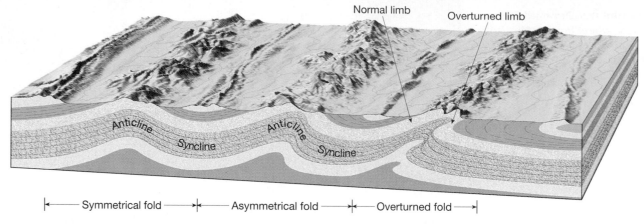

FIGURE 17.6 Block diagram of principal types of folded strata. The upfolded or arched structures are anticlines. The downfolds or troughs are synclines. Notice that the limb of an anticline is also the limb of the adjacent syncline.

Although most folds are caused by horizontal compression, which squeezes strata, some are a consequence of *vertical* displacement. When *upwarping* produces a circular or somewhat elongated structure, the feature is called a **dome** (Figure 17.9A). The Black Hills of western South Dakota are a domal structure. Erosion has stripped away the upwarped sedimentary beds, exposing older igneous and metamorphic rocks in the center (See Figure 17.18).

Structures of similar shape, except *downwarped,* are termed **basins** (Figure 17.9B). Several basins exist in the United States. The basins of Michigan and Illinois, each roughly the size of its state, have very gently sloping beds, similar to vast saucers.

Because large basins contain sedimentary beds sloping at such low angles, they are usually identified by the age of the rocks composing them. The

FIGURE 17.7 Sheep Mountain, a doubly plunging anticline. Note that erosion has cut the flanking sedimentary beds into low ridges that make a "V" pointing in the direction of plunge. (Photo by John S. Shelton)

youngest rocks are found near the center and the oldest rocks are at the flanks. This is just the opposite order of a domal structure such as the Black Hills, where the oldest rocks form the core.

Faults and Joints

Faults are fractures in the crust along which appreciable displacement has occurred, on a scale from centimeters to kilometers. Occasionally, small faults can be recognized in road cuts where sedimentary beds have been offset a few meters, as shown in Figure 17.10. Faults of this scale usually occur as single discrete breaks. By contrast, large faults, like the infamous San Andreas fault in California, have displacements of hundreds of kilometers. These large faults consist of multiple interconnecting fault surfaces, forming *fault zones* that can be several kilometers wide. Often they are easier to identify from high-altitude images than at ground level (see Figure 17.A in Box 17.1).

Movement along a fault can be horizontal, vertical, or a combination. Faults in which the movement is primarily *vertical* are called **dip-slip faults,** because the displacement is along the inclination *(dip)* of the fault plane. Movement along dip-slip faults can be either up or down the fault plane—that is, rocks on one side of the fault can be either pushed up or dropped down with respect to the other side. Thus, two types of dip-slip faults are recognized. To distinguish between the two, we call the rock that is higher than the fault surface the *hanging wall,* and the rock below the fault surface the *footwall.* This nomenclature arose from prospectors and miners who excavated shafts along fault zones because such zones are frequently sites of ore deposition. During these operations, the miners would walk on the rocks below the fault plane (hence, footwall), and hang their lanterns on the rocks above (hence, hanging wall).

FIGURE 17.8 Plunging folds.
A. Idealized view of plunging folds in which a horizontal surface has been added. **B.** View of plunging folds as they might appear after extensive erosion. Notice that in a plunging anticline the outcrop pattern "points" in the direction of the plunge, while the opposite is true of plunging synclines.

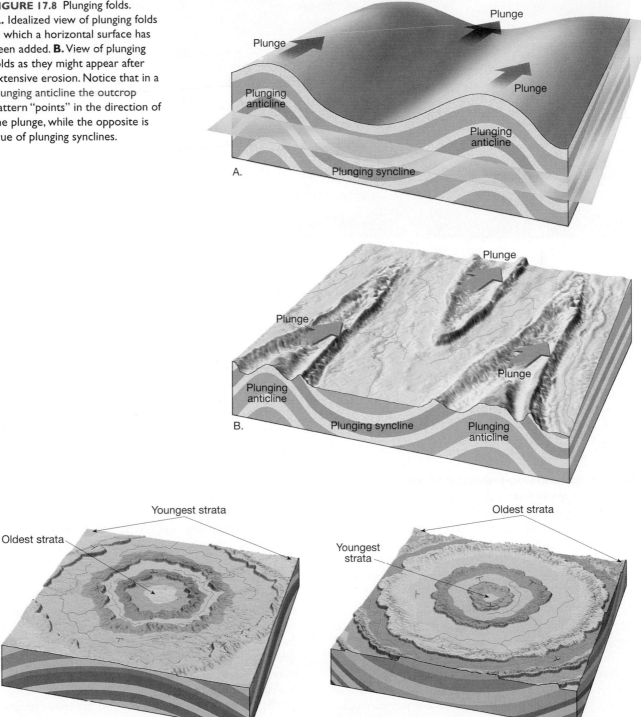

FIGURE 17.9 Gentle upwarping and downwarping of crustal rocks produce domes (**A**) and basins (**B**). Erosion of these structures results in an outcrop pattern that is roughly circular or elongated.

Dip-slip faults are classified as **normal faults** when the hanging wall moves downward relative to the footwall (Figure 17.11A). Conversely, **reverse faults** occur when the hanging wall moves upward relative to the footwall (Figure 17.11B). Reverse faults having a very low angle are also called **thrust faults** (Figure 17.11C). In mountainous regions such as the Alps and the Appalachians, thrust faults have displaced rock as far as 50 kilometers over adjacent strata. Thrust faults of this type result from strong compressional stresses.

FIGURE 17.10 Faulting caused the displacement of these beds. Arrows show relative motion of rock units. Exposure located along a roadcut near Kanab, Utah. (Photo by Tom Bean)

Faults in which the dominant displacement is along the *strike*, or trend, of the fault are called **strike-slip faults** (Figure 17.11D). Many large strike-slip faults are associated with plate boundaries and are called *transform faults*. Transform faults have nearly vertical dips and connect large structures, such as segments of an ocean ridge. The San Andreas fault in California is a well-known transform fault in which the displacement is several hundred kilometers (see Box 17.1). When faults have both vertical and horizontal movement, they are called **oblique-slip faults.**

Extensional and Compressional Forces Fault motion assists the geologist in determining the direction of forces at work within Earth. Normal faults indicate *extensional* stresses that pull the crust apart (Figure 17.11A). This "pulling apart" can be accomplished either by uplifting that causes the surface to stretch and break, or by horizontal forces that actually rip the crust apart.

Normal faulting is known to occur at spreading centers, where plates are diverging. Here a central block called a **graben,** bounded by normal faults, drops as the plates separate (Figure 17.12). These grabens produce an elongated valley bounded by upfaulted structures called **horsts.** The Great Rift Valley of East Africa consists of several large grabens, above which tilted horsts produce a linear mountainous topography (see Figure 16.11). This valley, nearly 6000 kilometers (3700 miles) long, contains the discovery sites of some of the earliest human fossils. Other rift valleys include the Rhine Valley in Germany and the valley of the Dead Sea in the Middle East.

In reverse and thrust faulting, the sections of crust are displaced toward one another, so geologists conclude that *compressional forces* are at work (Figure 17.11 B, C). The primary regions of such activity are convergent zones where plates are colliding.

Compressional forces generally produce folds as well as faults and result in a general thickening and shortening of the crust. As you shall see, most orogenesis occurs in compressional situations.

Joints Among the most common rock structures are fractures called **joints.** Unlike faults, joints are fractures along which *no appreciable displacement* has occurred. Although some joints have random orientation, most occur in roughly parallel groups.

We have already considered two types of joints. Earlier we learned that *columnar joints* form when igneous rocks cool and develop shrinkage fractures, producing elongated, pillarlike columns. Also recall that *sheeting* produces a pattern of gently curved joints that develop more or less parallel to the surface of large exposed igneous bodies such as batholiths. Here the jointing is thought to result from the gradual expansion that occurs as erosion removes the pressure of the overlying load.

In contrast to the situations just described, most joints are produced when rocks are deformed by the stresses associated with crustal movements during mountain building. Extensive joint patterns can also develop in response to relatively subtle and often barely perceptible regional upwarping and downwarping of the crust.

Many rock units are broken by two or even three sets of intersecting joints that divide the rock into numerous regularly shaped blocks. These joint sets often exert a strong influence on other geologic processes. For example, chemical weathering tends to concentrate along joints, and in many areas groundwater movement and the resulting solution activity in soluble rocks is controlled by the joint pattern (Figure 17.13). Moreover, a system of joints can influence the path taken by a stream. The rectangular drainage pattern shown in Figure 9.19 is such a case.

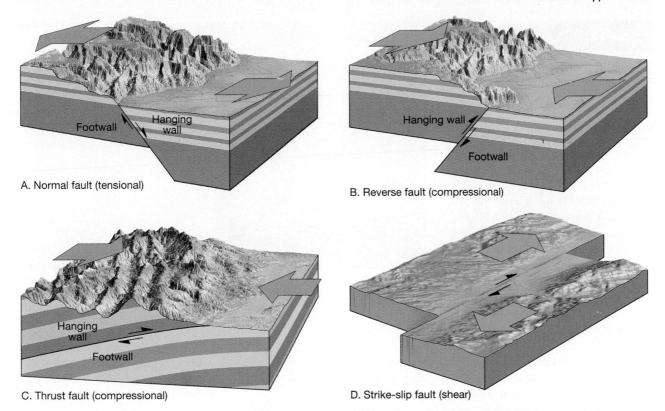

A. Normal fault (tensional)

B. Reverse fault (compressional)

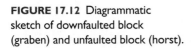

C. Thrust fault (compressional)

D. Strike-slip fault (shear)

FIGURE 17.11 Block diagrams of four types of faults. **A.** Normal fault. **B.** Reverse fault. **C.** Thrust fault. **D.** Strike-slip fault.

FIGURE 17.12 Diagrammatic sketch of downfaulted block (graben) and unfaulted block (horst).

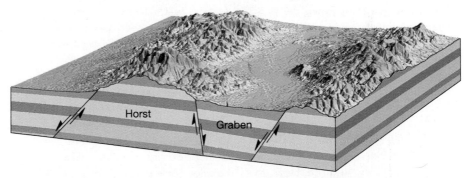

Mountain Types

Now that you have some knowledge of crustal uplift and rock deformation, we will examine different types of mountains and how they come to be.

Mountain Types

No two mountain ranges are exactly alike, but they can be classified according to their structural characteristics. Using this approach, four main categories result: (1) fault-block mountains, (2) upwarped mountains, (3) folded mountains (complex mountains), and (4) volcanic mountains.

Mountain ranges of the same structural type commonly occur in close proximity, forming a *mountain system*. For example, nearly the entire state of Nevada is composed of numerous elongated fault-block mountains that are separated by faulted basins

(Figure 17.14). Further, within any mountainous belt, such as that portion of the American Cordillera in the western United States, mountain ranges representing each of the four groups can be found.

In addition to these basic varieties, some regions exhibit mountainous topography that was produced *without appreciable crustal deformation*. For example, plateaus (areas of high-standing rocks that are essentially horizontal) can become deeply dissected by erosion into rugged, mountainlike terrains. Although these highlands have the topographical expression of mountains, they lack the structures associated with orogenesis.

In the following sections we will examine three of the four basic mountain types: fault-block, upwarped, and, the most common, folded mountains. (Volcanic mountains are treated in detail in Chapter 4.)

Box 17.1 The San Andreas Fault System

The great San Francisco earthquake and fire of 1906 first attracted the public's attention to what is now California's best-known fault: the San Andreas (Figure 17.A). It is North America's largest fault system. Geologic studies after the quake discovered that a displacement up to 5 meters (16 feet) along the fault caused the tremor. It is now known that this event was just one of thousands caused by repeated slippage along the San Andreas throughout its 29-million-year history.

The San Andreas runs for nearly 1300 kilometers (780 miles) through much of western California (Figure 17.B). To the south, it connects with a spreading center in the Gulf of California. To the north, the fault enters the Pacific Ocean at Point Arena, where it is thought to continue northwestward, eventually joining the Mendocino fracture zone.

In its central section, the San Andreas is relatively simple and straight. However, at its north and south ends, several branches spread from the main trace, so that in some areas the fault zone exceeds 100 kilometers (60 miles) in width.

With the development of the plate tectonics theory, geologists began to recognize the significance of this great fault system. The San Andreas fault is a transform boundary separating two crustal plates. The Pacific plate, on the

FIGURE 17.A Aerial view showing the displacement of stream channels along the trace of the San Andreas fault. (Photo by Michael Collier)

west, is slipping and lurching northwestward relative to the North American plate on the east, causing earthquakes along the fault.

Over much of its extent, a linear trough reveals the presence of the San Andreas fault. When viewed from the air, linear scars, offset stream channels, and elongated ponds strikingly mark its trace (Figure 17.A). On the ground, however, surface expressions of the fault are much more difficult to detect.

The San Andreas is undoubtedly the most studied fault system on Earth. Geologists have learned that each fault segment exhibits somewhat different behavior. Some portions creep slowly with little noticeable seismic activity. Other segments regularly slip, producing small earthquakes. Still other segments seem to store elastic energy for

hundreds of years and then rupture to generate great earthquakes. This knowledge is useful when assigning a potential earthquake hazard to a given segment of the fault zone.

Because of its great length and complexity, the San Andreas is more appropriately called a "fault system." It consists of the San Andreas fault and several major branches, including the Hayward, Calaveras, San Jacinto, and Elsinore faults. These major segments, plus a vast number of smaller faults that include the Imperial, San Fernando, and Santa Monica faults, collectively accommodate the relative motion between the North American and Pacific plates. Geologists have determined that the total accumulated displacement from earthquakes and creep exceeds 560 kilometers (340 miles).

Fault-Block Mountains

Most orogenesis occurs in compressional environments. Evidence for this is the predominance of large thrust faults and folded strata in mountainous areas. However, some mountain building is associated with extensional stresses (Figure 17.15). The mountains that form under such circumstances, termed **fault-block mountains,** are bounded on at least one side by normal faults of high to moderate angle.

Some fault-block mountains form in response to broad crustal uplifting, which causes elongation

(stretching) and faulting. Such a situation is exemplified by the fault blocks that rise above the rift valleys of East Africa. Other fault-block mountains appear to develop where isolated blocks are tilted, or lifted nearly vertically, high above adjacent underformed valleys.

The *Basin and Range Province,* a region that encompasses Nevada and portions of Utah, New Mexico, Arizona, and California, contains excellent examples of fault-block mountains. Here the crust has been broken into hundreds of blocks, giving rise to

FIGURE 17.B The San Andreas fault system, part of the boundary between the Pacific and North American plates. Inset shows a few of the many splinter faults that are part of this great fault system.

nearly parallel mountain ranges, averaging about 80 kilometers in length, which rise precipitously above the adjacent sediment-laden basins (Figure 17.15).

No single explanation for the development of the Basin and Range is universally accepted. It is clear, however, that important geological events preceded the period of stretching and faulting. During this earlier time span, a nearly horizontal oceanic plate was subducted eastward, beneath western North America

(Figure 17.16A). This caused compressional stresses in the overlying plate, thickening the crust as far inland as the Rocky Mountains of Colorado.

Then, about 30 million years ago, the western portion of the North American plate began to override a spreading center in the Pacific Basin. This attached the southwestern portion of the United States to the Pacific plate, which was moving northwesterly. This movement continues today and is

FIGURE 17.13 Chemical weathering is enhanced along joints in granitic rocks near the top of Lembert Dome, Yosemite National Park. (Photo by E. J. Tarbuck)

most pronounced along the San Andreas fault, where portions of southern California and Mexico's Baja Peninsula have migrated several hundred kilometers toward the northwest.

In addition to creating movement along the San Andreas fault system, this change in the direction of plate motion also may have stretched and fractured the crust in the region now occupied by the Basin and Range Province, generating its fault-block topography.

Some geologists disagree with part of the preceding scenario. Instead, they suggest that extension in the Basin and Range was triggered when the horizontal oceanic slab that had been thrust under the continent began to sink (Figure 17.16B). Sinking of the oceanic slab produced an upwelling of magma from the asthenosphere. The buoyancy of the magma caused upwarping and tensional fracturing in the overlying crust. This event was associated with volcanism and east-west stretching of the crust.

Regardless of which explanation is correct, if either, there is general agreement that the fault-block mountains in this region resulted from tensional forces that stretched the crust by nearly 150 kilometers.

Other fault-block mountains in the United States are the Teton Range of Wyoming (see Figure 17.17) and the Sierra Nevada of California. Both are faulted along their eastern flanks, which were uplifted as the blocks tilted downward to the west. Looking west from Jackson Hole, Wyoming, and Owens Valley, California, the eastern fronts of these ranges (the Tetons and the Sierra Nevada, respectively) rise over 2 kilometers, making them two of the most precipitous mountain fronts in the United States (Figure 17.17).

Upwarped Mountains

Upwarped mountains are caused by a broad arching of the crust or, in some instances, because of great vertical displacement along a high-angle fault. The Black Hills in western South Dakota and the Adirondack Mountains in upstate New York are examples of arching. They consist of older igneous and metamorphic bedrock that was once eroded flat and subsequently mantled with sediment. As these regions were upwarped, erosion removed the veneer of sedimentary strata, leaving a core of igneous and metamorphic rocks standing above the surrounding terrain (Figure 17.18).

Another example of upwarped mountains is that portion of the Rockies extending from southern Montana through Colorado and into New Mexico.

Folded Mountains

The largest and most complex mountain systems are **folded mountains** (also called **complex mountains**). Although folding is often more conspicuous, faulting, metamorphism, and igneous activity are always present in varying degrees. Most major mountain belts, including the Alps, Urals, Himalayas, Appalachians, and northern Rockies, are of this type (see Box 17.2. p. 361). Folded mountains represent the world's major mountain systems, so the process of mountain building is usually described in terms of their formation. Thus, the next section is devoted to the evolution these majestic and always complex mountain systems.

Mountain Building

Mountains are spectacular features that often rise several hundred meters or more above the surrounding terrain. All mountain systems show evidence of enormous forces that have folded, faulted, and generally deformed large sections of Earth's crust (Figure 17.19). Although this folding and faulting have contributed to the majestic structure of mountains, much of the credit for their beauty must be given to erosion by running water and glacial ice, which sculpture these uplifted masses in an unending effort to lower them to sea level. **Mountain building,** however, refers to the processes that uplift the mountains.

Mountain building has operated during the recent geologic past in several locations around the world. These relatively young mountainous belts include: the American Cordillera, which runs along the western margin of the Americas from Cape Horn to Alaska (Rockies, Andes); the Alpine-Himalayan chain, which extends from the Mediterranean through Iran to northern India and into Indochina;

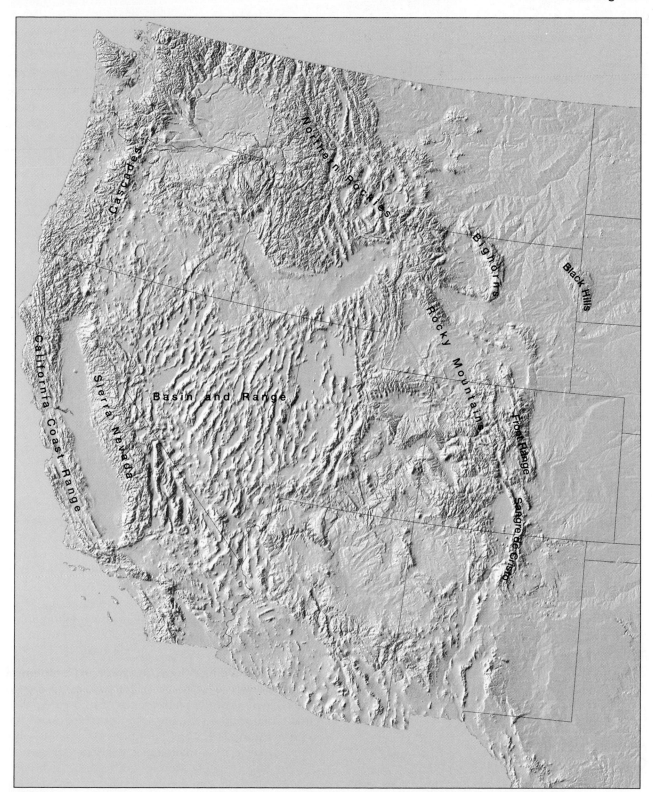

FIGURE 17.14 Map of major mountainous landforms of the western United States. (After Thelin and Pike, USGS)

and the mountainous terrains of the western Pacific, which include mature island arcs such as Japan, the Philippines, and Sumatra. Most of these young mountain belts have come into existence within the last 100 million years. Some, including the Himalayas, began their growth as recently as 45 million years ago.

In addition to these recently formed mountains, several chains of much older mountains exist.

FIGURE 17.15 Normal faulting in the Basin and Range Province. Here, tensional stresses have elongated and fractured the crust into numerous blocks. Movement along these fractures has tilted the blocks, producing parallel mountain ranges called fault-block mountains. (Photo by Michael Collier)

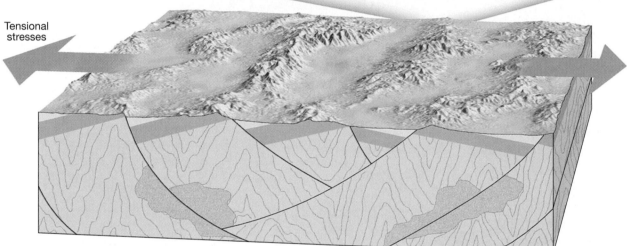

Tensional stresses

Although these structures are deeply eroded and topographically less prominent, they clearly possess the same structural features found in younger mountains. Typical of this older group are the Appalachians in the eastern United States and the Urals in Russia.

The first encompassing explanation of mountain building came a little over three decades ago as part of plate tectonic theory. It is now known that Earth's major mountain systems formed along convergent plate boundaries. Recall that convergence can occur between one oceanic and one continental plate, between two oceanic plates, or between two continental plates. We will consider these sites of mountain building in the following sections.

Mountain Building at Convergent Boundaries

To unravel the events that produce mountains, many studies have been conducted along active subduction zones where plates are converging. At most modern-day subduction zones, volcanic arcs are forming. This situation is typified by Alaska's Aleutian Islands and by the Andes Mountains of western South America. Although all volcanic arcs are similar,

Aleutian-type subduction zones occur where *two oceanic plates* converge (Figure 17.20). *Andean-type subduction zones* are situated where *oceanic crust is being thrust beneath a continental mass.* Consequently, the events that are generating these volcanic arcs follow different evolutionary paths. Further, although the development of a volcanic arc does result in the formation of mountainous topography, this activity is viewed as just one of the phases in the development of a major mountain belt.

Mountain Building Where Oceanic and Continental Crust Converge Mountain building along continental margins involves the convergence of an oceanic plate and a plate whose leading edge contains continental crust. Exemplified by the Andes Mountains, this *Andean type* of convergence generates structures resembling those of a developing volcanic island arc.

The first stage in the development of an Andean-type mountain belt occurs prior to the formation of the subduction zone. During this period the continental margin is *passive*. It is not a plate boundary but a part of the same plate as the adjoining oceanic crust. The East Coast of the United States is a present-day example of a passive continental margin.

FIGURE 17.16 One proposal for the formation of the Basin and Range Province. Top: Nearly horizontal subduction of an oceanic plate produced compressional stresses that generally thickened the crust in the Basin and Range. Bottom: Sinking of this oceanic slab allowed for the upwelling of hot material from the asthenosphere. The buoyancy of the warm material caused upwarping and tensional fracturing in the crust above. This event was associated with volcanism and east-west extension of the crust by nearly 150 kilometers.

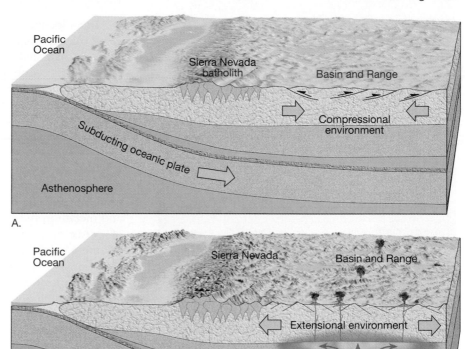

Here, as at other passive continental margins surrounding the Atlantic, deposition of sediment along the continental margin is producing a thick wedge of sandstones, limestones, and shales (Figure 17.21A).

At some point the continental margin becomes active. A subduction zone forms and the deformation process begins (Figure 17.21B). A good place to examine such an active continental margin is the western coast of South America. Here, an oceanic plate is being subducted eastward beneath the South American plate along the Peru–Chile trench (see Figure 16.8).

FIGURE 17.17 The Teton Range in Grand Teton National Park, Wyoming, is an example of fault-block mountains. (Photo by Carr Clifton)

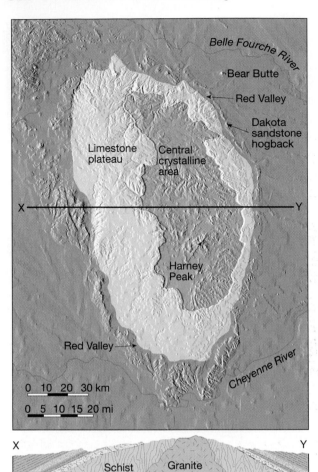

FIGURE 17.18 The Black Hills of South Dakota, a large domal structure with resistant igneous and metamorphic rocks exposed in the core. (After Arthur N. Strahler, *Introduction to Physical Geography,* 3rd ed., New York: John Wiley & Sons, 1973. Reprinted by permission.)

In an idealized Andean-type subduction, convergence of the continental block and the subducting oceanic plate leads to deformation and metamorphism of the continental margin. Once the oceanic plate descends to about 100 kilometers, partial melting generates magma that slowly migrates upward, intruding and further deforming these strata (Figure 17.21B). During the development of the volcanic arc, sediment derived from the land as well as that scraped from the subducting plate becomes plastered against the landward side of the trench. This chaotic accumulation of sedimentary and metamorphic rocks, including occasional scraps of ocean crust, is called an **accretionary wedge** (Figure 17.21B). Prolonged subduction can build an accretionary wedge that is large enough to stand above sea level (Figure 17.21C).

Andean-type mountain belts, like island arcs, are composed of two roughly parallel zones. The landward segment is the volcanic arc, made up of volcanoes and large bodies of intrusive igneous rock, intermixed with high-temperature metamorphic rocks. The seaward segment is the accretionary wedge. It consists of folded, faulted, and metamorphosed sediments and volcanic debris (Figure 17.21C).

One of the best examples of an inactive Andean-type mountain belt is found in the western United States. It includes the Sierra Nevada and the Coast Ranges in California (Figure 17.22). These parallel mountain belts were produced by the subduction of a portion of the Pacific Basin under the western edge of the North American plate. The Sierra Nevada batholith is a remnant of a portion of the volcanic arc that was produced by several surges of magma over

FIGURE 17.19 Highly deformed sedimentary strata in the Rocky Mountains of British Columbia. These sedimentary rocks are continental shelf deposits that were displaced toward the interior of Canada by low-angle thrust faults. (Photo by John Montagne)

Box 17.2   The Rocky Mountains

The portion of the Rocky Mountains from southern Montana to New Mexico was produced during a deformation period known as the *Laramide Orogeny.* This event, which created some of the most picturesque United States scenery, peaked about 60 million years ago. The mountain ranges it generated include the Front Range of Colorado, the Sangre de Cristo of New Mexico and Colorado, and the Bighorns of Wyoming (Figure 17.C).

These mountains are structurally much different from the northern Rockies (Idaho, western Wyoming and Montana, and Canada). The northern Rockies are *folded mountains,* thick sequences of sedimentary rocks that were deformed by folding and low-angle thrust faulting. The middle and southern Rockies are *upwarped mountains,* pushed almost vertically upward as part of broad upwarping of the crust.

In general, these mountains consist of ancient basement rocks overlain by relatively thin layers of younger strata. Since the time of deformation, much of the sedimentary cover has been eroded from the highest portions of the uplifted blocks, exposing their igneous and metamorphic cores. Examples include a number of granitic outcrops that project as steep summits, such as Pikes Peak and Longs Peak in Colorado's Front Range. In many areas, remnants of the sedimentary strata that once covered this region are visible as prominent angular ridges called *hogbacks.*

Although the Rockies have been extensively studied for over a century,

FIGURE 17.C The spectacular Maroon Bells are part of the Colorado Rockies. (Photo by Larry Ulrich/Tony Stone Images)

much debate continues over their uplift mechanisms. According to the plate tectonics model, most mountain belts are produced along continental margins at convergent plate boundaries. Here, crustal buckling generates thick sequences of folded, faulted, and metamorphosed strata that are often intruded by massive igneous bodies.

However, this model does not fit the middle and southern Rocky Mountains, which consist of elongated, uplifted blocks of Precambrian basement rocks separated by sediment-filled basins.

One popular hypothesis is that a subducted plate of oceanic lithosphere remained nearly horizontal as it pushed eastward under North Amer-

ica as far inland as the Black Hills of South Dakota (see Figure 17.14). As the subducted slab scraped beneath North America, compression stresses shortened and thickened the rocks in the lower crust. Furthermore, this event locally uplifted blocks of ancient basement rocks along high-angle faults to produce the Rockies and intermountain basins.

Thus, the Laramide Orogeny may be associated with a special type of convergent plate boundary. Whereas plate subduction along a typical Andean-type plate boundary occurs at a steep angle, the middle and southern Rocky Mountains may have been produced by the nearly horizontal subduction of an oceanic plate.

tens of millions of years. Subsequent uplifting and erosion have removed most evidence of past volcanic activity and exposed a core of crystalline metamorphic and igneous rocks.

In the trench region, sediments scraped from the subducting plate, plus those provided by the eroding

volcanic arc, were intensely folded and faulted into the accretionary wedge, which presently constitutes part of California's Coast Ranges. Uplifting of the Coast Ranges took place only recently, as evidenced by the unconsolidated sediments that still mantle portions of these highlands.

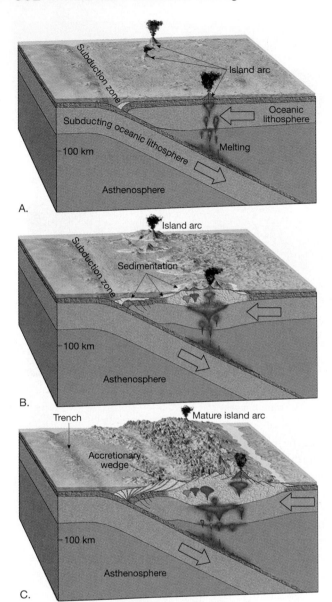

FIGURE 17.20 The development of a mature volcanic island arc by the convergence of two oceanic plates. These Aleutian-type subduction zones produce a mountainous topography that is similar to the volcanic arc that comprises the Aleutian islands of Alaska.

Mountain Building Where Continents Converge

So far, we have discussed the formation of mountain belts where the leading edge of just one of the two converging plates contained continental crust. However, it is possible for two converging plates to be carrying continental crust. Because continental lithosphere is too buoyant to undergo any appreciable subduction, a collision between the continental fragments eventually results (Figure 17.23).

An example of such a collision occurred about 45 million years ago when India collided with the Eurasian plate. India was once part of Antarctica but split from that continent and moved a few thousand kilometers due north. The result of this collision was the formation of the spectacular Himalaya Mountains and the Tibetan Highlands.

Although most of the oceanic crust that separated these landmasses prior to the collision was subducted, some was caught up in the squeeze, along with sediment that lay offshore (Figure 17.23). Today it is elevated high above sea level. Geologists think that, following such a collision, the subducted oceanic plate decouples from the rigid continental plate and continues its descent into the mantle.

A similar but much older collision is believed to have taken place when the European continent collided with the Asian continent to produce the Ural Mountains, which extend north-south through Russia. Prior to the discovery of plate tectonics, geologists had difficulty explaining the existence of mountain ranges, like the Urals, which are located deep within continental interiors. How could thousands of meters of marine sediment be deposited and then become highly deformed while situated in the middle of a large landmass?

Other mountain ranges showing evidence of continental collisions are the Alps and the Appalachians. The Appalachians resulted from collisions between North America, Europe, and northern Africa. Although they have since separated, these landmasses were juxtaposed as part of the supercontinent Pangaea less than 200 million years ago. Detailed studies in the southern Appalachians indicate that the formation of this mountain belt was more complex than once thought. Rather than forming during a single continental collision, the Appalachians resulted from several distinct episodes of mountain building that occurred over a period of nearly 300 million years.

Mountain Building and Continental Accretion

Plate tectonics theory originally suggested two mechanisms for mountain building: (1) continental collisions were proposed to explain the formation of such mountainous terrains as the Appalachians, Himalayas, and Urals; and (2) subduction of oceanic lithosphere was thought to be the underlying tectonic process for many circum-Pacific mountain chains, as typified by the Andes. Recent investigations, however, indicate a third mechanism of mountain building: *Smaller crustal fragments collide and accrete to continental margins.* Through this process of collision and accretion, many of the mountainous regions rimming the Pacific have been generated.

What is the nature of these small crustal fragments, and where do they come from? Researchers

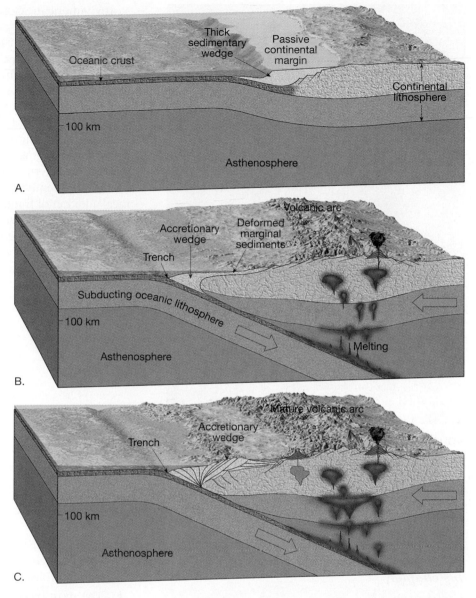

FIGURE 17.21 Mountain building along an Andean-type subduction zone. **A.** Passive continental margin with extensive wedge of sediments. **B.** Plate convergence generates a subduction zone, and partial melting produces a developing volcanic arc. **C.** Continued convergence and igneous activity further deform and thicken the crust, elevating the mountain belt, while the accretionary wedge grows.

suggest that, prior to their accretion to a continental block, some of the fragments may have been microcontinents similar to the present-day island of Madagascar near Africa. Others were island arcs like Japan, the Philippines, and the Aleutian Islands. Still others may have been submerged crustal fragments that extended high above the ocean floor. It is believed that these fragments originated as submerged continental fragments, extinct volcanic island arcs, or submerged volcanic chains associated with hot-spot activity.

Accretion of Foreign Terranes The widely accepted view today is that, as oceanic plates move, they carry the embedded oceanic plateaus or microcontinents to a subduction zone. Here the upper portions of these thickened zones are peeled from the descending plate and thrust in relatively thin sheets onto the adjacent

continental block. This newly added material increases the width of the continent. The material may later be overridden and displaced further inland by colliding with other fragments.

Geologists refer to these accreted crustal blocks as terranes. Simply, the term **terrane** designates any crustal fragment whose geologic history is distinct from that of the adjoining terranes. (Do not confuse the term *terrane* with the word *terrain,* which indicates the topography or lay of the land.) Terranes come in varied shapes and sizes; some are no larger than volcanic islands. Others, such as the one composing the entire Indian subcontinent, are very large.

Accretion and Mountain Building The idea that mountain building occurs in association with the accretion of small crustal fragments to a continental mass arose principally from studies conducted in the

FIGURE 17.22 Map of the California Coast Ranges and the Sierra Nevada.

North American Cordillera (Figure 17.24). Some mountainous areas, principally those of Alaska and British Columbia, contain fossil and magnetic evidence that these strata formed much nearer the equator.

It is now assumed that many other terranes found in the North American Cordillera were once scattered throughout the eastern Pacific, much as we find island arcs and oceanic plateaus distributed in the western Pacific today. Over the last 200 million years, these fragments migrated toward and collided with the western coast of North America (Figure 17.24). Apparently, this activity resulted in the piecemeal addition of fragments to the entire Pacific Coast, from the Baja Peninsula to northern Alaska. In a like manner, many modern microcontinents will eventually be accreted to active continental margins, thus resulting in the formation of new mountainous belts.

The Chapter in Review

The following statements are intended to help you review the primary objectives presented in this chapter.

- The name for the processes that collectively produce a mountain system is *orogenesis*. Earth's less dense crust is believed to float on top of the denser rocks in the mantle, much like wooden blocks floating in water. This concept of a floating crust in gravitational balance is called *isostasy*. As erosion lowers the peaks of mountains, *isostatic adjustment* gradually raises the mountains in response.

- Rocks that are subjected to stresses greater than their own strength begin to deform, usually by *folding* or *fracturing*. During *elastic deformation* a rock behaves like a rubber band and will return to nearly its original size and shape when the stress is removed. Once the *elastic limit* is surpassed, rocks either deform plastically or

fracture. *Plastic deformation* results in permanent changes in the size and shape of a rock through folding and flowing.

- The most basic geologic structures associated with rock deformation are *folds* (flat-lying sedimentary and volcanic rocks bent into a series of wavelike undulations) and *faults*. The two most common types of folds are *anticlines*, formed by the upfolding, or arching, of rock layers, and *synclines*, which are downfolds. Most folds are the result of horizontal *compressional stresses*. Folds can be *symmetrical, asymmetrical*, or, if one limb has been tilted beyond the vertical, *overturned*. *Domes* (upwarped structures) and *basins* (downwarped structures) are circular or somewhat elongated folds formed from vertical displacements of rocks.

- Faults are fractures in the crust along which appreciable displacement has occurred. Faults in which the movement is primarily vertical are

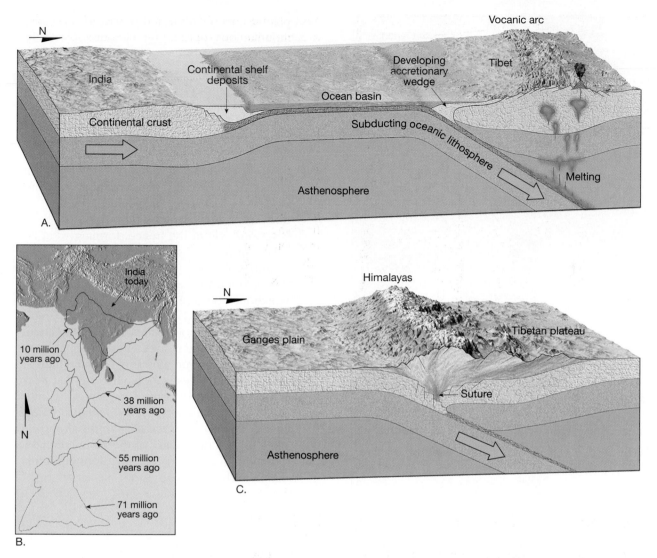

FIGURE 17.23 Simplified diagrams showing the northward migration and collision of India with the Eurasian plate. **A.** Converging plates generated a subduction zone, while partial melting of the subducting oceanic slab produced a volcanic arc. Sediments scraped from the subducting plate were added to the accretionary wedge. **B.** Position of India in relation to Eurasia at various times. (Modified after Peter Molnar) **C.** Eventually the two landmasses collided, deforming and elevating the accretionary wedge and continental shelf deposits. In addition, slices of the Indian crust were thrust up onto the Indian plate.

called *dip-slip faults*. Dip-slip faults include both *normal* and *reverse faults*. Low-angle reverse faults are also called *thrust faults*. In *strike-slip faults*, horizontal movement causes displacement along the trend, or *strike*, of the fault. Normal faults indicate *tensional stresses* that pull the crust apart. Along the spreading centers of plates, divergence can cause a central block called a *graben*, bounded by normal faults, to drop as the plates separate. Reverse and thrust faulting indicate that *compressional forces* are at work. A *joint* is a fracture along which no appreciable displacement has occurred.

• Mountains can be classified according to their structural characteristics. The four main mountain types are (1) *fault-block mountains* (e.g., Basin and Range Province and Sierra Nevada of California) are associated with tensional stresses and are bounded at least on one side by normal faults; (2) *folded mountains* (e.g., Alps, Urals, Himalayas, Appalachians, and northern Rockies) are the most complex and form most of the major mountain belts; (3) *upwarped mountains* (e.g., Black Hills and Adirondack Mountains) are caused by broad arching; and (4) *volcanic mountains*. Some regions, such as

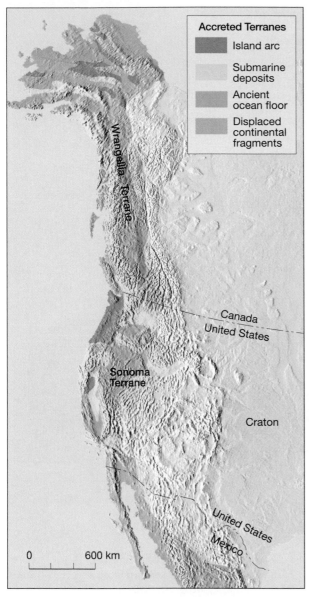

Accreted Terranes

Island arc

Submarine deposits

Ancient ocean floor

Displaced continental fragments

Wrangellia Terrane

Canada
United States

Sonoma Terrane

Craton

United States
Mexico

0   600 km

FIGURE 17.24 Map showing terranes thought to have been added to western North America during the past 200 million years. (Redrawn after D. R. Hutchinson and others)

plateaus deeply dissected by erosion, exhibit mountainous topography without appreciable crustal deformation.

- *Major mountain systems form along convergent plate boundaries. Andean-type mountain building* along continental margins involves the convergence of an oceanic plate and a plate whose leading edge contains continental crust. At some point in the formation of Andean-type mountains a *subduction zone* forms along with a *volcanic arc*. Sediment from the land as well as material scraped from the subducting plate become plastered against the landward side of the trench, forming what is called an *accretionary wedge*. One of the best examples of an inactive Andean-type mountain belt is found in the western United States, which includes the Sierra Nevada and the Coast Ranges in California. *Continental collisions,* in which both plates are carrying continental crust, have resulted in the formation of the Himalaya Mountains and Tibetan Highlands. Recent investigations indicate that *accretion,* a third mechanism of orogenesis, takes place where *smaller crustal fragments collide and accrete to continental margins* along some plate boundaries. Many of the mountainous regions rimming the Pacific have formed in this manner. The accreted crustal blocks are referred to as *terranes*.

Key Terms

accretionary wedge (p. 360)
anticline (p. 349)
basin (p. 350)
dip-slip fault (p. 350)
dome (p. 350)
fault (p. 350)
fault-block mountains (p. 354)

fold (p. 349)
folded mountains (complex mountains) (p. 356)
graben (p. 352)
horst (p. 352)
isostasy (p. 347)
isostatic adjustment (p. 347)
joint (p. 352)

mountain building (p. 356)
normal fault (p. 351)
oblique-slip fault (p. 352)
orogenesis (p. 346)
reverse fault (p. 351)
strike-slip fault (p. 352)
syncline (p. 349)

terrane (p. 363)
thrust fault (p. 351)
upwarped mountains (p. 356)

Questions for Review

1. State two lines of evidence that support the concept of crustal uplift. Can you think of another?

2. What happens to a floating object when weight is added? Subtracted? How do these principles apply to changes in the elevations of mountains? What term is applied to the adjustment that causes crustal uplift of this type?

3. List one line of evidence that supports the idea of a floating crust.

4. What conditions favor rock deformation by folding? By faulting?

5. The San Andreas fault is an excellent example of a _____ fault.

6. Compare the movement of normal and reverse faults. What type of force produces each?

7. At which of the three types of plate boundaries does normal faulting predominate? Reverse faulting? Strike-slip faulting?

8. Describe a horst and a graben. Explain how a graben valley forms and name one.

9. Compare and contrast anticlines and synclines. Domes and basins. Anticlines and domes.

10. Although we classify many mountains as folded, why might this description be misleading?

11. What type of faults are associated with fault-block mountains?

12. During the formation of fault-block mountains, are the forces that act upon the region compressional or extensional?

13. There are four types of mountain ranges. Name the type of mountain range exemplified by each of the following:
    a. Black Hills   **b.** Basin and Range
    c. Adirondacks   **d.** Cascades
    e. Appalachians   **f.** Tetons
    g. Bighorns   **h.** Himalayas
    i. Front Range

14. What do we call the site where sediments are deposited along the margin of the continent where they have a good chance of being squeezed into a mountain range?

15. Which type of plate boundary is most directly associated with mountain building?

16. Describe an accretionary wedge and explain its formation.

17. Suppose a sliver of oceanic crust was discovered 1000 kilometers into the interior of a continent. Would this support or refute the theory of plate tectonics? Why?

18. Why might it have been difficult for geologists to conclude that the Appalachian Mountains were formed by plate collision if examples like the Himalayas did not exist?

19. How does the plate tectonics theory explain the existence of fossil marine life on top of the Ural Mountains?

Testing What You Have Learned

To test your knowledge of the material presented in this chapter, answer the following questions:

Multiple-Choice Questions

1. Faults in which the dominant displacement is along the trend of the fault are called _____ faults.
   a. strike-slip   **c.** reverse   **e.** inclined-slip
   b. normal   **d.** complex

2. Which one of the following is NOT one of the four main types of mountains?
   a. divergent   **c.** upwarped   **e.** folded
   b. fault-block   **d.** volcanic

3. Faults in which the movement is primarily vertical are called _____ faults.
   a. strike-slip   **c.** overturned   **e.** folded
   b. complex   **d.** dip-slip

4. The _____ Mountains formed as the result of continental collision.
   a. Sierra Nevada   **c.** Ethiopian   **e.** Rocky
   b. Andes   **d.** Himalaya

5. Dip slip faults in which the hanging wall moves upward relative to the footwall are called _____ faults.
   a. strike-slip   **c.** reverse   **e.** oblique-slip
   b. normal   **d.** complex

6. Normal faults indicate _____ stresses that pull the crust apart.
   a. vertical   **c.** complex   **e.** compressional
   b. extensional   **d.** reverse

7. The chaotic accumulation of sedimentary and metamorphic rocks that are plastered against the leeward side of a trench during the development of a volcanic arc is referred to as a(n) _____ wedge.
   a. accretionary   **c.** mixed   **e.** deformed
   b. subduction   **d.** compressional

8. A(n) _____ is the most common structure formed by the upfolding, or arching, of rock layers.
   a. normal fault   **c.** anticline   **e.** reverse fault
   b. syncline   **d.** graben

9. The Alps, Urals, Himalayas, and Appalachians are all examples of _____ mountains.
   a. upwarped   **c.** folded   **e.** displaced
   b. volcanic   **d.** fault-block

10. Reverse faults having a very low angle are also called _____ faults.
   a. complex   **c.** anticlinal   **e.** thrust
   b. synclinal   **d.** displaced

Fill-In Questions

11. The concept of _____ refers to a floating crust in gravitational balance.
12. At spreading centers, where plates are diverging, a central block called a _____, bounded by normal faults, drops as the plates separate.
13. Most folds are the result of _____ stresses.
14. The name for the processes that collectively produce a mountain system is _____.
15. _____ are fractures in the crust along which appreciable displacement has occurred.

True/False Questions

16. Changes resulting from plastic deformation of rock are reversible. _____
17. In a basin. the youngest rocks are located near the center and the oldest rocks are at the flanks. _____
18. Andean-type subduction zones are situated where oceanic crust is being thrust beneath a continental mass. _____
19. When the weight of a mountain is removed by erosion, crustal subsidence occurs. _____
20. The outcrop pattern of an eroded, plunging anticline points in the direction it is plunging. _____

Answers

1.a; 2.a; 3.d; 4.d; 5.c; 6.b; 7.a; 8.c; 9.c; 10.e; 11. isostasy; 12. graben; 13.compressional; 14. orogenesis; 15. faults; 16.F; 17.T; 18.T; 19.F; 20.T

Geologic Time

Focus on Learning

To assist you in learning the important concepts in this chapter, you will find it helpful to focus on the following questions:

- What are the two types of dates used by geologists to interpret Earth history?

- What are the laws, principles, and techniques used to establish relative dates?

- What are fossils? What are the conditions that favor the preservation of organisms as fossils?

- How are fossils used to correlate rocks of similar ages that are in different places?

- What is radioactivity and how are radioactive isotopes used in radiometric dating?

- What is the geologic time scale and what are its principal subdivisions?

- Why is it difficult to assign reliable absolute dates to samples of sedimentary rock?

The strata exposed in the Grand Canyon contain clues to hundreds of millions of years of Earth history. Sunset viewed from Yaki Point. (Photo by Tom Bean)

In 1869, John Wesley Powell, who was later to head the U.S. Geological Survey, led a pioneering expedition down the Colorado River and through the Grand Canyon (Figure 18.1). Writing about the rock layers that were exposed by the downcutting of the river, Powell said that "the canyons of this region would be a Book of Revelations in the rock-leaved Bible of geology." He was undoubtedly impressed with the millions of years of Earth history exposed along the walls of the Grand Canyon.

Powell realized that the evidence for an ancient Earth is concealed in its rocks. Like the pages in a long and complicated history book, rocks record the geological events and changing life-forms of the past. The book, however, is not complete. Many pages, especially in the early chapters, are missing. Others are tattered, torn, or smudged. Yet enough of the book remains to allow much of the story to be deciphered.

Interpreting Earth history is a prime goal of the science of geology. Like a modern-day sleuth, the geologist must interpret clues found preserved in the rocks. By studying rocks, especially sedimentary rocks, and the features they contain, geologists can unravel the complexities of the past.

Geological events by themselves, however, have little meaning until they are put into a time perspective. Studying history, whether it be the Civil War or the Age of Dinosaurs, requires a calendar. Among geology's major contributions is a calendar called the *geologic time scale* and the discovery that Earth history is exceedingly long.

Relative Dating—Key Principles

During the late 1800s and early 1900s, various attempts were made to determine the age of Earth. Although some of the methods appeared promising at the time, none proved to be reliable. What these scientists were seeking was an **absolute date.** Such dates pinpoint the time in history when something took place—for example, the extinction of the dinosaurs about 66 million years ago. Today our understanding of radioactivity allows us to accurately determine absolute dates for rock units that represent important events in Earth's distant past. We will study radioactivity later in this chapter. Prior to the discovery of radioactivity, geologists had no accurate and dependable method of absolute dating and had to rely solely on relative dating.

Relative dating means placing rocks in their proper *sequence of formation*—which formed first, second, third, and so on. Relative dating cannot tell us how long ago something took place, only that it followed one event and preceded another. The relative dating techniques that were developed are valuable and still widely used. Absolute dating methods

A.

B.

FIGURE 18.1 A. Start of the expedition from Green River station in Wyoming, shown in a drawing from Powell's 1875 book. **B.** Major John Wesley Powell, pioneering geologist and the second director of the U.S. Geological Survey. (Courtesy of the U.S. Geological Survey)

did not replace these techniques; they simply supplemented them. To establish a relative time scale, a few simple principles or rules had to be discovered and applied. Although they may seem obvious to us today, they were major breakthroughs in thinking at the time, and their discovery and acceptance was an important scientific achievement.

Law of Superposition

Nicolaus Steno, a Danish anatomist, geologist, and priest (1636–1686), is credited with being the first to recognize a sequence of historical events in an outcrop of sedimentary rock layers. Working in the mountains of western Italy, Steno applied a very simple rule that has come to be the most basic principle of relative dating—the **law of superposition.** The law simply states that in an underformed sequence of sedimentary rocks, each bed is older than the one above it and younger than the one below. Although it may seem obvious that a rock layer could not be deposited unless it had something older beneath it for support, it was not until 1669 that Steno clearly stated the principle. This rule also applies to other surface-deposited materials such as lava flows and beds of ash from volcanic eruptions. Applying the law of superposition to the beds exposed in the upper portion of the Grand Canyon (Figure 18.2), you can easily place the layers in their proper order. Among those that are shown, the sedimentary rocks

in the Supai Group must be the oldest, followed in order by the Hermit Shale, Coconino Sandstone, Toroweap Formation, and Kaibab Limestone.

Principle of Original Horizontality

Steno is also credited with recognizing the **principle of original horizontality.** This principle simply states that most layers of sediment are deposited in a horizontal position. Thus, if we observe rock layers that are flat, it means they have not been disturbed, and still have their *original* horizontality (Figure 18.3A). But if they are folded or inclined at a steep angle, they must have been moved into that position by crustal disturbances sometime after their deposition (Figure 18.3B).

Principle of Cross-Cutting Relationships

When a fault cuts through other rocks, or when magma intrudes and crystallizes, we can assume that the fault or intrusion is younger than the rocks affected. For example, in Figure 18.4, the faults and dikes clearly must have occurred *after* the sedimentary layers were deposited.

This is the **principle of cross-cutting relationships.** By applying the cross-cutting principle you can see that fault A occurred *after* the sandstone layer was deposited, because it "broke" the layer. However, fault A occurred *before* the conglomerate was laid down, because that layer is unbroken.

A.

B.

FIGURE 18.2 Applying the law of superposition to these layers exposed in the upper portion of the Grand Canyon, the Supai Group is oldest and the Kaibab Limestone is youngest. (Photo by E. J. Tarbuck)

A.

B.

FIGURE 18.3 A. The principle of original horizontality states that most layers of sediment are deposited in a nearly horizontal position. **B.** When we see folded rock layers such as these, we can assume they must have been moved into that position by crustal disturbances *after* their deposition. (Photos by E. J. Tarbuck)

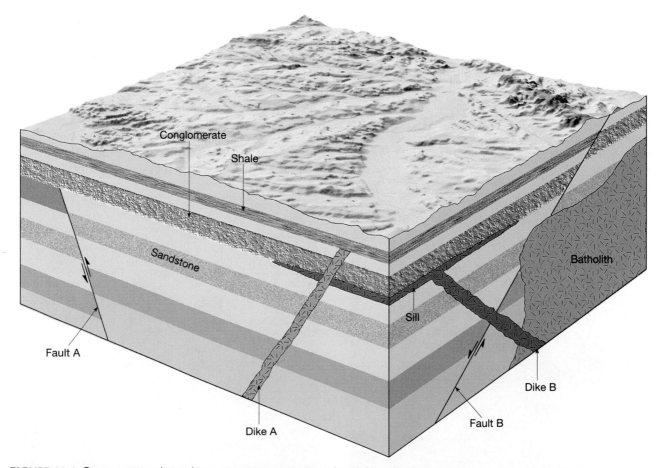

FIGURE 18.4 Cross-cutting relationships are an important principle used in relative dating. An intrusive rock body is younger than the rocks it intrudes. A fault is younger than the rock layers it cuts.

We can also state that dike B and its associated sill are older than dike A, because dike A cuts the sill. In the same manner, we know that the batholith was emplaced after movement occurred along fault B, but before dike B was formed. This is true because the batholith cuts across fault B and dike B cuts across the batholith.

Inclusions

Sometimes inclusions can aid the relative dating process. **Inclusions** are pieces of one rock unit that are contained within another. The basic principle is logical and straightforward. The rock mass adjacent to the one containing the inclusions must have been there first in order to provide the rock fragments. Therefore, the rock mass containing inclusions is the younger of the two. Figure 18.5 provides an example. Here the inclusions of granite in the adjacent sedimentary layer indicate the sedimentary layer was deposited on top of a weathered mass of granite rather than being intruded from below by a mass of magma that later crystallized.

Unconformities

When we observe layers of rock that have been deposited essentially without interruption, we call them **conformable.** Many areas exhibit conformable beds representing certain spans of geologic time. However, no place on Earth has a complete set of conformable strata.

Throughout Earth history, the deposition of sediment has been interrupted again and again. All such breaks in the rock record are termed unconformities.

An **unconformity** represents a long period during which deposition ceased, erosion removed previously formed rocks, and then deposition resumed. In each case Earth's crust has undergone uplift and erosion, followed by subsidence and renewed sedimentation. Unconformities are important features because they represent significant geologic events in Earth history. Moreover, their recognition helps us identify what intervals of time are not represented by strata and thus are missing from the geologic record.

The rocks exposed in the Grand Canyon of the Colorado River represent a tremendous span of geologic history. It is a wonderful place to take a trip through time. The canyon's colorful strata record a long history of sedimentation in a variety of environments—advancing seas, rivers and deltas, tidal flats, and sand dunes. But the record is not continuous. Unconformities represent vast amounts of time that have not been recorded in the canyon's layers. Figure 18.6 is a geologic cross section of the Grand Canyon. Refer to it as you read about the three basic types of unconformities: angular unconformities, disconformities, and nonconformities.

Angular Unconformity Perhaps the most easily recognized unconformity is an **angular unconformity.** It consists of tilted or folded

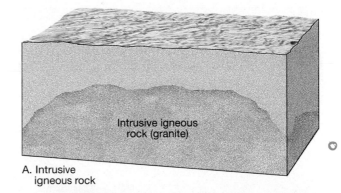

A. Intrusive igneous rock

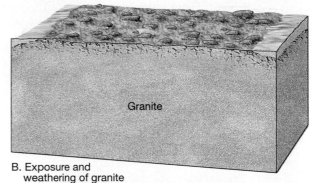

B. Exposure and weathering of granite

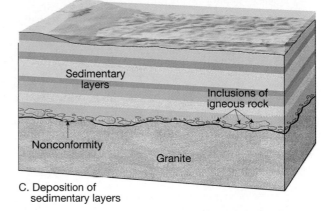

C. Deposition of sedimentary layers

FIGURE 18.5 Because pieces of granite (*inclusions*) are contained within the overlying sedimentary bed, we know the granite must be older. When older intrusive igneous rocks are overlain by younger sedimentary layers, a type of unconformity termed a *nonconformity* is said to exist.

sedimentary rocks that are overlain by younger, more flat-lying strata. An angular unconformity indicates that during the pause in deposition, a period of deformation (folding or tilting) and erosion occurred (Figure 18.7).

Disconformity When contrasted with angular unconformities, **disconformities** are more common, but usually far less conspicuous because the strata on either side are essentially parallel. For example, look at the disconformities in the cross section of the Grand Canyon in Figure 18.6. Many disconformities are difficult to identify because the rocks above and

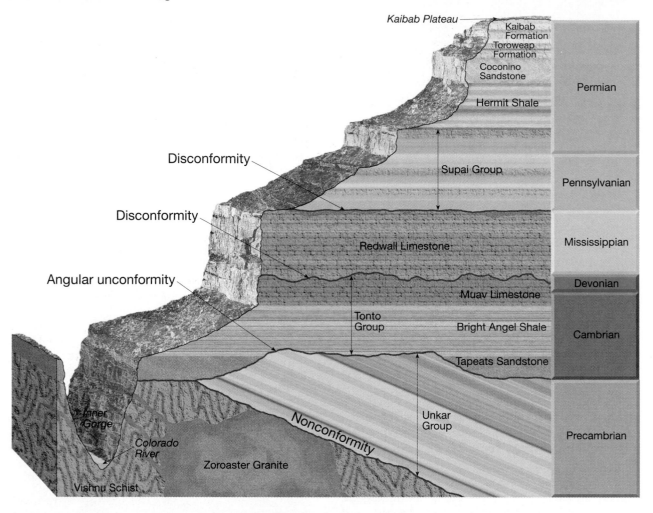

FIGURE 18.6 This cross section through the Grand Canyon illustrates the three basic types of unconformities. An angular unconformity can be seen between the tilted Precambrian Unkar Group and the Cambrian Tapeats Sandstone. Two disconformities are marked, above and below the Redwall Limestone. A nonconformity occurs between the igneous and metamorphic rocks of the Inner Gorge and the sedimentary strata of the Unkar Group.

below are similar and there is little evidence of erosion. Such a break often resembles an ordinary bedding plane. Other disconformities are easier to identify because the ancient erosion surface is cut deeply into the older rocks below.

Nonconformity The third basic type of unconformity is a **nonconformity.** Here the break separates older metamorphic or intrusive igneous rocks from younger sedimentary strata (Figure 18.5). Just as angular unconformities and disconformities imply crustal movements, so too do nonconformities. Intrusive igneous masses and metamorphic rocks originate far below the surface. Thus, for a nonconformity to develop, there must be a period of uplift and the erosion of overlying rocks. Once exposed at the surface, the igneous or metamorphic rocks are

subjected to weathering and erosion prior to subsidence and the renewal of sedimentation.

Using Relative Dating Principles

If you apply the principles of relative dating to the hypothetical geologic cross section in Figure 18.8, you can place in proper sequence the rocks and the events they represent. The statements within the figure summarize the logic used to interpret the cross section.

In this example, we establish a relative time scale for the rocks and events in the area of the cross section. Remember that this method gives us no idea of how many years of Earth history are represented, for we have no absolute dates. Nor do we know how this area compares to any other.

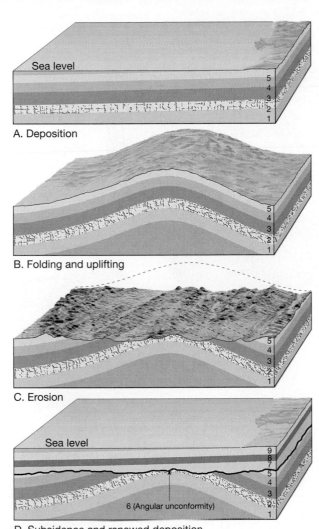

A. Deposition

B. Folding and uplifting

C. Erosion

Sea level

6 (Angular unconformity)

D. Subsidence and renewed deposition

FIGURE 18.7 Formation of an angular unconformity. An angular unconformity represents a period during which deformation and erosion occurred.

Correlation of Rock Layers

To develop a geologic time scale that applies to the whole Earth, rocks of similar age in different regions must be matched up. Such a task is referred to as **correlation.** Within a limited area correlating the rocks of one locality with those of another may be done simply by walking along the outcropping edges. However, this may not be possible when the rocks are mostly concealed by soil and vegetation. Correlation over short distances is often achieved by noting the position of a distinctive rock layer in a sequence of strata. Or, a layer may be identified in another location if it is composed of very distinctive or uncommon minerals.

By correlating the rocks from one place to another, a more comprehensive view of the geologic history of a region is possible. Figure 18.9, for example, shows the correlation of strata at three sites on the Colorado Plateau. No single locale exhibits the entire sequence, but correlation reveals a more complete picture of the sedimentary rock record.

Many geologic studies involve relatively small areas. Such studies are important in their own right, but their full value is realized only when the rocks are correlated with those of other regions. Although the methods just described are sufficient to trace a rock formation over relatively short distances, they are not adequate for matching rocks that are separated by great distances. When correlation between widely separated areas or between continents is the objective, the geologist must rely on fossils.

Fossils: Evidence of Past Life

Fossils, the remains or traces of prehistoric life, are important inclusions in sediment and sedimentary rocks. They are important tools for interpreting the geologic past. Knowing the nature of the life-forms that existed at a particular time helps researchers understand past environmental conditions. Further, fossils are important time indicators and play a key role in correlating rocks of similar ages that are from different places.

Types of Fossils

Fossils are of many types. The remains of relatively recent organisms may not have been altered at all. Such objects as teeth, bones, and shells are common examples. Far less common are entire animals, flesh included, that have been preserved because of rather unusual circumstances. Remains of prehistoric elephants called mammoths that were frozen in the Arctic tundra of Siberia and Alaska are examples, as are the mummified remains of sloths preserved in a dry cave in Nevada.

Given enough time, the remains of an organism are likely to be modified. Often fossils become *petrified* (literally, "turned into stone"), meaning that the small internal cavities and pores of the original structure are filled with precipitated mineral matter (Figure 18.10A). In other instances *replacement* may occur. Here the cell walls and other solid material are removed and replaced with mineral matter. Sometimes the microscopic details of the replaced structure are faithfully retained.

Molds and casts constitute another common class of fossils. When a shell or other structure is buried in sediment and then dissolved by underground water, a *mold* is created. The mold faithfully

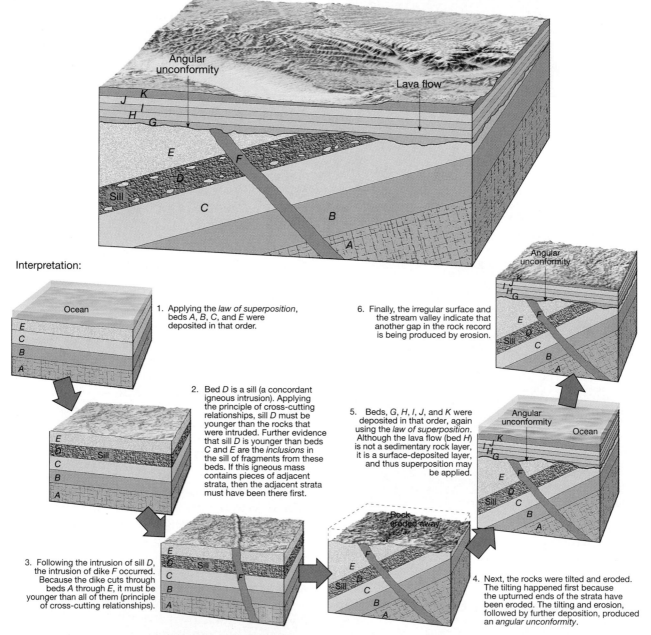

FIGURE 18.8 Geologic cross section of a hypothetical region.

reflects only the shape and surface marking of the organism; it does not reveal any information concerning its internal structure. If these hollow spaces are subsequently filled with mineral matter, *casts* are created (Figure 18.10B)

A type of fossilization called *carbonization* is particularly effective in preserving leaves and delicate animal forms. It occurs when fine sediment encases the remains of an organism. As time passes, pressure squeezes out the liquid and gaseous components and leaves behind a thin residue of carbon

(Figure 18.10C). Black shales deposited as organic-rich mud in oxygen-poor environments often contain abundant carbonized remains. If the film of carbon is lost from a fossil preserved in fine-grained sediment, a replica of the surface, called an *impression,* may still show considerable detail (Figure 18.10D).

Delicate organisms, such as insects, are difficult to preserve and consequently are relatively rare in the fossil record. Not only must they be protected from decay, but they must not be subjected to any pressure that would crush them. One way in which

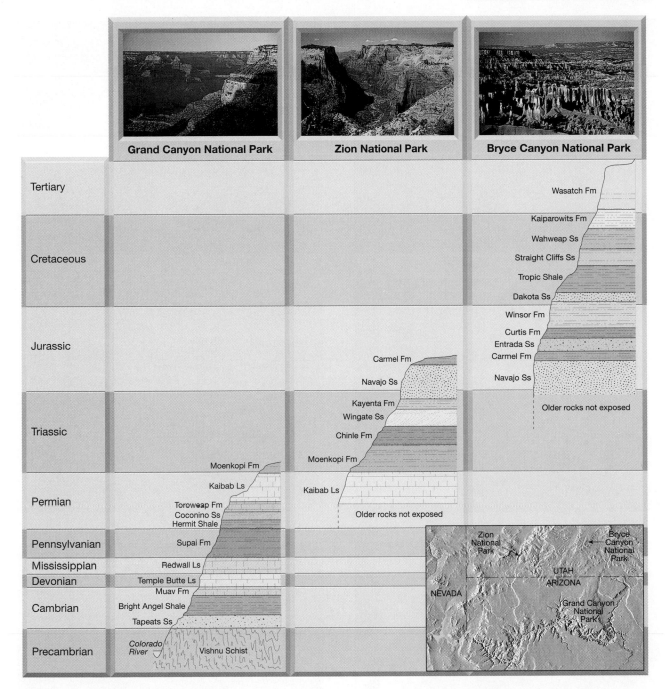

FIGURE 18.9 Correlation of strata at three locations on the Colorado Plateau reveals the total extent of sedimentary rocks in the region. (After U.S. Geological Survey; Photos by E. J. Tarbuck)

some insects have been preserved is in *amber*, the hardened resin of ancient trees. The fly in Figure 18.10E was preserved after being trapped in a drop of sticky resin. Resin sealed off the insect from the atmosphere and protected the remains from damage by water and air. As the resin hardened, a protective, pressure-resistant case was formed.

In addition to the fossils already mentioned, there are numerous other types, many of them only traces of prehistoric life. Examples of such indirect evidence include:

1. Tracks—animal footprints made in soft sediment that was later lithified (Figure 18.10F).

A.

B.

C.

D.

E.

F.

FIGURE 18.10 There are many types of fossilization. Six examples are shown here. **A.** Petrified wood in Petrified Forest National Park, Arizona. **B.** Natural casts of shelled invertebrates. **C.** A fossil bee preserved as a thin carbon film. **D.** Impressions are common fossils and often show considerable detail. **E.** Insect in amber. **F.** Dinosaur footprint in fine-grained limestone near Tuba City, Arizona. (Photo A by David Muench; Photos B, D, and F by E. J. Tarbuck; Photo C courtesy of the National Park Service; Photo E by Breck P. Kent)

2. Burrows—tubes in sediment, wood, or rock made be an animal. These holes may later become filled with mineral matter and preserved. Some of the oldest-known fossils are believed to be worm burrows.

3. Coprolites—fossil dung and stomach contents that can provide useful information pertaining to food habits of organisms.

4. Gastroliths—highly polished stomach stones that were used in the grinding of food by some extinct reptiles.

Conditions Favoring Preservation

Only a tiny fraction of the organisms that have lived during the geologic past have been preserved as fossils. Normally the remains of an animal or plant are destroyed. Under what circumstances are they preserved? Two special conditions appear to be necessary: rapid burial and the possession of hard parts.

When an organism perishes, its soft parts usually are quickly eaten by scavengers or decomposed by bacteria. Occasionally, however, the remains are buried by sediment. When this occurs, the remains are protected from the environment where destructive processes operate. Rapid burial therefore is an important condition favoring preservation.

In addition, animals and plants have a much better chance of being preserved as part of the fossil record if they have hard parts. Although traces and imprints of soft-bodied animals such as jellyfish, worms, and insects existed, they are rare. Flesh usually decays so rapidly that preservation is exceedingly unlikely. Hard parts such as shells, bones, and teeth predominate in the record of past life.

Because preservation is contingent on special conditions, the record of life in the geologic past is biased. The fossil record of those organisms with hard parts that lived in areas of sedimentation is quite abundant. However, we get only an occasional glimpse of the vast array of other life-forms that did not meet the special conditions favoring preservation.

Fossils and Correlation

The existence of fossils had been known for centuries, yet it was not until the late 1700s and early 1800s that their significance as geologic tools was made evident. During this period an English engineer and canal builder, William Smith, discovered that each rock formation in the canals he worked on contained fossils unlike those in the beds either above or below. Further, he noted that sedimentary strata in widely separated areas could be identified—and correlated—by their distinctive fossil content.

Based on Smith's classic observations and the findings of many geologists who followed, one of the most important and basic principles in historical geology was formulated: *Fossil organisms succeed one another in a definite and determinable order, and therefore any time period can be recognized by its fossil content.* This has come to be known as the **principle of fossil succession.** In other words, when fossils are arranged according to their age by applying the law of superposition to the rocks in which they are found, they do not present a random or haphazard picture. To the contrary, fossils show progressive change documenting the evolution of life through time.

For example, an Age of Trilobites is recognized quite early in the fossil record. Then, in succession, paleontologists recognize an Age of Fishes, an Age of Coal Swamps, an Age of Reptiles, and an Age of Mammals. These "ages" pertain to groups that were especially plentiful and characteristic during particular time periods. Within each of the "ages," there are many subdivisions based, for example, on certain species of trilobites and certain types of fish, reptiles, and so on. This same succession of dominant organisms, never out of order, is found on every major landmass.

Once fossils were recognized as time indicators, they became the most useful means of correlating rocks of similar age in different regions. Geologists pay particular attention to certain fossils called **index fossils.** These fossils are widespread geographically and are limited to a short span of geologic time, so their presence provides an important method of matching rocks of the same age. Rock formations, however, do not always contain a specific index fossil. In such situations, groups of fossils are used to establish the age of the bed. Figure 18.11 illustrates how a group of fossils can be used to date rocks more precisely than could be accomplished by the use of only one of the fossils.

In addition to being important and often essential tools for correlation, fossils are important environmental indicators. Although much can be deduced about past environments by studying the nature and characteristics of sedimentary rocks, a close examination of any fossils present can usually provide a great deal more information. For example, when the remains of certain clam shells are found in limestone, the geologist can assume that the region was once covered by a shallow sea, because that is where clams live today. Also, by using what we know of living organisms, we can conclude that fossil animals with thick shells capable of withstanding pounding and surging waves must have inhabited

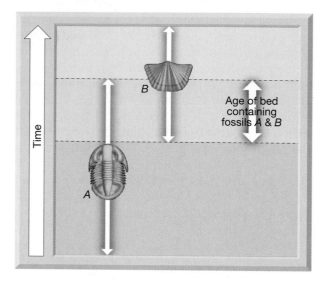

FIGURE 18.11 Overlapping ranges of fossils help date rocks more exactly than using a single fossil.

shorelines. On the other hand, animals with thin, delicate shells probably indicate deep, calm offshore waters. Hence, by looking closely at the types of fossils, the approximate position of an ancient shoreline may be identified.

Further, fossils can indicate the former temperature of the water. Certain present-day corals require warm and shallow tropical seas like those around Florida and the Bahamas. When similar corals are found in ancient limestones, they indicate that a Florida-like marine environment must have existed when they were alive. These are just a few examples of how fossils can help unravel the complex story of Earth history.

Absolute Dating with Radioactivity

In addition to establishing relative dates by using the principles described in the preceding sections, it is also possible to obtain reliable absolute dates for events in the geologic past. For example, we know that Earth is about 4.6 billion years old and that the dinosaurs became extinct about 66 million years ago. Dates that are expressed in millions and billions of years truly stretch our imagination because our personal calendars involve time measured in hours, weeks, and years. Nevertheless, the vast expanse of geologic time is a reality and it is radiometric dating that allows us to measure it accurately. In this section you will learn about radioactivity and its application in radiometric dating.

Recall from Chapter 1 that each atom has a *nucleus* containing protons and neutrons, and that the nucleus is orbited by electrons. *Electrons* have a negative electrical charge, and *protons* have a positive

charge. A *neutron* is actually a proton and an electron combined, so it has no charge (it is neutral).

The *atomic number* (each element's identifying number) is the number of protons in the nucleus. Every element has a different number of protons, and thus a different atomic number (hydrogen = 1, carbon = 6, oxygen = 8, uranium = 92, etc.) Atoms of the same element always have the same number of protons, so the atomic number stays constant.

Practically all of an atom's mass (99.9%) is in the nucleus, indicating that electrons have virtually no mass at all. So, by adding the protons and neutrons in an atom's nucleus, we derive the atom's *mass number*. The number of neutrons can vary, and these variants, or *isotopes,* have different numbers.

To summarize with an example, uranium's nucleus always has 92 protons, so its atomic number always is 92. But its neutron population varies, so uranium has three isotopes: uranium-234 (mass of protons + neutrons = 234), uranium-235, and uranium-238. All three isotopes are mixed in nature. They look the same and behave the same in chemical reactions.

Radioactivity

The forces that bind protons and neutrons together in the nucleus usually are strong. However, in some isotopes, the nuclei are unstable because the forces binding protons and neutrons together are not strong enough. As a result, the nuclei spontaneously break apart (decay), a process called **radioactivity.**

What happens when unstable nuclei break apart? Three common types of radioactive decay are illustrated in Figure 18.12 and are summarized as follows:

1. Alpha particles (α particles) may be emitted from the nucleus. An alpha particle is composed of 2 protons and 2 neutrons. Consequently, the emission of an alpha particle means (a) the mass number of the isotope is reduced by 4, and (b) the atomic number is decreased by 2.

2. When a beta particle (β particle), or electron, is given off from a nucleus, the mass number remains unchanged, because electrons have practically no mass. However, because the electron has come from a neutron (remember, a neutron is a combination of a proton and an electron), the nucleus contains one more proton than before. Therefore, the atomic number increases by 1.

3. Sometimes an electron is captured by the nucleus. The electron combines with a proton and forms an additional neutron. As in the last

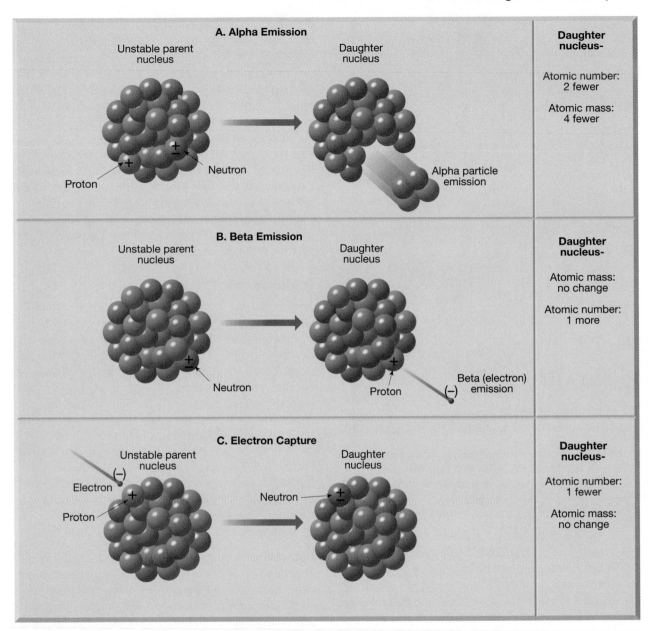

FIGURE 18.12 Common types of radioactive decay. Notice that in each case the number of protons (atomic number) in the nucleus changes, thus producing a different element.

example, the mass number remains unchanged. However, as the nucleus now contains one less proton, the atomic number decreases by 1.

An unstable (radioactive) isotope of an element is called the *parent*. The isotopes resulting from the decay of the parent are the *daughter products*. Figure 18.13 provides an example of radioactive decay. Here it can be seen that when the radioactive parent, uranium-238 (atomic number 92, mass number 238), decays, it follows a number of steps, emitting 8 alpha particles and 6 beta particles before finally becoming

the stable daughter product lead-206 (atomic number 82, mass number 206). One of the unstable daughter products produced during this decay series is radon. Box 18.1 examines the hazards associated with this radioactive gas.

Certainly among the most important results of the discovery of radioactivity is that it provided a reliable means of calculating the ages of rocks and minerals that contain particular radioactive isotopes. The procedure is called **radiometric dating.** Why is radiometric dating reliable? Because the rates of decay for many isotopes have been precisely measured and do

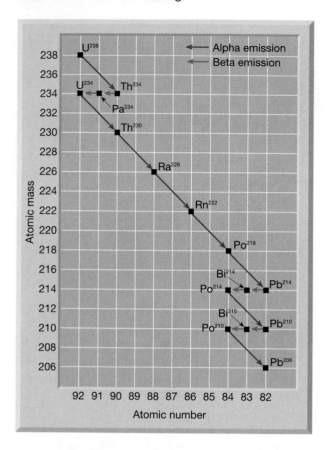

FIGURE 18.13 The most common isotope of uranium (U-238) is an example of a radioactive decay series. Before the stable end product (Pb-206) is reached, many different isotopes are produced as intermediate steps.

not vary under the physical conditions that exist in Earth's outer layers. Therefore, each radioactive isotope used for dating has been decaying at a fixed rate ever since the formation of the rocks in which it occurs, and the products of decay have been accumulating at a corresponding rate.

For example, when uranium is incorporated into a mineral that crystallizes from magma, there is no lead (the stable daughter product) from previous decay. The radiometric "clock" starts at this point. As the uranium in this newly formed mineral disintegrates, atoms of the daughter product are trapped and measurable amounts of lead eventually accumulate.

Half-Life

The time required for one-half of the nuclei in a sample to decay is called the **half-life** of the isotope. Half-life is a common way of expressing the rate of radioactive disintegration. Figure 18.14 illustrates what occurs when a radioactive parent decays directly into its stable daughter product. When the quantities of parent and daughters are equal (ratio

1:1), we know that one half-life has transpired. When one-quarter of the original parent atoms remain and three-quarters have decayed to the daughter product, the parent/daughter ratio is 1:3 and we know that two half-lives have passed. After three half-lives, the ratio of parent atoms to daughter atoms is 1:7 (one parent for every seven daughter atoms).

If the half-life of a radioactive isotope is known and the parent/daughter ratio can be measured, the age of the sample can be calculated. For example, assume that the half-life of a hypothetical unstable isotope is one million years and the parent/daughter ratio in a sample is 1:15. Such a ratio indicates that four half-lives have passed and that the sample must be four million years old.

Radiometric Dating

Notice that the *percentage* of radioactive atoms that decay during one half-life is always the same: 50 percent. However, the *actual number* of atoms that decay with the passing of each half-life continually decreases. Thus, as the percentage of radioactive parent atoms declines, the proportion of stable daughter atoms rises, with the increase in daughter atoms just matching the drop in parent atoms. This fact is the key to radiometric dating.

Of the many radioactive isotopes that exist in nature, five have proven particularly useful in providing radiometric ages for ancient rocks (Table 18.1). Rubidium-87, thorium-232, and the two isotopes of uranium are used only for dating rocks that are millions of years old, but potassium-40 is more versatile.

Potassium-Argon Although the half-life of potassium-40 is 1. 3 billion years, analytical techniques make possible the detection of tiny amounts of its stable daughter product, argon-40, in some rocks that are younger than 100,000 years. Another important reason for its frequent use is that potassium is an abundant constituent of many common materials, particularly micas and feldspars.

Although potassium (K) has three natural isotopes K^{39}, K^{40}, and K^{41}, only K^{40} is radioactive. When

TABLE 18.1 Radioactive isotopes frequently used in radiometric dating.

Radioactive Parent	Stable Daughter Product	Currently Accepted Half-life Values
Uranium-238	Lead-206	4.5 billion years
Uranium-235	Lead-207	713 million years
Thorium-232	Lead-208	14.1 billion years
Rubidium-87	Strontium-87	47.0 billion years
Potassium-40	Argon-40	1.3 billion years

Box 18.1   Radon[*]   Richard L. Hoffmann, Ph.D.

Radioactivity is defined as the spontaneous emission of atomic particles and/or electromagnetic waves from unstable atomic nuclei. For example, in a sample of uranium-238, unstable nuclei decay and produce a variety of radioactive progeny or "daughter" products as well as energetic forms of radiation (Table 18.A). One of its radioactive decay products is radon—a colorless, odorless, invisible gas.

Radon gained public attention in 1984 when a worker in a Pennsylvania nuclear power plant set off radiation alarms—not when he left work, but as he entered. His clothing and hair were contaminated with radon decay products. Investigation revealed that his basement at home had a radon level 2800 times the average level in indoor air. The home was located along a geological formation known as the Reading Prong—a mass of uranium-bearing black shale that runs from near Reading, Pennsylvania, to near Trenton, New Jersey.

Originating in the radio-decay of traces of uranium and thorium found in almost all soils, radon isotopes (Rn-222 and Rn-220) are continually renewed in an ongoing, natural process. Geologists estimate that the top six feet of soil from an average acre of land contains about fifty pounds of uranium (about 2 to 3 parts per million); some types of rocks contain more. Radon is continually generated by the gradual decay of this uranium. Because uranium has

TABLE 18.A   Decay Products of Uranium-238.

Some Decay Products of Uranium-238	Decay Particle Produced	Half-Life
Uranium-238	alpha	4.5 billion years
Radium-226	alpha	1600 years
Radon-222	alpha	3.82 days
Polonium-218	alpha	3.1 minutes
Lead-214	beta	26.8 minutes
Bismuth-214	beta	19.7 minutes
Polonium-214	alpha	1.6×10^{-4} second
Lead-210	beta	20.4 years
Bismuth-210	beta	5.0 days
Polonium-210	alpha	138 days
Lead-206	none	stable

a half-life of about 4.5 billion years, radon will be with us forever.

Radon itself decays, having a half-life of only about four days. Its decay products (except lead-206) are all radioactive solids that adhere to dust particles, many of which we inhale. During prolonged exposure to a radon-contaminated environment, some decay will occur while the gas is in the lungs, thereby placing the radioactive radon progeny in direct contact with delicate lung tissue. Steadily accumulating evidence indicates radon to be a significant cause of lung cancer second only to smoking.

A house with a radon level of 4.0 picocuries per liter of air has about 8 to 9 atoms of radon decaying every minute in every liter of air. The EPA suggests indoor radon levels be kept below this level. EPA risk estimates are conservative—they are based on an assumption that one would spend 75 percent of a 70-year time span (about 52 years) in the contaminated space, which most people would not.

Once radon is produced in the soil, it diffuses throughout the tiny spaces between soil particles. Some radon ultimately reaches the soil surface where it dissipates into the air. Radon enters buildings and homes through holes and cracks in basement floors and walls. Radon's density is greater than air, so it tends to remain in basements during its short decay cycle.

The source of radon is as enduring as its generation mechanism within Earth; radon will never go away. However, cost-effective mitigation strategies are available to reduce radon to acceptable levels, generally without great expense.

[*]Dr. Hoffmann is Professor of Chemistry, Illinois Central College.

K^{40} decays, it does so in two ways. About 11 percent changes to argon-40 (Ar^{40}) by means of electron capture (see Figure 18.12C). The remaining 89 percent of K^{40} decays to calcium-40 (Ca^{40}) by beta emission (see Figure 18.12B). The decay of K^{40} to Ca^{40}, however, is not useful for radiometric dating, because the Ca^{40} produced by radioactive disintegration cannot be distinguished from calcium that may have been present when the rock formed.

The potassium-argon clock begins when potassium-bearing minerals crystallize from a magma or form within a metamorphic rock. At this point the new minerals will contain K^{40} but will be free of Ar^{40}, because this element is an inert gas that does not chemically combine with other elements. As time passes, the K^{40} steadily decays by electron capture. The Ar^{40} produced by this process remains trapped within the mineral's crystal lattice. Because no Ar^{40}

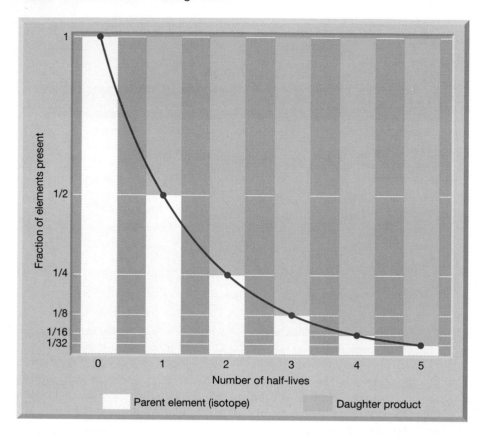

FIGURE 18.14 The radioactive decay curve shows change that is exponential. Half of the radioactive parent remains after one half-life. After a second half-life one-quarter of the parent remains, and so forth.

was present when the mineral formed, all of the daughter atoms trapped in the mineral must have come from the decay of K^{40}. To determine a sample's age, the K^{40}/Ar^{40} ratio is measured precisely and the known half-life for K^{40} applied.

Sources of Error It is important to realize that an accurate radiometric date can be obtained only if the mineral remained a closed system during the entire period since its formation; that is, a correct date is possible only if neither addition nor loss of parent or daughter isotopes occurred. This is not always the case. In fact, an important limitation of the potassium-argon method arises from the fact that argon is a gas and may leak from the minerals in which it forms. Indeed, losses can be significant if the rock is subjected to relatively high temperatures. Of course, a reduction in the amount of Ar^{40} leads to an underestimation of the rock's age. Sometimes temperatures are high enough for a sufficiently long period that all argon escapes. When this happens, the potassium-argon clock is reset and dating the sample will give only the time of thermal resetting, not the true age of the rock. For other radiometric clocks, a loss of daughter atoms can occur if the rock has been subjected to weathering or leaching. To avoid such a problem, one simple safeguard is to use only fresh, unweathered material and not samples that may have been chemically altered.

If parent/daughter ratios are not always reliable, how can meaningful radiometric dates be obtained? One common precaution against unknown errors is the use of cross checks. Often this simply involves subjecting a sample to two different radiometric methods. If the two dates agree, the likelihood is high that the date is reliable. If, on the other hand, there is an appreciable difference between the two dates, other cross checks must be employed to determine which, if either, is correct.

Carbon-14 Dating

To date very recent events, carbon-14 is used. Carbon-14 is the radioactive isotope of carbon. Because the half-life of carbon-14 is only 5730 years, it can be used for dating events from the historic past as well as those from recent geologic history. In some cases carbon-14 can be used to date events as far back as 75,000 years.

Carbon-14 is continuously produced in the upper atmosphere as a consequence of cosmic ray bombardment. Cosmic rays, which are high-energy particles, shatter the nuclei of gas atoms, releasing neutrons. Some of the neutrons are absorbed by

FIGURE 18.15 **A.** Production and **B.** decay of carbon-14. These sketches represent the nuclei of the respective atoms.

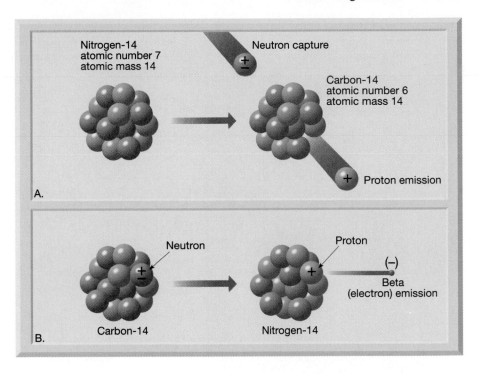

nitrogen atoms (atomic number 7) causing their nuclei to emit a proton. As a result, the atomic number decreases by 1 (to 6), and a different element, carbon-14, is created (Figure 18.15A). This isotope of carbon quickly becomes incorporated into carbon dioxide, which circulates in the atmosphere and is absorbed by living matter. As a result, all organisms contain a small amount of carbon-14, including yourself.

While an organism is alive, the decaying radiocarbon is continually replaced, and the proportions of carbon-14 and carbon-12 remain constant. Carbon-12 is the stable and most common isotope of carbon. However, when any plant or animal dies, the amount of carbon-14 gradually decreases as it decays to nitrogen-14 by beta emission (Figure 18.15B). By comparing the proportions of carbon-14 and carbon-12 in a sample, radiocarbon dates can be determined.

Although carbon-14 is useful in dating only the last small fraction of geologic time, it has become a very valuable tool for anthropologists, archeologists, and historians, as well as for geologists who study very recent Earth history. In fact, the development of radiocarbon dating was considered so important that the chemist who discovered this application, Willard F. Libby, received a Nobel prize.

Importance of Radiometric Dating

Bear in mind that, although the basic principle of radiometric dating is simple, the actual procedure is quite complex. The analysis that determines the quantities of parent and daughter must be painstakingly precise. In addition, some radioactive materials do not decay directly into the stable daughter product. As you saw in Figure 18.13, uranium-238 produces thirteen intermediate unstable daughter products before the fourteenth and final daughter product, the stable isotope lead-206, is produced.

Radiometric dating methods have produced literally thousands of dates for events in Earth history. Rocks from several localities have been dated at more than three billon years, and geologists realize that still-older rocks exist. For example, a granite from South Africa has been dated at 3.2 billion years—and it contains inclusions of quartzite that must be older. Quartzite itself is a metamorphic rock that originally was the sedimentary rock sandstone. Sandstone, in turn, is the product of the lithification of sediments produced by the weathering of existing rocks. Thus, we have a positive indication that much older rocks existed.

Radiometric dating has vindicated the ideas of Hutton, Darwin, and others who inferred that geologic time must be immense. Indeed, modern dating methods have proven that there has been enough time for the processes we observe to have accomplished tremendous tasks.

The Geologic Time Scale

Geologists have divided the whole of geologic history into units of varying magnitude. Together they comprise the **geologic time scale** of Earth history

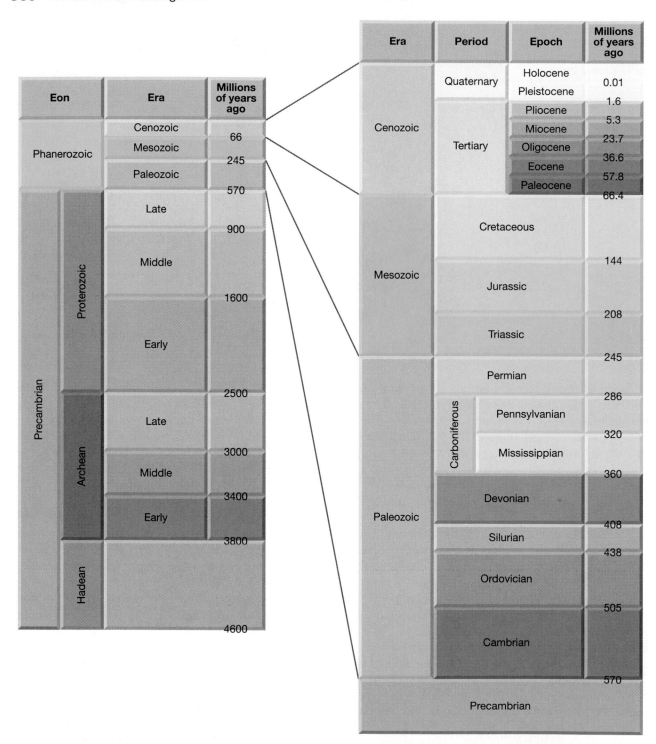

FIGURE 18.16 The geologic time scale. The absolute dates were added long after the time scale had been established using relative dating techniques. (Data from Geological Society of America)

(Figure 18.16). The major units of the time scale were delineated during the 1800s, principally by workers in Western Europe and Great Britain. Because absolute dating was unavailable at that time, the entire time scale was created using methods of relative dating. It has been only in this century that radiometric dating permitted absolute dates to be added.

The geologic time scale subdivides the 4.6-billion-year history of Earth into many different units and provides a meaningful time frame within which the events of the geologic past are arranged. As shown in Figure 18.16, **eons** represent the greatest expanses of time. The eon that began about 570 million years ago is the **Phanerozoic,** meaning *visible life.* It is an appropriate description because the rocks and deposits of the Phanerozoic eon contain abundant fossils that document major evolutionary trends.

Another glance at the time scale reveals that the Phanerozoic eon is divided into **eras.** The three eras within the Phanerozoic are the **Paleozoic** ("ancient life"), the **Mesozoic** ("middle life"), and the **Cenozoic** ("recent life"). As the names imply, the eras are bounded by profound worldwide changes in life-forms. Each era is subdivided into **periods.** The Paleozoic has seven, the Mesozoic three, and the Cenozoic two. Each of these twelve periods is characterized by a somewhat less profound change in life-forms as compared with the eras.

Finally, periods are divided into still smaller units called **epochs.** As you can see in Figure 18.16, seven epochs have been named for the periods of the Cenozoic. The epochs of other periods, however, are not usually referred to by specific names. Instead, the terms *early, middle,* and *late* are generally applied to the epochs of these earlier periods.

Notice that the detail of the geologic time scale does not begin until about 570 million years ago, the date for the beginning of the Cambrian period. The more than four billion years prior to the Cambrian is divided into three eons, the *Hadean,* the *Archean,* and the *Proterozoic.* It is also common for this vast expanse of time to simply be referred to as the **Precambrian.** Although it represents more than 85 percent of Earth history, the Precambrian is not divided into nearly as many smaller time units as the Phanerozoic eon.

The quantity of information geologists have deciphered about Earth's past is somewhat analogous to the detail of human history. The farther back we go, the less we know. Certainly more data and information exist about the past ten years than for the first decade of the twentieth century; the events of the nineteenth century have been documented much better than the events of the first century A.D.; and so on. So it is with Earth history. The more recent past has the freshest, least disturbed, and most observable record. The farther back in time the geologist goes, the more fragmented the record and clues become.

Difficulties in Dating the Geologic Time Scale

Although reasonably accurate absolute dates have been worked out for the periods of the geologic time scale (see Figure 18.16), the task is not without difficulty. The primary problem in assigning absolute dates to units of time is the fact that not all rocks can be dated radiometrically. Recall that for a radiometric date to be useful, all minerals in the rock must have formed at approximately the same time. For this reason, radioactive isotopes can be used to determine when minerals in an igneous rock crystallized and when pressure and heat created new minerals in a metamorphic rock.

However, samples of sedimentary rock can only rarely be dated directly by radiometric means. A sedimentary rock may include particles that contain radioactive isotopes, but the rock's age cannot be accurately determined because the grains, making up the rock are not the same age as the rock in which they occur. Rather, the sediments have been weathered from rocks of diverse ages.

Radiometric dates obtained from metamorphic rocks may also be difficult to interpret, because the age of a particular mineral in a metamorphic rock does not necessarily represent the time when the rock initially formed. Instead, the date may indicate any one of a number of subsequent metamorphic phases.

If samples of sedimentary rocks rarely yield reliable radiometric ages, how can absolute dates be assigned to sedimentary layers? Usually the geologist must relate them to datable igneous masses, as in Figure 18.17. In this example, radiometric dating has determined the ages of the volcanic ash bed within the Morrison Formation and the dike cutting the Mancos Shale and Mesaverde Formation. The sedimentary beds below the ash are obviously older than the ash, and all the layers above the ash are younger (principle of superposition). The dike is younger than the Mancos Shale and the Mesaverde Formation but older than the Wasatch Formation because the dike does not intrude the Tertiary rocks (cross-cutting relationships).

From this kind of evidence, geologists estimate that a part of the Morrison Formation was deposited about 160 million years ago, as indicated by the ash bed. Further, they conclude that the Tertiary period began after the intrusion of the dike, 66 million years ago. This is one example of literally thousands that illustrates how datable materials are used to *bracket* the various episodes in Earth history within specific time periods. It shows the necessity of combining laboratory dating methods with field observations of rocks.

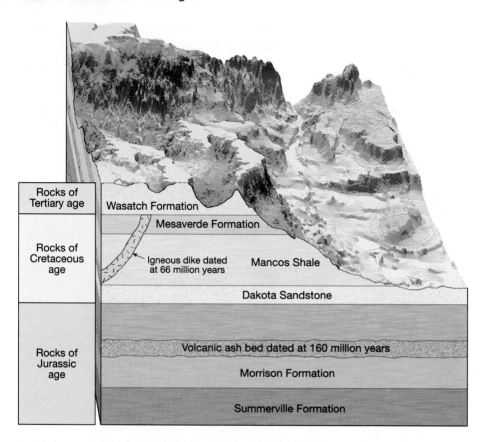

The Chapter in Review

The following statements are intended to help you review the primary objectives presented in this chapter.

- The two types of dates used by geologists to interpret Earth history are (1) *relative dates,* which put events in their *proper sequence of formation,* and (2) *absolute dates,* which pinpoint the *time in years* when an event took place.

- Relative dates can be established using the *law of superposition* (in an underformed sequence of sedimentary rocks or surface-deposited igneous rocks, each bed is older than the one above, and younger than the one below), *principle of original horizontality* (most layers are deposited in a horizontal position), *principle of cross-cutting relationships* (when a fault or intrusion cuts through another rock, the fault or intrusion is younger than the rocks cut through), *inclusions* (the rock mass containing the inclusion is younger than the rock that provided the inclusion), and *unconformities* (which represent a long period during which deposition ceased, erosion removed previously formed rocks, and then deposition resumed).

- The three basic types of unconformities are *angular unconformities* (tilted or folded sedimentary rocks that are overlain by younger, more flat-lying strata), *disconformities* (the strata on either side of the unconformity are essentially parallel), and *nonconformities* (where a break separates older metamorphic or intrusive igneous rocks from younger sedimentary strata).

- *Correlation,* the matching up of two or more geologic phenomena in different areas, is used to develop a geologic time scale that applies to the whole Earth.

- *Fossils* are the remains or traces of prehistoric life. Some fossils form by *petrification, replacement,* or *carbonization.* Others are *impressions, casts,* or organisms that have been *preserved in amber.* Fossils can also be indirect evidence of prehistoric life such as *tracks* or *burrows.* The special conditions that favor preservation are *rapid burial* and the possession of *hard parts* such as shells, bones, or teeth.

- Fossils are used to *correlate* sedimentary rocks that are from different regions by using the rocks' distinctive fossil content and applying the *principle of fossil succession.* The principle of fossil succession, which is based on the work of *William Smith* in the late 1700s, states that fossil organisms succeed one another in a definite and determinable order, and therefore any time period can be recognized by its fossil content. The use of *index fossils,* those that are wide-spread geographically and are limited to a short span of geologic time, provides an important method for matching rocks of the same age.

- Each atom has a nucleus containing *protons* (positively charged particles) and *neutrons* (neutral particles). Orbiting the nucleus are negatively charged *electrons.* The *atomic number* of an atom is the number of protons in the nucleus. The *mass number* is the number of protons plus the number of neutrons in an atom's nucleus. *Isotopes* are variants of the same atom, but with a different number of neutrons, and hence a different mass number.

- *Radioactivity* is the spontaneous breaking apart (decay) of certain unstable atomic nuclei. Three common forms of radioactive decay are (1) emission of *alpha particles* from the nucleus, (2) emission of *beta particles* (or electron) from the nucleus, and (3) *capture of an electron* by the nucleus.

- An unstable *radioactive isotope,* called the *parent,* will decay and form *daughter products.* The length of time for one-half of the nuclei of a radioactive isotope to decay is called the *half-life* of the isotope. Using a procedure called *radiometric dating,* if the half-life of the isotope is known, and the parent/daughter ratio can be measured, the age of a sample can be calculated. An accurate radiometric date can only be obtained if the mineral containing the radioactive isotope remained in a closed system during the entire period since its formation. *Carbon-14,* the radioactive isotope of carbon that is absorbed by living matter, is used to date very recent events.

- The *geologic time scale* divides Earth's history into units of varying magnitude. It is commonly presented in chart form, with the oldest time and event at the bottom and the youngest at the top. The principal subdivisions of the geologic time scale, called *eons,* include the *Hadean, Archean, Proterozoic* (together, these three eons are commonly referred to as the *Precambrian*), and, beginning about 570 million years ago, the *Phanerozoic.* The Phanerozoic (meaning "visible life") eon is divided into the following *eras: Paleozoic* ("ancient life"), *Mesozoic* ("middle life"), and *Cenozoic* ("recent life").

- The primary problem in assigning absolute dates to units of time is that *not all rocks can be radiometrically dated.* A sedimentary rock may contain particles of many ages that have been weathered from different rocks that formed at various times. One way geologists assign absolute dates to sedimentary rocks is to relate them to datable igneous masses, such as volcanic ash beds.

Key Terms

absolute date (p. 370)

angular conformity (p. 373)

Cenozoic era (p. 387)

conformable (p. 373)

correlation (p. 375)

cross-cutting relationships, principle of (p. 371)

disconformity (p. 373)

eon (p. 387)

epoch (p. 387)

era (p. 387)

fossil (p. 375)

fossil succession, principle of (p. 379)

geologic time scale (p. 385)

half-life (p. 382)

inclusion (p. 373)

index fossil (p. 379)

Mesozoic era (p. 387)

nonconformity (p. 374)

original horizontality, principle of (p. 371)

Paleozoic era (p. 387)

period (p. 383)

Phanerozoic eon (p. 387)

Precambrian (p. 387)

radioactivity (p. 380)

radiometric dating (p. 381)

relative dating (p. 370)

superposition, law of (p. 371)

unconformity (p. 373)

Questions for Review

1. Distinguish between absolute and relative dating.

2. What is the law of superposition? How are cross-cutting relationships used in relative dating?

3. When you observe an outcrop of steeply inclined sedimentary layers, what principle allows you to assume that the beds were tilted after they were deposited?

4. Refer to Figure 18.4 and answer the following questions:

    a. Is fault A older or younger than the sandstone layer?
    b. Is dike A older or younger than the sandstone layer?
    c. Was the conglomerate deposited before or after fault A?
    d. Was the conglomerate deposited before or after fault B?
    e. Which fault is older, A or B?
    f. Is dike A older or younger than the batholith?

5. A mass of granite is in contact with a layer of sandstone. Using a principle described in this chapter, explain how you might determine whether the sandstone was deposited on top of the granite, or whether it was intruded from below after the sandstone was deposited.

6. Distinguish among angular unconformity, disconformity, and nonconformity.

7. What is meant by the term *correlation?*

8. Describe several types of fossils. What organisms have the best chance of being preserved as fossils?

9. Describe William Smith's important contribution to the science of geology.

10. Why are fossils such useful tools in correlation?

11. In addition to being important aids in dating and correlating rocks, how else are fossils helpful in geologic investigations?

12. If a radioactive isotope of thorium (atomic number 90, mass number 232) emits 6 alpha particles and 4 beta particles during the course of radioactive decay, what is the atomic number and mass number of the stable daughter product?

13. Why is radiometric dating the most reliable method of dating the geologic past?

14. A hypothetical radioactive isotope has a half-life of 10,000 years. If the ratio of radioactive parent to stable daughter product is 1:3, how old is the rock containing the radioactive material?

15. Why is potassium-40 used more frequently in radiometric dating than other isotopes?

16. In order to provide a reliable radiometric date, a mineral must remain a closed system from the time of its formation until the present. Why is this true?

17. What precautions are taken to ensure reliable radiometric dates?

18. To make calculations easier, let us round the age of Earth to 5 billion years.

    a. What fraction of geologic time is represented by recorded history (assume 5000 years for the length of recorded history)?
    b. The first abundant fossil evidence does not appear until the beginning of the Cambrian period (570 million years ago). What percentage of geologic time is represented by abundant fossil evidence?

19. What general types of subdivisions make up the geologic time scale?

20. Briefly describe the difficulties in assigning absolute dates to layers of sedimentary rock.

21. Figure 18.18 is a block diagram of a hypothetical area in the American Southwest. Place the lettered features in the proper sequence, from oldest to youngest. Identify an angular unconformity and a nonconformity.

Testing What You Have Learned

To test your knowledge of the material presented in this chapter, answer the following questions:

Multiple-Choice Questions

1. The entire geologic time scale was originally created using only methods of _____ dating.
    a. absolute
    b. relative
    c. reverse
    d. inclusive
    e. precise

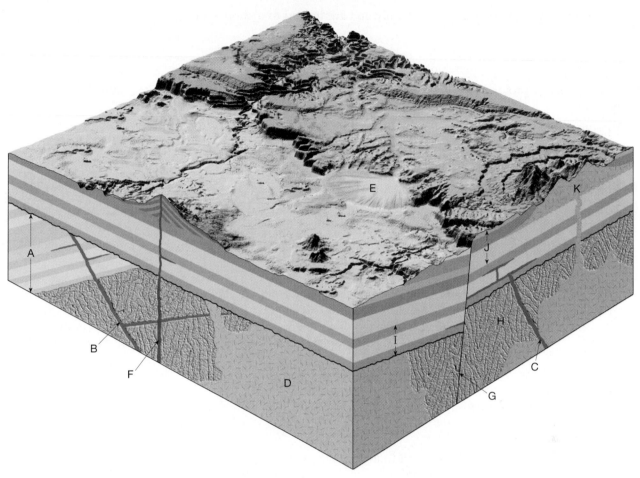

FIGURE 18.18 Figure for Review Questions

2. Within a sequence of undisturbed sedimentary rocks, the bed at the bottom can be assumed to be the oldest rock because of the law of _____.
a. superposition   **c.** inclusions   **e.** constancy
b. cross-cutting   **d.** unconformities

3. Isotopes of the same atom will have a different number of _____.
a. protons   **c.** electrons   **e.** neutrons
b. ions   **d.** molecules

4. What type of unconformity consists of folded sedimentary rocks overlain by younger, more flat-lying strata?
a. nonconformity   **c.** reversed   **e.** angular unconformity
b. conformity   **d.** disconformity

5. Which one of the following is NOT a method or type of fossilization?
a. cast   **c.** carbonization   **e.** replacement
b. petrification   **d.** conformation

6. The time required for one-half of the nuclei in a radioactive isotope to decay is referred to as the atom's _____.
a. period   **c.** emission   **e.** era
b. radioactivity   **d.** half-life

7. Fossils that are widespread geographically and are limited to a short span of geologic time are called —— fossils.
   a. base   **c.** term   **e.** connecting
   b. index   **d.** matching

8. The number of protons plus neutrons in an atom's nucleus is the atom's _____ number.
   a. mass   **c.** form   **e.** atomic
   b. isotope   **d.** key

9. The task of matching rocks of similar age in different regions is called _____.
   a. inclusion   **c.** dating   **e.** connecting
   b. conforming   **d.** correlation

10. Which one of the following is NOT used to determine relative dates?
    a. superposition   **c.** inclusions   **e.** radioactivity
    b. cross-cutting   **d.** fossils

Fill-In Questions

11. _____ dates pinpoint the time when something took place, while _____ dates put events in their proper sequence.
12. The principle of _____ states that layers of sediment are deposited in a horizontal position.
13. A(n) _____ represents a long time span during which deposition ceased, erosion removed previously formed rocks, and then deposition resumed.
14. The fact that fossil organisms succeed one another in a definite and determinable order is known as the principle of _____.
15. Rock layers that have been deposited essentially without interruption are said to be _____.

True/False Questions

16. Inclusions are pieces of one rock unit found within another. _____
17. Rapid burial is the only condition necessary for an organism to be preserved as a fossil. _____
18. The spontaneous breaking apart of unstable atomic nuclei is called radioactivity. _____
19. Together, the Hadean, Archean, and Proterozoic eons are often referred to simply as the Precambrian. _____
20. Radiometric dates are more reliable for igneous rocks than for samples of sedimentary rocks. _____

Answers

1.b; 2.a; 3.e; 4.e; 5.d; 6.d; 7.b; 8.a; 9.d; 10.e; 11. Absolute, relative; 12. original horizontality; 13. unconformity; 14. fossil succession; 15. conformable; 16.T; 17.F; 18.T; 19.T; 20.T.

Earth History: A Brief Summary

Focus on Learning

To assist you in learning the important concepts in this chapter, you will find it helpful to focus on the following questions:

- How did the planets in the solar system originate?

- How did Earth's atmosphere evolve and change through time?

- What were the principal geological and biological events of Earth history?

Dinosaur tracks near Cameron, Arizona. (Photo by Tom Bean)

arth has a long and complex history. The splitting and colliding of continents has resulted in
the formation of new ocean basins and the creation of Earth's great mountain ranges. Furthermore,
biological evolution has spurred a constant changing
of Earth's life-forms.

Changes on planet Earth occur at a "snail's pace,"
generally too slowly for people to perceive. Thus,
human awareness of evolutionary change is fairly
recent. Evolution is not confined to life-forms, for all of
Earth's "spheres" have evolved together: the atmosphere, hydrosphere, lithosphere, and biosphere. Examples are evolutionary changes in the air we breathe,
evolution of the world oceans, the rise of mountains,
ponderous movements of crustal plates, the comings
and goings of vast ice sheets, and the evolution of a
vast array of life-forms. As each facet of Earth has
evolved, it has powerfully influenced the others.

As we saw in Chapter 18, geologists have many
tools at their disposal for interpreting the clues about
Earth's past. Using these tools, and clues that are
contained in the rock record, geologists have been
able to unravel many of the complex events of the
geological past. The goal of this chapter is to provide
a brief overview of the history of our planet and its
life-forms (Figure 19.1). We will begin nearly five billion years ago with the formation of Earth and its
atmosphere. Then we will describe how our physical
world assumed its present form and how Earth's
inhabitants changed through time.

Origin of Earth

At the center of the solar system is the Sun. Around
it revolve nine planets, and around seven of the
planets revolve one or more moons. This orderly
nature of our solar system led most astronomers to
conclude that its members formed at essentially the
same time and from the same primordial material.
This proposal, known as the **nebular hypothesis**,
suggests that the bodies of our solar system condensed from an enormous cloud. It was composed
mostly of hydrogen and helium, with only a small
percentage of all other heavier elements.

FIGURE 19.1 Paleontologists study
ancient life. This scientist is working
with the remains of a dinosaur
known as *Tyrannosaurus Rex*. (Photo
by Rich Frishman/Tony Stone Images)

FIGURE 19.2 Nebular hypothesis. **A.** The nebula, a huge rotating cloud of dust and gases, began to contract. **B.** Most of the material was gravitationally swept toward the center, producing the Sun. However, owing to rotational motion, some dust and gases remained orbiting the central body as a flattened disk. **C.** The planets began to accrete from the material that was orbiting within the flattened disk. **D.** In time, most of the remaining debris either coalesced to form the nine planets and their moons, or was swept into space by the solar wind.

About five billion years ago, this huge cloud of minute rocky fragments and gases began to contract under its own gravitational influence (Figure 19.2A). The contracting material somehow began to rotate. Like a spinning ice skater pulling in her arms, the cloud rotated faster and faster as it contracted. This rotation in turn caused the nebular cloud to flatten into a disk (Figure 19.2B). Within the rotating disk, smaller condensations formed nuclei from which the planets eventually coalesced. However, the great concentration of material was pulled toward the center of this rotating mass. As it packed inward upon itself, it gravitationally heated, forming the hot *protosun* (sun in the making).

After the protosun formed, the temperature out in the rotating disk dropped significantly. This cooling caused substances with high melting points to condense into small particles, perhaps the size of sand grains. Iron and nickel solidified first. Next to condense were the elements of which rocky substances are composed. As these fragments collided over a few tens of millions of years, they accreted into the planets (Figure 19.2C,D). In the same manner, but on a lesser scale, the processes of condensation and accretion acted to form the moons and other small bodies of the solar system.

As the *protoplanets* (planets in the making) accumulated more and more debris, the space within the solar system began to clear. This removal of debris allowed sunlight to reach planetary surfaces unimpeded and to heat them. The resulting high surface temperatures of the inner planets (Mercury, Venus, Earth, Mars), coupled with their comparatively weak gravitational fields, meant that Earth and its neighbors were unable to retain appreciable amounts of the lighter components of the primordial cloud. These light materials, which included hydrogen, helium, ammonia, methane, and water, vaporized from their surfaces and were eventually whisked from the inner solar system by streams of solar particles called the *solar winds*.

At distances beyond Mars, temperatures were much cooler. Consequently, the large outer planets (Jupiter, Saturn, Uranus, and Neptune) accumulated huge amounts of hydrogen and other light materials from the primordial cloud. The accumulation of these gaseous substances is thought to account for the comparatively large sizes and low densities of the outer planets.

Shortly after Earth formed, the decay of radioactive elements, coupled with heat released by colliding particles, produced at least some melting of the interior. Melting allowed the denser elements, principally, iron and nickel, to sink to Earth's center, while the lighter rocky components floated outward, toward the surface. The sorting of material by density, which began early in Earth's history, is believed to be occurring still, but on a much smaller scale. As a result of this differentiation, Earth's interior is not homogeneous. Rather, it consists of shells or spheres of materials that have different properties.

An important consequence of this period of differentiation is that gaseous materials were allowed to escape from Earth's interior, just as gases are given off today during volcanic eruptions. By this process, an atmosphere gradually evolved, composed chiefly of gases expelled from within the planet.

Earth's Atmosphere Evolves

Today, the air you breathe is a stable mixture of 79 percent nitrogen, 20 percent oxygen, about 1 percent argon (an inert gas), and trace gases like carbon dioxide and water vapor. But our planet's original atmosphere, several billion years ago, was far different.

Earth's very earliest atmosphere probably was swept into space by the *solar wind,* a vast stream of particles emitted by the Sun. As Earth slowly cooled, a more enduring atmosphere formed. The molten surface solidified into a crust, and gases that had been dissolved in the molten rock were gradually released, a process called **outgassing**. Outgassing continues today from hundreds of active volcanoes worldwide. Thus, geologists hypothesize that Earth's original atmosphere was made up of gases similar to those released in volcanic emissions today: water vapor, carbon dioxide, nitrogen, and several trace gases.

As the planet continued to cool, the water vapor condensed to form clouds, and great rains commenced. At first, the water evaporated in the hot air before reaching the ground, or quickly boiled away upon contacting the surface, just like water sprayed on a hot grill. This accelerated the cooling of Earth's crust. When the surface had cooled below water's boiling point (100ºC or 212ºF), torrential rains slowly filled low areas, forming the oceans. This reduced not only the water vapor in the air, but the amount of carbon dioxide as well, for it became dissolved in the water. What remained was a nitrogen-rich atmosphere.

If Earth's primitive atmosphere resulted from volcanic outgassing, we have a problem, because volcanoes do not emit free oxygen. Where did the very significant percentage of oxygen in our present atmosphere (20 percent) come from?

The major source of oxygen is green plants. Put another way, *life itself* has strongly influenced the composition of our present atmosphere. Plants did not just adapt to their environment; they actually influenced it, dramatically altering the composition of the entire planet's atmosphere by using carbon dioxide and releasing oxygen. This is a good example of how Earth operates as a giant system in which living things interact with their environment.

How did plants come to alter the atmosphere? The key is the way in which plants create their own food. They employ *photosynthesis,* in which they use light energy to synthesize food sugars from carbon dioxide and water. The process releases a waste gas, oxygen. Those of us in the animal kingdom rely on oxygen to metabolize our food, and we in turn exhale carbon dioxide as a waste gas. The plants use this carbon dioxide for more photosynthesis, and so on, in a continuing system.

The first life-forms on Earth, probably bacteria, did not need oxygen. Their life processes were geared to the earlier, oxygenless atmosphere. Even today, many *anaerobic* bacteria thrive in environments that lack free oxygen. Later, primitive plants evolved that used photosynthesis and released oxygen. Slowly, the oxygen content of Earth's atmosphere increased. The Precambrian rock record suggests that much of the first free oxygen did not remain free because it combined with (oxidized) other substances dissolved in water, especially iron. Iron has tremendous affinity for oxygen, and the two elements combine to form iron oxides (rust) at any opportunity.

Then, once the available iron satisfied its need for oxygen, substantial quantities of oxygen accumulated in the atmosphere. By the beginning of the Paleozoic era, about four billion years into Earth's existence (after seven-eighths of Earth's history had transpired), the fossil record reveals abundant ocean-dwelling organisms that require oxygen to live. Hence, the composition of Earth's atmosphere has evolved together with its lifeforms, from an oxygenless envelope to today's oxygen-rich environment.

With this background on the evolution of the solar system, Sun, Earth, and atmosphere, we will look at Earth's history.

Precambrian Time: Vast and Enigmatic

The Precambrian encompasses immense geological time, from Earth's distant beginnings 4.6 billion years ago until the start of the Cambrian period, some four billion years later. Thus, the Precambrian spans about 87 percent of Earth's history. Our knowledge of this ancient time is sketchy, for much of the early rock record has been obscured by the very Earth processes you have been studying, especially plate tectonics, erosion, and deposition.

Untangling the long, complex Precambrian rock record is a formidable task, and we are far from done. Most Precambrian rocks are devoid of fossils, which hinders correlation of rocks. Rocks of this great age are metamorphosed and deformed, extensively eroded, and obscured by overlying strata. Consequently, this least-understood span of Earth's history has not been successfully divided into briefer time units, as have later intervals. Indeed, Precambrian history is written in scattered, speculative episodes, like a long book with many missing chapters.

Precambrian Rocks

Looking upon Earth from the Space Shuttle, astronauts see plenty of ocean (71 percent) and much less land area (29 percent). Over large expanses of the continents, the orbiting space scientists gaze upon many Paleozoic, Mesozoic, and Cenozoic rock surfaces, but fewer Precambrian surfaces. This demonstrates the law of superposition: Precambrian rocks in these regions are buried from view beneath varying thicknesses of more recent rocks. Here, Precambrian rocks peek through the surface where younger strata are extensively eroded, as in the Grand Canyon and in some mountain ranges. However, on each

continent, large "core areas" of Precambrian rocks dominate the surface, mostly as deformed metamorphic rocks. These areas are called **shields** because they roughly resemble a warrior's shield in shape.

Figure 19.3 shows these shield areas of Precambrian rocks worldwide. In North America (including Greenland), the Canadian Shield encompasses 7.2 million square kilometers (2.8 million square miles), the equivalent of about ten states of Texas put together.

Much of what we know about Precambrian rocks comes from mining the ores that some contain. The mining of iron, nickel, gold, silver, copper, chromium, uranium, and diamonds has provided Precambrian rock samples for study, and surveys to locate valuable ore deposits have revealed much about the rocks.

Noteworthy are extensive iron ore deposits. Rocks from the middle Precambrian (1.2–2.5 billion years ago) contain most of Earth's iron ore, mainly as the mineral hematite (Fe_2O_3). These iron-rich sedimentary rocks probably represent the time when oxygen became sufficiently abundant to react with iron dissolved in shallow lakes and seas. Later, after much of the iron was oxidized and deposited on lake and sea bottoms, formation of these iron-rich deposits declined and oxygen levels in the ocean and atmosphere began to increase. Because most of Earth's free oxygen results from plant photosynthesis, the formation of extensive Precambrian iron ore deposits is linked to life in the sea.

Notably absent in the Precambrian are fossil fuels (coal, oil, natural gas). The reason is clear—a virtual absence of land plants to form coal swamps

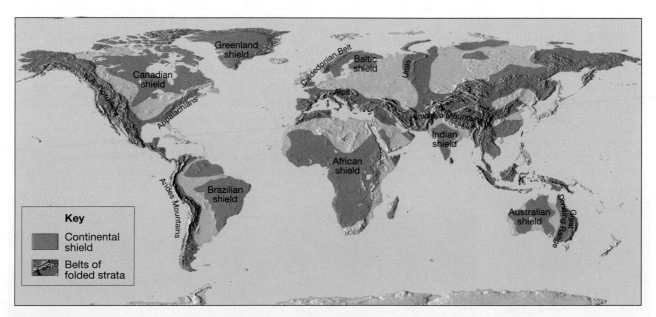

FIGURE 19.3 Today's continents look very different from the Precambrian. Remnants of Precambrian rocks are the continental shields, composed largely of metamorphic rocks.

and of certain animals to form petroleum. Fossil fuels are from a later time.

Precambrian Fossils

A century ago, the earliest fossils known dated from the Cambrian period, about 570 million years ago. None were known from the Precambrian. This created a major problem for science at that time: How could complex organisms like trilobites abruptly appear in the geologic record? The answer, of course, was that there were fossils in Precambrian rocks, but they were rare, small, obscure, and simply had not been discovered yet. Today, our knowledge of Precambrian life, although far from complete, is quite extensive.

Precambrian fossils are disappointing if you are expecting to see fascinating plants and large animals, for these had not yet evolved. Instead, the most common Precambrian fossils are **stromatolites.** These are distinctively layered mounds or columns of calcium carbonate (Figure 19.4). Stromatolites are not the remains of actual organisms, but are material deposited by algae. They are indirect evidence of algae because they closely resemble similar deposits made by modern algae.

Stromatolites did not become common until the middle Precambrian, about two billion years ago. Stromatolites are large, but most actual organisms preserved in Precambrian rocks are microscopic. Well-preserved remains of many tiny organisms have been discovered, extending the record of life back beyond 3.5 billion years.

Many of these most ancient fossils are preserved in *chert,* a hard, dense chemical sedimentary rock. Chert must be very thinly sliced and studied under powerful microscopes to observe bacteria and algae fossils within it.

Microfossils have been found at several locations worldwide. Two notable areas are in southern Africa, where the rocks date to more than 3.1 billion years, and in the Gunflint Chert (named for its use in flintlock rifles) of Lake Superior, which dates to 1.7 billion years. In both places, bacteria and blue-green algae have been discovered. The fossils are of the most primitive organisms, *prokaryotes.* Their cells lack organized nuclei, and they reproduce asexually.

More advanced organisms, *eukaryotes,* have cells that contain nuclei. Eukaryotes are among billion-year-old fossils discovered at Bitter Springs in Australia, such as green algae. Unlike prokaryotes, eukaryotes reproduce sexually, which means that genetic material is exchanged between organisms. This reproductive mode permits greatly increased genetic variation. Thus, development of eukaryotes may have dramatically increased the rate of evolutionary change.

Plant fossils date from the middle Precambrian, but animal fossils came a bit later, in the late Precambrian. Many of these fossils are *trace fossils,* meaning that they are not of the animals themselves, but of their activities, such as trails and worm holes. Areas in Australia and Newfoundland have yielded hundreds of fossil impressions of soft-bodied creatures. Most, if not all, of the Precambrian fauna lacked shells, which would develop as protective armor during the Paleozoic.

As the Precambrian came to a close, the fossil record disclosed diverse and complete multicelled organisms. This set the stage for more complex plants and animals to evolve at the dawn of the Paleozoic era.

Paleozoic Era: Life Explodes

Following the long Precambrian, the most recent 570 million years of Earth history are divided into three eras: Paleozoic, Mesozoic, and Cenozoic. The Paleozoic era encompasses more than 345 million years and is by far the longest of the three. Seven periods make up the Paleozoic era (Figure 19.5).

Before the Paleozoic, life-forms possessed no hard parts—shells, scales, bones, or teeth. The beginning of the Paleozoic is marked by the appearance

FIGURE 19.4 Stromatolites are among the most common Precambrian fossils. **A.** Precambrian fossil stromatolites composed of calcium carbonate deposited by algae in the Helena Formation, Glacier National Park. (Photo by Ken M. Johns/Photo Researchers, Inc.) **B.** Modern stromatolites growing in shallow saline area, western Australia. (Photo by Bill Bachman/Photo Researchers, Inc.)

A.

B.

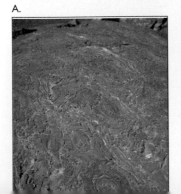

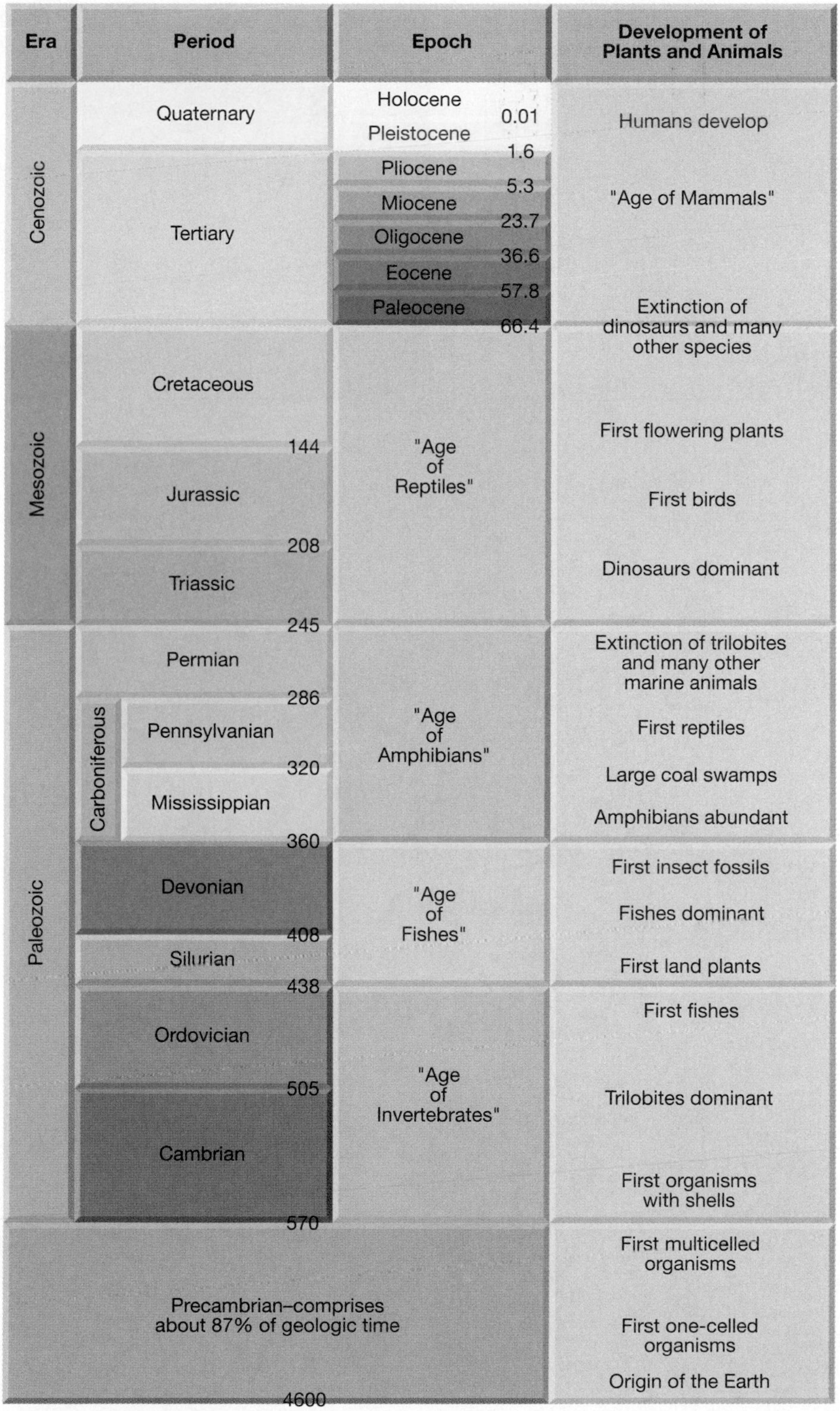

FIGURE 19.5 The geologic time scale. Numbers on the time scale represent time in millions of years before the present. These dates were added long after the time scale had been established using relative dating techniques. The Precambrian accounts for about 87 percent of geologic time. (Data from Geological Society of America)

Era	Period	Epoch	Development of Plants and Animals
Cenozoic	Quaternary	Holocene 0.01 / Pleistocene 1.6	Humans develop
Cenozoic	Tertiary	Pliocene 5.3 / Miocene 23.7 / Oligocene 36.6 / Eocene 57.8 / Paleocene 66.4	"Age of Mammals" / Extinction of dinosaurs and many other species
Mesozoic	Cretaceous 144	"Age of Reptiles"	First flowering plants
Mesozoic	Jurassic 208	"Age of Reptiles"	First birds
Mesozoic	Triassic 245	"Age of Reptiles"	Dinosaurs dominant
Paleozoic	Permian 286	"Age of Amphibians"	Extinction of trilobites and many other marine animals
Paleozoic	Carboniferous – Pennsylvanian 320	"Age of Amphibians"	First reptiles
Paleozoic	Carboniferous – Mississippian 360	"Age of Amphibians"	Large coal swamps / Amphibians abundant
Paleozoic	Devonian 408	"Age of Fishes"	First insect fossils / Fishes dominant
Paleozoic	Silurian 438	"Age of Fishes"	First land plants
Paleozoic	Ordovician 505	"Age of Invertebrates"	First fishes
Paleozoic	Cambrian 570	"Age of Invertebrates"	Trilobites dominant / First organisms with shells
	Precambrian—comprises about 87% of geologic time 4600		First multicelled organisms / First one-celled organisms / Origin of the Earth

of the first life-forms with hard parts (Figure 19.6). Hard parts greatly enhanced their chance of being preserved as part of the fossil record. Therefore, our knowledge of life's diversification improves greatly from the Paleozoic onward. This diversity is demonstrated in Figure 19.7.

Abundant Paleozoic fossils have allowed geologists to construct a far more detailed time scale for

FIGURE 19.6 Fossils of common Paleozoic life-forms. **A.** Natural cast of a trilobite. Trilobites dominated the Paleozoic ocean, scavenging food from the bottom. **B.** Extinct coiled cephalopods. Like their modern descendants, these were highly developed marine organisms. (Photos by E. J. Tarbuck)

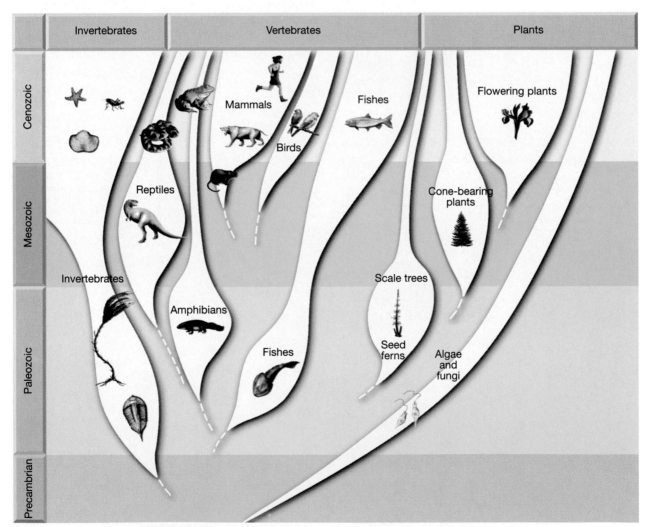

FIGURE 19.7 This chart indicates the times of appearance and relative abundance of major groups of organisms. The wider the band, the more dominant the group.

the last one-eighth of geologic time than for the preceding seven-eighths, the Precambrian. Moreover, because every organism is associated with a particular environment, the greatly improved fossil record provided invaluable information for deciphering ancient environments. To facilitate our brief tour of the Paleozoic, we divide it into Early Paleozoic (Cambrian, Ordovician, Silurian periods) and Late

Paleozoic (Devonian, Mississippian, Pennsylvanian, Permian periods).

Early Paleozoic History

The early Paleozoic consists of a 162-million-year span that embraces the Cambrian, Ordovician and Silurian periods. Anyone approaching Earth from space at this time would have seen the familiar blue planet with plentiful white clouds, but the arrangement of continents would have looked very different from today (Figure 19.8). At this time, the vast southern continent of Gondwanaland encompassed five continents (South America, Africa, Australia, Antarctica, India, and perhaps China). Evidence of an extensive continental glaciation places western Africa near the South Pole!

Landmasses that were not part of Gondwanaland existed as five separate units and some scattered fragments. Although the exact position of these "northern" continents is uncertain, ancestral North America and Europe are thought to have been near the equator and separated by a narrow sea, as shown on the map in Figure 19.8.

As the Paleozoic opened, North America was a land with no living things, plant or animal. There were no Appalachian or Rocky Mountains; rather, the continent was largely a barren lowland. Several times during the Cambrian and Ordovician periods, shallow seas moved inland and then receded from the interior of the continent. Deposits of clean sandstones, used today to make glass, mark the edge of these shallow seas in the midcontinent.

Early in the Paleozoic, a mountain-building event affected eastern North America from the present-day central Appalachians to Newfoundland. The mountains produced during this event, the Taconic orogeny, have since eroded away, leaving behind deformed strata and a large volume of detrital sedimentary rocks that were derived from the weathering of these mountains.

During the Silurian period, much of North America was once again inundated by shallow seas. This time large barrier reefs restricted circulation between shallow marine basins and the open ocean. Water in these basins evaporated, causing deposition of large quantities of rock salt and gypsum. Today these thick *evaporite beds* are important resources for the chemical, rubber, plasterboard, and photographic industries in Ohio, Michigan, and western New York State.

Early Paleozoic Life

Life in early Paleozoic time was restricted to the seas. Vertebrates had not yet evolved, so life consisted of several invertebrate groups (shown in Figure 19.9). The Cambrian period was the golden age of *trilobites.* More than 600 genera of these mud-burrowing scavengers flourished worldwide. By Ordovician times,

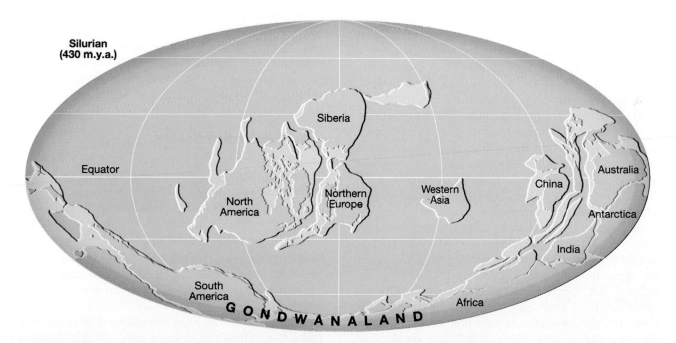

FIGURE 19.8 Reconstruction of Earth as it may have appeared in early Paleozoic time. The southern continents were joined into a single landmass called Gondwanaland. Four of the five landmasses that would later join to form the northern continent of Laurasia lay scattered roughly along the equator. (After C. Scotese, R.K. Bambach, C. Barton, R. VanderVoo, and A. Ziegler)

FIGURE 19.9 During the Ordovician period (505 million to 438 million years ago), the shallow waters of an inland sea over central North America contained an abundance of marine invertebrates. Shown in this reconstruction are straight-shelled cephalopods, trilobites, brachiopods, snails, and corals. (© The Field Museum, Neg. #GEO80820c, Chicago)

brachiopods outnumbered the trilobites. Brachiopods are among the most widespread Paleozoic fossils and, except for one modern group, are now extinct. Although the adults lived attached to the seafloor, the young larvae were free-swimming. This mobility accounts for the group's wide geographic distribution.

The Ordovician also marked the appearance of abundant *cephalopods,* mobile, highly developed mollusks that became the major predators of their time. The descendants of cephalopods include the modern squid, octopus, and nautilus. Cephalopods were the first truly large organisms on Earth (Figure 19.9). Whereas the largest trilobites seldom exceeded 30 centimeters (12 inches) in length and the biggest brachiopods were no more than about 20 centimeters (8 inches) across, one species of cephalopod reached a length of nearly 10 meters (30 feet).

The beginning of the Cambrian period marks an important event in animal evolution. For the first time, organisms appeared that secreted material which formed *hard parts,* such as shells. Why several diverse life-forms began to develop hard parts about the same time remains unanswered. One proposal suggests that, because an external skeleton provides protection from predators, hard parts evolved for survival. Yet, the fossil record does not seem to support this hypothesis. Organisms with hard parts were plentiful in the Cambrian period, whereas predators such as cephalopods were not abundant until the Ordovician period, some 70 million years later.

Whatever the answer, hard parts clearly served many useful purposes and aided adaptations to new ways of life. Sponges, for example, developed a network of fine interwoven silica spicules that allowed them to grow larger and more erect, capable of extending above the surface in search for food. Mollusks (clams and snails) secreted external shells of calcium carbonate that protected them and allowed body organs to function in a more controlled environment. The successful trilobites developed an exoskeleton of a protein called *chitin,* which permitted them to burrow through soft sediment in search of food (Figure 19.10).

Late Paleozoic History

The late Paleozoic consists of four periods—the Devonian, Mississippian, Pennsylvanian, and Permian—that span about 160 million years. Tectonic forces reorganized Earth's landmasses during this time, culminating with the formation of the supercontinent *Pangaea* (Figure 19.11).

Forming Pangaea, the Supercontinent As ancestral North America collided with Africa, the narrow sea that separated these landmasses began to close slowly (compare Figure 19.11B and 19.11C). Strong compressional forces from this collision deformed the rocks to produce the original northern Appalachian Mountains of eastern North America.

During the fusion of North America and Africa, the other northern continents began to converge

FIGURE 19.10 These trilobites date from the Devonian period, 408 million to 360 million years ago. (Photo by Phil Degginger/Tony Stone Images)

(Figure 19.11). By the Permian period, this newly formed landmass had collided with western Asia and the Siberian landmass along the line of the Ural Mountains. Through this union, the northern continent of *Laurasia* was born, encompassing present-day North America, Europe, western Asia, Siberia, and perhaps China.

As Laurasia was forming, Gondwanaland migrated northward. By the Pennsylvanian period, Gondwanaland collided with Laurasia, forming a mountainous belt through central Europe. Simultaneously, a collision between the African fragment of Gondwanaland and the southeastern edge of North America rumpled up the southern Appalachian Mountains.

By the close of the Paleozoic, all the continents had fused into the supercontinent of Pangaea (Figure 19.11). With only a single vast continent, the world's climate became very seasonal, having extremes far greater than those we experience today. These altered climatic conditions caused one of the most dramatic biological declines in all of Earth history.

Late Paleozoic Life

During most of the late Paleozoic, organisms diversified dramatically. Some 400 million years ago, plants that had adapted to survive at the water's edge began to move inland, becoming *land plants*. These earliest land plants were leafless, vertical spikes about the size of your index finger. However, by the end of the Devonian, 40 million years later, the fossil record indicates the existence of forests with trees tens of meters high.

In the oceans, armor-plated fishes that had evolved during the Ordovician continued to adapt. Their armor plates thinned to lightweight scales that increased their speed and mobility (Figure 19.12).

Other fishes evolved during the Devonian, including primitive sharks that had a skeleton made of cartilage and bony fishes, the groups to which virtually all modern fishes belong. Because of this, the Devonian period is often called the "age of fishes."

By late Devonian time, two groups of bony fishes, the lung fish and the lobe-finned fish, became adapted to land environments. Not unlike their modern relatives, these fishes had primitive lungs that supplemented their breathing through gills. It is believed that the lobe-finned fish occupied tidal flats or small ponds and that in times of drought they may have used their bone fins to "walk" from dried-up pools in search of other ponds. Through time, the lobe-finned fish began to rely more on their lungs and less on their gills. By late Devonian time, they had evolved into true air-breathing amphibians with fishlike heads and tails. It should be noted that insects had already invaded the land.

Modern amphibians, like frogs, toads, and salamanders, are small and occupy limited biological niches. But conditions during the remainder of the Paleozoic were ideal for these newcomers to the land. Plants and insects, which were their main diet, already were very abundant and large. Having only minimal competition from other land dwellers, the amphibians rapidly diversified. Some groups took on roles and forms that were more similar to modern reptiles, such as crocodiles, than to modern amphibians (Figure 19.13).

By the Pennsylvanian period, large tropical swamps extended across North America, Europe, and Siberia (Figure 19.14). Trees approached 30 meters (100 feet), with trunks over 1 meter across. The coal deposits that fueled the Industrial Revolution, and which provide a substantial portion of our electric power today, originated in these vast

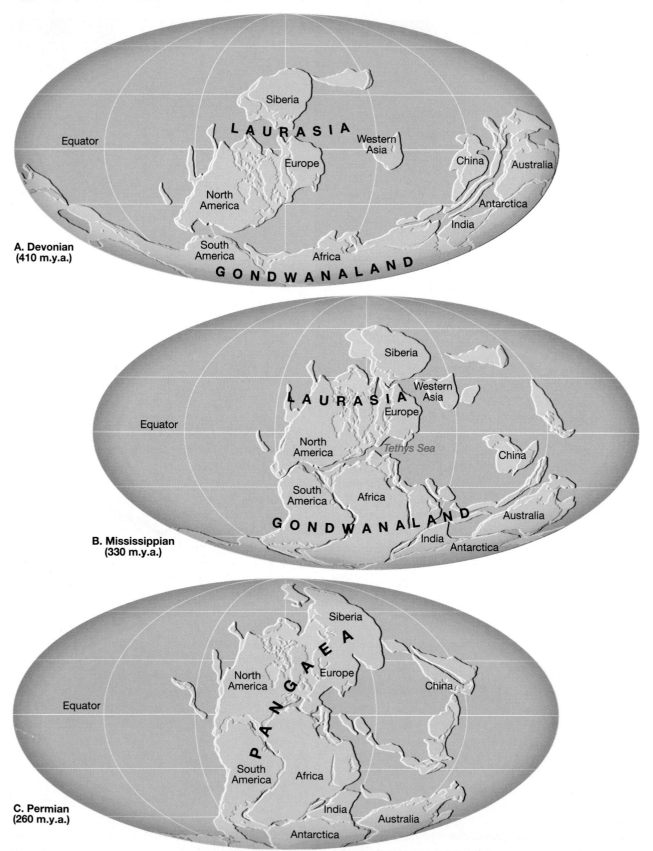

FIGURE 19.11 During the late Paleozoic, plate movements were joining together the major landmasses to produce the supercontinent of Pangaea. (After C. Scotese, R.K. Bambach, C. Barton, R. VanderVoo, and A. Ziegler)

FIGURE 19.12 These placoderms or "plate-skinned" fish were abundant during the Devonian (408 million to 360 million years ago). (Drawing after A.S. Romer)

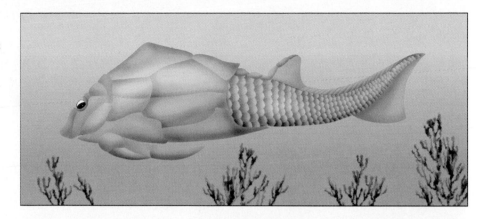

FIGURE 19.13 Permian scene (286 million to 245 million years ago) shows the large amphibian *Eryops*, which exceeded 2 meters (6 feet) in length. Eryops was a carnivore, as you can see from the teeth.

swamps. Further, it was in the lush coal swamp environment of the late Paleozoic that the amphibians evolved quickly into a variety of species.

Mesozoic Era: Age of the Dinosaurs

Spanning nearly 180 million years, the Mesozoic era is divided into three periods: the Triassic, Jurassic, and Cretaceous. The Mesozoic era witnessed the beginning of the breakup of the supercontinent Pangaea. Also in this era, organisms that had survived the great Permian extinction began to diversify in spectacular ways (see Box 19.1). On land, dinosaurs became dominant and remained unchallenged for

over 100 million years. Because of this, the Mesozoic Era is often called the "age of dinosaurs."

Early geologists recognized a profound difference between the fossils in Permian strata and those in younger Triassic rocks, as if someone had drawn a bold line to separate the two time periods. Clearly one-half of the fossil groups that occurred in late Paleozoic rocks were missing in Mesozoic rocks. On this basis, it was decided to separate the Paleozoic and Mesozoic at the Permian-Triassic boundary.

Mesozoic History

The Mesozoic era began with much of the world's land above sea level. In fact, in North America no

FIGURE 19.14 Restoration of a Pennsylvania coal swamp (320 million to 286 million years ago). Shown are scale trees (left), seed ferns (lower left), and scouring rushes (right). Also note the large dragonfly. (© The Field Museum, Neg. #GEO85637, Chicago. Photographer John Weinstein.)

period exhibits a more meager marine sedimentary record than the Triassic period. Of the exposed Triassic strata, most are red sandstones and mudstones that lack fossils and contain features indicating a terrestrial environment.

As the second period opened, the Jurassic, the sea invaded western North America. Adjacent to this shallow sea, extensive continental sediments were deposited on what is now the Colorado Plateau. The most prominent is the Navajo Sandstone, a windblown, white quartz sandstone that in places approaches a thickness of 300 meters (1000 feet). These massive dunes indicate that a major desert occupied much of the American Southwest during early Jurassic times.

A well-known Jurassic deposit is the Morrison Formation, within which is preserved the world's richest storehouse of dinosaur fossils. Included are fossilized bones of huge dinosaurs such as *Apatosaurus* (formerly *Brontosaurus*), *Brachiosaurus,* and *Stegosaurus*.

As the Jurassic period gave way to the Cretaceous, shallow seas once again invaded much of western North America, the Atlantic, and Gulf coastal

regions. This created great swamps like those of the Paleozoic era, forming Cretaceous coal deposits that are very important economically in the western United States and Canada. For example, on the Crow Indian reservation in Montana, there exists nearly 20 billion tons of high-quality coal of Cretaceous age.

A major event of the Mesozoic era was the breakup of Pangaea (Figure 19.15). A rift developed between what is now the eastern United States and western Africa, marking the birth of the Atlantic Ocean. It also represents the beginning of the breakup of Pangaea, a process that continued for 200 million years, through the Mesozoic and into the Cenozoic.

As Pangaea fragmented, the westward-moving North American plate began to override the Pacific plate. This tectonic event began a continuous wave of deformation that moved inland along the entire western margin of the continent. By Jurassic times, subduction of the Pacific plate had begun to produce the chaotic mixture of rocks that exists today in the Coast Ranges of California. Further inland, igneous activity was widespread, and for nearly 60 million years, huge

Box 19.1   The Great Paleozoic Extinction

The Paleozoic ended with the Permian period, a time when Earth's major landmasses joined to form the supercontinent Pangaea. This redistribution of land and water and changes in the elevations of landmasses brought pronounced changes in world climates. Broad areas of the northern continents became elevated above sea level and the climate grew drier. These changes apparently triggered extinctions of many species on land and sea.

By the close of the Permian, 75 percent of the amphibian families had disappeared, and plants had declined in number and variety. Although many amphibian groups became extinct, their descendants, the reptiles, would become the most successful and advanced animals on Earth. Marine life was not spared. At least 80 percent, and perhaps as much as 95 percent, of marine life disappeared. Many marine invertebrates that had been dominant during the Paleozoic, including all the remaining trilobites as well as some types of corals and brachiopods, failed to adapt to the widespread environmental changes.

The late Paleozoic extinction was the greatest of at least five mass extinctions to occur over the past 600 million years. Each extinction wreaked havoc with the existing biosphere, wiping out large numbers of species. In each case, however, the survivors formed new biological communities that were more diverse than their predecessors. Thus, mass extinctions actually invigorated life on Earth, as the few hardy survivors eventually filled more niches than the ones left by the victims.

The cause of the great Paleozoic extinction is uncertain. The climate changes from the formation of Pangaea and the associated drop in sea level undoubtedly stressed many species. In addition, at least two million cubic kilometers of lava flowed across Siberia to produce what is called the Siberian Traps. Perhaps debris from these eruptions blocked incoming sunlight, or perhaps enough sulfuric acid was emitted to make the seas virtually uninhabitable. Whatever caused the late Paleozoic extinction, it is clear that, without it, a very different population of organisms would today inhabit this planet.

---

masses of magma rose to within a few miles of the surface, where they cooled and solidified. The remnants of this intrusive activity included the granitic rocks of the Sierra Nevada, the Idaho batholith, and the Coast Range batholith of British Columbia.

Tectonic activity that began in the Jurassic continued throughout the Cretaceous, ultimately forming the vast mountains of western North America (Figure 19.16). Compressional forces moved huge rock units in a shingle-like fashion toward the east. Throughout much of the western margin of North America, older rocks were thrust eastward over younger strata, for a distance exceeding 150 kilometers (90 miles).

Toward the end of the Mesozoic, the middle and southern ranges of the Rocky Mountains formed. This mountain-building event, called the Laramide Orogeny, resulted when tectonic forces uplifted large blocks of Precambrian rocks in Colorado and Wyoming.

Mesozoic Life

As the Mesozoic era dawned, its life-forms were survivors of the great Paleozoic extinction. These survivors diversified in many new ways to fill the biological voids created at the close of the Paleozoic. On land, conditions favored those that could adapt to drier climates. Among plants, the gymnosperms were one such group. Unlike the first plants to invade the land, the seed-bearing gymnosperms did not depend on free-standing water for fertilization. Consequently, these plants were not restricted to a life near the water's edge.

The gymnosperms quickly became the dominant trees of the Mesozoic. They included the cycads, the conifers, and the ginkgoes. The cycads resembled a large pineapple plant. The ginkgoes had pan-shaped leaves, much like their modern relatives. Largest were the conifers, whose modern descendants include the pines, firs, and junipers. The best-known fossil occurrence of these ancient trees is in northern Arizona's Petrified Forest National Park. Here, huge petrified logs lie exposed at the surface, having been weathered from rocks of the Triassic Chinle Formation (Figure 19.17).

The Shelled Egg Among the animals, reptiles readily adapted to the drier Mesozoic environment. They were the first true terrestrial animals. Unlike amphibians, reptiles have shell-covered eggs that can be laid on land. The elimination of a water-dwelling stage (like the tadpole stage in frogs) was an important evolutionary step. Of interest is the fact that the watery fluid within the reptilian egg closely resembles seawater in chemical composition. Because the reptile

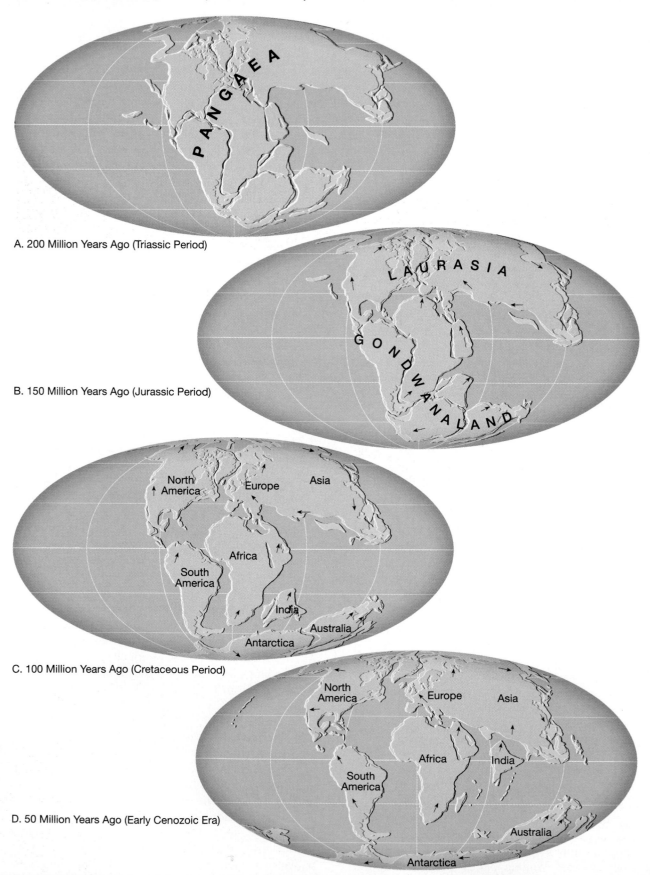

A. 200 Million Years Ago (Triassic Period)

B. 150 Million Years Ago (Jurassic Period)

C. 100 Million Years Ago (Cretaceous Period)

D. 50 Million Years Ago (Early Cenozoic Era)

FIGURE 19.15 The breakup of the supercontinent of Pangaea began and continued throughout the Mesozoic era. (After D. Walsh and C. Scotese)

FIGURE 19.16 Uplifted and deformed sedimentary strata in the Canadian Rockies near Lake Louise, Alberta, Canada. (Photo by Carr Clifton)

embryo develops in this watery environment, the shelled egg has been characterized as a "private aquarium" in which the embryos of these land vertebrates spend their water-dwelling stage of life.

Dinosaurs Dominate With the perfection of the shelled egg, reptiles quickly became the dominant land animals. They continued this dominance for more than 160 million years. Most awesome of the Mesozoic reptiles were the dinosaurs. Some of the huge dinosaurs were carnivorous (*Tyrannosaurus*), whereas others were herbivorous (like the ponderous *Apatosaurus* formerly *Brontosaurus*). The extremely long neck of *Apatosaurus* may have been an adaptation for feeding on tall conifer trees. However, not all dinosaurs were large. In fact, certain small forms closely resembled modern fleet-footed lizards. Further, evidence indicates that some dinosaurs, unlike their present-day reptile relatives, were warm blooded.

The reptiles made one of the most spectacular adaptive radiations in all of Earth history (Figure 19.18). One group, the pterosaurs, took to the air. These "dragons of the sky" possessed huge membranous wings that allowed them rudimentary flight (Figure 19.19). Another group of reptiles, exemplified by the fossil *Archaeopteryx,* led to more successful flyers: the birds. Whereas some reptiles took to the skies, others returned to the sea, including the fish-eating plesiosaurs and ichthyosaurs. These reptiles became proficient swimmers, but retained their reptilian teeth and breathed by means of lungs.

At the close of the Mesozoic, many reptile groups became extinct. Only a few types survived to recent times, including the turtles, snakes, crocodiles, and lizards. The huge land-dwelling dinosaurs, the marine plesiosaurs, and the flying pterosaurs all are known only through the fossil record. What caused this great extinction? (See Box 19.2.)

Cenozoic Era: Age of Mammals

The Cenozoic era, or "era of recent life," encompasses the past 66 million years of Earth history. It is the "post-dinosaur" era, the time of mammals, including humans. It is during this span that the physical landscapes and life-forms of our modern world came into being. The Cenozoic era represents a much smaller fraction of geologic time than either the Paleozoic or the Mesozoic. Although shorter, it nevertheless possesses a rich history, because the completeness of the geologic record

FIGURE 19.17 Petrified logs that have been weathered out of the Triassic Chinle Formation in the Petrified Forest of Arizona. (Photo by David Muench)

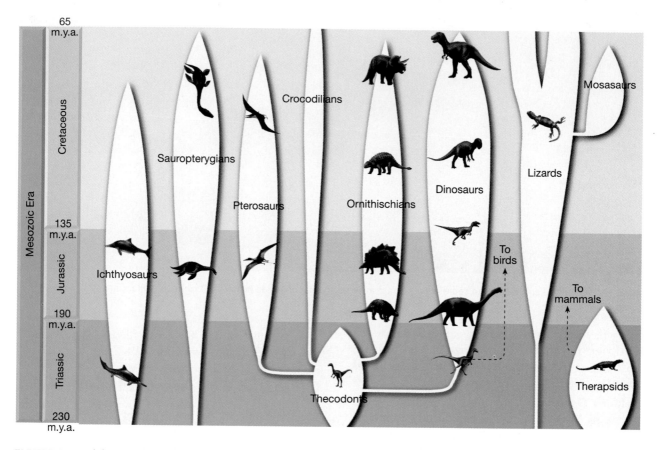

FIGURE 19.18 Adaptive radiation of major reptile groups during the Mesozoic era—"the golden age of reptiles." Dinosaurs, pterosaurs (flying reptiles), and large marine reptiles all became extinct by the end of the Mesozoic. (After Robert W. Tope)

FIGURE 19.19 Fossils of the great flying Pteranodon have been recovered from Cretaceous chalk deposits located in Kansas. Although Pteranodon had a wingspan of 7 meters (22 feet), flying reptiles with twice this wingspan have been discovered in strata of west Texas.

improves as time approaches the present. The rock formations of this time span are more widespread and less disturbed than those of any preceding time.

The Cenozoic era is divided into two periods of very unequal duration, the Tertiary period and the Quaternary period. The Tertiary period includes five epochs and embraces about 64 million years, practically all of the Cenozoic era. The Quaternary period consists of two epochs that represent only the last two million years of geologic time.

Cenozoic North America

Most of North America was above sea level throughout the Cenozoic era. However, the eastern and western margins of the continent experienced markedly contrasting events, because of their different relationships with plate boundaries. The Atlantic and Gulf coastal regions, far removed from an active plate boundary, were tectonically stable. Western North America, on the other hand, was the leading edge of the North American plate. As a result, plate interactions during the Cenozoic gave rise to many events of mountain building, volcanism, and earthquakes in the West.

Eastern North America The stable continental margin of eastern North America was the site of abundant marine sedimentation. The most extensive deposition surrounded the Gulf of Mexico, from the Yucatan Peninsula to Florida. Here, the great buildup of sediment caused the crust to downwarp and produced numerous faults. In many instances, the faults created traps in which oil and natural gas accumulated. Today, these and other petroleum traps are the most economically important resource in Cenozoic strata of the Gulf Coast, as evidenced by the Gulf's well-known off-shore drilling platforms.

By early Cenozoic time, most of the original Appalachians had been eroded to a low plain. Then, by the mid-Cenozoic, isostatic adjustments raised the region once again, rejuvenating its rivers. Streams eroded with renewed vigor, gradually sculpturing the surface into its present-day topography (Figure 19.20). The sediments from all of this erosion were deposited along the eastern margin of the continent, where they attained a thickness of many kilometers. Today, portions of the strata deposited during the Cenozoic are exposed as the gently sloping Atlantic and Gulf coastal plains.

Western North America In the West, the Laramide orogeny that built the Rocky Mountains was coming to an end. As erosion lowered the mountains, the basins between uplifted ranges filled with sediments. Eastward, a great wedge of sediment from the eroding Rockies was building, creating the Great Plains.

Beginning in the Miocene epoch, a broad region from northern Nevada into Mexico experienced crustal movements that formed more than 150 fault-block mountain ranges. They rise abruptly above the adjacent basins, creating the Basin and Range Province (see Chapter 17).

As the Basin and Range Province was forming, the entire western interior of the continent was gradually uplifted. This uplift re-elevated the Rockies and rejuvenated many of the West's major rivers. As the rivers became entrenched, many spectacular gorges were formed, including the Grand Canyon of the Colorado River, the Grand Canyon of the Snake River, and the

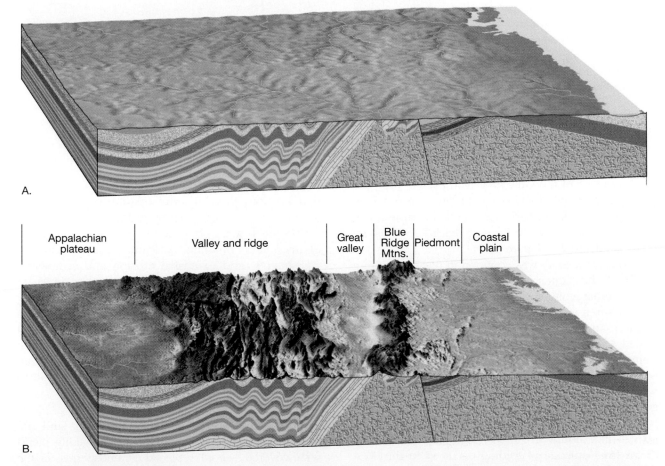

FIGURE 19.20 The formation of the modern Appalachian Mountains. **A.** The original Appalachians eroded to a low plain. **B.** The recent upwarping and erosion of the Appalachians began nearly 30 million years ago to produce the present topography.

Box 19.2 The KT Extinction

Review for a moment the geologic time scale (Figure 19.5). You can see that its divisions represent times of significant geological and/or biological change. Of special interest is the boundary between the Mesozoic era ("middle life") and Cenozoic era ("recent life"), about 66 million years ago. Around this time, more than half of all plant and animal species died out in a *mass extinction*. This boundary marks the end of the era in which dinosaur and other reptiles dominated the landscape and the beginning of the era when mammals became very important (Figure 19.A).

Because the last period of the Mesozoic is the Cretaceous (abbreviated K to avoid confusion with other "C" periods), and the first period of the Cenozoic is Tertiary (abbreviated T), the time of this mass extinction is called the Cretaceous-Tertiary or *KT boundary*.

The dinosaurs met their demise around the KT boundary, along with large numbers of other animal and plant groups, both terrestrial and marine. More important, of course, is

FIGURE 19.A Dinosaurs dominated the Mesozoic landscape until their extinction at the close of the Cretaceous period. This skeleton of *Tyrannosaurus* stands on display in Chicago's Natural History Museum. (Photo by Don and Pat Valenti/DRK Photo)

that many species survived. Human beings are descended from these survivors. Perhaps this explains why an event that occurred 66 million years ago has captured the interest of so many people.

Black Canyon of the Gunnison. The present topography of the Rocky Mountains is in large measure the result of this late Tertiary uplift and the subsequent excavation of the early Tertiary basin deposits by rejuvinated streams (Figure 19.21).

Volcanic activity was common in the West during much of the Cenozoic. Beginning in the Miocene epoch, great volumes of fluid basaltic lava flowed from fissures in portions of present-day Washington, Oregon, and Idaho. These eruptions built the extensive (1.3 million square kilometers) Columbia Plateau. Immediately west of the Columbia Plateau, volcanic activity was quite different. Here, thicker magmas with a higher silica content erupted explosively, creating a chain of stratovolcanoes from northern California to the Canadian border, some of which remain active—like Mount St. Helens (Figure 19.22).

A final episode of folding occurred in the West in late Tertiary time, creating the Coast Ranges that stretch along the Pacific Coast. Meanwhile, the Sierra Nevada became fault-block mountains as they were uplifted along their eastern flank, creating the imposing mountain front we know today.

As the Tertiary period drew to a close, the effects of mountain building, volcanic activity, isostatic adjustments, and extensive erosion and sedimentation had created a physical landscape very similar to the configuration of today. All that remained of Cenozoic time was the final two-million-year episode called the Quaternary period. During this most recent (and current) phase of Earth history, in which humans evolved, the action of glacial ice and other erosional agents added the finishing touches.

Cenozoic Life

Mammals replaced reptiles as the dominant land animals in the Cenozoic. Angiosperms (flowering plants with covered seeds) replaced gymnosperms as the dominant land plants. Marine invertebrates took on a modern look. Microscopic animals called *foraminifera*

The extinction of the great reptiles is generally attributed to this group's inability to adapt to some radical change in environmental conditions. What event could have triggered the rapid extinction of the dinosaurs—the most successful group of land animals ever to have lived?

One view proposes that, approximately 66 million years ago, a large asteroid or comet about 10 kilometers in diameter collided with Earth. The impact of such a body would have produced a dust cloud thousands of times larger than that released during the 1980 eruption of Mount St. Helens. For many months the dust would have greatly restricted the sunlight reaching Earth's surface. With insufficient sunlight for photosynthesis, delicate food chains would collapse. Large, plant-eating dinosaurs would be affected more adversely than would smaller life-forms because of the tremendous volume of vegetation they consumed. When the sunlight returned, more than half of the species on Earth, including numerous marine organisms, had become extinct.

What evidence points to such a catastrophic collision 66 million years ago? First, a thin layer of sediment nearly 1 centimeter thick has been discovered at the KT boundary, worldwide. This sediment contains a high level of the element *iridium*, rare in Earth's crust but found in high proportions in stony meteorites. Could this layer be the scattered remains of an asteroid that was responsible for the environmental changes that led to the demise of many reptile groups?

The second piece of evidence is that this period of mass extinction appears to have affected all land animals larger than dogs. Supporters of this catastrophic-event scenario suggest that small, ratlike mammals could survive a breakdown of food chains lasting perhaps several months. Large animals, proponents argue, could not have survived such an event.

Although the impact hypothesis is widely held, some scientists disagree. They claim that what appears to be a mass extinction over a short period in fact occurred over a much broader time span. Based on the fossil record at the KT boundary, these geologists conclude that the decline of the dinosaurs was gradual.

Those who disagree with the impact hypothesis have suggested an upsurge in volcanism. This is based on enormous outpourings of basaltic lavas in India approximately 65 million years ago. Volcanism advocates argue that the consequences of extensive volcanism would be quite similar to those from an asteroid impact: large quantities of dust and ash in the atmosphere that reduced sunlight, causing food chains to collapse. Further, some eruptions emit large quantities of sulfur gases that could form toxic sulfuric acid rains.

Volcano advocates point to the 1783 eruption at Laki, Iceland. Although small, this activity killed 75 percent of all livestock and ultimately 24 percent of the inhabitants of Iceland.

Whatever the cause, the KT extinction provided habitat vacancies for the mammals, allowing mammals to attain dominance during the Cenozoic era.

became especially important. Today, foraminifera are among the most intensely studied of all fossils because their widespread occurrence makes them invaluable in correlating Tertiary sediments. Tertiary strata are very important to the modern world, for they yield more oil than rocks of any other age.

The Cenozoic is often called the "age of mammals," because these animals came to dominate land life. It could also be called the "age of flowering plants," for the angiosperms enjoy a similar status in the plant world. As a result of advances in seed fertilization and dispersal, angiosperms experienced rapid development and expansion as the Mesozoic drew to a close. Thus, as the Cenozoic era began, angiosperms were already the dominant land plants.

Development of the flowering plants, in turn, strongly influenced the evolution of both birds and mammals. Birds that feed on seeds and fruits, for example, evolved rapidly during the Cenozoic in close association with the flowering plants. During the middle Tertiary, grasses developed rapidly and spread over the plains. This fostered the emergence of herbivorous (plant-eating) mammals that were mainly grazers. In turn, the development and spread of grazing animals established the setting for the evolution of the carnivorous mammals that preyed upon them.

Mammals Replace Reptiles Back in the Mesozoic, an important evolutionary event was the appearance of primitive mammals in the late Triassic, about the same time that the dinosaurs emerged. Yet throughout the period of dinosaur dominance, mammals remained in the background as small and inconspicuous animals. By the close of the Mesozoic era, dinosaurs and other reptiles no longer dominated the land. It was only after these large reptiles became extinct that mammals came into their own as the dominant land animals. The transition is a major example in the fossil record of the replacement of one large group by another.

Mammals are distinct from reptiles in important respects. Mammalian young are born live and

FIGURE 19.21 Wyoming's spectacular Teton Range is part of the Rocky Mountains. They were formed during the late Tertiary uplift about 10 million years ago. (Photo by Tom Till)

mammals maintain a steady body temperature, that is, they are "warm blooded." This latter adaptation allowed mammals to lead more active and diversified lives than reptiles because they could survive in cold regions and search for food during any season or time of day. Other mammalian adaptations included the development of insulating body hair and more efficient heart and lungs.

It is worth noting that perfect boundaries cannot be drawn between mammalian and reptilian traits. One minor group of mammals, the monotremes, still lay eggs. The two species in this group, the duck-billed platypus and the spiny anteater, are found only in Australia. Moreover, although modern reptiles are "cold-blooded," some paleontologists believe that dinosaurs may have been "warm blooded."

With the demise of most Mesozoic reptiles, Cenozoic mammals diversified rapidly. The many forms that exist today evolved from small primitive mammals that were characterized by short legs, flat five-toed feet, and small brains. Their development and specialization took four principal directions: (1) increase in size, (2) increase in brain capacity, (3) specialization of teeth to better accommodate a particular diet, and (4) specialization of limbs to better equip the animal for life in a particular environment.

Marsupials, Placentals, and Diversity Following the reptilian extinctions at the close of the Mesozoic, two groups of mammals, the marsupials and the placentals, evolved and expanded to dominate the Cenozoic. The groups differ principally in their modes of reproduction. Young marsupials are born live but at a very early stage of development. After birth, the tiny and immature young crawl into the mother's external stomach pouch to complete their development. Examples are kangaroos, opossums, and koala bears.

Placental mammals, on the other hand, develop within the mother's body for a much longer period, so that birth occurs after the young are relatively mature and independent. Most mammals are placental, including humans.

Today, marsupials are found primarily in Australia, where they went through a separate evolutionary expansion during the Cenozoic, largely isolated from placental mammals.

In South America, both primitive marsupials and placentals coexisted before that landmass became completely isolated during the break-up of Pangaea. Evolution and specialization of both groups continued undisturbed for approximately 40 million years until the close of the Pliocene epoch, when the Central American land bridge emerged, connecting the two American continents. Then an invasion of advanced carnivores from North America brought the extinction of many hoofed mammals that had persisted in South America for millions of years. The marsupials, except for opossums, also could not compete and became extinct. Both Australia and South America provide excellent examples of how isolation caused by the separation of continents increased the diversity of animals in the world.

Large Mammals and Extinction As we have seen, mammals diversified quite rapidly during the Cenozoic era. One tendency was for some groups to became very large. For example, by the Oligocene

FIGURE 19.22 Mount Hood, Oregon. This dormant volcano is one of several large composite cones that comprise the Cascade Range. (Photo by John M. Roberts/The Stock Market)

epoch a hornless rhinoceros that stood nearly 5 meters high (16 feet) had evolved. It is the largest land mammal known to have existed. As time approached the present, many other types evolved to a large size as well; more, in fact, than now exist. Many of these large forms were common as recently as 11,000 years ago. However, a wave of late Pleistocene extinctions rapidly eliminated these animals from the landscape.

In North America, the mastodon and mammoth, both huge relatives of the elephant, became extinct. In addition, saber-toothed cats, giant beavers, large ground sloths, horses, camels, giant bison, and others died out (Figure 19.23). In Europe, late Pleistocene extinctions included woolly rhinos, large cave bears, and the Irish elk. The reason for this recent wave of large animal extinctions puzzles scientists. These animals had survived several major glacial advances and interglacial periods, so it is difficult to ascribe these extinctions to climatic change. Some scientists believe that early humans hastened the decline of these mammals by selectively hunting large forms. Although this hypothesis is preferred by many, it is not yet accepted by all.

FIGURE 19.23 Partially exposed skeleton of an extinct sabre-toothed cat. This lion-sized cat roamed the White River Badlands of South Dakota during the Oligocene epoch. (Photo by T.A. Wiewandt/DRK Photo)

The Chapter in Review

The following statements are intended to help you review the primary objectives presented in this chapter.

- The *nebular hypothesis* describes the formation of the solar system. The planets and Sun began forming about five billion years ago from a large cloud of dust and gases composed of hydrogen and helium, with only a small percentage of all the other heavier elements. As the cloud contracted, it began to rotate and assume a disk shape. Material that was gravitationally pulled toward the center became the *protosun.* Within the rotating disk, small centers, called *protoplanets,* swept up more and more of the cloud's debris. Because of their high temperatures and weak gravitational fields, the inner planets (Mercury, Venus, Earth, and Mars) were unable to accumulate and retain many of the lighter components (hydrogen, helium, ammonia, methane, and water). However, because of the very cold temperatures existing far from the Sun, the fragments from which the large outer planets (Jupiter, Saturn, Uranus, and Neptune) formed consisted of huge amounts of lighter materials. These gaseous substances account for the comparatively large sizes and low densities of the outer planets.

- The decay of radioactive elements and heat released by colliding particles aided the melting of Earth's interior, allowing the denser elements, principally iron and nickel, to sink to its center. As a result of this *differentiation,* Earth's interior consists of shells or spheres of materials, each having distinct properties.

- Earth's primitive atmosphere probably consisted of water vapor, carbon dioxide, nitrogen, and several trace gases that were released in volcanic emissions, a process called *outgassing.* The first life-forms on Earth, probably *anaerobic bacteria,* did not need oxygen. As life evolved, plants, through the process of *photosynthesis,* used carbon dioxide and water and released oxygen into the atmosphere. Once the available iron on Earth was oxidized (combined with oxygen), substantial quantities of oxygen accumulated in the atmosphere. About four billion years into Earth's existence, the fossil record reveals abundant ocean-dwelling organisms that require oxygen to live.

- The *Precambrian* spans about 87 percent of Earth's history, beginning with the formation of Earth about 4.6 billion years ago and ending approximately 570 million years ago with the diversification of life that marks the start of the Paleozoic era. It is the least understood span of Earth's history. On each continent there is a "core area" of Precambrian rocks called a *shield.* The iron ore deposits of Precambrian age represent the time when oxygen became abundant and combined with iron to form iron oxide. The most common middle Precambrian fossils are *stromatolites.* Microfossils of bacteria and blue-green algae, both primitive *prokaryotes* whose cells lack organized nuclei, have been found in chert, a hard, dense, chemical sedimentary rock in southern Africa (3.1 billion years of age) and near Lake Superior (1.7 billion years of age). *Eukaryotes,* with cells containing organized nuclei, are among billion-year-old fossils discovered in Australia. Plant fossils date from the middle Precambrian, but animal fossils came a bit later, in the late Precambrian. Many of these fossils are *trace fossils.*

- The *Paleozoic era* extends from 570 million years ago to about 245 million years ago. The beginning of the Paleozoic is marked by the *appearance of the first life-forms with hard parts* such as shells. Therefore, abundant Paleozoic fossils occur and a far more detailed record of Paleozoic events can be constructed. During the early Paleozoic (the Cambrian, Ordovician, and Silurian periods) the vast southern continent of *Gondwanaland* existed. Seas inundated and receded from North America several times, leaving thick evaporate beds of rock salt and gypsum. Life in the early Paleozoic was restricted to the seas and consisted of several invertebrate groups. During the late Paleozoic (the Devonian, Mississippian, Pennsylvanian, and Permian periods), ancestral North America collided with Africa to produce the original northern Appalachian Mountains, and the northern continent of *Laurasia* formed. By the close of the Paleozoic, all the continents had fused into the supercontinent of *Pangaea.* During most of the late Paleozoic, organisms diversified dramatically. Insects and plants moved onto the land, and amphibians evolved and diversified quickly. By the Pennsylvanian period, large tropical swamps, which became the major coal deposits of today, extended across North America, Europe, and Siberia. At the close of the Paleozoic, altered climatic conditions caused one of the most dramatic biological declines in all of Earth's history.

- The *Mesozoic era,* literally the era of *middle life,* is often called the "age of dinosaurs." It began about 245 million years ago and ended approximately 66 million years ago. Early in the Mesozoic much of the land was above sea level. However, by the middle Mesozoic, seas invaded western North America. As Pangaea began to break up, the westward-moving North American plate began to override the Pacific plate, causing crustal deformation along the entire western margin of the continent. Organisms that had survived extinction at the end of the Paleozoic began to diversify in spectacular ways. *Gymnosperms* (cycads, conifers, and ginkgoes) became the dominant trees of the Mesozoic because they could adapt to the drier climates. Reptiles became the dominant land animals, with one group eventually becoming the birds. The most awesome of the Mesozoic reptiles were the *dinosaurs.* At the close of the Mesozoic, many reptile groups, including the dinosaurs, became extinct.

- The *Cenozoic era,* or *"era of recent life,"* begins approximately 66 million years ago and continues today. It is the *time of mammals,* including

humans. The widespread, less disturbed rock formations of the Cenozoic provide a rich geologic record. Most of North America was above sea level throughout the Cenozoic. Owing to their different relations with tectonic plate boundaries, the eastern and western margins of the continent experienced contrasting events. The stable eastern margin was the site of abundant sedimentation as isostatic adjustment raised the Appalachians, causing streams to erode with renewed vigor and deposit their sediment along the continental margin. In the West, building of the Rocky Mountains (the *Laramide orogeny*) was coming to an end, the Basin and Range Province was forming, and volcanic activity was extensive. The Cenozoic is often called *"the age of mammals"* because these animals replaced the reptiles as the dominant land life. Two groups of mammals, the marsupials and the placentals, evolved and expanded to dominate the era. One tendency was for some mammal groups to become very large. However, a wave of late *Pleistocene* extinctions rapidly eliminated these animals from the landscape. Some scientists believe that early humans hastened their decline by selectively hunting the larger animals. The Cenozoic could also be called the *"age of flowering plants."* As a source of food, flowering plants (angiosperms) strongly influenced the evolution of both birds and herbivorous (plant-eating) mammals throughout the Cenozoic era.

Key Terms

nebular hypothesis (p. 394)

outgassing (p. 396)

shields (p. 397)

stromatolites (p. 398)

Questions for Review

1. Briefly describe the events believed to have led to the formation of the solar system.

2. What is the major source of free oxygen in Earth's atmosphere?

3. Explain why Precambrian history is more difficult to decipher than more recent geological history.

4. Match the following words and phrases to the most appropriate time span. Select among the following: Precambrian, early Paleozoic, late Paleozoic, Mesozoic, Cenozoic.

(a) Pangaea came into existence.
(b) Bacteria and blue-green algae preserved in chert.
(c) The era that encompasses the least amount of time.
(d) Shields.
(e) "Age of dinosaurs."
(f) Formation of the original northern Appalachian mountains.
(g) Mastodons and mammoths.
(h) Extensive deposits of rock salt.

(i) Triassic, Jurassic, and Cretaceous.

(j) Coal swamps extended across North America, Europe, and Siberia.

(k) Gulf Coast oil deposits formed.

(l) Formation of most of the world's major iron ore deposits.

(m) Massive sand dunes covered large portions of the Colorado Plateau region.

(n) The "age of fishes" occurred during this span.

(o) Cambrian, Ordovician, and Silurian.

(p) Pangaea began to break apart.

(q) "Age of mammals."

(r) Animals with hard parts first appeared in abundance.

(s) Gymnosperms were the dominant trees.

(t) Columbia Plateau formed.

(u) Stromatolites are among its more common fossils.

(v) "Golden age of trilobites" occurred during this span.

(w) Fault-block mountains form in the Basin and Range Region

5. Briefly discuss two proposals that attempt to explain why several groups developed hard parts at the beginning of the Cambrian period. Do these proposals appear to provide a satisfactory explanation? Why or why not?

6. List some differences between amphibians and reptiles. List differences between reptiles and mammals.

7. Describe one hypothesis that attempts to explain the extinction of the dinosaurs. Cite a major objection to this proposal.

8. Contrast the eastern and western margins of North America during the Cenozoic era in terms of their relationships to plate boundaries.

Testing What You Have Learned

To test your knowledge of the material presented in this chapter, answer the following questions:

Multiple Choice Questions

1. Earth is about _____ years old.
a. 4000   **c.** 5.8 million   **e.** 6.7 billion
b. 4.6 million   **d.** 4.6 billion

2. Life on Earth began during _____ time.
a. Cenozoic   **c.** Paleozoic   **e.** Pleistocene
b. Mesozoic   **d.** Precambrian

3. Which era of the geologic time scale is often referred to as the "age of mammals"?
a. Hadean   **c.** Cenozoic   **e.** Precambrian
b. Paleozoic   **d.** Mesozoic

4. One group of _____, exemplified by the fossil *Archaeopteryx*, led to the birds.
a. reptiles   **c.** rodents   **e.** angiosperms
b. amphibians   **d.** fishes

5. Large core areas of Precambrian rocks, called _____, exist on each continent.
a. plateaus   **c.** roots   **e.** terranes
b. shields   **d.** complexes

6. Which span of geologic time encompasses the greatest percentage of Earth's history?
a. Precambrian   **c.** Cenozoic   **e.** Paleozoic
b. Mesozoic   **d.** Pleistocene

7. The beginning of the Paleozoic era is marked by the appearance of the first life-forms with _____.
a. legs   **c.** fins   **e.** lungs
b. cells   **d.** hard parts

8. With the perfection of the shelled egg, reptiles quickly became the dominant animals during the _____ era.
   - **a.** Mesozoic
   - **c.** Cenozoic
   - **e.** Paleozoic
   - **b.** Devonian
   - **d.** Precambrian

9. The supercontinent of Pangaea began forming during the late _____ era.
   - **a.** Precambrian
   - **c.** Devonian
   - **e.** Mesozoic
   - **b.** Cenozoic
   - **d.** Paleozoic

10. Which one of the following was NOT a characteristic of primitive mammals?
    - **a.** small size
    - **c.** short legs
    - **b.** large brain
    - **d.** flat, five-toed feet

Fill-In Questions

11. Earth's atmosphere consists of gases emitted from the planet during a process called _____.
12. During the Mesozoic, _____, such as the cycads and conifers, quickly became the dominant trees; however, during the Cenozoic, flowering plants called _____ were the primary land plants.
13. At the close of the Mesozoic, two groups of mammals, the _____ and _____, evolved and expanded to dominate the Cenozoic.
14. Plants, employing the process called _____, are the major contributors of oxygen to the atmosphere.
15. The proposal that suggests that the bodies of our solar system condensed from an enormous cloud is referred to as the _____ hypothesis.

True/False Questions

16. The Devonian period is often called the "age of fishes." _____
17. The ancestral Appalachian Mountains formed during the late Paleozoic as North America and Africa collided. _____
18. Melting of Earth's interior allowed denser elements to rise to the surface. _____
19. Most of North America was below sea level during the Cenozoic era. _____
20. Evidence indicates that some dinosaurs, unlike their present-day reptile relatives, were warm blooded. _____

Answers

1.d; 2.d; 3.c; 4.a; 5.b; 6.a; 7.d; 8.a; 9.d; 10.b; 11. outgassing; 12. gymnosperms, angiosperms; 13. marsupials, placentals; 14. photosynthesis; 15. nebular; 16.T; 17.T; 18.T; 19.F; 20.T

Metric and English Units Compared

Units

1 kilometer (km)	= 1000 meters (m)
1 meter (m)	= 100 centimeters (cm)
1 centimeter (cm)	= 0.39 inch (in.)
1 mile (mi)	= 5280 feet (ft)
1 foot (ft)	= 12 inches (in.)
1 inch (in.)	= 2.54 centimeters (cm)
1 square mile (mi^2)	= 640 acres (a)
1 kilogram (kg)	= 1000 grams (g)
1 pound (lb)	= 16 ounces (oz)
1 fathom	= 6 feet (ft)

Conversions

When you want to convert:	Multiply by:	To find:

Length

inches	2.54	centimeters
centimeters	0.39	inches
feet	0.30	meters
meters	3.28	feet
yards	0.91	meters
meters	1.09	yards
miles	1.61	kilometers
kilometers	0.62	miles

Area

square inches	6.45	square centimeters
square centimeters	0.15	square inches
square feet	0.09	square meters
square meters	10.76	square feet
square miles	2.59	square kilometers
square kilometers	0.39	square miles

Volume

cubic inches	16.38	cubic centimeters
cubic centimeters	0.06	cubic inches
cubic feet	0.028	cubic meters
cubic meters	35.3	cubic feet
cubic miles	4.17	cubic kilometers
cubic kilometers	0.24	cubic miles
liters	1.06	quarts
liters	0.26	gallons
gallons	3.78	liters

Masses and Weights

ounces	28.33	grams
grams	0.035	ounces
pounds	0.45	kilograms
kilograms	2.205	pounds

Temperature

When you want to convert degrees Fahrenheit (°F) to degrees Celsius (°C), subtract 32 degrees and divide by 1.8.

When you want to convert degrees Celsius (°C) to degrees Fahrenheit (°F), multiply by 1.8 and add 32 degrees.

When you want to convert degrees Celsius (°C) to kelvins (K), delete the degree symbol and add 273. When you want to convert kelvins (K) to degrees Celsius (°C), add the degree symbol and subtract 273.

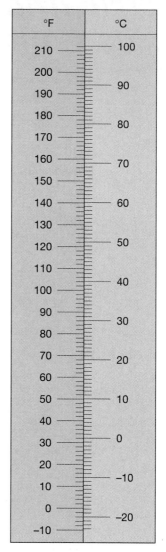

FIGURE A.1 A comparison of Fahrenheit and Celsius temperature scales.

Appendix B

Periodic Table of the Elements

Key:

2
He
4.003
Helium

- Atomic number
- Symbol of element
- Atomic weight
- Name of element

Legend:
- Metals
- Transition metals
- Nonmetals
- Noble gases
- Lanthanide series
- Actinide series

Period	IA	IIA	IIIB	IVB	VB	VIB	VIIB	VIIIB	VIIIB	VIIIB	IB	IIB	IIIA	IVA	VA	VIA	VIIA	VIIIA
1	1 H 1.0080 Hydrogen																	2 He 4.003 Helium
2	3 Li 6.939 Lithium	4 Be 9.012 Beryllium											5 B 10.81 Boron	6 C 12.011 Carbon	7 N 14.007 Nitrogen	8 O 15.9994 Oxygen	9 F 18.998 Fluorine	10 Ne 20.183 Neon
3	11 Na 22.990 Sodium	12 Mg 24.31 Magnesium											13 Al 26.98 Aluminum	14 Si 28.09 Silicon	15 P 30.974 Phosphorus	16 S 32.064 Sulfur	17 Cl 35.453 Chlorine	18 Ar 39.948 Argon
4	19 K 39.102 Potassium	20 Ca 40.08 Calcium	21 Sc 44.96 Scandium	22 Ti 47.90 Titanium	23 V 50.94 Vanadium	24 Cr 52.00 Chromium	25 Mn 53.94 Manganese	26 Fe 55.85 Iron	27 Co 58.93 Cobalt	28 Ni 58.71 Nickel	29 Cu 63.54 Copper	30 Zn 65.37 Zinc	31 Ga 69.72 Gallium	32 Ge 72.59 Germanium	33 As 74.92 Arsenic	34 Se 78.96 Selenium	35 Br 79.909 Bromine	36 Kr 83.80 Krypton
5	37 Rb 85.47 Rubidium	38 Sr 87.62 Strontium	39 Y 88.91 Yttrium	40 Zr 91.22 Zirconium	41 Nb 92.91 Niobium	42 Mo 95.94 Molybdenum	43 Tc (99) Technetium	44 Ru 101.1 Ruthenium	45 Rh 102.90 Rhodium	46 Pd 106.4 Palladium	47 Ag 107.87 Silver	48 Cd 112.40 Cadmium	49 In 114.82 Indium	50 Sn 118.69 Tin	51 Sb 121.75 Antimony	52 Te 127.60 Tellurium	53 I 126.90 Iodine	54 Xe 131.30 Xenon
6	55 Cs 132.91 Cesium	56 Ba 137.34 Barium	57 TO 71	72 Hf 178.49 Hafnium	73 Ta 180.95 Tantalum	74 W 183.85 Tungsten	75 Re 186.2 Rhenium	76 Os 190.2 Osmium	77 Ir 192.2 Iridium	78 Pt 195.09 Platinum	79 Au 197.0 Gold	80 Hg 200.59 Mercury	81 Tl 204.37 Thallium	82 Pb 207.19 Lead	83 Bi 208.98 Bismuth	84 Po (210) Polonium	85 At (210) Astantine	86 Rn (222) Radon
7	87 Fr (223) Francium	88 Ra 226.05 Radium	89 TO 103															

Lanthanide series

57 LA 138.91 Lanthanum	58 Ce 140.12 Cerium	59 Pr 140.91 Praseodymium	60 Nd 144.24 Neodymium	61 Pm (147) Promethium	62 Sm 150.35 Samarium	63 Eu 151.96 Europium	64 Gd 157.25 Gadolinium	65 Tb 158.92 Terbium	66 Dy 162.50 Dysprosium	67 Ho 164.93 Holmium	68 Er 167.26 Erbium	69 Tm 168.93 Thulium	70 Yb 173.04 Ytterbium	71 Lu 174.97 Lutetium

Actinide series

89 Ac (227) Actinium	90 Th 232.04 Thorium	91 Pa (231) Protactinium	92 U 238.03 Uranium	93 Np (237) Neptunium	94 Pu (242) Plutonium	95 Am (243) Americium	96 Cm (247) Curium	97 Bk (249) Berkelium	98 Cf (251) Californium	99 Es (253) Einsteinium	100 Fm (253) Fermium	101 Md (256) Mendelevium	102 No (254) Nobelium	103 Lw (257) Lawrencium

FIGURE B.1 Periodic table of the elements.

Mineral Identification Key

Group 1 Metallic Luster			
Hardness	**Streak**	**Other Diagnostic Properties**	**Name (Chemical Composition)**
Harder than glass	Black streak	Black; magnetic; hardness = 6; specific gravity = 5.2; often granular	Magnetite (Fe_3O_4)
	Greenish-black streak	Brass yellow; hardness = 6; specific gravity = 5.2; generally an aggregate of cubic crystals	Pyrite (FeS_2)—fool's gold
	Red-brown streak	Gray or reddish brown; hardness = 5–6; specific gravity = 5; platy appearance	Hematite (Fe_2O_3)
Softer than glass	Greenish-black streak	Golden yellow; hardness = 4; specific gravity = 4.2; massive	Chalcopyrite ($CuFeS_2$)
	Gray-black streak	Silvery gray; hardness = 2.5; specific gravity = 7.6 (very heavy); good cubic cleavage	Galena (PbS)
	Yellow-brown streak	Yellow brown to dark brown; hardness variable (1–6); specific gravity = 3.5–4; often found in rounded masses; earthy appearance	Limonite ($Fe_2O_3 \cdot H_2O$)
	Gray-black srteak	Black to bronze; tarnishes to purples and greens; hardness = 3; specific gravity = 5; massive	Bornite (Cu_5FeS_4)
Softer than your fingernail	Dark gray streak	Silvery gray; hardness = 1 (very soft); specific gravity = 2.2; massive to platy; writes on paper (pencil lead); feels greasy	Graphite (C)

		Group II Nonmetallic Luster (dark colored)	
Hardness	**Cleavage**	**Other Diagnostic Properties**	**Name (Chemical Composition)**
Harder than glass	Cleavage present	Black to greenish black; hardness = 5–6; specific gravity = 3.4; fair cleavage, two directions at nearly 90 degrees	Augite (Ca, Mg, Fe, Al silicate)
		Black to greenish black; hardness = 5–6; specific gravity = 3.2; fair cleavage, two directions at nearly 60 degrees and 120 degrees	Hornblende (Ca, Na, Mg, Fe, OH, Al silicate)
		Red to reddish brown; hardness = 6.5-7.5; conchoidal fracture; glassy luster	Garnet (Fe, Mg, Ca, Al silicate)
	Cleavage not prominent	Gray to brown; hardness = 9; specific gravity = 4; hexagonal crystals common	Corundum (Al_2O_3)
		Dark brown to black; hardness = 7; conchoidal fracture; glassy luster	Smoky quartz (SiO_2)
		Olive green; hardness = 6.5–7; small glassy grains	Olivine $(Mg, Fe)_2SiO_4$
Softer than glass	Cleavage present	Yellow brown to black; hardness = 4; good cleavage in six directions, light yellow streak that has the smell of sulfur	Sphalerite (ZnS)
		Dark brown to black; hardness = 2.5–3, excellent cleavage in one direction; elastic in thin sheets; black mica	Biotite (K, Mg, Fe, OH, Al silicate)
	Cleavage absent	Generally tarnished to brown or green; hardness = 2.5; speciic gravity = 9; massive	Native copper (Cu)
Softer than your fingernail	Cleavage not prominent	Reddish brown; hardness = 1–5; specific gravity = 4–5; red streak; earthy appearance	Hematite (Fe_2O_3)
		Yellow brown; hardness = 1–3; specific gravity = 3.5; earthy appearance; powders easily	Limonite ($Fe_2O_3 \cdot H_2O$)

		Group III Nonmetallic Luster (light colored)	
Hardness	**Cleavage**	**Other Diagnostic Properties**	**Name (Chemical Composition)**
Harder than glass	Cleavage present	Salmon colored or white to gray; hardness = 6; specific gravity = 2.6; two directions of cleavage at nearly right angles	Potassium feldspar ($KAlSi_3O_8$) Plagioclase feldspar ($NaAlSi_3O_8$ to $CaAl_2Si_2O_8$)
	Cleavage absent	Any color; hardness = 7; specific gravity = 2.65; conchoidal fracture; glassy appearance; varieties; milky, rose, smoky, amethyst (violet)	Quartz (SiO_2)
Softer than glass	Cleavage present	White, yellowish to colorless; hardness = 3; three directions of cleavage at 75 degrees (rhombohedral); effervesces in HCl; often transparent	Calcite ($CaCO_3$)
		White to colorless; hardness = 2.5; three directions of cleavage at 90 degrees (cubic); salty taste	Halite (NaCl)
		Yellow, purple, green, colorless; hardness = 4; white streak; translucent to transparent; four directions of cleavage	Fluorite (CaF_2)
Softer than your fingernail	Cleavage present	Colorless; hardness = 2–2.5; transparent and elastic in thin sheets; excellent cleavage in one direction; light mica	Muscovite (K, OH, Al silicate)
		White to transparent, hardness = 2; when in sheets, is flexible but not elastic; varieties: selenite (transparent, three directions of cleavage); satin spar (fibrous, silky luster); alabaster (aggregate of small crystals)	Gypsum ($CaSO_4 \cdot 2H_2O$)
	Cleavage not prominent	White, pink, green; hardness = 1; forms in thin plates; soapy feel; pearly luster	Talc (Mg silicate)
		Yellow; hardness = 1–2.5	Sulfur (S)
		White; hardness = 2; smooth feel; earthy odor when moistened, has typical clay texture	Kaolinite (Hydrous Al silicate)
		Green; hardness = 2.5; fibrous; variety of serpentine	Asbestos (Mg, Al silicate)
		Pale to dark reddish brown; hardness = 1–3; dull luster; earthy; often contains spheroidal-shaped particles; not a true mineral	Bauxite (Hydrous Al oxide)

Topographic Maps

A map is a representation on a flat surface of all or a part of Earth's surface drawn to a specific scale. Maps are often the most effective means for showing the locations of both natural and human structures, their sizes, and their relationships to one another. Like photographs, maps readily display information that would be impractical to express in words.

While most maps show only the two horizontal dimensions, geologists, as well as other map users, often require that the third dimension, elevation, be shown on maps. Maps that show the shape of the land are called **topographic maps.** Although various techniques may be used to depict elevations, the most accurate method involves the use of contour lines.

Contour Lines

A **contour line** is a line on a map representing a corresponding imaginary line on the ground that has the same elevation above sea level along its entire length. While many map symbols are pictographs, resembling the objects they represent, a contour line is an abstraction, that has no counterpart in nature. It is, however, an accurate and effective device for representing the third dimension on paper.

Some useful facts and rules concerning contour lines are listed as follows. This information should be studied in conjunction with Figure D.l.

1. Contour lines bend upstream or upvalley. The contours form Vs that point upstream, and in the upstream direction the successive contours represent higher elevations. For example, if you were standing on a stream bank and wished to get to the point at the same elevation directly opposite you on the other bank, without stepping up or down, you would need to walk upstream along the contour at that elevation to where it crosses the stream bed, cross the stream, and then walk back downstream along the same contour.

2. Contours near the upper parts of hills form closures. The top of a hill is higher than the highest closed contour.

3. Hollows (depressions) without outlets are shown by closed, hatched contours. Hatched contours are contours with short lines on the inside pointing downslope.

4. Contours are widely spaced on gentle slopes.

5. Contours are closely spaced on steep slopes.

6. Evenly spaced contours indicate a uniform slope.

7. Contours usually do not cross or intersect each other, except in the rare case of an overhanging cliff.

8. All contours eventually close, either on a map or beyond its margins.

9. A single high contour never occurs between two lower ones, and vice versa. In other words, a change in slope direction is always determined by the repetition of the same elevation either as two different contours of the same value or as the same contour crossed twice.

10. Spot elevations between contours are given at many places, such as road intersections, hill summits, and lake surfaces. Spot elevations differ from control elevation stations, such as bench marks, in not being permanently established by permanent markers.

Relief

Relief refers to the difference in elevation between any two points. Maximum relief refers to the difference in elevation between the highest and lowest points in the area being considered. Relief determines the **contour interval,** which is the difference in elevation between succeeding contour lines that is used on topographic maps. Where relief is low, a small contour interval, such as 10 or 20 feet, may be used. In flat areas, such as wide river valleys or broad, flat uplands, a contour interval of 5 feet is often used. In rugged mountainous terrain, where

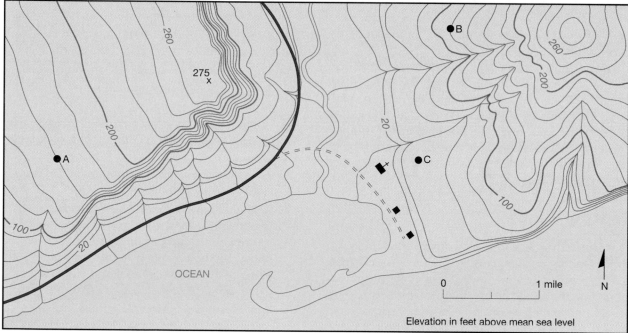

FIGURE D.I Perspective view of an area and a contour map of the same area. These illustrations show how features are depicted on a topographic map. The upper illustration is a perspective view of a river valley and the adjoining hills. The river flows into a bay, which is partly enclosed by a hooked sandbar. On either side of the valley are terraces through which streams have cut gullies. The hill on the right has a smoothly eroded form and gradual slopes, whereas the one on the left rises abruptly in a sharp precipice, from which it slopes gently, and forms an inclined plateau traversed by a few shallow gullies. A road provides access to a church and the two houses situated across the river from a highway that follows the seacoast and curves up the river valley. The lower illustration shows the same features represented by symbols on a topographic map. The contour interval (vertical distance between adjacent contours) is 20 feet. (After U.S. Geological Survey)

relief is many hundreds of feet, contour intervals as large as 50 or 100 feet are used.

Scale

Map **scale** expresses the relationship between distance or area on the map to the true distance or area on Earth's surface. This is generally expressed as a ratio or fraction, such as 1:24,000 or 1/24,000. The numerator, usually 1, represents map distance, and the denominator, a large number, represents ground distance. Thus, 1:24,000 means that a distance of 1 unit on the map represents a distance of 24,000 such units on the surface of Earth. It does not matter what the units are.

Often, the graphic or bar scale is more useful than the fractional scale, because it is easier to use for measuring distances between points. The graphic scale (Figure D.2) consists of a bar divided into equal segments, which represent equal distances on the map. One segment on the left side of the bar is usually divided into smaller units to permit more accurate estimates of fractional units.

Topographic maps, which are also referred to as *quadrangles,* are generally classified according to publication scale. Each series is intended to fulfill a specific type of map need. To select a map with the proper scale for a particular use, remember that large-scale maps show more detail and small-scale maps show less detail. The sizes and scales of topographic maps published by the U.S. Geological Survey are shown in Table D. l.

Color and Symbol

Each color and symbol used on U.S. Geological Survey topographic maps has significance. Common topographic map symbols are shown in Figure D.3. The meaning of each color is as follows:

Blue — water features
Black — works of humans, such as homes, schools, churches, roads, and so forth
Brown — contour lines
Green — woodlands, orchards, and so forth
Red — urban areas, important roads, public land subdivision lines

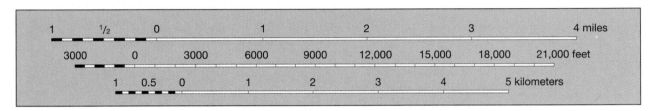

FIGURE D.2 Graphic scale.

TABLE D.3 National Topographic Maps

Series	Scale	1 inch Represents	Standard Quadrangle Size (latitude-longitude)	Quadrangle Area (square miles)	Paper Size E-W N-S Width Length (inches)
7 ¹/₂-minute	1:24,000	2000 feet	7 ¹/₂′ x 7 ¹/₂′	49–70	22 x 27
Puerto Rico 7 ¹/₂-minute	1:20,000	about 1667 feet	7 ¹/₂′ x 7 ¹/₂′	71	29 ¹/₂ x 32 ¹/₂
15-minute	1:62,500	nearly 1 mile	15′ x 15′	197–282	17 x 21
Alaska 1:63,360	1:63,360	1 mile	15′ x 20′–36′	207–281	18 x 21
U.S. 1:250,000	1:250,000	nearly 4 miles	1° x 2°	4580–8669	34 x 22
U.S. 1:1,000,000	1:1,000,000	nearly 16 miles	4° x 6°	73,734–102,759	27 x 27

SOURCE: U.S. Geological Survey

TOPOGRAPHIC MAP SYMBOLS

VARIATIONS WILL BE FOUND ON OLDER MAPS

Primary highway, hard surface	
Secondary highway, hard surface	
Light-duty road, hard or improved surface	
Unimproved road	
Road under construction, alinement known	
Proposed road	
Dual highway, dividing strip 25 feet or less	
Dual highway, dividing strip exceeding 25 feet	
Trail	

Railroad: single track and multiple track	
Railroads in juxtaposition	
Narrow gage: single track and multiple track	
Railroad in street and carline	
Bridge: road and railroad	
Drawbridge: road and railroad	
Footbridge	
Tunnel: road and railroad	
Overpass and underpass	
Small masonry or concrete dam	
Dam with lock	
Dam with road	
Canal with lock	

Buildings (dwelling, place of employment, etc.)	
School, church, and cemetery	
Buildings (barn, warehouse, etc.)	
Power transmission line with located metal tower	
Telephone line, pipeline, etc. (labeled as to type)	
Wells other than water (labeled as to type)	o Oil .... o Gas
Tanks: oil, water, etc. (labeled only if water)	Water
Located or landmark object; windmill	
Open pit, mine, or quarry; prospect	
Shaft and tunnel entrance	

Horizontal and vertical control station:

Tablet, spirit level elevation	BM △ 5653
Other recoverable mark, spirit level elevation	△ 5455
Horizontal control station: tablet, vertical angle elevation	VABM △ 95/9
Any recoverable mark, vertical angle or checked elevation	△ 3775
Vertical control station: tablet, spirit level elevation	BM × 957
Other recoverable mark, spirit level elevation	× 954
Spot elevation	× 7369 × 7369
Water elevation	670 670

Boundaries: National	
State	
County, parish, municipio	
Civil township, precinct, town, barrio	
Incorporated city, village, town, hamlet	
Reservation, National or State	
Small park, cemetery, airport, etc.	
Land grant	
Township or range line, United States land survey	
Township or range line, approximate location	
Section line, United States land survey	
Section line, approximate location	
Township line, not United States land survey	
Section line, not United States land survey	
Found corner: section and closing	
Boundary monument: land grant and other	
Fence or field line	

Index contour		Intermediate contour	
Supplementary contour		Depression contours	
Fill		Cut	
Levee		Levee with road	
Mine dump		Wash	
Tailings		Tailings pond	
Shifting sand or dunes		Intricate surface	
Sand area		Gravel beach	

Perennial streams		Intermittent streams	
Elevated aqueduct		Aqueduct tunnel	
Water well and spring	o ...... o~	Glacier	
Small rapids		Small falls	
Large rapids		Large falls	
Intermittent lake		Dry lake bed	
Foreshore flat		Rock or coral reef	
Sounding, depth curve	10	Piling or dolphin	
Exposed wreck		Sunken wreck	
Rock, bare or awash; dangerous to navigation			

Marsh (swamp)		Submerged marsh	
Wooded marsh		Mangrove	
Woods or brushwood		Orchard	
Vineyard		Scrub	
Land subject to controlled inundation		Urban area	

FIGURE D.3 U.S. Geological Survey topographic map symbols. (Variations will be found on older maps.)

Landforms of the Conterminous United States

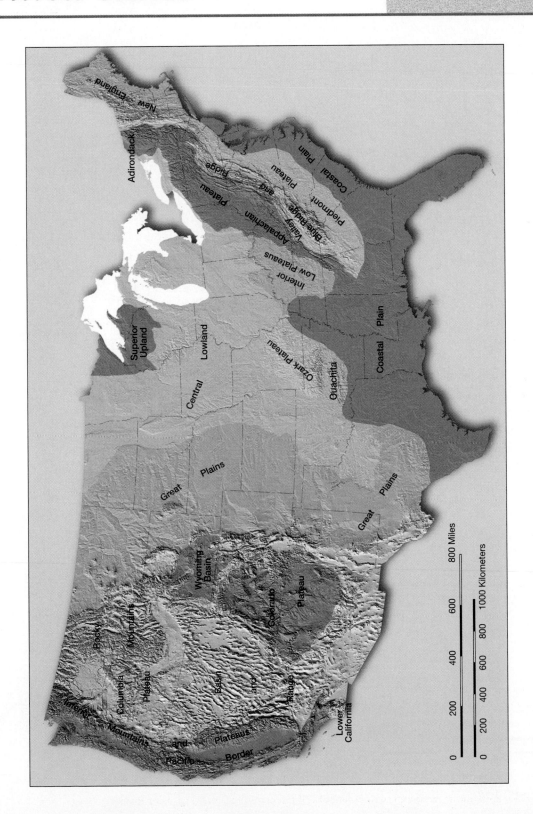

FIGURE E.1 Outline map showing major physiographic provinces of the United States.

0   200   400   600   800 Miles

0   200   400   600   800   1000 Kilometers

Landforms of the Conterminous United States

FIGURE E.2 Digital shaded relief landform map of the United States. (Data provided by the U.S. Geological Survey)

Glossary

Aa A type of lava flow that has a jagged, blocky surface.

Ablation A general term for the loss of ice and snow from a glacier.

Abrasion The grinding and scraping of a rock surface by the friction and impact of rock particles carried by water, wind, or ice.

Absolute dating Determination of the number of years since the occurrence of a given geologic event.

Abyssal plain Very level area of the deep-ocean floor, usually lying at the foot of the continental rise.

Accretionary prism A large wedge-shaped mass of sediment that accumulates in subduction zones. Here sediment is scraped from the subducting oceanic plate and accreted to the overriding crustal block.

Active layer The zone above the permafrost that thaws in summer and refreezes in winter.

Aftershock A smaller earthquake that follows the main earthquake.

Alluvial fan A fan-shaped deposit of sediment formed when a stream's slope is abruptly reduced.

Alluvium Unconsolidated sediment deposited by a stream.

Alpine glacier See *Valley glacier.*

Angle of repose The steepest angle at which loose material remains stationary without sliding downslope.

Angular unconformity An unconformity in which the older strata dip at an angle different from that of the younger beds.

Anthracite A hard, metamorphic form of coal that burns clean and hot.

Anticline A fold in sedimentary strata that resembles an arch.

Aphanitic texture A texture of igneous rocks in which the crystals are too small for individual minerals to be distinguished with the unaided eye.

Aquifer Rock or sediment through which groundwater moves easily.

Aquitard An impermeable bed that hinders or prevents groundwater movement.

Archean eon The earliest time unit on the geologic time scale. The eon ending 2500 million years ago that preceded the Proterozoic eon.

Arête A narrow, knifelike ridge separating two adjacent glaciated valleys.

Arkose A feldspar-rich sandstone.

Artesian well A well in which the water rises above the level where it was initially encountered.

Asthenosphere A subdivision of the mantle situated below the lithosphere. This zone of weak material exists below a depth of about 100 kilometers and in some regions extends as deep as 700 kilometers. The rock within this zone is easily deformed.

Atmosphere The gaseous portion of a planet; the planet's envelope of air. One of the traditional subdivisions of Earth's physical environment.

Atoll A continuous or broken ring of coral reef surrounding a central lagoon.

Atom The smallest particle that exists as an element.

Atomic mass unit A mass unit equal to exactly one twelfth the mass of a carbon-12 atom.

Atomic number The number of protons in the nucleus of an atom.

Atomic weight The average of the atomic masses of isotopes of a given element.

Aureole A zone or halo of contact metamorphism found in the country rock surrounding an igneous intrusion.

Back swamp A poorly drained area on a floodplain resulting when natural levees are present.

Barchan dune A solitary sand dune shaped like a crescent with its tips pointed downwind.

Barchanoid dunes Dunes forming scalloped rows of sand oriented at right angles to the wind. This form is intermediate between isolated barchans and extensive waves of transverse dunes.

Barrier island A low, elongated ridge of sand that parallels the coast.

Basal slip A mechanism of glacial movement in which the ice mass slides over the surface below.

Basalt A fine-grained igneous rock of mafic composition.

Base level The level below which a stream cannot erode.

Basin A circular downfolded structure.

Batholith A large mass of igneous rock that formed when magma was emplaced at depth, crystallized, and was subsequently exposed by erosion.

Baymouth bar A sandbar that completely crosses a bay, sealing it off from the main body of water.

Beach drift The transport of sediment in a zigzag pattern along a beach. It is caused by the uprush of water from obliquely breaking waves.

Beach nourishment Large quantities of sand are added to the beach system to offset losses caused by wave erosion. By building beaches seaward, beach quality and storm protection are both improved.

Bed See *Srata.*

Bedding plane A nearly flat surface separating two beds of sedimentary rock. Each bedding plane marks the end of one deposit and the beginning of another having different characteristics.

Bed load Sediment rolled along the bottom of a stream by moving water, or particles robed along the ground surface by wind.

Belt of soil moisture. A zone in which water is held as a film on the surface of soil particles and may be used by plants or withdrawn by evaporation. The uppermost subdivision of the zone of aeration.

Biochemical Describing a type of chemical sediment that forms when material dissolved in water is precipitated by water-dwelling organisms. Shells are common examples.

Biogenous sediment Seafloor sediments consisting of material of marine-organic origin.

Bituminous coal The most common form of coal, often called soft, black coal.

Blowout (deflation hollow) A depression excavated by wind in easily eroded materials.

Body wave A seismic wave that travels through Earth's interior.

Bottomset bed A layer of fine sediment deposited beyond the advancing edge of a delta and then buried by continuous delta growth.

Bowen's reaction series A concept proposed by N. L. Bowen that illustrates the relationship between magma and the minerals crystallizing from it during the formation of igneous rock.

Braided stream A stream consisting of numerous intertwining channels.

Breakwater A structure protecting a near shore area from breaking waves.

Breccia A sedimentary rock composed of angular fragments that were lithified.

Caldera A large depression typically caused by collapse of the summit area of a volcano following a violent eruption.

Caliche A hard layer, rich in calcium carbonate, that forms beneath the *B* horizon in soils of arid regions.

Calving Wastage of a glacier that occurs when large pieces of ice break off into water.

Capacity The total amount of sediment a stream is able to transport.

Capillary fringe A relatively narrow zone at the base of the zone of aeration. Here water rises from the water table in tiny threadlike openings between grains of soil or sediment.

Cap rock A necessary part of an oil trap. The cap rock is impermeable and hence keeps upwardly mobile oil and gas from escaping at the surface.

Catastrophism The concept that Earth was shaped by catastrophic events of a short-term nature.

Cavern A naturally formed underground chamber or series of chambers most commonly produced by solution activity in limestone.

Cementation One way in which sedimentary rocks are lithified. As material precipitates from water that percolates through the sediment, open spaces are filled and particles are joined into a solid mass.

Cenozoic era A time span on the geologic calendar beginning about 66 million years ago following the Mesozoic era.

Chemical sedimentary rock Sedimentary rock consisting of material that was precipitated from water by either inorganic or organic means.

Chemical weathering The processes by which the internal structure of a mineral is altered by the removal and/or addition of elements.

Cinder cone A rather small volcano built primarily of pyroclastics ejected from a single vent.

Cirque An amphitheater-shaped basin at the head of a glaciated valley produced by frost wedging and plucking.

Clastic A sedimentary rock texture consisting of broken fragments of pre-existing rock.

Cleavage The tendency of a mineral to break along planes of weak bonding.

Col A pass between mountain valleys where the headwalls of two cirques intersect.

Color A phenomenon of light by which otherwise identical objects may be differentiated.

Column A feature found in caves that is formed when a stalactite and stalagmite join.

Columnar joints A pattern of cracks that forms during cooling of molten rock to generate columns.

Compaction A type of lithification in which the weight of overlying material compresses more deeply buried sediment. It is most important in fine-grained sedimentary rocks such as shale.

Competence A measure of the largest particle a stream can transport; a factor dependent on velocity.

Composite cone A volcano composed of both lava flows and proclastic material.

Compound A substance formed by the chemical combination of two or more elements in definite proportions and usually having properties different from those of its constituent elements.

Concordant A term used to describe intrusive igneous masses that form parallel to the bedding of the surrounding rock.

Cone of depression A cone-shaped depression immediately surrounding a well.

Conformable layers Rock layers that were deposited without interruption.

Conglomerate A sedimentary rock composed of rounded, gravel-sized particles.

Contact metamorphism Changes in rock caused by the heat of a nearby magma body.

Continental drift A hypothesis, credited largely to Alfred Wegener, that suggested all present continents once existed as a single supercontinent. Further, beginning about 200 million years ago, the supercontinent began breaking into smaller continents, which then "drifted" to their present positions.

Continental margin That proportion of the sea floor adjacent to the continents. It may include the continental shelf, continental slope, and continental rise.

Continental rise The gently sloping surface at the base of the continental slope.

Continental shelf The gently sloping submerged portion of the continental margin extending from the shoreline to the continental slope.

Continental slope The steep gradient that leads to the deep-ocean floor and marks the seaward edge of the continental shelf.

Convergent boundary A boundary in which two plates move together, causing one of the slabs of lithosphere to be consumed into the mantle as it descends beneath the overriding plate.

Core Located beneath the mantle, it is Earth's innermost layer. The core is divided into an outer core and an inner core.

Correlation Establishing the equivalence of rocks of similar age in different areas.

Covalent bond A chemical bond produced by the sharing of electrons.

Crater The depression at the summit of a volcano, or that which is produced by a meteorite impact.

Creep The slow downhill movement of soil and regolith.

Crevasse A deep crack in the brittle surface of a glacier.

Cross-bedding Structure in which relatively thin layers are inclined at an angle to the main bedding. Formed by currents of wind or water.

Cross-cutting A principle of relative dating. A rock or fault is younger than any rock (or fault) through which it cuts.

Crust The very thin, outermost layer of Earth.

Crystal An orderly arrangement of atoms.

Crystal form The external appearance of a mineral as determined by its internal arrangement of atoms.

Crystallization The formation and growth of a crystalline solid from a liquid or gas.

Curie point The temperature above which a material loses its magnetization.

Cut bank The area of active erosion on the outside of a meander.

Cutoff A short channel segment created when a river erodes through the narrow neck of land between meanders.

Daughter product An isotope resulting from radioactive decay.

Deep-focus earthquake An earthquake focus at a depth of more than 300 kilometers.

Deep-ocean basin The portion of sea floor that lies between the continental margin and the oceanic ridge system. This region compromises almost 30 percent of Earth's surface.

Deep-ocean trench A narrow, elongated depression of the seafloor.

Deflation The lifting and removal of loose material by wind.

Delta An accumulation of sediment formed where a stream enters a lake or ocean.

Dendritic pattern A stream system that resembles the pattern of a branching tree.

Density The weight per unit volume of a particular material.

Desert One of the two types of dry climate; the drier of the dry climates.

Desert pavement A layer of coarse pebbles and gravel created when wind removes the finer material.

Detrital sedimentary rocks Rocks that form from the accumulation of materials that originate and are transported as solid particles derived from both mechanical and chemical weathering.

Dike A tabular-shaped intrusive igneous feature that cuts through the surrounding rock.

Dip The angle at which a rock layer is inclined from the horizontal. The direction of dip is at a right angle to the strike.

Dip-slip fault A fault in which the movement is parallel to the dip of the fault.

Discharge The quantity of water in a stream that passes a given point in a period of time.

Disconformity A type of unconformity in which the beds above and below are parallel.

Discontinuity A sudden change with depth in one or more of the physical properties of the material making up Earth's interior. The boundary between two dissimilar materials in Earth's interior as determined by the behavior of seismic waves.

Discordant A term used to describe plutons that cut across existing rock structures, such as bedding planes.

Disseminated deposit Any economic mineral deposit in which the desired mineral occurs as scattered particles in the rock but in sufficient quantity to make the deposit an ore.

Dissolved load The portion of a stream's load carried in solution.

Distributary A section of a stream that leaves the main flow.

Diurnal tide A tide characterized by a single high and low water height each tidal day.

Divergent boundary A boundary in which two plates move apart, resulting in upwelling of material from the mantle to create new seafloor.

Divide An imaginary line that separates the drainage of two streams; often found along a ridge.

Dome A roughly circular, upfolded structure.

Drainage basin The land area that contributes water to a stream.

Drawdown The difference in height between the bottom of a cone of depression and the original height of the water table.

Drift The general term for any glacial deposit.

Drumlin A streamlined asymmetrical hill composed of glacial till. The steep side of the hill faces the direction from which the ice advanced.

Dry climate A climate in which the yearly precipitation is less than the potential loss of water by evaporation.

Dune A hill or ride of wind-deposited sand.

Earthflow The downslope movement of water-saturated, clay-rich sediment. Most characteristic of humid regions.

Earthquake Vibration of Earth produced by the rapid release of energy.

Echo sounder An instrument used to determine the depth of water by measuring the time interval between emission of a sound signal and the return of its echo from the bottom.

Effluent stream A stream channel that intersects the water table. Consequently, groundwater feeds into the stream.

Elastic deformation Nonpermanent deformation in which rock returns to its original shape when the stress is released.

Elastic rebound The sudden release of stored strain in rocks that results in movement along a fault.

Electron A negatively charged subatomic particle that has a negligible mass and is found outside the atom's nucleus.

Element A substance that cannot be decomposed into simpler substances by ordinary chemical or physical means.

Eluviation The washing out of fine soil components from the *A* horizon by downward-percolating water.

Emergent coast A coast where land formerly below the sea level has been exposed either by crustal uplift or a drop in sea level or both.

End moraine A ridge of till marking a former position of the front of the glacier.

Energy-level shell The region occupied by electrons with a specific energy level.

Entrenched meander A meander cut into bedrock when uplifting rejuvenated a meandering stream.

Eon The largest time unit on the geologic time scale, next in order of magnitude above era.

Epicenter The location on Earth's surface that lies directly above the focus of an earthquake.

Epoch A unit of the geological calendar that is a subdivision of a period.

Era A major division on the geologic calendar; eras are divided into shorter units called periods.

Erosion The incorporation and transportation of material by a mobile agent, such as water, wind, or ice.

Esker Sinuous ridge composed largely of sand gravel deposited by a stream flowing in a tunnel beneath a glacier near its terminus.

Estuary A funnel-shaped inlet of the sea that formed when a rise in sea level or subsidence of land caused the mouth of a river to be flooded.

Evaporite A sedimentary rock formed of material deposited from solution by evaporation of water.

Evapotranspiration The combined effect of evaporation and transpiration.

Exfoliation dome Large, dome-shaped structure, usually composed of granite, formed by sheeting.

Exotic stream A permanent stream that traverses a desert and has its source in well-watered areas outside the desert.

Extrusive Igneous activity that occurs at Earth's surface.

Fall A type of movement common to mass-wasting processes that refers to the free falling of detached individual pieces of any size.

Fault A break in a rock mass along which movement has occurred.

Fault-block mountain A mountain formed by the displacement of rock along a fault.

Fetch The distance that the wind has traveled across the open water.

Fiord A steep-sided inlet of the sea formed when a glacial trough was partially submerged.

Fissure eruption An eruption in which lava is extruded from narrow fractures or cracks in the crust.

Flood basalts Flows of basaltic lava that issue from numerous cracks or fissures and commonly cover extensive areas to thicknesses of hundreds of meters.

Floodplain The flat, low-lying portion of a stream valley subject to periodic inundation.

Flow A type of movement common to mass-wasting processes in which water-saturated material moves downslope as a viscous fluid.

Fluorescence The absorption of ultraviolet light, which is re-emitted as visible light.

Focus (earthquake) The zone within Earth where rock displacement produces an earthquake.

Fold A bent layer or series of layers that were originally horizontal and subsequently deformed.

Foliated A texture of metamorphic rocks that gives the rock a layered appearance.

Foliation A term for a linear arrangement of textural features often exhibited by metamorphic rocks.

Foreset bed An inclined bed deposited along the front of a delta.

Foreshocks Small earthquakes that often precede a major earthquake.

Fossil The remains or traces of organisms preserved from the geologic past.

Fossil fuel General term for any hydrocarbon that may be used as a fuel, including coal, oil, natural gas, bitumen from tar sands, and shale oil.

Fossil succession Fossil organisms succeed one another in a definite and determinable order, and any time period can be recognized by its fossil content.

Fractional crystallization The process that separates magma into components having varied compositions and melting points.

Fracture (mineral) One of the basic physical properties of minerals. It relates to the breakage of minerals when there are no planes of weakness in the crystalline structure. Examples include conchoidal, irregular, and splintery.

Fracture (rock) Any break or rupture in rock along which no appreciable movement has taken place.

Frost wedging The mechanical breakup of rock caused by the expansion of freezing water in cracks and crevices.

Fumarole A vent in a volcanic area from which fumes or gases escape.

Geology The science that examines Earth, its form and composition, and the changes which it has undergone and is undergoing.

Geosyncline A large linear downwarp in Earth's crust in which thousands of meters of sediment have accumulated.

Geothermal energy Natural steam used for power generation.

Geothermal gradient The gradual increase in temperature with depth in the crust. The average is 30ºC per kilometer in the upper crust.

Geyser A fountain of hot water ejected periodically from the ground.

Glacial budget The balance, or lack of balance, between accumulation at the upper end of a glacier, and loss at the other end.

Glacial erratic An ice-transported boulder that was not derived from the bedrock near its present site.

Glacial striations Scratches and grooves on bedrock caused by glacial abrasion.

Glacial trough A mountain valley that has been widened, deepened, and straightened by a glacier.

Glacier A thick mass of ice originating on land from the compaction and recrystallization of snow. The ice shows evidence of past or present flow.

Glass (volcanic) Natural glass produced when molten lava cools too rapidly to permit crystallization. Volcanic glass is a solid composed of unordered atoms.

Glassy texture A term used to describe the texture of certain igneous rocks, such as obsidian, that contain no crystals.

Gneissic texture The texture displayed by the metamorphic rock *gneiss* in which dark and light silicate minerals have separated, giving the rock a banded appearance.

Gondwanaland The southern portion of Pangaea consisting of South America, Africa, Australia, India, and Antarctica.

Graben A valley formed by the downward displacement of a fault-bounded block.

Graded bed A sediment layer characterized by a decrease in sediment size from bottom to top.

Graded stream A stream that has the correct channel characteristics to maintain the exact velocity required to transport the material supplied to it.

Gradient The slope of a stream; generally measured in feet per mile.

Granitization The process of converting country rock into granite. The process is thought to occur when hot, ion-rich fluids migrate through a rock and chemically alter its composition.

Greenhouse effect Carbon dioxide and water vapor in a planet's atmosphere absorb and re-radiate infrared wavelengths, effectively trapping solar energy and raising the temperature.

Groin A short wall built at a right angle to the seashore to trap moving sand.

Groundmass The matrix of smaller crystals within an igneous rock that has porphyritic texture.

Ground moraine An undulating layer of till deposited as the ice front retreats.

Groundwater Water in the zone of saturation.

Guyot A submerge flat-topped seamount.

Half-life The time required for one-half of the atoms of a radioactive substance to decay.

Hanging valley A tributary valley that enters a glacial trough at a considerable height above the floor of the trough.

Hardness A mineral's resistance to scratching and abrasion.

Head The vertical distance between the recharge and discharge points of a water table. Also the source area or beginning of a valley.

Headward erosion The extension upslope of the head of a valley due to erosion.

Historical geology A major division of geology that deals with the origin of Earth and its development through time. Usually involves the study of fossils and their sequence in rock beds.

Hogback A narrow, sharp-crested ridge formed by the upturned edge of a steeply dipping bed of resistant rock.

Horn A pyramid-like peak formed by glacial action in three or more cirques surrounding a mountain summit.

Horst An elongate, uplifted block of crust bounded by faults.

Hot spot A proposed concentration of heat in the mantle capable of introducing magma that, in turn, extrudes onto Earth"s surface. The intraplate volcanism that produced the Hawaiian Islands is one example.

Hot spring A spring in which the water is 6º–9ºC (10–15ºF) warmer than the mean annual air temperature of its locality.

Humus Organic matter in soil produced by the decomposition of plants and animals.

Hydrogenous sediment Seafloor sediments consisting of minerals that crystallize from seawater. The principal example is manganese nodules.

Hydrologic cycle The unending circulation of Earth's water supply. The cycle is powered by energy from the Sun and is characterized by continuous exchanges of water among the oceans, the atmosphere, and the continents.

Hydrolysis A chemical weathering process in which minerals are altered by chemically reacting with water and acids.

Hydrosphere The water portion of our planet; one of the traditional subdivisions of Earth's physical environment.

Hydrothermal solution The hot, watery solution that escapes from a mass of magma during the latter stages of crystallization. Such solutions may alter the surrounding country rock and are frequently the source of significant ore deposits.

Hypothesis A tentative explanation that is then tested to determine if it is valid.

Ice cap A mass of glacial ice covering a high upland or plateau and spreading out radially.

Ice-contact deposit An accumulation of stratified drift deposited in contact with a supporting mass of ice.

Ice sheet A very large, thick mass of glacial ice flowing outward in all directions from one or more accumulation centers.

Igneous rock A rock formed by the crystallization of molten magma.

Immature soil A soil lacking horizons.

Inclusion A piece of one rock unit contained within another. Inclusions are used in relative dating. The rock mass adjacent to the one containing the inclusion must have been there first in order to provide the fragment.

Index fossil A fossil that is associated with a particular span of geologic time.

Index mineral A mineral that is a good indicator of the metamorphic environment in which it formed. Used to distinguish different zones of regional metamorphism.

Inertia Objects at rest tend to remain at rest and objects in motion tend to stay in motion unless either is acted upon by an outside force.

Infiltration The movement of surface water into rock or soil through crack and pore spaces.

Infiltration capacity The maximum rate at which soil can absorb water.

Influent stream A stream channel that is above the water table level. Water seeps downward from the channel to the zone of saturation to produce an upward bulge in the water table.

Inner core The solid, innermost layer of Earth, about 1216 kilometers (754 miles) in radius.

Inselberg An isolated mountain remnant characteristic of the late stage of erosion in a mountainous and region.

Intensity (earthquake) An indication of the destructive effects of an earthquake at a particular place. Intensity is affected by such factors as distance to the epicenter and the nature of the surface materials.

Interior drainage A discontinuous pattern of intermittent streams that do not flow to the ocean.

Intermediate focus An earthquake focus at a depth of between 60 and 300 kilometers.

Intrusive rock Igneous rock that formed below Earth's surface.

Ion An atom or molecule that possesses an electrical charge.

Ionic bond A chemical bond between two oppositely charged ions formed by the transfer of valence electrons from one atom to another.

Island arc A chain of volcanic islands generally located a few hundred kilometers from a trench where active subduction of one oceanic slab beneath another is occurring.

Isostasy The concept that Earth's crust is "floating" in gravitational balance upon the material of the mantle.

Isotopes Varieties of the same element that have different mass numbers; their nuclei contain the same number of protons but different numbers of neutrons.

Joint A fracture in rock along which there has been no movement.

Kame A steep-sided hill composed of sand and gravel originating when sediment collected in openings in stagnant glacial ice.

Kame terrace A narrow, terrace-like mass of stratified drift deposited between a glacier and an adjacent valley wall.

Karst A topography consisting of numerous depressions called sinkholes.

Kettle holes Depressions created when blocks of ice become lodged in glacial deposits and subsequently melt.

Laccolith A massive, concordant igneous body intruded between pre-existing strata.

Lahar Mudflows on the slopes of volcanoes that result when unstable layers of ash and debris become saturated and flow downslope, usually following stream channels.

Laminar flow The movement of water particles in straight line paths that are parallel to the channel. The water particles move downstream without mixing.

Lateral moraine A ridge of till along the sides of a valley glacier composed primarily of debris that fell to the glacier from the valley walls.

Laterite A red, highly leached soil type found in the tropics and rich in oxides of iron and aluminum.

Laurasia The northern portion of Pangaea consisting of North America and Eurasia.

Lava Magma that reaches Earth's surface.

Lava dome A bulbous mass associated with an old-age volcano, produced when thick lava is slowly squeezed from the vent. Lava domes may act as plugs to deflect subsequent gaseous eruptions.

Law of superposition In any undeformed sequence of sedimentary rocks or surface-deposited igneous materials, each layer is older than the one above it and younger than the one below.

Leaching The depletion of soluble materials from the upper soil by downward-percolating water.

Lithification The process, generally cementation and/or compaction, of converting sediments to solid rock.

Lithogenous sediment Seafloor sediment consisting primarily of mineral grains that were weathered from continental rocks and transported to the ocean. Also called terrigenous sediment.

Lithosphere The rigid outer layer of Earth, including the crust and upper mantle.

Loess Deposits of windblown silt, lacking visible layers, generally buff colored, and capable of maintaining a nearly vertical cliff.

Longitudinal dunes Long ridges of sand oriented parallel to the prevailing wind; these dunes form where sand supplies are limited.

Longitudinal profile A cross section of a stream channel along its descending course from the head to the mouth.

Longshore current A near shore current that flows parallel to the shore.

Luster The appearance or quality of light reflected from the surface of a mineral.

Magma A body of molten rock found at depth, including any dissolved gases and crystals.

Magnetometer A sensitive instrument used to measure the intensity of Earth's magnetic field at various points.

Magnitude (earthquake) The total amount of energy released during an earthquake.

Manganese nodules A type of hydrogenous sediment scattered on the ocean floor, consisting mainly of manganese and iron, and usually containing small amounts of copper, nickel, and cobalt.

Mantle The 2885-kilometer (1789-mile) thick layer of Earth located below the crust.

Mass number The sum of the number of neutrons and protons in the nucleus of an atom.

Mass wasting The downslope movement of rock, regolith, and soil under the direct influence of gravity.

Meander A looplike bend in the course of a stream.

Meander scar A floodplain feature created when an oxbow lake becomes filled with sediment.

Mechanical weathering The physical disintegration of rock, resulting in smaller fragments.

Medial moraine A ridge of till formed when lateral moraines from two coalescing valley glaciers join.

Melt The liquid portion of magma excluding the solid crystals.

Mercalli intensity scale A 12-point scale originally developed to evaluate earthquake intensity based upon the amount of damage to various types of structures.

Mesozoic era A time span on the geologic calendar between the Paleozoic and Cenozoic eras—from about 245 million to 66 million years ago.

Metallic bond A chemical bond present in all metals that may be characterized as an extreme type of electron sharing in which the electrons move freely from atom to atom.

Metamorphic rock Rock formed by the alteration of pre-existing rock deep within Earth (but still in the solid state) by heat, pressure, and/or chemically active fluids.

Metamorphism The changes in mineral composition and texture of a rock subjected to high temperature and pressure within earth.

Mid-ocean ridge A continuous mountainous ridge on the floor of all major ocean basins and varying in width from 500 to 5000 kilometers (300 to 3000 miles). The rifts at the crests of these ridges represent divergent plate boundaries.

Migmatite A rock exhibiting both igneous and metamorphic rock characteristics. Such rocks may form when light-colored silicate minerals melt and then crystallize, while the dark silicate minerals remain solid.

Mineral A naturally occurring, inorganic crystalline material with a unique chemical structure.

Mineral resource All discovered and undiscovered deposits of a useful mineral that can be extracted now or at some time in the future.

Mohorovičić discontinuity (Moho) The boundary separating the crust and the mantle, discernible by an increase in seismic velocity.

Mohs scale A series of ten minerals used as a standard in determining hardness.

Monocline A one-limbed flexure in strata. The strata are usually flat lying or very gently dipping on both sides of the monocline.

Mouth The point downstream where a river empties into another stream or water body.

Mud crack A feature in some sedimentary rocks that forms when wet mud dries out, shrinks, and cracks.

Mudflow The flow of debris containing a large amount of water; most characteristic of canyons and gullies in dry, mountainous regions.

Natural levees The elevated landforms composed of alluvium that parallel some streams and act to confine their waters, except during floodstage.

Neap tide The lowest tidal range, occurring near the times of the first and third quarters of the Moon.

Nebular hypothesis A model for the origin of the solar system that assumes a rotating nebula of dust and gases that contracted to produce the sun and planets.

Neutron A subatomic particle found in the nucleus of an atom. The neutron is electrically neutral with a mass approximately equal to that of a proton.

Nonclastic A term for the texture of sedimentary rocks in which the minerals form a pattern of interlocking crystals.

Nonconformity An unconformity in which older metamorphic or intrusive igneous rocks are overlain by younger sedimentary strata.

Nonfoliated Metamorphic rocks that do not exhibit foliation.

Nonmetallic mineral resource Mineral resource that is not a fuel or processed for the metals it contains.

Nonrenewable resource Resource that forms or accumulates over such long time spans that it must be considered as fixed in total quantity.

Normal fault A fault in which the rock above the fault plane has moved down relative to the rock below.

Normal polarity A magnetic field the same as that which presently exists.

Nucleus The small, heavy core of an atom that contains all of its positive charge and most of its mass.

Nuée ardente Incandescent volcanic debris that is buoyed up by hot gases and moves downslope in an avalanche fashion.

Oblique-slip fault A fault having both vertical and horizontal movement.

Octet rule Atoms combine in order that each may have the electron arrangement of a noble gas; that is, the outer energy level contains eight electrons.

Oil trap A geologic structure that allows for significant amounts of oil and gas to accumulate.

Ore Usually a useful metallic mineral that can be mined at a profit. The term is also applied to certain nonmetallic minerals such as fluorite and sulfur.

Original horizontality Layers of sediment are generally deposited in a horizontal or nearly horizontal position.

Orogenesis The processes that collectively result in the formation of mountains.

Outer core A layer beneath the mantle about 2270 kilometers (1407 miles) thick that has the properties of a liquid.

Outgassing The release of gases dissolved in molten rock.

Outwash Sediments deposited by glacial meltwater.

Outwash plain A relatively flat, gently sloping plain consisting of materials deposited by meltwater streams in front of the margin of an ice sheet.

Oxbow lake A curved lake produced when a stream cuts off a meander.

Oxidation The removal of one or more electrons from an atom or ion. So named because elements commonly combine with oxygen.

Pahoehoe A lava flow with a smooth-to-ropy surface.

Paleomagnetism The natural remnant magnetism in rock bodies. The permanent magnetization acquired by rock that can be used to determine the location of the magnetic poles and the latitude of the rock at the time it became magnetized.

Paleontology The systematic study of fossils and the history of life on earth.

Paleozoic era A time span on the geologic calendar between the Precambrian and Mesozoic eras—from about 570 million to 245 million years ago.

Pangaea The proposed supercontinent that 200 million years ago began to break apart and form the present landmasses.

Parabolic dune A sand dune similar in shape to a barchan dune except that its tips point into the wind. These dunes often form along coasts that have strong onshore winds, abundant sand, and vegetation that partly covers the sand.

Parasitic cone A volcanic cone that forms on the flank of a larger volcano.

Parent material The material upon which a soil develops.

Partial melting The process by which most igneous rocks melt. Because individual minerals have different melting points, most igneous rocks melt over a temperature range of a few hundred degrees. If the liquid is squeezed out after some melting has occurred, a melt with a higher silica content results.

Pater noster lakes A chain of small lakes in a glacial trough that occupy basins created by glacial erosion.

Pedalfer Soil of humid regions characterized by the accumulation of iron oxides and aluminum-rich clays in the *B* horizon.

Pedocal Soil associated with drier regions and characterized by an accumulation of calcium carbonate in the upper horizons.

Pegmatite A very coarse-grained igneous rock (typically granite) commonly found as a dike associated with a large mass of plutonic rock that has smaller crystals. Crystallization in a water-rich environment is believed to be responsible for the very large crystals.

Peneplain In the idealized cycle of landscape evolution in a humid region, an undulating plain near base level associated with old age.

Perched water table A localized zone of saturation above the main water table created by an impermeable layer (aquitard).

Peridotite An igneous rock of ultramafic composition thought to be abundant in the upper mantle.

Period A basic unit of the geologic calendar that is a subdivision of an era. Periods may be divided into smaller units called epochs.

Permafrost Any permanently frozen subsoil. Usually found in the subarctic and arctic regions.

Permeability A measure of a material's ability to transmit water.

Phaneritic texture An igneous rock texture in which the crystals are roughly equal in size and large enough so that individual minerals can be identified with the unaided eye.

Phanerozoic eon That part of geologic time represented by rocks containing abundant fossil evidence. The eon extending from the end of the Proterozoic eon (570 million years ago) to the present.

Phenocryst Conspicuously large crystals in a porphyry that are imbedded in a matrix of finer-grained crystals (the groundmass).

Physical geology A major division of geology that examines the materials of Earth and seeks to understand the processes and forces acting beneath upon Earth's surface.

Piedmont glacier A glacier that forms when one or more valley glaciers emerge from the confining walls of mountain valleys and spread out to create a broad sheet in the lowlands at the base of the mountains.

Pillow lava Basaltic lava that solidifies in an underwater environment and develops a structure that resembles a pile of pillows.

Pipe A vertical conduit through which magmatic materials have passed.

Placer Deposit formed when heavy minerals are mechanically concentrated by currents, most commonly streams and waves. Placers are sources of gold, tin, platinum, diamonds, and other valuable minerals.

Plastic deformation Permanent deformation that results in a change in size and shape through folding or flowing.

Plastic flow A type of glacial movement that occurs within the glacier, below a depth of approximately 50 meters, in which the ice is not fractured.

Plate One of numerous rigid sections of the lithosphere that moves as a unit over the material of the asthenosphere.

Plate tectonics The theory that proposes Earth's outer shell consists of individual plates, which interact in various ways and thereby produce earthquakes, volcanoes, mountains, and the crust itself.

Playa The flat central area of an undrained desert basin.

Playa lake A temporary lake in a playa.

Pleistocene epoch An epoch of the Quaternary period beginning about 1.6 million years ago and ending about 10,000 years ago. Best known as a time of extensive continental glaciation.

Plucking The process by which pieces of bedrock are lifted out of place by a glacier.

Pluton A structure that results from the emplacement and crystallization of magma beneath the surface of earth.

Pluvial lake A lake formed during a period of increased rainfall. For example, this occurred in many nonglaciated areas during periods of ice advance elsewhere.

Point bar A crescent-shaped accumulation of sand and gravel deposited on the inside of a meander.

Polar wandering hypothesis As the result of paleomagnetic studies in the 1950s, researchers proposed that either the magnetic poles migrated greatly through time or the continents had gradually shifted their positions.

Polymorphs Two or more minerals having the same chemical composition but different crystalline structures. Exemplified by the diamond and graphite forms of carbon.

Porosity The volume of open spaces in rock or soil.

Porphyritic texture An igneous rock texture characterized by two distinctively different crystal sizes. The larger crystals are called *phenocrysts* and the matrix of smaller crystals is termed the *groundmass*.

Porphyry An igneous rock with a porphyritic texture.

Pothole A depression formed in a stream channel by the abrasive action of the water's sediment load.

Precambrian All geologic time prior to the Paleozoic era.

Principle of fossil succession Fossil organisms succeed one another in a definite and determinable order, and any time period can be recognized by its fossil content.

Principle of original horizontality Layers of sediment are generally deposited in a horizontal or nearly horizontal position.

Proterozoic eon The eon following the Archean and preceding the Phanerozoic eon. It extends between 2500 million and 570 million years ago.

Proton A positively charged subatomic particle found in the nucleus of an atom.

P wave The fastest earthquake wave; travels by compression and expansion of the medium.

Pyroclastic flow A highly heated mixture, largely of ash and pumice fragments, traveling down the flanks of a volcano or along the surface of the ground.

Pyroclastic material The volcanic rock ejected during an eruption. Pyroclastics include ash, bombs, and blocks.

Pyroclastic texture An igneous rock texture resulting from the consolidation of individual rock fragments that are ejected during a violent eruption.

Radial drainage A system of streams running in all directions away from a central elevated structure, such as a volcano.

Radioactivity The spontaneous decay of certain unstable atomic nuclei.

Radiocarbon (carbon-14) The radioactive isotope of carbon produced continuously in the atmosphere and used in dating events as far back as 75,000 years.

Radiometric dating The procedure of calculating the absolute ages of rocks and minerals containing certain radioactive isotopes.

Rainshadow desert A dry area on the lee side of a mountain range. Many middle-latitude deserts are of this type.

Rapids A part of a stream channel in which the water suddenly begins flowing more swiftly and turbulently because of an abrupt steepening of the gradient.

Recessional moraine An end moraine formed as the ice front stagnated during glacial retreat.

Rectangular pattern A drainage pattern that develops on jointed or fractured bedrock and is characterized by numerous right-angle bends.

Refraction A change in direction of waves as they enter shallow water. The portion of the wave in shallow water is slowed, which causes the wave to bend an align with the underwater contours.

Regional metamorphism Metamorphism associated with large-scale mountain building.

Regolith The layer of rock and mineral fragments that nearly everywhere covers Earth's land surface.

Rejuvenation A change in relation to base level, often caused by regional uplift, that causes the forces of erosion to intensify.

Relative dating Rocks are placed in their proper sequence or order. Only the chronological order of events is determined.

Renewable resource A resource that is virtually inexhaustible or that can be replenished over relatively short time spans.

Reserve Already identified deposits from which minerals can be extracted profitably.

Reservoir rock The porous, permeable portion of an oil trap that yields oil and gas.

Residual soil Soil developed directly from the weathering of the bedrock below.

Reverse fault A fault in which the material above the fault plane moves up in relation to the material below.

Reverse polarity A magnetic field opposite to that which presently exists.

Richter scale A scale of earthquake magnitude based on the motion of a seismograph.

Rift A region of Earth's crust along which divergence is taking place.

Ripple marks Small waves of sand that develop on the surface of a sediment layer by the action of moving water or air.

Roche moutonnée An asymmetrical knob of bedrock formed when glacial abrasion smoothes the gentle slope facing the advancing ice sheet and plucking steepens the opposite side as the ice overrides the knob.

Rock A consolidated mixture of minerals.

Rock avalanche The very rapid downslope movement of rock and debris. These rapid movements may be aided by a layer of air trapped beneath the debris, and they have been known to reach speeds in excess of 200 kilometers per hour.

Rock cleavage The tendency of rock to split along parallel, closely spaced surfaces. These surfaces are often highly inclined to the bedding planes in the rock.

Rock cycle A model that illustrates the origin of the three basic rock types and the interrelatedness of Earth's materials and processes.

Rock flour Ground-up rock produced by the grinding effect of a glacier.

Rockslide The rapid slide of a mass of rock downslope along planes of weakness.

Runoff Water that flows over the land rather than infiltrating into the ground.

Saltation Transportation of sediment through a series of leaps or bounces.

Salt flat A white crust on the ground produced when water evaporates and leaves its dissolved materials behind.

Schistosity A type of foliation characteristic of coarser-grained metamorphic rocks. Such rocks have a parallel arrangement of platy minerals such as the micas.

Scoria Hardened lava as retained the vesicles produced by escaping gases.

Sea arch An arch formed by wave erosion when caves opposite sides of a headland unite.

Seafloor spreading The hypothesis first proposed in the 1960s by Harry Hess that suggested that new oceanic crust is produced at the crests of mid-ocean ridges, which are the sites of divergence.

Seamount An isolated volcanic peak that rises at least 1000 meters (3300 feet) above the deep-ocean floor.

Sea stack An isolated mass of rock standing just offshore, produced by wave erosion of a headland.

Seawall A barrier constructed to prevent waves from reaching the area behind the wall. Its purpose is to defend property from the force of breaking waves.

Secondary enrichment The concentration of minor amounts of metals that are scattered through unweathered rocks into economically valuable concentrations by weathering processes.

Secondary (S) wave A seismic wave that involves oscillation perpendicular to the direction of propagation.

Sediment Unconsolidated particles created by the weathering and erosion of rock, by chemical precipitations from solution in water, or from the secretions of organisms, and transported by water, wind, or glaciers.

Sedimentary rock Rock formed from the weathered products of pre-existing rocks that have been transported, deposited, and lithified.

Seiche The rhythmic sloshing of water in lakes, reservoirs, and other smaller enclosed basins. Some seiches are initiated by earthquake activity.

Seismic sea wave A rapidly moving ocean wave generated by earthquake activity and capable of inflicting heavy damage in coastal regions.

Seismogram The record made by a seismograph.

Seismograph An instrument that records earthquake waves.

Seismology The study of earthquakes and seismic waves.

Settling velocity The speed at which a particle falls through a still fluid. The size, shape, and specific gravity of particles influence settling velocity.

Shadow zone The zone between 105 and 140 degrees distance from an earthquake epicenter that direct waves do not penetrate because of refraction by Earth's core.

Shallow-focus earthquake An earthquake focus at a depth of less than 60 kilometers.

Shear Stress that causes two adjacent parts of a body to slide past one another.

Sheet flow Runoff moving in unconfined thin sheets.

Sheeting A mechanical weathering process characterized by the splitting off of slablike sheets of rock.

Shelf break The point at which a rapid steepening of the gradient occurs, marking the outer edge of the continental shelf and the beginning of the continental slope.

Shield A large, relatively flat expanse of ancient metamorphic rock within the stable continental interior.

Shield volcano A broad, gently sloping volcano built from fluid basaltic lavas.

Silicate Any one of numerous minerals that have the silicon-oxygen tetrahedron as their basic structure.

Silicon-oxygen tetrahedron A structure composed of four oxygen atoms surrounding a silicon atom that constitutes the basic building block of silicate minerals.

Sill A tabular igneous body that was intruded parallel to the layering of pre-existing rock.

Sinkhole A depression produced in a region where soluble rock has been removed by groundwater.

Slaty cleavage The type of foliation characteristic of slates in which there is a parallel arrangement of finegrained metamorphic minerals.

Slide A movement common to mass-wasting processes in which the material moving downslope remains fairly coherent and moves along a well-defined surface.

Slip face The steep, leeward surface of a sand dune that maintains a slope of about 34 degrees.

Slump The downward slipping of a mass of rock or unconsolidated material moving as a unit along a curved surface.

Snowfield An area where snow persists year-round.

Snowline Lower limit of perennial snow.

Soil A combination of mineral and organic matter, water, and air; that portion of the regolith that supports plant growth.

Soil horizon A layer of soil that has identifiable characteristics produced by chemical weathering and other soil-forming processes.

Soil profile A vertical section through a soil showing its succession of horizons and the underlying parent material.

Solifluction Slow, downslope flow of water-saturated materials common to permafrost areas.

Solum The *O, A,* and *B* horizons in a soil profile. Living roots and other plant and animal life are largely confined to this zone.

Solution The change of matter from the solid or gaseous state into the liquid state by its combination with a liquid.

Sorting The degree of similarity in particle size in sediment or sedimentary rock.

Specific gravity The ratio of a substance's weight to the weight of an equal volume of water.

Speleothem A collective term for the dripstone features found in caverns.

Spheroidal weathering Any weathering process that tends to produce a spherical shape from an initially blocky shape.

Spit An elongate ridge of sand that projects from the land into the mouth of an adjacent bay.

Spring A flow of groundwater that emerges naturally at the ground surface.

Spring tide The highest tidal range. Occurs near the times of the new and full moons.

Stalactite The iciclelike structure that hangs from the ceiling of a cavern.

Stalagmite The columnlike form that grows upward from the floor of a cavern.

Star dune Isolated hill of sand that exhibits a complex form and develops where wind conditions are variable.

Steppe One of the two types of dry climate. A marginal and more humid variant of the desert that separates the desert from bordering humid climates.

Stock A pluton similar to but smaller than a batholith.

Strata Parallel layers of sedimentary rock.

Stratovolcano See *Composite cone.*

Streak The color of a mineral in powdered form.

Stream A general term to denote the flow of water within any natural channel. Thus, a small creek and a large river are both streams.

Stress The force per unit area acting on any surface within a solid. Also known as *directed pressure*.

Striations The multitude of fine parallel lines found on some cleavage faces of plagioclase feldspars but not present on orthoclase feldspar.

Striations (glacial) Scratches or grooves in a bedrock surface caused by the grinding action of a glacier and its load of sediment.

Strike The compass direction of the line of intersection created by a dipping bed or fault and a horizontal suite face. Strike is always perpendicular to the direction of dip.

Strike-slip fault A fault along which the movement is horizontal.

Stromatolite Structures are deposited by algae and that consist of layered mounds or columns of calcium carbonate.

Subduction The process of thrusting oceanic lithosphere into the mantle along a convergent zone.

Subduction zone A long, narrow zone where one lithospheric plate descends beneath another.

Submarine canyon A seaward extension of a valley that was cut on the continental shelf during a time when sea level was lower, or a canyon carved into the outer continental shelf, slope, and rise by turbidity currents.

Submergent coast A coast whose form is largely the result of the partial drowning of a former land surface either due to a rise of sea level or subsidence of the crust, or both.

Subsoil A term applied to the *B* horizon of a soil profile.

Superposition, law of In any undeformed sequence of sedimentary rocks, each bed is older than the one above and younger than the one below.

Surf A collective term for breakers; also the wave activity in the area between the shoreline and the outer limit of breakers.

Surface soil The upper portion of a soil profile consisting of the *O* and *A* horizons.

Surface waves Seismic waves that travel along the outer layer of Earth.

Surge A period of rapid glacial advance. Surges are typically sporadic and short-lived.

Suspended load The fine sediment carried within the body of flowing water or air.

S wave An earthquake wave, slower than a P wave, that travels only in solids.

Swells Wind-generated waves that have moved into an area of weaker winds or calm.

Syncline A linear downfold in sedimentary strata; the opposite of anticline.

Talus An accumulation of rock debris at the base of a cliff.

Tarn A small lake in a cirque.

Tectonics The study of the large-scale processes that collectively deform Earth's crust.

Temporary (local) base level The level of a lake, resistant rock layer, or any other base level that stands above sea level.

Terminal moraine The end moraine marking the farthest advance of a glacier.

Terrace A flat, benchlike structure produced by a stream, which was left elevated as the stream cut downward.

Terrane A crustal block bounded by faults, whose geologic history is distinct from the histories of adjoining crustal blocks.

Texture The size, shape, and distribution of the particles that collectively constitute a rock.

Theory A well-tested and widely accepted view that explains certain observable facts.

Thrust fault A low-angle reverse fault.

Tide Periodic change in the elevation of the ocean's surface.

Till Unsorted sediment deposited directly by a glacier.

Tillite A rock formed when glacial till is lithified.

Tombolo A ridge of sand that connects an island to the mainland or to another island.

Topset bed An essentially horizontal sedimentary layer deposited on top of a delta during floodstage.

Transform boundary A boundary in which two plates slide past one another without creating or destroying lithosphere.

Transpiration The release of water vapor to the atmosphere by plants.

Transported soil Soils that form on unconsolidated deposits.

Transverse dunes A series of long ridges oriented at right angles to the prevailing wind; these dunes form where vegetation is sparse and sand is very plentiful.

Travertine A form of limestone ($CaCO_3$) that is deposited by hot springs or as a cave deposit.

Trellis drainage A system of streams in which nearly parallel tributaries occupy valleys cut in folded strata.

Trench An elongate depression in the seafloor produced by bending of oceanic crust during subduction.

Truncated spurs Triangular-shaped cliffs produced when spurs of land that extend into a valley are removed by the great erosional force of a valley glacier.

Tsunami The Japanese word for a seismic sea wave.

Turbidite Turbidity current deposit characterized by graded bedding.

Turbidity current A downslope movement of dense, sediment-laden water created when sand and mud on the continental shelf and slope are dislodged and thrown into suspension.

Turbulent flow The movement of water in an erratic fashion often characterized by swirling, whirlpool-like eddies. Most streamflow is of this type.

Ultimate base level Sea level; the lowest level to which stream erosion could lower the land.

Unconformity A surface that represents a break in the rock record; caused by erosion or nondeposition.

Uniformitarianism The concept that the processes that have shaped Earth in the geologic past are essentially the same as those operating today.

Valence electron The electrons involved in the bonding process; the electrons occupying the highest principal energy level of an atom.

Valley glacier A glacier confined to a mountain valley, which in most instances had previously been a stream valley.

Valley train A relatively narrow body of stratified drift deposited on a valley floor by meltwater streams that issue from the terminus of a valley glacier.

Vein deposit A mineral filling a fracture or fault in a host rock. Such deposits have a sheetlike, or tabular, form.

Ventifact A cobble or pebble polished and shaped by the sandblasting effect of wind.

Vesicles Spherical or elongated openings on the outer portion of a lava flow that were created by escaping gases.

Vesicular A term applied to igneous rocks that contain small cavities called vesicles, which are formed when gases escape from lava.

Viscosity A measure of a fluid's resistance to flow.

Volcanic Pertaining to the activities, structures, or rock types of a volcano.

Volcanic arc Mountains formed in part by igneous activity associated with the subduction of oceanic lithosphere beneath a continent. Examples include the Andes and the Cascades.

Volcanic bomb A streamlined pyroclastic fragment ejected from a volcano while molten.

Volcanic neck An isolated, steep-sided, erosional remnant consisting of lava that once occupied the vent of a volcano.

Volcano A mountain formed from lava and/or pyroclastics.

Wash A desert stream course that is typically dry except for brief periods immediately following rainfall.

Water gap A pass through a ridge or mountain in which a stream flows.

Water table The upper level of the saturated zone of groundwater.

Wave-cut cliff A seaward-facing cliff along a steep shore line formed by wave erosion at its base and mass wasting.

Wave-cut platform A bench or shelf along a shore at sea level, cut by wave erosion.

Wave height The vertical distance between the trough and crest of a wave.

Wave length The horizontal distance separating successive crests or troughs.

Wave of oscillation A water wave in which the wave form advances as the water particles move in circular orbits.

Wave of translation The turbulent advance of water created by breaking waves.

Wave period The time interval between the passage of successive crests at a stationary point.

Weathering The disintegration and decomposition of rock at or near the surface of the Earth.

Welded tuff A pyroclastic deposit composed of particles fused together by the combination of heat still contained in the deposit after it has come to rest and the weight of overlying material.

Well An opening bored into the zone of saturation.

Wind gap An abandoned water gap. These gorges typically result from stream piracy.

Xenolith An inclusion of unmelted country rock in an igneous pluton.

Xerophyte A plant highly tolerant of drought.

Yazoo tributary A tributary that flows parallel to the main stream because a natural levee is present.

Zone of accumulation The part of a glacier characterized by snow accumulation and ice formation. The outer limit of this zone is the snowline.

Zone of aeration Area above the water table where openings in soil, sediment, and rock are not saturated but are filled mainly with air.

Zone of fracture The upper portion of a glacier consisting of brittle ice.

Zone of saturation Zone where all open spaces in sediment and rock are completely filled with water.

Index